D0858123

DISCRETE COSINE TRANSFORM

Algorithms, Advantages, Applications

DISCRETE COSINE TRANSFORM

Algorithms, Advantages Applications

K. R. RAO

The University of Texas at Arlington
Department of Electrical Engineering
Arlington, Texas

P. YIP

McMaster University
Department of Mathematics and Statistics
and Communications Research Laboratory
Hamilton, Ontario
Canada

ACADEMIC PRESS, INC

Harcourt Brace Jovanovich, Publishers

Boston San Diego New York
London Sydney Tokyo Toronto

This book is printed on acid-free paper. ∞

ACADEMIC PRESS, INC.
1250 Sixth Avenue, San Diego, CA 92101

United Kingdom Edition published by
ACADEMIC PRESS LIMITED
24–28 Oval Road, London NW1 7DX

Library of Congress Cataloging-in-Publication Data

Rao, K. R., 1990
Discrete cosine transform : algorithms, advantages, and applications / K. R. Rao, P. Yip.
p. cm.
Includes bibliographical references.
ISBN 0-12-580203-X
1. Signal processing—Mathematics. 2. Transformations (Mathematics) I. Yip, P. (Ping), 1956– II. Title.
TK5102.5.R335 1990 89-29800
621.382'2'0151—dc20 CIP

PRINTED IN THE UNITED STATES OF AMERICA

90 91 92 93 9 8 7 6 5 4 3 2 1

CONTENTS

PREFACE

As indicated in the Introduction (Chapter 1) discrete cosine transform (DCT) has become the industry standard in image coding. It has attracted the attention of engineers, researchers, and academicians, leading to various developments including the dedicated DCT VLSI chips. However, it is surprising that no book on DCT has emerged so far. It is fervently hoped that this book will fill this gap. Now that the CCITT specialists group on coding for visual telephony has finalized the draft recommendations for $p \times 64$ KBPS ($p = 1, 2, \ldots, 30$) video codec based on the DCT, it is speculated that there is an extensive market in consumer electronics (videophone) with DCT as the main compression tool.

Although this book is primarily aimed at the graduate student, an advanced senior student in engineering or sciences should be able to grasp the basics of the DCT and proceed further toward the algorithms and applications. It can also serve as an excellent reference, since a wide variety of DCT applications, supported by an extensive list of references, is furnished. In most cases, these applications are described conceptually rather than in detail. These concepts are complimented by flowgraphs and block diagrams. The objective is to limit the book to a reasonable size but at the same time provide the reader with the power and the practical utility of the DCT. The computer programs and the list of DCT VLSI chip manufacturers augmented by the fundamentals and fast algorithms should familiarize the engineer and researcher with the necessary background for delving more deeply into the applications.

A major problem during the preparation of this book was the rapid pace of development, both in software and hardware, relating to the DCT. Many application-specific ICs have been realized in the past few years. The authors have tried hard to keep pace by including many of these latest developments. In this way, it is hoped that the book is timely and that it will appeal to a wide audience in the engineering, scientific, and technical community so that additional DCT applications may emerge.

ACKNOWLEDGMENTS

It is a pleasure to acknowledge the invaluable help received from a number of people associated with universities, industry, and research labs. This help was in the form of technical papers, discussions, processed images, tables, brochures, the review of various sections of the manuscript, computer programs, etc. Special thanks are due to the following persons:

S. Acharya	Telephoto Communications, Inc.
A. Artieri	SGS-Thomson Microelectronics
M. Barbero	Telettra
R. C. Brainard	AT&T Bell Labs
W. K. Cham	The Chinese University of Hong Kong
M. Chelehmal	U.T.A.
A. M. Chiang	Mass. Institute of Technology
B. Chitprasert	Compression Labs, Inc.
N. Demassieux	Telecom Paris University
P. Duhamel	CNET
J. Duran	Video Telecom Corpn.
T. Fjällbrant	Linköping University
R. Forchheimer	Linköping University
H. Gharavi	Bellcore
P. Guiseppe	Consiglio Naxionale Ricerche
Y. Hatori	Kokusai Denshin Denwa Co., Ltd.
Y. S. Ho	University of California, Santa Barbara
H. Hölzlwimmer	SIEMENS
A. Jalali	Bellcore
M. Khan	Plessey Company
J. K. Kim	Korea Advanced Institute of Science and Technology
T. Koga	NEC
B. G. Lee	Seoul National University

S. U. Lee	Seoul National University
A. Lèger	STM/TSA CCETT
W. Li	Lehigh University
W. Liebsch	Heinrich-Hertz Institut für Nachrichtentechnik
C. Ma	Comdisco System, Inc.
H. Malvar	Universidale de Brasilia
K. Matsuda	Fujitsu Labs
B. T. McSweeney	Philips Research Labs
J. L. Mitchell	IBM
F. Molo	Telettra
J. Y. Nam	U. T. A.
K. N. Ngan	National University of Singapore
G. Nunan	INMOS, Ltd.
N. Ohta	NTT
K. Ohzeki	Toshiba
S. Okubo	NTT
F. Pellandini	University of Neuchatel
P. Pirsch	University of Hannover
K. Prabhu	Rockwell Intl.
S. Sabri	BNR Recherches Bell-Northern
K. Sawada	NTT
H. Shu	Zoran Corp.
H. Sobol	U. T. A.
R. W. Sovinee	Telephoto Communications, Inc.
M. T. Sun	Bellcore
Z. M. Sun	U. T. A.
S. Takahashi	Toshiba
K. H. Tzou	Bellcore
K. Vanhoof	Katholieke Universiteit Leuven
M. Vetterli	Columbia University
G. K. Wallace	Digital Equipment Corp.
L. Wang	University of Ottawa
Welzenbach	ANT Nachrichtentechnik
K. M. Yang	Bellcore
H. Yasuda	NTT

The authors wish to thank several people who contributed to the manuscript preparation (graphics, typing, word processing, etc.). The majority of this arduous task was undertaken by Ms. Bui, Mr. H. Dao, Ms. S. P. Lam, and Ms. A. Tatum. The authors particularly appreciate their patience and willingness to help far beyond the call of duty.

The encouragement and understanding of our families during the preparation of this book is gratefully acknowledged. The time and effort spent on writing this book must certainly have been reflected in the neglect of our families, whom we thank for their forbearance.

LIST OF ACRONYMS

ABAC	Adaptive binary arithmetic coding
ABTC	Adaptive block truncation coding
ACTV	Advanced compatible TV
A/D	Analog-to-digital
ADPCM	Adaptive DPCM
AGHMM	Autoregressive Gaussian hidden Markov model
AOT-AR	Adaptive orthogonal transform based on the autoregressive model
ASIC	Application-specific integrated circuit
ATC	Adaptive transform coding
ATC/HS	Adaptive transform coding/harmonic scaling
ATSC	Advanced television systems committee
ATV	Advanced TV
ATVQ	Adaptive transform VQ
AVTQ	Adaptive vector transform quantization
BIFORE	Binary Fourier representation
BMA	Block matching algorithms
BPF	Bandpass filter
BPP	Bits per pel
BPS	Bits per second
BRO	Bit-reversed order
BSPC	Block separated component progressive coding
BTC	Block truncation coding
CATV	Cable TV
CBP	Coded block pattern
CCD	Charge-coupled device
CCF	Cross-correlation function
CCIR	International Radio Consultative Committee
CCITT	International Telegraph and Telephone Consultative Committee
CDS	Conjugate direction search
CEPT	European Conference of Postal and Telecommunication Administration
CMT	C-matrix transform
CMTT	Committee for Mixed Telephone and Television
CODEC	Coder docoder
col VQ	Color VQ
COPD	Chronic obstructive pulmonary disease
CPD	Coefficient power distribution
CRT	Cathode ray tube
CSIF	Common source intermediate format
CT	Computer tomography
CTCC	Cosine transform complex cepstrum
CTRC	Cosine transform real cepstrum
CTVQ	Classified transform VQ
CU	Control unit
CVQ	Classified VQ
DA	Distributed arithmetic
DAT	Digital audio tape
DBS	Direct broadcast satellite
DCT	Discrete cosine transform
DDP	1D DCT processor
DFT	Discrete Fourier transform
DHT	Discrete Hartley transform
DIF	Decimation in frequency

DIR	Directional features
DIT	Decimation in time
DLT	Discrete Legendre transform
DM	Delta modulation
DPCM	Differential pulse code modulation
DSM	Digital storage media
DSP	Digital signal processing
DST	Discrete sine transform
DWL	Dual word length
D/A	Digital to analog
ECG	Electrocardiogram
EDTV	Enhanced definition TV
EIA	Electronics Industries Association
em	Essential maximum
EOB	End of block
EPE	Energy packing efficiency
EQTV	Extended quality TV
FC	Fractional correlation
FCC	Federal Communications Commission
FCT	Fourier cosine transform
FDM	Frequency division multiplexing
FEC	Forward error correction
FFT	Fast Fourier transform
FIN	Fineness of a subimage
FIR	Finite impulse response
FPR	Fast progressive reconstruction
FST	Fourier sine transform
GBPS	Giga BPS
GBTC	Generalized BTC
GOB	Group of blocks
HCT	High correlation transform
HDTV	High-definition TV
HHT	Hadamard–Haar transform
HIVITS	High-quality videotelephone and high definition television system
HPC	Hierarchical predictive coding
HT	Haar transform
HVQ	Hierarchical VQ
HVS	Human visual system
IC	Integrated circuit
ICT	Integer cosine transform
IDCT	Inverse DCT
IDTV	Improved definition TV
IEEE	Institute of Electrical and Electronics Engineers
IRM	Intermediate result memory
ISDN	Integrated Services Digital Network
ISO	International Standards Organization
IVQ	Inverse vector quantization, interpolative vector quantization
JPEG	Joint photographic experts group
KBPS	Kilo BPS
KLT	Karhunen–Loeve transform
LBG	Linde Buzo Gray
LBR	Low bit rate
LCT	Low correlation transform
LMS	Least mean square
LOT	Lapped orthogonal transform
LPC	Linear predictive coding
LSP	Least significant part
MAC	Multiplex analog component
MACE	Middle frequency ac energy
MAE	Mean absolute error
MB	Macro block
MBA	Macro block addressing
MBPS	Mega BPS
MC	Motion compensation
MI	Myocardial infarction
MOS	Mean opinion score
MPEG	Moving picture experts group
MRI	Magnetic resonance imaging
mrb	Maximum reducible bits
ms	Millisecond
MSCT	Modified symmetric cosine transform
MSDCT	Modified symmetric DCT
MSE	Mean square error
MSP	Most significant part
MSQE	Mean square quantization error
MSVQ	Multistage VQ
MTF	Modulation transfer function
MVD	Motion vector data
MVZS	Maximum variance zonal sampling
NAB	National Association of Broadcasters
NCC	Normalized correlation coefficient
NIC	New image communications
NMAE	Normalized mae
NMSE	Normalized mse
NTSC	National Television Systems Committee

OTS One-at-a-time search
PACS Picture archiving and communication systems
PAL Phase alternating line
PCM Pulse code modulation
PCS Progressive coding scheme
PE Processing element
PFA Prime-factor algorithm
PIT Progressive image transmission
PLL Phase lock loop
pp Peak-to-peak
PRA Pel recursive algorithms
PSTN Public Service Telephone Network
QMF Quadrature mirror filter
RAC ROM and accumulator
RACE Research in Advanced Communication Technologies for Europe
RBC Recursive block coding
RBN Recursive binary nesting
RELP Residual excited LPC
RICA Raster image communication architecture
RICAPS Raster image communication application profiles
RM Reference model
ROM Read-only memory
RPV Remotely piloted vehicle
RT Rapid transform
RUC Region of uncertainty
SAR Synthetic aperture radar
SBC Sub-band
SC Subcarrier
SCT Symmetric cosine transform
SDFT Symmetrical DFT
SECAM Sequential couleur a memoire
SEGSNR Segmental SNR
SHT Slant–Haar transform
SMPTE Society of Motion Picture and Television Engineers
SNR Signal-to-noise ratio
SPA Significant pel area
SQ Scalar quantization
SR Search region
SRB Shift register bank
ST Slant transform
STA Surface texture analysis
SVD Singular-value decomposition
TCAD Circuit de transformé en cosinus dicrete en arithmetique distribué
TCWVQ Transform coder weighted vector quantization
TDHS Time domain harmonic scaling
TDM Time division multiplexing
TSBC Transform based split band coder
TSVQ Tree search VQ
TTC Transform trellis code
VAWTPCVQ Vector-adaptive wideband transform product code vector quantization
VCG Vectorcardiogram
VCR Videocassette recorder
VDP Videodisc player
VLC Variable length coding
VLIW Very long instruction word
VLSI Very large scale integration
VMP Vector-matrix multiplication
VQ Vector quantization
VQPS VQ using peano scan
VTR Videotape recorder
VWL Variable word length
WFTA Winograd Fourier transform algorithm
WHT Walsh–Hadamard transform

NOTATION

Symbol	Meaning
$[A]$	Auto-covariance matrix, see (3.2.9)
$[A]_{ik}$	ik^{th} element of $[A]$
$[A_N]$	$(N \times N)$ matrix, see (4.3.1)
$[A_{\text{tr}}]$	Auto-covariance matrix in the transform domain
$[C^{\text{I}}_{N+1}]$	DCT-I matrix of size $(N+1) \times (N+1)$
C^{I}_k	$\cos(i\pi/k)$
$[C^k_N]$	DCT-k matrix of size $(N \times N)$, $k = 2,3,4$
$\mathbf{D}^{c(k)}$	$\mathbf{X}^{c(k)}_{+} - \mathbf{X}^{c(k)}$ $k = 1,2,3,4$, see (2.7.2)
$D(\theta)$	Distortion, see (6.5.2)
$\mathbf{d}$	$\mathbf{x}_{+} - \mathbf{x}$, see (2.7.1b)
$d(n)$	$x(n+1) - x(n)$, see (2.7.1a)
$E[\ ]$	Expectation operator
$e_p(n)$	Prediction error
$e_{\text{pq}}(n)$	Quantized prediction error
$F[\ ]$	Fourier transform of [ ], see (2.2.1a)
$F^{-1}[\ ]$	Inverse Fourier transform of [ ], see (2.2.1b)
$F_C[\ ]$	Fourier cosine transform of [ ], see (2.2.3)
$F_C^{-1}[\ ]$	Inverse cosine transform of [ ], see (2.2.4)
$F_{Ck}[\ ]$	DCT $-$ k of [ ], $k = 1,2,3,4$
$F_{Ck}^{-1}[\ ]$	IDCT $-$ k of [ ], $k = 1,2,3,4$
$F_F[\ ]$	DFT of [ ]
$F_F^{-1}[\ ]$	IDFT of [ ]
$F_s[\ ]$	Fourier sine transform of $x(t)$, see (2.2.8b)
$F_s^{-1}[\ ]$	Inverse Fourier sine transform of $x(t)$
$F_{sk}[\ ]$	DST $-$ k of [ ], k = 1, 2, 3, 4
$F_{sk}^{-1}[\ ]$	IDST $-$ k of [ ], k = 1, 2, 3, 4
$\hat{F}_F(k)$	$2N$-point DFT of $\hat{f}(n)$, see (2.8.4)
f	Frequency in Hz or radial frequency in cycles/degree
$f(n)$	$f(t)$ sampled at $n\Delta t$
$f(t)$	Continuous function of time
f_{sc}	Subcarrier frequency
$\mathbf{g}_k$	$[g_{1k}, g_{2k}, \ldots, g_{Nk}]^{\text{T}}$
$\langle \mathbf{g}_k, \mathbf{g}_m \rangle$	Inner product of $\mathbf{g}_k$ and $\mathbf{g}_m$, see (2.4.1)
$H(f)$	Modulation transfer function, see (7.14.1)
$[H_w]$	Walsh-ordered $(N \times N)$ WHT matrix
$\text{Im}[\ ]$	Imaginary part of [ ]
I_{N}	$(N \times N)$ identity matrix
$\tilde{\mathrm{I}}_{\mathrm{N}}$	(N × N) opposite diagonal identity matrix e.g., $\tilde{I}_4 = \begin{bmatrix} 0 & 0 & 0 & 1 \\ 0 & 0 & 1 & 0 \\ 0 & 1 & 0 & 0 \\ 1 & 0 & 0 & 0 \end{bmatrix}$
j	$\sqrt{-1}$
$\mathbf{n}$	Additive noise vector
$[P_N]$	Permutation matrix, see (4.3.4)
$p(x)$	Probability density function of random variable x
q_N	Quantization error
$\bar{R}$	Average bit rate in BPS
$R(\theta, D)$	Rate distortion, see (6.5.1)
$\text{Re}[\ ]$	Real part of [ ]
R_i	Number of bits assigned to the ith transform coefficient

Symbol	*Meaning*
S_k^i	Sin $(i\pi/k)$
$[S_N^k]$	DST-k matrix of size $(N \times N)$, $k = 1, 2, 3, 4$
T	Duration in time for a sequence of length N − 1 or N + 1, see (2.5.2b)
$T_n(x)$	n^{th} Tchebyshev polynomial of the first kind, see (3.3.6)
t	Time
$tr[\ ]$	Trace of a matrix
W_N	$\text{Exp}(-j\,2\pi/N)$
w	Radian frequency
$\mathbf{X}_{\text{H}}$	DHT of $\mathbf{x}$
$\mathbf{X}_{\text{w}}$	Walsh ordered WHT of $\mathbf{x}$, see (4.5.1)
$\mathbf{X}^{C(k)}$	DCT-k of $\mathbf{x}$, $k = 1, 2, 3, 4$
$\mathbf{X}^{\text{F}}$	DFT of $\mathbf{x}$
$\mathbf{X}^{S(k)}$	DST-k of $\mathbf{x}$, $k = 1, 2, 3, 4$
$X(\omega)$	Fourier transform of $x(t)$, see (2.2.1a)
$X_{\text{C}}(\omega)$	Fourier cosine transform of $x(t)$, see (2.2.3)
$X_{\text{S}}(\omega)$	Fourier sine transform of $x(t)$
$X^{C(2)}(u, v)$	DCT-II coefficient in row u and column v
$X^{C(k)}(m)$	m^{th} coefficient of $\mathbf{X}^{C(k)}$, $k = 1, 2, 3, 4$
$X^{\text{F}}(m)$	m^{th} DFT coefficient
$X^{SC(2)}(m)$	m^{th} coefficient of SDCT-II
$X^{s(k)}(m)$	m^{th} coefficient of $\mathbf{X}^{s(k)}$
$\mathbf{X}_+^{c(k)}$	DCT-k of $\mathbf{x}_+$, $k = 1, 2, 3, 4$
$\mathbf{X}_+^{s(k)}$	DST-k of $\mathbf{x}_+$, $k = 1, 2, 3, 4$
$X_+^{s(k)}(m)$	m^{th} coefficient of $\mathbf{X}_+^{s(k)}$
$\mathbf{x}$	Sampled data vector
$x_e(t)$	Even extension of function $x(t)$
$x(m, n)$	Intensity of original image in row m and column n

Symbol	*Meaning*
$x_{\text{p}}(n)$	Predicted value of $x(n)$
x_{pp}	Peak-to-peak image intensity
$x(t)$	Function of time
$\mathbf{x}_+$	Shifted sampled data vector
$\mathbf{x} * \mathbf{y}$	Circular convolution of $\mathbf{x}$ and $\mathbf{y}$
$\hat{x}(m, n)$	Intensity of reconstructed image in row m and column n
α, β	Constants
$\alpha(N)$	Number of additions required for an N-point discrete transform
Δf	Sampling interval in the frequency domain
Δt	Sampling interval in the time domain
ε	MSE between $\mathbf{x}$ and $\hat{\mathbf{x}}$, see (3.2.5)
μ_i	i^{th} eigenvalue
$\mu(N)$	Number of multiplications required for an N-point discrete transform
δ_{lm}	Kronecker delta
ρ	Adjacent correlation coefficient
$\tilde{\sigma}^2(u, v)$	Variance of $X^{c(2)}(u, v)$
$[\Sigma]$	Correlation matrix in the data domain
$[\tilde{\Sigma}]$	Correlation matrix in the DCT-II domain
$\mathbf{\Phi}_i$	Set of linearly independent vectors $i = 0, 1, \ldots, N - 1$, see (3.2.2)
$\Phi_m(n)$	n^{th} component of mth eigenvector
$[\phi]$	matrix of eigenvectors $[\Phi_0, \Phi_1, \ldots, \Phi N - 1]$, see (3.2.11)
$[\phi]_{mn}$	mn-element of matrix ϕ
$\otimes$	Kronecker or direct product

CHAPTER 1

DISCRETE COSINE TRANSFORM

1.1 Introduction

As is well known, the development of fast algorithms for efficient implementation of the discrete Fourier transform (DFT) by Cooley and Tukey [M-34] in 1965 has led to phenomenal growth in its applications in digital signal processing (DSP). Various improvements, modifications, and enhancements of the basic Cooley–Tukey DFT algorithm [M-24, M-30] (some of these are tailored for hardware realization), coupled with the rapid rise in digital device technology, have further enhanced the DFT application arena. Similarly, the discovery of the discrete cosine transform (DCT) in 1974 [G-1] has provided a significant impact in the DSP field. While the original DCT algorithm is based on the FFT, a real arithmetic and recursive algorithm, developed by Chen, Smith, and Fralick in 1977 [FRA-1], was the major breakthrough in the efficient implementation of the DCT. A less well-known but equally efficient algorithm was developed by Corrington [FRA-6]. Subsequently, other algorithms, such as the decimation-in-time (DIT) [FRA-8]; decimation-in-frequency (DIF) [FRA-3, FRA-4, FRA-9, FRA-14]; split radix [FRA-29]; DCT via other discrete transforms such as the discrete Hartley transform (DHT) [FRA-16, FRA-19, FRA-21, FRA-28] or the Walsh–Hadamard transform (WHT) [DW-1, DW-2, DW-3, DW-5]; prime factor algorithm (PFA) [FRA-10], a fast recursive algorithm [FRA-13]; and planar rotations [DP-22, DP-27, FRA-23, FRA-26], which concentrate on reducing the computational complexity and/or improving the structural simplicity (recursiveness and repetitive structure), have been developed. Although all of these algorithms can be extended to multidimensional DCT, since this is a separable transform (see Figs. 5.1 and 5.2), additional algorithms for direct implementation of the 2D (two-dimensional) DCT have also been developed [FID-1 through FID-7]. The latter concept can possibly be extended to the 3D DCT. Practical applications of the 3D DCT, however, appear to be limited. The architecture of the 2D DCT chips developed by the industry [IC-48, IC-55, HDTV-10] is based on the "row-column" (separability) decomposition. While the DCT has been implemented in hardware using either discrete components [T-1, T-2, LBR-3 through LBR-6, LBR-18, DP-5, DP-19] or the DSP chips [LBR-11, LBR-15, LBR-16, LBR-45 through LBR-47, DP-9, DP-14, DP-32, DP-40, DP-50, DP-55, SC-17, SC-20, SC-23 through SC-26], the evolution of DCT chips only began in 1984 [DP-1, DP-13, DP-23]. Subsequently, industry [DP-15, DP-17, DP-33, DP-35, DP-38, DP-44, IC-53, IC-67, IC-73, HDTV-9, DP-52], research labs [DP-22, DP-27, DP-29, DP-30, DP-34, DP-37, DP-39, DP-41, DP-46, DP-53, DP-56], and universities [DP-26, DP-31, DP-45, DP-47, DP-48] have developed ASICs (application-specific ICs) for real-time implementation of the DCT at rates up to 14.32 MHz, which is four times the color subcarrier frequency of the NTSC TV signal. Versatility, in terms of the 2D DCT of block sizes ranging from 4×4 to 16×16 and increased accuracy of the internal logic, has also been accomplished. More recently, an 8×8 DCT chip for real time processing of video at 27 MHz has been developed [DP-44]. DCT chips for real-time processing of high-definition TV (HDTV) signals [DP-52, DP-53] (see Section 7.11 on HDTV coding) are also being developed. (Appendix B.1 provides a list of DCT VLSI chip manufacturers.) Also, image compression boards and PC compatible multiprocessor workstations based on DCT (see Appendix B.2) that facilitate the accelerated transmission over narrow bandwidth channels and increased database storage have been developed by industry [PIT-29, DP-40, IC-34, IC-68, LBR-47].

This dramatic development of DCT-based DSP is by no means an accident. Besides being real, separable, and orthogonal, DCT approaches the statistically optimal transform, KLT (Karhunen–Loeve transform) (see Section 3.2), using a number of performance criteria, as illustrated in Chapter 6. KLT, however, suffers from computational problems, not to mention the problem of generating the KLT, as it is dependent on signal statistics. The appeal and attractiveness of the DCT are further accentuated by fast algorithms. There are a number of other discrete transforms [B-1 through B-3, B-12], such as the Walsh–Hadamard transform (WHT), slant transform (ST), Haar transform (HT), discrete sine transform (DST), discrete Legendre transform (DLT), and hybrid transforms such as the slant-Haar (SHT) and Hadamard–Haar (HHT) that also possess properties similar to those of the DCT. DCT, on the other hand, has shown its superiority in bandwidth compression (redundancy reduction) of a wide range of signals such as speech, TV signals, color print images, infrared images, synthetic aperture radar (SAR) images, photovideotex, surface texture, etc. (Chapter 7 details these applications.) Because of the frequency decomposition and simple convolution-multiplication properties, DFT is in the forefront, and rightly so, of all the DSP applications. After DFT, DCT appears to take the limelight. Recently, some other powerful tools, such as vector quantization (VQ) (see Section 7.9) and sub-band coding, although not new in concept, have emerged. In many DSP applications, especially speech and image coding, combination of these tools (DCT, VQ, prediction, sub-band, etc.) has culminated in cumulative compression. Also, a new transform (real, discrete, and orthogonal) having fast algorithms with very low arithmetic complexity and a simple structure has been developed by Guillemot and Duhamel [PC-16]. Its performance in image coding appears to be similar to that of the DCT.

Chapter 7 presents the basic concepts of the compression techniques and illustrates these with applications in diverse disciplines. It must be cautioned, both in speech and image coding, that the DCT, although significant, is only one of the many ingredients in the compression cuisine. For example, bit allocation, quantizer, variable word length (VWL) coder, buffer, error detection/correction codes, multiplexing various signals (sync, audio, data, etc.), and network interface can complement and complete the coding process. Also the codec (coder–decoder) operations start with preprocessing (low pass filtering, A/D, generating various clocks) and end with postprocessing (interpolation when the original signal is subsampled, inserting sync signals, D/A, etc.). Further evidence of DCT popularity may be observed from the draft recommendations being formulated by various committees of international standards organizations such as the CCITT/SGXV, Working Party XV/1, Specialists Group on Coding for Visual Telephony [LBR-23 through LBR-27, LBR-35, LBR-37, LBR-38, LBR-44], and Joint Photographic Experts Group (JPEG) formed from the CCITT and ISO for still picture image transmission [PIT-13, PIT-14, PIT-18, PIT-20 through PIT-27, PIT-29, PIT-30, PIT-33 through PIT-36]. (See Fig. 7.186 for an organizational chart for recommending these standards.) In both applications, DCT has been recommended in the overall compression scheme. Using the recommendations of the Specialists Group on coding for visual telephony, several companies have developed the codec for testing, evaluation, and refinement of the basic compression algorithm (Fig. 7.123). Also a software-based 64KBPS video codec [LBR-45, LBR-46] using a modular hardware architecture for videophone application is being built (picture format is 1/4 CSIF) (see Table 7.16). Videoconferencing codecs using DCT as the main compression algorithm have been built and marketed by many companies such as Compression Labs, Video Telecom, and Picturetel. Similarly, DCT-based broadcast-quality codecs (see Figs. 7.58 and 7.59) that can transmit and receive TV signals in various formats (NTSC, PAL, 4:2:2, RGB, etc.) at two different rates (34.368 MBPS and 44.736 MBPS) have been developed and marketed by Telettra [IC-53, IC-75, IC-76, IC-78, HDTV-18]. Also, the CMTT/2 (Committee for Mixed Telephone and Television) has set up an experts group to define the main parameters of a single worldwide standard based on hybrid coding (DPCM/DCT) (see Fig. 7.61) for both contribution and distribution of 4:2:2 TV signals (CCIR Rec. 601 [G-21], see Table 7.1) at both 32.768 MBPS (European hierarchy) and 44.736 MBPS (North American hierarchy). The primary objective in highlighting these milestones, apart from describing the multitude of the DCT applications, is to attract the attention of the researcher so that other application areas may be explored. For example, DCT has been successfully used in transcoding (conversion of TV signal from a 4:2:2 format to

3:1:1 format) [F-6], decimation and interpolation [F-3, F-4, F-8] and in progressive transmission of medical images [PIT-4, PIT-6, PIT-8 through PIT-10]. It is hoped that the book needles the novice into new frontiers, entices and excites the expert into further explorations, and seduces the student into investigating additional applications.

1.2 Organization of the Book

The book is organized into several chapters with the last chapter (Chapter 7) focusing on the applications of the DCT. Chapter 2 introduces the DCT originally developed in [G-1] along with other versions defined by Wang [FD-1]. This family of DCTs and their inverses are explicitly described in (2.4.11). This is followed by a discussion on some basic properties of the DCTs, such as orthogonality, time or spatial shift, and convolution. Although the convolution theorem for the DCT is not as simple as the convolution-multiplication relationship of the DFT, the former has been applied to image enhancement [IE-1], filtering [F-1, F-7], and progressive image transmission [PIT-12]. The Karhunen–Loeve transform (KLT), which is statistically optimal based on a number of criteria, is developed in Chapter 3. This is followed by the asymptotic equivalence of the family of DCTs with respect to the KLT, for a Markov process in terms of the transform size and the adjacent correlation coefficient. The asymptotic behavior of the DCT, as related to the KLT, is the key to its powerful performance in the bit-rate reduction of a variety of signals. It is tacitly assumed throughout the book that the DCT implies DCT-II [G-1], while other forms of the DCT are explicitly expressed as DCT-I, DCT-III, etc. (2.4.11). There are a number of other versions of the DCT, including the symmetric cosine transform (SCT) [S-1], modified symmetric cosine transform (MSCT) [DP-24], and integer cosine transform (ICT) [DP-42, DP-43].

With the definitions, properties, and asymptotic behavior of the DCT described in Chapters 2 and 3, it is only logical that the fast algorithms for its efficient implementation should follow. This development can be observed in Chapter 4, where several algorithms are outlined. It is interesting to note that the algorithmic history of the DCT parallels that of the DFT, i.e., algorithms based on sparse factorization of the transform matrix, decimation-in-time (DIT) and decimation-in-frequency (DIF) formats, prime factor algorithms (PFAs), algorithms based on different radices, index mapping, a fast recursive algorithm, and via other discrete transforms. Some of these algorithms formed the basis for the ASICs. For example, the DCT chip developed by SGS–Thomson Microelectronics [DP-28, DP-33, DP-41] based on Lee's algorithm [FRA-3, FRA-4] can process video in real time and also has multiple block size capability. A table summarizing the computational complexities of these algorithms concludes Chapter 4.

Extension of the 1D DCT to multidimensions and the development of corresponding algorithms form the basis of Chapter 5. While the DCT chips, when applied to image coding, have utilized the separability property (Fig. 5.1), several algorithms for direct implementation of the 2D DCT have also been developed. The published literature relates the DCT/VLSI activity to Linkoping University [DP-13, DP-23, DP-24], MIT Lincoln Labs [DP-26, DP-31], AT&T Bell Labs [DP-15, DP-22, DP-27, FRA-24], BELLCORE [DP-29, DP-34, DP-37, DP-39], CCETT-CNET-XCOM-ENST-SGS Thomson Microelectronics (France) [DP-17, DP-28, DP-33, DP-41, DP-44], HHI Berlin GmbH [DP-53], Telettra [IC-53, IC-67, IC-73], Philips Research Labs [DP-46], Siemens AG [DP-52], Inmos, Ltd., Matsushita Electric [DP-54], Fujitsu Labs, Toshiba R&D Center, NEC, Universite de Neuchatel [DP-45], University of California at Davis [DP-35], and Stanford University [DP-47]. Also, DCT-based image compression boards [DP-40, PIT-29] (Appendix B.2) have been developed. As indicated in the introduction, the incentive and impetus for the development of the dedicated DCT chips came from the recommendations of the standards groups for the low bit rate codecs (videophone/videoconference), progressive image transmission, etc.

The performance of the DCT is compared with several other discrete transforms [B-1 through B-3, B-12] in Chapter 6. This comparison is based on variance distribution, residual correlation, rate distortion, and Wiener filtering. The results of the comparison reinforce the superior performance of the DCT. With the background on the definition, properties, fast algorithms, and performance of the DCT espoused in Chapters 2–6, the stage is now set for its applications. This aspect is addressed in Chapter 7. A wide spectrum of applications ranging from filtering to image coding and covering diverse disciplines (such as speech, video, printed image, medical imaging, transmultiplexers, surface texture, etc.) demonstrates the significance of the DCT in the DSP. Because of the extensive nature of these applications, the presentation is based on the block-diagram-concepts-properties format. Detailed design of a codec, for example, requires extensive exploration of the literature, classified into different categories, at the end of the book.

1.3 Appendices

As is customary with all the contemporary books, computer programs for implementing the DCT and other operations, such as the quantizer design, are provided. Documentation describes these programs in adequate detail. Additional appendices provide lists of manufacturers of DCT chips, image compression boards, and motion estimation chips.

1.4 References

While there is no optimal way of categorizing the references, the objective here is to classify them according to their applications. Some references may overlap the disciplines, and they may be cited more than once, the overriding concern being to make the list of references as extensive and exhaustive as possible. No claim as to the completeness of this list is made.

CHAPTER 2

DEFINITIONS AND GENERAL PROPERTIES

2.1 Introduction

Transforms, and in particular integral transforms, are used primarily for the reduction of complexity in mathematical problems. Differential equations and integral equations may, by judicious application of appropriate transforms, be changed into algebraic equations, whose solutions are more easily obtained. It is thus important to derive the basic mathematical properties of these transforms before applications are considered. Transform analysis, as applied in digital signal processing, bears a similar aim. The Fourier transform, which decomposes a signal into its frequency components, and the Karhunen–Loeve transform (KLT), which decorrelates a signal sequence, are well-known examples in the digital signal processing area. Here the mathematical properties are also important.

In discussing the discrete cosine transforms (DCTs), we shall first consider the Fourier cosine transform (FCT), whose properties are well known. These are recalled in Section 2.2. It is tempting to treat DCTs as discretized approximations of the continuous Fourier cosine transform. It would, however, be quite wrong. For in reality one has to deal with samples, measurements, and time instants. The continuum is merely an idealization to permit the use of calculus. Thus, in the remaining sections of this chapter, we will derive the various properties of the DCTs directly. We shall attempt to derive properties similar to those possessed by the continuous Fourier cosine transform, which will serve as a convenient and familiar starting point.

2.2 The Fourier Cosine Transform

We will start by recalling the definition of the Fourier transform. Given a function $x(t)$ for $-\infty < t < \infty$, its Fourier transform is given by (see, for example, Elliott and Rao [B-2] or Sneddon [M-1])

$$X(\omega) \equiv F[x(t)] = \left(\frac{1}{2\pi}\right)^{1/2} \int_{-\infty}^{\infty} x(t)e^{-j\omega t}\,dt, \tag{2.2.1a}$$

subject to the usual existence conditions for the integral. Here, $j = \sqrt{-1}$, $\omega = 2\pi f$ is the radian frequency and f is the frequency in Hertz. The function $x(t)$ can be recovered by the inverse Fourier transform, i.e.,

$$x(t) \equiv F^{-1}[X(\omega)] = \left(\frac{1}{2\pi}\right)^{1/2} \int_{-\infty}^{\infty} X(\omega)e^{j\omega t}\,d\omega. \tag{2.2.1b}$$

In (2.2.1), $F[\cdot]$ and $F^{-1}[\cdot]$ denote forward and inverse Fourier transforms of the functions enclosed. If $x(t)$ is defined only for $t \geqslant 0$, we can construct a function $y(t)$ given by:

$$\begin{aligned} y(t) = x(t) \qquad & t \geqslant 0, \\ x(-t) \qquad & t \leqslant 0. \end{aligned}$$

Then

$$\begin{aligned} F[y(t)] &= \left(\frac{1}{2\pi}\right)^{1/2} \left\{\int_0^{\infty} x(t)e^{-j\omega t}\,dt + \int_{-\infty}^{0} x(-t)e^{-j\omega t}\,dt\right\} \\ &= \left(\frac{1}{2\pi}\right)^{1/2} \int_0^{\infty} x(t)[e^{-j\omega t} + e^{j\omega t}]dt \\ &= \left(\frac{2}{\pi}\right)^{1/2} \int_0^{\infty} x(t)\cos(\omega t)dt. \end{aligned} \tag{2.2.2}$$

We can now define this as the Fourier cosine transform (FCT) of $x(t)$ given by

$$X_c(\omega) \equiv F_c[x(t)] = \left(\frac{2}{\pi}\right)^{1/2} \int_0^{\infty} x(t)\cos(\omega t)\,dt. \tag{2.2.3}$$

Noting that $X_c(\omega)$ is an even function of ω, we can apply the Fourier inversion to (2.2.2) to obtain

$$y(t) = x(t) \equiv F_c^{-1}[X_c(\omega)] = \left(\frac{2}{\pi}\right)^{1/2} \int_0^{\infty} X_c(\omega)\cos(\omega t)\,d\omega \qquad (t \geqslant 0) \tag{2.2.4}$$

Equations (2.2.3) and (2.2.4) define a FCT pair. Some of the properties are immediately clear.

(a) *Inversion:*

$$F_c \equiv F_c^{-1}. \tag{2.2.5}$$

It is clear from (2.2.3) and (2.2.4) that

$$F_c\{F_c[x(t)]\} = x(t) \qquad t \geqslant 0.$$

(b) *Linearity:*

$$F_c[\alpha x(t) + \beta y(t)] = \alpha X_c(\omega) + \beta Y_c(\omega), \tag{2.2.6}$$

where α and β are constants. F_c is clearly a linear operator.

(c) *Scaling in time:*

$$\begin{aligned} F_c[x(at)] &= \left(\frac{2}{\pi}\right)^{1/2} \int_0^\infty x(at)\cos(\omega t)\,dt \\ &= a^{-1}\left(\frac{2}{\pi}\right)^{1/2} \int_0^\infty x(\tau)\cos\left(\frac{\omega\tau}{a}\right)d\tau \\ &= \left(\frac{1}{a}\right)X_c\left(\frac{\omega}{a}\right) \qquad \text{for } a > 0. \end{aligned} \tag{2.2.7}$$

(d) *Shift in time:*

$$\begin{aligned} F_c[x(t-\alpha)] &= \left(\frac{2}{\pi}\right)^{1/2} \int_0^\infty x(t-\alpha)\cos(\omega t)\,dt \\ &= \left(\frac{2}{\pi}\right)^{1/2} \int_0^\infty x(\tau)\cos[\omega(\tau+\alpha)]\,d\tau \\ &= \cos(\omega\alpha)F_c[x(t)] - \sin(\omega\alpha)F_s[x(t)], \end{aligned} \tag{2.2.8a}$$

where F_s denotes the Fourier sine transform (FST) given by

$$F_s[x(t)] \equiv \left(\frac{2}{\pi}\right)^{1/2} \int_0^\infty x(t)\sin(\omega t)\,dt. \tag{2.2.8b}$$

We have also assumed that $x(t) = 0$ for $t < 0$.

(e) *Differentiation in time:*

$$F_c\left[\frac{d}{dt}x(t)\right] = \left(\frac{2}{\pi}\right)^{1/2} \int_0^\infty \left[\frac{d}{dt}x(t)\right]\cos(\omega t)\,dt$$

$$= \left(\frac{2}{\pi}\right)^{1/2} \left\{ [x(t)\cos(\omega t)]|_0^\infty + \omega \int_0^\infty x(t)\sin(\omega t)\, dt \right\}$$

$$= -\left(\frac{2}{\pi}\right)^{1/2} x(0) + \omega F_s[x(t)]. \tag{2.2.9}$$

We assume that $x(t)$ vanishes as t tends to infinity. Transforms of higher order derivatives may be obtained in a similar way (See Problem 2.1).

(f) *The convolution property:* Recall that the convolution of two functions $x(t)$ and $y(t)$ is given by

$$x(t) * y(t) \equiv \int_0^\infty x(t-u)y(u)\, du. \tag{2.2.10}$$

The Fourier cosine transform of $x * y$ is given by

$$F_c[x(t) * y(t)] = \left(\frac{2}{\pi}\right)^{1/2} \int_0^\infty \int_0^\infty x(t-u)y(u)\cos(\omega t)\, du\, dt$$

$$= \left(\frac{\pi}{2}\right)^{1/2} [X_c(\omega)Y_c(\omega) - X_s(\omega)Y_s(\omega)]. \tag{2.2.11}$$

In the following section, we shall derive properties for the discrete cosine transforms in parallel to the above properties for the Fourier cosine transform.

2.3 Definitions

It is seen in (2.2.3) that the Fourier cosine transform has a kernel given by

$$K_c(\omega, t) = \cos(\omega t) \tag{2.3.1a}$$

up to a normalization constant. Equation (2.2.5) shows that this kernel is involutary if proper normalization is taken into account. Let $\omega_m = 2\pi m\delta f$ and $t_n = n\delta t$ be the sampled angular frequency and time, where δf and δt represent the unit sample intervals for frequency and time, respectively; m and n are integers. Equation (2.3.1a) may now be written in the form

$$K_c(\omega_m, t_n) = K_c(2\pi m\delta f, n\delta t) = \cos(2\pi mn\delta f\delta t) = K_c(m, n). \tag{2.3.1b}$$

If we further let $\delta f\delta t = 1/(2N)$ where N is an integer, we have

$$K_c(m, n) = \cos\left(\frac{\pi mn}{N}\right). \tag{2.3.2}$$

Equation (2.3.2) represents a discretized Fourier cosine kernel. It is tempting to discretize the various properties from (a) to (f) in the previous section in a similar fashion. But, it is far simpler, both mathematically and conceptually, to regard the discrete kernel (2.3.2) as elements in an $(N + 1)$ by $(N + 1)$ transform matrix. Let this matrix be denoted by $[M]$. Then, the mnth element of this matrix is

$$[M]_{mn} = \cos\left(\frac{\pi mn}{N}\right) \qquad m, n = 0, 1, \ldots, N. \tag{2.3.3}$$

When $[M]$ is applied to a column vector $\mathbf{x} = [x(0), x(1), \ldots, x(N)]^T$, we obtain the vector $\mathbf{X} = [X(0), X(1), \ldots, X(N)]^T$ such that

$$\mathbf{X} = [M]\mathbf{x}, \tag{2.3.4}$$

where

$$X(m) = \sum_{n=0}^{N} \cos\left(\frac{\pi mn}{N}\right) x(n) \qquad m = 0, 1, \ldots, N. \tag{2.3.5}$$

The vector $\mathbf{x}$ is said to have undergone a discrete transform. This discrete cosine transform in (2.3.5) was first reported by Kitajima in 1980 [S-1], and was named the symmetric cosine transform (SCT).

Since inversion is part of transform processing in general, nonsingular matrices are usually preferred. However, there are discrete transforms that are singular or noninvertible. An example of this is the rapid transform (RT) (see Reitboeck and Brody [M-23]). In practice, if the nonsingular matrix is real and orthogonal, its inverse is easily obtained as its transpose. Such unitary transform matrices (with the proper normalizations) have a preeminent place in the digital signal processing field. We shall now present the definitions for the four discrete cosine transforms as classified by Wang [FD-1].

(1) DCT-I:

$$[C^{\mathrm{I}}_{N+1}]_{mn} = \left(\frac{2}{N}\right)^{1/2}\left[k_m k_n \cos\left(\frac{mn\pi}{N}\right)\right] \qquad m, n = 0, 1, \ldots, N;$$

(2) DCT-II:

$$[C^{\mathrm{II}}_{N}]_{mn} = \left(\frac{2}{N}\right)^{1/2}\left[k_m \cos\left(\frac{m(n + \frac{1}{2})\pi}{N}\right)\right] \qquad m, n = 0, 1, \ldots, N - 1;$$

(3) DCT-III:

$$[C^{\mathrm{III}}_{N}]_{mn} = \left(\frac{2}{N}\right)^{1/2}\left[k_n \cos\left(\frac{(m + \frac{1}{2})n\pi}{N}\right)\right] \qquad m, n = 0, 1, \ldots, N - 1;$$

(4) DCT-IV:

$$[C^{\mathrm{IV}}_{N}]_{mn} = \left(\frac{2}{N}\right)^{1/2}\left[\cos\left\{\frac{(m + \frac{1}{2})(n + \frac{1}{2})\pi}{N}\right\}\right] \qquad m, n = 0, 1, \ldots, N - 1;$$

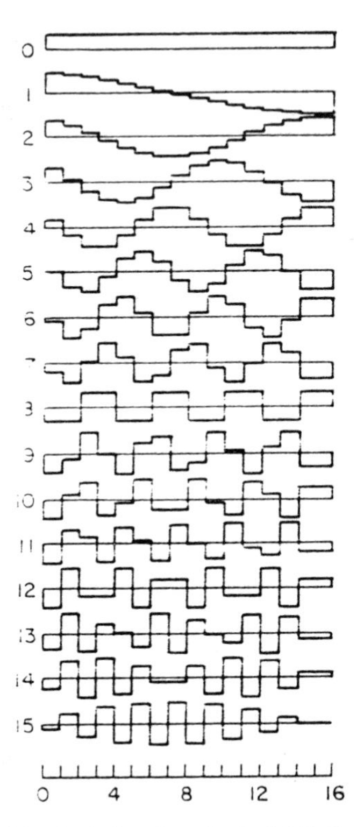

Fig. 2.1 Basis functions for DCT-II, $N = 16$.

where

$$k_j = 1 \qquad \text{if } j \neq 0 \text{ or } N$$
$$= \frac{1}{\sqrt{2}} \qquad \text{if } j = 0 \text{ or } N. \tag{2.3.6}$$

Here, we note that DCT-II is the discrete cosine transform first reported by Ahmed, Natarajan, and Rao [G-1]. DCT-III is obviously the transpose of DCT-II, and DCT-IV is the shifted version of DCT-I. The basis functions for DCT-II for $N = 16$ are shown in Fig. 2.1.

2.4 The Unitarity of DCTs

In the original derivations of these discrete transforms (see Ahmed *et al.* [G-1], Kitajima, [S-1]), a diagonalization problem is considered and the unitarity of the transform matrices is assured, since they are the similarity transform matrices in the diagonalization problem.

Here, we recall the definition of an inner product for two N-dimensional real vectors $\mathbf{g}_k$ and $\mathbf{g}_m$:

$$\langle \mathbf{g}_k, \mathbf{g}_m \rangle = \sum_{n=1}^{N} g_{nk} g_{nm}, \tag{2.4.1}$$

where $\mathbf{g}_k = [g_{1k}, g_{2k}, \ldots, g_{Nk}]^T$.

$\mathbf{g}_k$ and $\mathbf{g}_m$ are said to be orthogonal if their inner product vanishes for $k \neq m$. In addition, if $\langle \mathbf{g}_k, \mathbf{g}_k \rangle = 1$, the vectors are said to be normalized, and the corresponding real transform matrix is then unitary.

Instead of deriving the transforms by considering the diagonalization problem, we shall use the simple concept in (2.4.1) and investigate the unitarity property of the discrete transforms directly.

1. *DCT-I* Let $\mathbf{g}_k$ denote the kth column vector in the $(N+1)$ by $(N+1)$ transform matrix C^{I}_{N+1}, then

$$\begin{aligned}
\langle \mathbf{g}_k, \mathbf{g}_m \rangle &= \sum_{n=0}^{N} \left(\frac{2}{N}\right) k_n k_k \cos\left(\frac{nk\pi}{N}\right) k_n k_m \cos\left(\frac{nm\pi}{N}\right) \\
&= \left(\frac{k_k k_m}{N}\right)\left\{1 + (-1)^{m+k} + \sum_{n=1}^{N-1} \left(\cos\left[\frac{(m+k)n\pi}{N}\right] + \cos\left[\frac{(m-k)n\pi}{N}\right]\right)\right\} \\
&= \left(\frac{k_k k_m}{N}\right)\left\{\sum_{n=0}^{N-1} \cos\left[\frac{(m-k)n\pi}{N}\right] + \sum_{n=1}^{N} \cos\left[\frac{(m+k)n\pi}{N}\right]\right\} \\
&= \left(\frac{k_k k_m}{N}\right) \mathrm{Re}\left\{\sum_{n=0}^{N-1} W_{2N}^{-n(m-k)} + \sum_{n=1}^{N} W_{2N}^{-n(m+k)}\right\},
\end{aligned} \tag{2.4.2}$$

where we have defined W_{2N} as the primitive $2N$th root of 1, i.e.,

$$W_{2N} = \exp\left[-\frac{j\pi}{N}\right] = \left[\cos\left(\frac{\pi}{N}\right) - j\sin\left(\frac{\pi}{N}\right)\right]. \tag{2.4.3}$$

To examine the unitarity property, we consider

$$\begin{aligned}
\sum_{n=0}^{N-1} (W_{2N}^{-p})^n &= \frac{[1 - W_{2N}^{-Np}]}{[1 - W_{2N}^{-p}]}, \qquad \text{where } p = (m-k) \\
&= \left[2\left(1 - \cos\left[\frac{p\pi}{N}\right]\right)\right]^{-1} \{1 - W_{2N}^{-Np} - W_{2N}^{p} + W_{2N}^{-(N-1)p}\}.
\end{aligned} \tag{2.4.4}$$

Similarly, for $q = m + k$,

$$\sum_{n=1}^{N} (W_{2N}^{-q})^n = \left[2\left(1 - \cos\left[\frac{q\pi}{N}\right]\right)\right]^{-1} \{W_{2N}^{-q} - W_{2N}^{-(N+1)q} - 1 + W_{2N}^{-Nq}\}. \tag{2.4.5}$$

Thus for $m \neq k$, $p \neq 0$, we have

$$\operatorname{Re}\left[\sum_{n=0}^{N-1} W_{2N}^{-np}\right] = \frac{[1-(-1)^{p}]\left[1-\cos\left(\frac{p\pi}{N}\right)\right]}{\left[2\left(1-\cos\left(\frac{p\pi}{N}\right)\right)\right]}$$

and

$$\operatorname{Re}\left[\sum_{n=1}^{N} W_{2N}^{-nq}\right] = \frac{-[1-(-1)^{q}]\left[1-\cos\left(\frac{q\pi}{N}\right)\right]}{\left[2\left(1-\cos\left(\frac{q\pi}{N}\right)\right)\right]}.$$

Combining these two expressions, we see that since p and q differ by $2k$

$$\langle \mathbf{g}_k, \mathbf{g}_m \rangle = 0 \qquad \text{when } m \neq k \text{ (i.e., } p \neq 0). \tag{2.4.6}$$

For $m = k$, $k \neq 0$ or N, (2.4.2) reduces to

$$\langle \mathbf{g}_k, \mathbf{g}_k \rangle = \left(\frac{1}{N}\right) \operatorname{Re}\left[\sum_{n=0}^{N-1} 1 + \sum_{n=1}^{N} (W_{2N}^{-2k})^{n}\right] = 1. \tag{2.4.7}$$

For $m = k$, $k = 0$, or N, we have

$$\langle \mathbf{g}_k, \mathbf{g}_k \rangle = \left(\frac{1}{2N}\right) \operatorname{Re}\left[\sum_{n=0}^{N-1} 1 + \sum_{n=1}^{N} 1\right] = 1. \tag{2.4.8}$$

Combining (2.4.6), (2.4.7), and (2.4.8), we can now represent the unitarity property of DCT-I by

$$\langle \mathbf{g}_k, \mathbf{g}_m \rangle = \delta_{km}, \tag{2.4.9}$$

where

$$\delta_{km} = 1 \qquad \text{for } k = m,$$
$$= 0 \qquad \text{for } k \neq m.$$

Thus

$$[C_{N+1}^{\mathrm{I}}]^{-1} = [C_{N+1}^{\mathrm{I}}]^{T} = [C_{N+1}^{\mathrm{I}}]. \tag{2.4.10a}$$

The procedure used for demonstrating the unitarity of DCT-I can be applied to the other DCTs with little or no modifications (Problem 2.3). We shall simply state the results:

(2) DCT-II: $[C_N^{\mathrm{II}}]^{-1} = [C_N^{\mathrm{II}}]^{T} = [C_N^{\mathrm{III}}]$.

(3) DCT-III: $[C_N^{\mathrm{III}}]^{-1} = [C_N^{\mathrm{III}}]^{T} = [C_N^{\mathrm{II}}]$.

(4) DCT-IV: $[C_N^{\mathrm{IV}}]^{-1} = [C_N^{\mathrm{IV}}]^{T} = [C_N^{\mathrm{IV}}]$. (2.4.10b)

We note here that only DCT-I and DCT-IV are involutary, and thus property (2.2.5) of the Fourier cosine transform is retained only by these two DCTs. Based on (2.3.6) and (2.4.10), the family of forward and inverse DCTs can be defined as follows:

(1) DCT-I
Forward:

$$X^{C(1)}(m) = \left(\frac{2}{N}\right)^{1/2} k_m \sum_{n=0}^{N} k_n x(n) \cos\left(\frac{mn\pi}{N}\right), \qquad m = 0, \ldots, N;$$

Inverse:

$$x(n) = \left(\frac{2}{N}\right)^{1/2} k_n \sum_{m=0}^{N} k_m X^{C(1)}(m) \cos\left(\frac{mn\pi}{N}\right), \qquad n = 0, \ldots, N$$

(2) DCT-II
Forward:

$$X^{C(2)}(m) = \left(\frac{2}{N}\right)^{1/2} k_m \sum_{n=0}^{N-1} x(n) \cos\left[\frac{(2n+1)m\pi}{2N}\right], \qquad m = 0, \ldots, N-1;$$

Inverse:

$$x(n) = \left(\frac{2}{N}\right)^{1/2} \sum_{m=0}^{N-1} k_m X^{C(2)}(m) \cos\left[\frac{(2n+1)m\pi}{2N}\right], \qquad n = 0, \ldots, N-1;$$

(3) DCT-III
Forward:

$$X^{C(3)}(m) = \left(\frac{2}{N}\right)^{1/2} \sum_{n=0}^{N-1} k_n x(n) \cos\left[\frac{(2m+1)n\pi}{2N}\right], \qquad m = 0, \ldots, N-1;$$

Inverse:

$$x(n) = \left(\frac{2}{N}\right)^{1/2} k_n \sum_{m=0}^{N-1} X^{C(3)}(m) \cos\left[\frac{(2m+1)n\pi}{2N}\right], \qquad n = 0, \ldots, N-1;$$

(4) DCT-IV
Forward:

$$X^{C(4)}(m) = \left(\frac{2}{N}\right)^{1/2} \sum_{n=0}^{N-1} x(n) \cos\left[\frac{(2m+1)(2n+1)\pi}{4N}\right], \qquad m = 0, \ldots, N-1;$$

Inverse:

$$x(n) = \left(\frac{2}{N}\right)^{1/2} \sum_{m=0}^{N-1} X^{C(4)}(m) \cos\left[\frac{(2m+1)(2n+1)\pi}{4N}\right], \qquad n = 0, \ldots, N-1;$$

where

$$k_p = \frac{1}{\sqrt{2}} \qquad \text{when } p = 0 \text{ or } N$$

$$= 1 \qquad \text{when } p \neq 0 \text{ and } N. \tag{2.4.11a}$$

In (2.4.11a) $x(n)$ is the input (data) sequence and $X^{C(i)}(m)$ is the transform sequence corresponding to DCT-i, where $i = 1, 2, 3, 4$. It may be noted that the normalization factor $\sqrt{(2/N)}$ that appears in both the forward and the inverse transforms can be merged as $2/N$ and moved to either the forward or inverse transform. For example, DCT-II can also be defined as

$$X^{C(2)}(m) = \left(\frac{2}{N}\right) k_m \sum_{n=0}^{N-1} x(n) \cos\left[\frac{(2n+1)m\pi}{2N}\right], \qquad m = 0, \ldots, N-1,$$

and its inverse,

$$x(n) = \sum_{m=0}^{N-1} k_m X^{C(2)}(m) \cos\left[\frac{(2n+1)m\pi}{2N}\right], \qquad n = 0, \ldots, N-1. \tag{2.4.11b}$$

It is evident that the terms $X^{C(2)}(m)$ in (2.4.11a) and (2.4.11b) are related by the scale factor $\sqrt{(2/N)}$. By merging these normalization factors, the family of DCT matrices are no longer orthonormal. They are, however, still orthogonal.

The linearity property as expressed by (2.2.6) for the Fourier cosine transform is easily extended to the DCTs. Since matrix multiplication is a linear operation, i.e.,

$$[M]\{\alpha\mathbf{g} + \beta\mathbf{f}\} = \alpha[M]\mathbf{g} + \beta[M]\mathbf{f},$$

for a matrix $[M]$, constants α and β, and vectors $\mathbf{g}$ and $\mathbf{f}$, all four DCTs are linear transforms, retaining property (2.2.6).

2.5 Scaling in Time

Recall that in the discretization of the Fourier cosine transform, we consider

$$\delta f \delta t = \left(\frac{1}{2N}\right) \qquad \text{or} \tag{2.5.1}$$

$$\delta f = \left(\frac{1}{2N}\right)/\delta t. \tag{2.5.2a}$$

Since the DCTs deal with discrete sample points, a scaling in time has no effect in the transform, except in changing the unit frequency interval in the transform domain. Thus as δt changes to $a\delta t$, δf changes to $\delta f/a$, provided N remains constant.

Thus, a scaling in time leads to a scaling in frequency without a scaling of the transform, and property (2.2.7) is retained, except for the $1/a$ factor, which is absent in the DCTs case.

Equation (2.5.2a) may also be interpreted to produce the frequency resolution for a given sample length. Combining $N\delta t$ as T, we have

$$\delta f = \left(\frac{1}{2T}\right), \tag{2.5.2b}$$

where T is the duration in time for a sequence of length N.

2.6 Shift in Time

Let $\mathbf{x} = [x(0), \ldots, x(N)]^T$ and $\mathbf{x}_+ = [x(1), \ldots, x(N+1)]^T$ be two sampled data vectors of dimension $N + 1$. $\mathbf{x}_+$ is the data vector $\mathbf{x}$ shifted by one sample point. We shall detail here the shift property for the DCTs, corresponding to property (2.2.8) for the Fourier cosine transform. Again, we consider first the DCT-I.

(1) DCT-I

Using (2.3.6), we can define the DCT-I of both $\mathbf{x}$ and $\mathbf{x}_+$ as

$$X^{C(1)}(m) = \left(\frac{2}{N}\right)^{1/2} \sum_{n=0}^{N} k_m k_n \cos\left(\frac{mn\pi}{N}\right) x(n)$$

and

$$X_+^{C(1)}(m) = \left(\frac{2}{N}\right)^{1/2} \sum_{n=0}^{N} k_m k_n \cos\left(\frac{mn\pi}{N}\right) x(n+1), \tag{2.6.1}$$

where

$$\mathbf{X}^{C(1)} = [X^{C(1)}(0), \ldots, X^{C(1)}(N)]^T \qquad \text{and}$$

$$\mathbf{X}_+^{C(1)} = [X_+^{C(1)}(0), \ldots, X_+^{C(1)}(N)]^T$$

are, respectively, the DCT-I of the vectors $\mathbf{x}$ and $\mathbf{x}_+$.

Taking the mth element of $\mathbf{X}_+^{C(1)}$, and changing the summation index n to $(n + 1) - 1$, we obtain:

$$X_+^{C(1)}(m) = \left(\frac{2}{N}\right)^{1/2} \left\{ k_m \cos\left(\frac{m\pi}{N}\right) \sum_{n=0}^{N} k_n \cos\left[\frac{m(n+1)\pi}{N}\right] x(n+1) \right.$$
$$\left. + k_m \sin\left(\frac{m\pi}{N}\right) \sum_{n=0}^{N} k_n \sin\left[\frac{m(n+1)\pi}{N}\right] x(n+1) \right\}. \tag{2.6.2a}$$

We shall simplify this to

$$X_+^{C(1)}(m) = \left(\frac{2}{N}\right)^{1/2} k_m \left\{ \cos\left(\frac{m\pi}{N}\right) S_1 + \sin\left(\frac{m\pi}{N}\right) S_2 \right\}, \tag{2.6.2b}$$

where the sums S_1 and S_2 in (2.6.2b) can be further simplified. We show here the details for S_1:

$$S_1 = \sum_{n=0}^{N} k_n \cos\left[\frac{m(n+1)\pi}{N}\right] x(n+1)$$

$$= \left(\frac{1}{2}\right)^{1/2} \cos\left(\frac{m\pi}{N}\right) x(1) + \sum_{n=2}^{N} \cos\left(\frac{mn\pi}{N}\right) x(n)$$

$$+ \left(\frac{1}{2}\right)^{1/2} (-1)^m \cos\left(\frac{m\pi}{N}\right) x(N+1).$$

The second term in the above expression can be made into the DCT-I of the vector $\mathbf{x}$ by introducing the proper terms. With this, we obtain for S_1 the following result:

$$S_1 = \left(\frac{N}{2}\right)^{1/2} \frac{X^{C(1)}_{(m)}}{k_m} - \frac{x(0)}{\sqrt{2}} + \left(\frac{1}{\sqrt{2}} - 1\right) \cos\left(\frac{m\pi}{N}\right) x(1)$$

$$+ \left(1 - \frac{1}{\sqrt{2}}\right)(-1)^m x(N) + \left(\frac{1}{\sqrt{2}}\right)(-1)^m \cos\left(\frac{m\pi}{N}\right) x(N+1). \quad (2.6.3)$$

Similarly, we obtain for S_2:

$$S_2 = \left(\frac{N}{2}\right)^{1/2} X^{S(1)}(m) + \left(\frac{1}{\sqrt{2}} - 1\right) \sin\left(\frac{m\pi}{N}\right) x(1)$$

$$+ \left(\frac{1}{\sqrt{2}}\right)(-1)^m \sin\left(\frac{m\pi}{N}\right) x(N+1), \quad (2.6.4)$$

where we have defined

$$X^{S(1)}(m) = \left(\frac{2}{N}\right)^{1/2} \sum_{n=1}^{N-1} \sin\left(\frac{mn\pi}{N}\right) x(n), \quad (2.6.5)$$

which is the discrete sine transform of type I (DST-I) (see Jain [M-2]). Substituting (2.6.3) and (2.6.4) into (2.6.2b), we finally obtain the shift property for DCT-I:

$$\mathbf{X}^{C(1)}_{+}(m) = \cos\left(\frac{m\pi}{n}\right) X^{C(1)}(m) + k_m \sin\left(\frac{m\pi}{N}\right) X^{S(1)}(m)$$

$$+ \left(\frac{2}{N}\right)^{1/2} k_m \left\{\left(\frac{-1}{\sqrt{2}}\right) \cos\left(\frac{m\pi}{N}\right) x(0) + \left(\frac{1}{\sqrt{2}} - 1\right) x(1)\right.$$

$$\left. + (-1)^m \left(1 - \frac{1}{\sqrt{2}}\right) \cos\left(\frac{m\pi}{N}\right) x(N) + (-1)^m \frac{x(N+1)}{\sqrt{2}}\right\}. \quad (2.6.6)$$

We note that the first two terms are similar to the expression in (2.2.8). The remaining terms make the shift property for the DCTs much more complicated compared with that for the Fourier cosine transform.

However, it should be noted that when instantaneous DCTs have to be carried out on a continuous incoming data stream, (2.6.6) does represent a possible way of updating the instantaneous DCTs without having to perform complete DCTs at every instant. Such a situation may exist, for example, in the use of adaptive filtering in the transform domain.

Similar relations are available for the remaining DCTs (Problem 2.4). We list them here without detailed derivation (see Yip and Rao [FRA-14]).

(2) DCT-II

$$X_{+}^{C(2)}(m) = \cos\left(\frac{m\pi}{N}\right) X^{C(2)}(m) + \sin\left(\frac{m\pi}{N}\right) X_{-}^{S(2)}(m) + \left(\frac{2}{N}\right)^{1/2} k_m[(-1)^m x(N) - x(0)] \cos\left(\frac{m\pi}{2N}\right), \tag{2.6.7}$$

where

$$\mathbf{X}^{C(2)} = [C_N^{\text{II}}]\mathbf{x}, \; \mathbf{X}_{+}^{C(2)} = [C_N^{\text{II}}]\mathbf{x}_{+}$$

are the DCT-II for the data vector $\mathbf{x}$ and its time-shifted version $\mathbf{x}_{+}$,respectively. Also $X_{-}^{S(2)}(m)$ is used to denote the mth transform output of the DST (type II) for the sequence $x(0), \ldots, x(N-1)$, as defined by Wang (see Wang [FD-1]). We shall give the definitions for the DSTs later.

(3) DCT-III

$$X_{+}^{C(3)}(m) = \cos\left[\left(\frac{(2m+1)\pi}{2N}\right]X^{C(3)}(m) + \sin\left[\frac{(2m+1)\pi}{2N}\right] X^{S(3)}(m+1) + \left(\frac{2}{N}\right)^{1/2}\left\{\left(\frac{1}{\sqrt{2}} - 1\right)x(1) - \cos\left[\frac{(2m+1)\pi}{2N}\right]\frac{x(0)}{\sqrt{2}} + (-1)^m\left(1 - \frac{1}{\sqrt{2}}\right)\sin\left[\frac{(2m+1)\pi}{2N}\right] x(N)\right\}, \tag{2.6.8}$$

where

$$\mathbf{X}^{C(3)} = [C_N^{\text{III}}]\mathbf{x}, \; \mathbf{X}_{+}^{C(3)} = [C_N^{\text{III}}]\mathbf{x}_{+}$$

are the DCT-III of $\mathbf{x}$ and $\mathbf{x}_{+}$, respectively. Also $\mathbf{X}^{S(3)}$ is the DST-III of the data vector, to be defined later.

(4) DCT-IV

$$X_{+}^{C(4)}(m) = \cos\left[\frac{(2m+1)\pi}{2N}\right] X^{C(4)}(m) + \sin\left[\frac{(2m+1)\pi}{2N}\right] X^{S(4)}(m) + \left(\frac{2}{N}\right)^{1/2}\left\{-\cos\left[\frac{(2m+1)\pi}{4N}\right] x(0) + (-1)^m \sin\left[\frac{(2m+1)\pi}{4N}\right] x(N)\right\}, \tag{2.6.9}$$

where

$$\mathbf{X}^{C(4)} = [C_N^{\mathrm{IV}}]\mathbf{x}, \ \mathbf{X}_+^{C(4)} = [C_N^{\mathrm{IV}}]\mathbf{x}_+$$

are the DCT-IV of the data vector and its time-shifted version, respectively. Also, $\mathbf{X}^{S(4)}$ is the DST-IV of the data vector, to be defined later.

Equations (2.6.6) through (2.6.9) delineate the shift property of the four types of DCTs. It should be noted that none of them is as simple as the shift property (2.2.8) for the Fourier cosine transform. All of them involve the corresponding discrete sine transforms, which are defined in the following equations [FD-1]:

$$\text{DST-I:} \quad [S_{N-1}^{\mathrm{I}}]_{mn} = \left(\frac{2}{N}\right)^{1/2} \left\{\sin\left(\frac{mn\pi}{N}\right)\right\} \qquad m, n = 1, 2, \ldots, N-1$$

$$\text{DST-II:} \quad [S_N^{\mathrm{II}}]_{mn} = \left(\frac{2}{N}\right)^{1/2} \left\{k_m \sin\left[\frac{m(n-\frac{1}{2})\pi}{N}\right]\right\} \qquad m, n = 1, 2, \ldots, N$$

$$\text{DST-III:} \quad [S_N^{\mathrm{III}}]_{mn} = \left(\frac{2}{N}\right)^{1/2} \left\{k_n \sin\left[\frac{(m-\frac{1}{2})n\pi}{N}\right]\right\} \qquad m, n = 1, 2, \ldots, N$$

$$\text{DST-IV:} \quad [S_N^{\mathrm{IV}}]_{mn} = \left(\frac{2}{N}\right)^{1/2} \left\{\sin\left[\frac{(m+\frac{1}{2})(n+\frac{1}{2})\pi}{N}\right]\right\}$$

$$m, n = 0, 1, \ldots, N-1, \tag{2.6.10a}$$

where

$$k_n = 1 \qquad \text{for } n \neq N$$

$$= \frac{1}{\sqrt{2}} \qquad \text{for } n = N.$$

Similar to (2.4.11), the forward and inverse transforms for the DST family can be explicitly expressed. Also, the DST relations corresponding to (2.4.10) are [FD-1]:

$$\text{DST-I:} \quad [S_{N-1}^{\mathrm{I}}]^{-1} = [S_{N-1}^{\mathrm{I}}]^T = [S_{N-1}^{\mathrm{I}}],$$

$$\text{DST-II:} \quad [S_N^{\mathrm{II}}]^{-1} = [S_N^{\mathrm{II}}]^T = [S_N^{\mathrm{III}}], \tag{2.6.10b}$$

$$\text{DST-III:} \quad [S_N^{\mathrm{III}}]^{-1} = [S_N^{\mathrm{III}}]^T = [S_N^{\mathrm{II}}],$$

$$\text{DST-IV:} \quad [S_N^{\mathrm{IV}}]^{-1} = [S_N^{\mathrm{IV}}]^T = [S_N^{\mathrm{IV}}].$$

As in the case of the DCTs, the normalization factor $\sqrt{(2/N)}$, which appears in both the forward and the inverse transforms in (2.6.10a), can be merged as $(2/N)$ and moved to either the forward or the inverse transform. The basis functions of DST-I for $N = 16$ are shown in Fig. 2.2.

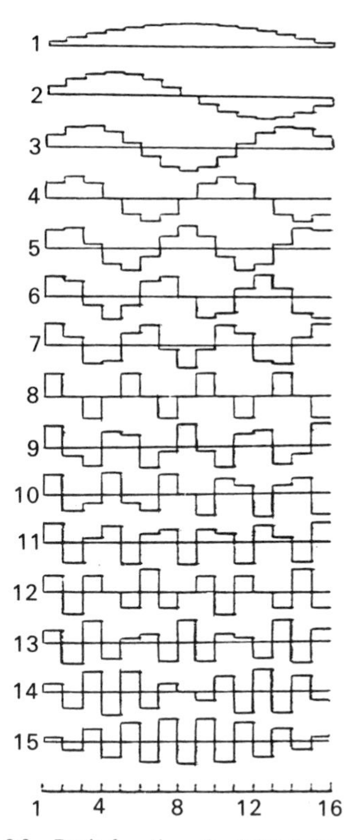

Fig. 2.2 Basis functions for DST-I, $N = 16$.

It is clear that the shift properties for the DSTs similar to those for the DCTs can be developed (Problem 2.5).

We conclude this section by noting that although the shift property is derived for a unit time shift, transforms of more general time shifts can be obtained by the repeated application of the appropriate formulae.

2.7 The Difference Property

Consider a sequence of data points, consisting of differences between adjacent samples, i.e.,

$$d(n) = x(n+1) - x(n) \qquad n = 0, 1, \ldots, N, \tag{2.7.1a}$$

which can also be written in terms of the data vectors:

$$\mathbf{d} = \mathbf{x}_{+} - \mathbf{x}. \tag{2.7.1b}$$

It is clear that the DCT of $d(n)$ can be obtained by applying the shift property derived in the last section.

(1) DCT-I: $\mathbf{D}^{C(1)} = \mathbf{X}_{+}^{C(1)} - \mathbf{X}^{C(1)},$ (2.7.2)

where $\mathbf{D}^{C(1)}$ denotes the DCT-I of the difference sequence defined in (2.7.1b). Other DCTs are similarly obtained.

$$\begin{aligned}
&(2)\quad \text{DCT-II:}\quad \mathbf{D}^{C(2)} = \mathbf{X}_{+}^{C(2)} - \mathbf{X}^{C(2)},\\
&(3)\quad \text{DCT-III:}\quad \mathbf{D}^{C(3)} = \mathbf{X}_{+}^{C(3)} - \mathbf{X}^{C(3)},\\
&(4)\quad \text{DCT-IV:}\quad \mathbf{D}^{C(4)} = \mathbf{X}_{+}^{C(4)} - \mathbf{X}^{C(4)}.
\end{aligned} \tag{2.7.3}$$

2.8 The Convolution Property

For discrete finite sequences $x(n)$ and $y(n)$, two types of convolutions are defined (see Elliott and Rao [B-2]). The circular convolution $a(i)$ is defined for sequences $x(n)$ and $y(n)$, which are both periodic with a period of N. $a(i)$ is given by

$$a(i) = x(n) * y(n) = \left(\frac{1}{\sqrt{N}}\right) \sum_{m=0}^{N-1} x(m)y(i-m), \qquad i = 0, 1, \ldots, N-1. \tag{2.8.1}$$

The linear convolution $b(i)$ for two nonperiodic sequences $x(n)$ and $y(n)$, of lengths L and M, respectively, is defined by

$$b(i) = x(n) * y(n) = \left(\frac{1}{\sqrt{N}}\right) \sum_{n=0}^{N-1} x(n)y(i-n), \qquad i = 0, 1, \ldots, N-1, \tag{2.8.2}$$

where $N \geqslant L + M - 1$. By augmenting $x(n)$ and $y(n)$ with zeros to equal lengths (N), $b(i)$ can be obtained as a portion of a circular convolution of the augmented sequences.

In this section, we derive the convolution property for DCT-II only. We shall see that because DCT-II can be easily related to DFT, the convolution property of DFT can be borrowed to more easily arrive at a comparable result for DCT-II. We shall also see that instead of the conventional convolution as expressed by (2.8.1) and (2.8.2), the convolution of even extension of given sequences must be considered.

Consider the sequence $f(n)$, $n = 0, 1, \ldots, N-1$. Construct a $2N$ sequence defined by

$$\begin{aligned}
\hat{f}(n) &= f(n) & n &= 0, 1, \ldots, N-1\\
&= f(2N-n-1) & &= N, N+1, \ldots, 2N-1.
\end{aligned} \tag{2.8.3}$$

Thus $\hat{f}(n)$ is an even extension of $f(n)$. Now the $2N$-point DFT of this sequence is given by

$$\hat{F}_F(k) = \left(\frac{1}{\sqrt{2N}}\right) \sum_{n=0}^{2N-1} \hat{f}(n) \exp\left[-\frac{j2\pi kn}{2N}\right]. \tag{2.8.4}$$

Using the symmetry in (2.8.3) we can now rewrite (2.8.4) as

$$\hat{F}_F(k) = \left(\frac{1}{\sqrt{2N}}\right) \sum_{n=0}^{N-1} f(n) \left\{ \exp\left[\frac{-j2\pi kn}{2N}\right] + \exp\left[\frac{j2\pi k(n+1)}{2N}\right] \right\}. \tag{2.8.5}$$

Let us now introduce a phase shift and a weight factor into (2.8.5),

$$k_k \hat{F}_F(k) \exp\left[\frac{-j\pi k}{2N}\right] = \sqrt{\left(\frac{2}{N}\right)} \sum_{n=0}^{N-1} k_k f(n) \cos\left\{\frac{k(n+\frac{1}{2})\pi}{N}\right\}, \tag{2.8.6}$$

where

$$\begin{aligned} k_m &= \frac{1}{\sqrt{2}} && m = 0, \\ &= 1 && m = 1, \ldots, N-1, \\ &= 0 && \text{otherwise.} \end{aligned}$$

We recognize the right-hand side of (2.8.6) as the definition (2.3.6) for DCT-II for the sequence $f(n)$. Therefore, using $F^{C(2)}(k)$ to denote the DCT-II for the sequence, we have

$$F^{C(2)}(k) = k_k \exp\left[\frac{-j\pi k}{2N}\right] \hat{F}_F(k), \qquad k = 0, 1, \ldots, N-1. \tag{2.8.7}$$

Equation (2.8.7) provides the relation between the DCT-II of the unappended sequence $f(n)$ and the Fourier transform of the appended sequence $\hat{f}(n)$. It is not difficult to show that the sequence $f(n)$ can be recovered from the transformed sequence $F^{C(2)}(k)$, and as a result we have

$$f(n) = \sqrt{\left(\frac{2}{N}\right)} \operatorname{Re}\left\{ \sum_{k=0}^{2N-1} k_k F^{C(2)}(k) \exp\left[\frac{j\pi k}{2N}\right] \exp\left[\frac{2j\pi kn}{2N}\right] \right\}$$
$$n = 0, 1, \ldots, N-1. \tag{2.8.8}$$

This equation states that the inverse N-point DCT-II of the sequence $F^{C(2)}(k)$ is given by twice the real part of a $(2N-1)$-point DFT inverse of the weighted and phase-shifted $F^{C(2)}(k)$.

Consider next the product of two DCT-II transforms $F^{C(2)}(k)$ and $G^{C(2)}(k)$ of the sequences $f(n)$ and $g(n)$, respectively,

$$W^{C(2)}(k) = F^{C(2)}(k) G^{C(2)}(k) \qquad k = 0, 1, \ldots, N-1. \tag{2.8.9}$$

Using (2.8.6), we get

$$W^{C(2)}(k) = k_k^2 \exp\left[\frac{-j\pi k}{N}\right] \hat{F}_F(k) \hat{G}_F(k), \tag{2.8.10}$$

where $\hat{F}_F(k)$ and $\hat{G}_F(k)$ are as defined in (2.8.5). To obtain the DCT-II inverse of (2.8.10), we apply the result in (2.8.8)

$$w(n) = 2\,\mathrm{Re}\left\{F_F^{-1}\left[k_k^3 \exp\left(-\frac{j\pi kn}{2N}\right)\hat{F}_F(k)\hat{G}_F(k)\right]\right\} \tag{2.8.11}$$

Here, F_F^{-1} denotes the inverse DFT operator. It is evident that the convolution theorem for the Fourier transform can now be applied. This results in the following:

$$w(n) = 2\,\mathrm{Re}\left\{F_F^{-1}\left[k_k^3 \exp\left(-\frac{j\pi kn}{2N}\right)\right] * \hat{f}(n) * \hat{g}(n)\right\}, \tag{2.8.12}$$

where $*$ denotes circular convolution. The remaining inverse DFT can be explicitly evaluated (see Chen and Fralick [IE-1]) to give

$$\begin{aligned} F_F^{-1}\left[k_k^3 \exp\left(\frac{-j\pi kn}{2N}\right)\right] &= \frac{\hat{h}(n)}{2} \\ &= \left\{\left(\frac{1}{2\sqrt{2}} - 1\right) + \exp\left[\frac{j(2n-1)(N-1)\pi}{4N}\right] \right. \\ &\quad \left. \times \frac{\sin\left[\dfrac{(2n-1)\pi}{4}\right]}{\sin\left[\dfrac{(2n-1)\pi}{4N}\right]}\right\} \Big/ \sqrt{(2N)}. \end{aligned} \tag{2.8.13a}$$

Since $\hat{f}(n)$ and $\hat{g}(n)$ are real, we get

$$w(n) = \hat{h}(n) * \hat{f}(n) * \hat{g}(n). \tag{2.8.13b}$$

With these results, we are now finally ready to state the convolution theorem for the DCT-II:

$$F^{C(2)}[\hat{h}(n) * \hat{f}(n) * \hat{g}(n)] = F^{C(2)}(k)G^{C(2)}(k), \qquad k = 0, 1, \ldots, N-1, \tag{2.8.14}$$

where $\hat{h}(n)$ is given by (2.8.13).

The result in (2.8.14) indicates that the product of the DCT-II transforms of two sequences is the DCT-II transform of the convolution of these two sequences and a third function given by $\hat{h}(n)$. In addition, the convolutions are circular convolutions of the evenly extended sequences (2.8.3). Similar convolution properties may be derived for the other members of the DCT family (Problem 2.11). Such properties are found to be useful for transform domain filtering (see Ngan and Clarke [F-1]). Another convolution property that is simpler than (2.8.14) has been developed for DCT-II by Chitprasert and Rao [F-7]. This property (Problem 2.18) has been utilized in progressive image transmission weighted by the human visual sensitivity [PIT-12].

2.9 Summary

We have briefly presented some relevant properties of the DCT. The discussion of the properties of the Fourier cosine transform forms the basis upon which corresponding properties for the DCT have been built. It is observed that both the shift property and the convolution property for the DCT are more complex than those for the Fourier cosine transform. However, there are situations under which these properties may be utilized. In the updating of an instantaneous DCT of a continuously incoming data stream, the shift property is considered to be useful. In applications that involve transform domain filtering, the convolution property is essential [F-1]. The asymptotic equivalence of the DCT to the statistically optimal Karhunen–Loeve transform (KLT), the development of the KLT, and other related topics will form the basis for Chapter 3.

PROBLEMS

2.1. In (2.2.9), the Fourier cosine transform of the derivative of a function is developed. Extend this to higher order derivatives and obtain a general formula for

$$F_c\left[\frac{d^n}{dt^n}x(t)\right].$$

2.2. In Section 2.2, the properties of Fourier cosine transform are developed. Obtain similar properties for the Fourier sine transform.

2.3. In Section 2.3, the unitarity of DCT-I is shown in detail [see (2.4.2) through (2.4.10)]. Prove this property for the remaining DCTs [see (2.4.11)].

2.4. In (2.6.6), the shift property of DCT-I is developed. In (2.6.7), (2.6.8), and (2.6.9), this property is listed for the other DCTs. Derive the same in detail.

2.5. Develop shift properties for the DSTs defined in (2.6.10).

2.6. Sketch the waveforms of DCTs for $N = 4$.

2.7. Repeat Problem 2.6 for DSTs.

2.8. In Section 2.7, the difference property for DCT-I is shown. Write down the expressions for the difference properties for all the DCTs, using (2.6.6) through (2.6.9).

2.9. Derive (2.8.5).

2.10. Derive (2.8.13).

2.11. The convolution property for DCT-II is derived in Section 2.8. Derive, if any, the same property for the other DCTs defined in (2.3.6).

2.12. Investigate if the convolution property exists for the family of DSTs defined in (2.6.10). If so, derive the same.

2.13. Derive (2.2.11).

2.14. Derive (2.4.4).

2.15. Derive (2.4.5).

2.16. Using (2.3.6), develop the matrices for all the DCTs for $N = 4$. Show that these matrices are orthogonal. (Hint: Compute the matrix multiplication of each DCT matrix by its transpose.)

2.17. Repeat Problem 2.16 for the DSTs using (2.6.10). Are any of the DSTs interrelated? If so, how? [See FD-1.]

2.18. A simple convolution property for DCT-II has been developed by Chitprasert and Rao [F-7]. The only restriction is that one of the functions to be convolved must be real and even. This property has been utilized in human visual weighted progressive image transmission [PIT-12]. Investigate if similar properties exist for the remaining DCTs defined in (2.3.6).

2.19. Repeat Problem 2.18 for the family of DSTs defined in (2.6.10).

CHAPTER 3

DCT AND ITS RELATIONS TO THE KARHUNEN–LOEVE TRANSFORM

3.1 Introduction

In Chapter 2, we have discussed some mathematical properties of the DCTs. We have also emphasized that they are not simply discretized versions of the Fourier cosine transform properties. However, it was only briefly mentioned that the transforms were obtained originally to diagonalize certain matrices. In this chapter, we shall discuss in more detail the "origin" of the DCTs and the reason why one of them, the DCT-II, has been regarded as one of the best tools in digital signal processing.

In Section 3.2, we present the Karhunen–Loeve transform (KLT), first discussed by Karhunen [M-3] and later by Loeve [M-4]. This is a series representation of a given random function. The orthogonal basis functions are obtained as the eigenvectors of the corresponding auto-covariance matrix. This transform is optimal in that it completely decorrelates the random function (i.e., the signal sequence) in the transform domain (see, for example, Ahmed and Rao [B-1]). It is also a canonical transformation that minimizes the mean square error (MSE) between a truncated representation and the actual signal (see, for example, Devijer and Kittler [M-5]). The KLT of a Markov-1 signal is then derived (Ray and Driver [M-6]). This forms the basis for the generation of the DCTs (Ahmed, Natarajan, and Rao [G-1]; Kitajima [S-1]), and the details are presented in Section 3.3.

In Section 3.4, we reflect on the formal treatment of asymptotic equivalence (Yemini and Pearl [M-7]) between classes of matrices and their orthonormal representations. Such considerations are then extended by using quadrature approximation to the actual generation of discrete unitary transforms.

3.2 The Karhunen–Loeve Transform

It is useful to consider the KLT from an intuitive point of view first. Suppose a signal in the form of a pure sinusoid is being transmitted over a medium of some sort. The signal can be transmitted as a sampled waveform, with each sampled data point being sent in a sequential manner. The number of points transmitted depends on how accurately we want to reconstruct the waveform. Intuitively, the more the number of points transmitted, the better is the reconstruction of the waveform. It is, however, well known that all that is required to construct a deterministic sinusoid are its magnitude, phase, frequency, starting time, and the fact that it is a sinusoid. This essentially implies that five pieces of information are all that is required to reconstruct the sinusoidal signal exactly. Thus in the transmission, assuming no degradation by the medium, only five parameters need be transmitted to reconstruct the sinusoid exactly at the receiver end. From an information theoretic point of view, the sampled values of the deterministic waveform are highly correlated and the information content of the transmitted signal is low. On the other hand, the five pieces of information regarding magnitude, phase, frequency, starting time, and shape are completely uncorrelated and have exactly the same amount of information content as the total number of sampled values being transmitted. This being the case, it is perhaps natural to pose the question of whether it is possible to take the N sampled points in the transmission and transform them to the five uncorrelated pieces of information. The KLT is just such a transformation when the signal being transmitted is a Markov-1 type signal.

Another way of formulating the KLT is to consider the representation of a random function. We seek the best representation of a given random function in the mean square error (MSE) sense. Let us consider N sampled points of a zero mean random vector $\mathbf{x}$,

$$\mathbf{x} = \{x(0), x(1), \ldots, x(N-1)\}^T. \tag{3.2.1}$$

If $\{\mathbf{\Phi}_i\}$ is a set of linearly independent vectors spanning the N-dimensional vector space, then $\mathbf{x}$ can be expanded in terms of $\mathbf{\Phi}_i$'s:

$$\mathbf{x} = \sum_{i=0}^{N-1} X_i \mathbf{\Phi}_i, \tag{3.2.2}$$

where X_i are the coefficients of expansion given by

$$X_i = \langle \mathbf{x}, \mathbf{\Phi}_i \rangle / \langle \mathbf{\Phi}_i, \mathbf{\Phi}_i \rangle, \qquad i = 0, 1, \ldots, N-1. \tag{3.2.3}$$

Here, $\langle \cdot, \cdot \rangle$ denotes inner product.

Thus, the vector $\mathbf{x}$ can be represented by the N numbers $x(i)$ or, just as accurately, the N numbers X_i in the space spanned by the basis functions $\mathbf{\Phi}_i$. Suppose in the solution of (3.2.3), only the first D coefficients ($D < N$) are significantly different from zero. Then, the vector $\mathbf{x}$ is well represented by the D coefficients in the $\{\mathbf{\Phi}_i\}$ space. This means that instead of N numbers, only D such numbers are needed to represent $\mathbf{x}$. In practice, this means that data compression and bandwidth reduction are possible. We now formulate the problem as follows:

Given a random vector $\mathbf{x} = \{x(0), x(1), \ldots, x(N-1)\}^T$, find a set of basis vectors $\mathbf{\Phi}_i$'s, $i = 0, \ldots, N-1$, such that the error of a truncated representation is minimized in the MSE sense.

We proceed by writing the truncated representation of $\mathbf{x}$ as $\tilde{\mathbf{x}}$ given by

$$\tilde{\mathbf{x}} = \sum_{i=0}^{D-1} X_i \mathbf{\Phi}_i. \tag{3.2.4}$$

The MSE in the truncation (3.2.4) is given by

$$\begin{aligned} \varepsilon &= E[(\mathbf{x} - \tilde{\mathbf{x}})^2] \\ &= E\left[\left\langle \sum_{i=D}^{N-1} X_i \mathbf{\Phi}_i, \sum_{i=D}^{N-1} X_i \mathbf{\Phi}_i \right\rangle\right], \end{aligned} \tag{3.2.5}$$

where E is the expectation operator.

Assuming that the basis functions are orthonormal, so that

$$\langle \mathbf{\Phi}_i, \mathbf{\Phi}_k \rangle = \delta_{ik}, \tag{3.2.6}$$

and that $\mathbf{x}$ is real, we obtain

$$\begin{aligned} \varepsilon &= E\left[\sum_{i=D}^{N-1} |X_i|^2\right] \\ &= E\left[\sum_{i=D}^{N-1} |\langle \mathbf{x}, \mathbf{\Phi}_i \rangle|^2\right]. \end{aligned} \tag{3.2.7}$$

Equation (3.2.7) can be further reduced to

$$\begin{aligned} \varepsilon &= E\left[\sum_{i=D}^{N-1} \mathbf{\Phi}_i^T \mathbf{x}\mathbf{x}^T \mathbf{\Phi}_i\right] \\ &= \sum_{i=D}^{N-1} \mathbf{\Phi}_i^T E[\mathbf{x}\mathbf{x}^T] \mathbf{\Phi}_i, \end{aligned} \tag{3.2.8}$$

where T denotes transposition as usual.

If one defines the auto-covariance matrix of the random vector $\mathbf{x}$ as

$$[A] = E[\mathbf{x}\mathbf{x}^T] \tag{3.2.9}$$

and minimizes the quantity in (3.2.8) by the proper choice of basis functions, subject to the orthonormality condition of (3.2.6), one obtains

$$\left(\frac{\delta}{\delta \mathbf{\Phi}_i}\right) \{\varepsilon - \mu_i \langle \mathbf{\Phi}_i, \mathbf{\Phi}_i \rangle\} = 0,$$

leading to the equation:

$$([A] - \mu_i [I_N]) \mathbf{\Phi}_i = 0, \qquad i = 0, 1, \ldots, N-1, \tag{3.2.10}$$

where we have introduced the Lagrange multiplier μ_i for the constraint and $[I_N]$ is the $N \times N$ identity matrix. We note also that $[A]$ is symmetric and positive semidefinite.

Thus, the minimization of the MSE in the truncated representation of $\mathbf{x}$ in (3.2.4) leads directly to the eigenvalue problem in (3.2.10). The set of basis vectors obtained will diagonalize the auto-covariance matrix $[A]$. Let $[\phi]$ be the matrix of the eigenvectors given by

$$[\phi] = [\mathbf{\Phi}_0, \mathbf{\Phi}_1, \ldots, \mathbf{\Phi}_{N-1}]. \tag{3.2.11}$$

Then

$$[\phi]^{-1}[A][\phi] = \mathrm{diag}[\mu_0, \mu_1, \ldots, \mu_{N-1}]. \tag{3.2.12}$$

Based on the foregoing, the MSE due to the truncation in (3.2.4) is given by

$$\varepsilon = \sum_{i=D}^{N-1} \mu_i, \tag{3.2.13}$$

and we note that ε is minimized by ranking the eigenvalues μ_i in a descending order.

The set of basis vectors $\{\mathbf{\Phi}_i\}$ form the bases for the KLT expansion, and the matrix in (3.2.11) is the K-L transform matrix. The KLT is said to be an optimal transform because it has the following properties:

(1) It completely decorrelates the signal in the transform domain.
(2) It minimizes the MSE in bandwidth reduction or data compression.
(3) It contains the most variance (energy) in the fewest number of transform coefficients.
(4) It minimizes the total representation entropy of the sequence.

These are, of course, direct consequences of the diagonalization problem (3.2.10). It is also immediately clear that the basis functions are dependent on the auto-covariance matrix $[A]$ and therefore cannot be predetermined. Furthermore, given an auto-covariance matrix, the solution of (3.2.10) is usually quite involved. There are only a few cases in which analytical solutions are available. Such is the case when the signal statistics produce an auto-covariance matrix of the form,

$$[A]_{ik} = \rho^{|i-k|} \qquad i, k = 0, 1, \ldots, N-1 \tag{3.2.14}$$

for $0 < \rho < 1$, where ρ is the adjacent correlation coefficient.

Such a signal is said to be a stationary Markov-1 signal. Davenport and Root [M-8] reported such an example in the continuous domain as a solution to an integral equation (see Chapter 3 of Davenport and Root [M-8]). In the discrete case, Ray and Driver [M-6] provided the solution as follows:

$$\begin{aligned}[\phi]_{nm} &= \Phi_m(n) \\ &= \left[\frac{2}{(N+\mu_m)}\right]^{1/2} \sin\left\{w_m\left[(n+1) - \frac{(N+1)}{2}\right] + (m+1)\frac{\pi}{2}\right\} \\ &\quad m, n = 0, 1, \ldots, N-1,\end{aligned} \tag{3.2.15}$$

where $\mu_m = (1 - \rho^2)/[1 - 2\cos(w_m) + \rho^2]$ are the eigenvalues and w_m's are the real positive roots of the transcendental equation:

$$\tan(Nw) = -\frac{(1 - \rho^2)\sin(w)}{[\cos(w) - 2\rho + \rho^2\cos(w)]}. \tag{3.2.16}$$

Here, we have used $[\phi]_{mn}$ to denote the mn-element of the matrix $[\phi]$ and $\Phi_m(n)$ is the nth component of the mth eigenvector, where $m, n = 0, 1, \ldots, N - 1$.

In Fig. 3.1 we show the basis functions in the K-L expansion for a stationary Markov-1 signal with $\rho = 0.95$ and $N = 16$. We note the intrinsic sinusoidal nature of the basis functions.

Practical implementation of KLT involves the estimation of the auto-covariance matrix of the data sequence, its diagonalization, and the construction of the basis vectors. The inability to predetermine the basis vectors in the transform domain has made KLT an ideal but impractical tool. A natural question to ask is whether there are predetermined basis vectors that are good approximations to the KLT. It is in answer to this question that attempts were made to examine the diagonalization of matrices that are asymptotically equivalent to (3.2.14). Stationary zero-mean Markov-1 signals are deemed to be

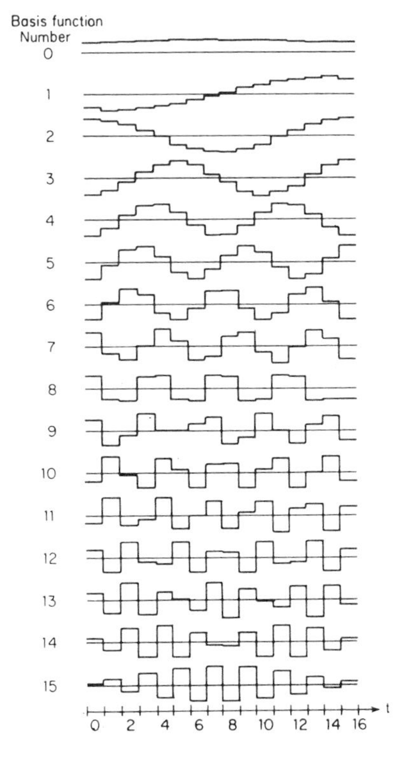

Fig. 3.1 KLT basis functions for $N = 16$ and $\rho = 0.95$ for a Markov-1 signal.

sufficiently general in signal processing to warrant detailed studies. Such studies have led to the construction of other discrete transforms. Although the KLT provides no easy solutions to the problem of actual decorrelation, it does provide a benchmark against which other discrete transforms may be judged. It may be noted that techniques for estimating the basis functions and the associated eigenvalues and their convergence to the desired values have been developed [M-35].

3.3 The DCT-I and DCT-II

In the case of a stationary Markov-1 signal, the auto-covariance matrix is a symmetric Toeplitz matrix. It is, as we have seen in the last section, one of the very few cases where its diagonalization can be obtained analytically. The symmetric Toeplitz matrix has a great deal of structure and its properties may be used. In fact, a Toeplitz matrix may be treated as asymptotically equivalent (i.e., as N, the size of the matrix, increases) to a circulant matrix (see, for example, Davis [M-9]), whose eigenvectors are the basis vectors for the discrete Fourier transform (DFT). From this point of view, DFT and KLT are asymptotically equivalent.

A parameter in the stationary Markov-1 signal is ρ, the correlation coefficient. One can therefore also examine the asymptotic behavior (as $\rho \to 0$ or as $\rho \to 1$) of the auto-covariance matrix. Thus, when one speaks of the asymptotic equivalence of the matrix $[A]$, it should be clearly understood whether it applies to N or ρ.

Yet another approach may be taken. Instead of examining the full covariance matrix, it is possible to examine an approximation to $[A]$, where only a few diagonals are retained and the diagonalization is considered for this approximate matrix.

The above approaches have been taken by Ahmed and Flickner [G-4], Jain [M-2], Kitajima [S-1], and Yemini and Pearl [M-7]. From these studies have come not only the DCTs and DSTs, but also other discrete unitary transforms. We shall present the derivations for the DCT-I and DCT-II in the following discussions.

3.3.1 *DCT-I*

In Section 2.3, we obtained the symmetric cosine transform or DCT-I by simply discretizing the Fourier cosine transform. This particular approach of obtaining the DCT-I bears no relation to the optimal KLT that we have discussed in the previous section. The original derivation by Kitajima [S-1], however, was motivated by an attempt to approximate the KLT basis functions by simpler basis functions. The derivation is based on a recursive relation of the Tchebyshev polynomial of the first kind, the diagonalization of a tridiagonal matrix, and its asymptotic equivalence (as N tends to infinity) to the inverse of the Markov-1 auto-covariance matrix. The derivation follows.

First, we recall a simple matrix property. If a nonsingular matrix $[A]$ is diagonalized by a similarity transformation $[S]$, i.e.,

$$[S]^{-1}[A][S] = [\cap] = \text{diag}[\mu_0, \mu_1, \ldots, \mu_{N-1}], \tag{3.3.1}$$

then its inverse is also diagonalized by $[S]$ so that

$$[S]^{-1}[A]^{-1}[S] = [\cap]^{-1} = \text{diag}[\mu_0^{-1}, \mu_1^{-1}, \ldots, \mu_{N-1}^{-1}]. \tag{3.3.2}$$

We note also that the symmetric Toeplitz matrix $[A]$ for the autocovariance of the Markov-1 signal has an inverse in tridiagonal form (see Theory and Applications of Toeplitz Matrices, Yip and Agrawal [M-10]) given by

$$[A]^{-1} = (1-\rho^2)^{-1} \begin{bmatrix} 1 & -\rho & 0 & 0 & \cdots & \cdots & \cdots \\ -\rho & 1+\rho^2 & -\rho & 0 & \cdots & \cdots & \cdots \\ 0 & -\rho & 1+\rho^2 & -\rho & 0 & \cdots & \cdots \\ \cdots & 0 & -\rho & 1+\rho^2 & -\rho & 0 & \cdots \\ \cdots & \cdots & \cdots & \cdots & \cdots & 1+\rho^2 & -\rho \\ \cdots & \cdots & \cdots & \cdots & \cdots & -\rho & 1 \end{bmatrix}. \tag{3.3.3}$$

It is therefore evident that the transformation that diagonalizes (3.3.3) will also diagonalize the matrix $[A]$. Instead of considering the diagonalization of (3.3.3), which will lead to the KLT basis functions, a matrix that is asymptotically equivalent (as $N \to \infty$) to $[A]^{-1}$ is considered.

Let $[A]^{-1}$ in (3.3.3) be decomposed into two parts,

$[A]^{-1} = [B] + [R]$, such that

$$[B] = (1-\rho^2)^{-1} \begin{bmatrix} 1+\rho^2 & -\sqrt{2}\rho & 0 & \cdots & \cdots \\ -\sqrt{2}\rho & 1+\rho^2 & -\rho & \cdots & \cdots \\ 0 & -\rho & 1+\rho^2 & -\rho & \cdots \\ \cdots & \cdots & \cdots & \cdots & \cdots \\ \cdots & \cdots & \cdots & -\sqrt{2}\rho & 1+\rho^2 \end{bmatrix} \tag{3.3.4}$$

and

$$[R] = (1-\rho^2)^{-1} \begin{bmatrix} -\rho^2 & (\sqrt{2}-1)\rho & 0 & \cdots & \cdots \\ (\sqrt{2}-1)\rho & 0 & 0 & \cdots & \cdots \\ \cdots & \cdots & \cdots & 0 & (\sqrt{2}-1)\rho \\ \cdots & \cdots & \cdots & (\sqrt{2}-1)\rho & -\rho^2 \end{bmatrix}.$$

We note that [R] is very close to being a null matrix. The matrix [B] can be further decomposed as follows,

$$[B] = \left\{\frac{-2\rho}{(1-\rho^2)}\right\}[B_1] + \left\{\frac{(1+\rho^2)}{(1-\rho^2)}\right\}[I_N],$$

where,

$$[B_1] = \begin{bmatrix} 0 & \frac{1}{\sqrt{2}} & 0 & 0 & \cdots & \cdots & \cdots \\ \frac{1}{\sqrt{2}} & 0 & \frac{1}{2} & 0 & \cdots & \cdots & \cdots \\ 0 & \frac{1}{2} & 0 & \frac{1}{2} & \cdots & \cdots & \cdots \\ \cdots & \cdots & \cdots & \cdots & \frac{1}{2} & 0 & \frac{1}{\sqrt{2}} \\ \cdots & \cdots & \cdots & \cdots & 0 & \frac{1}{\sqrt{2}} & 0 \end{bmatrix}. \tag{3.3.5}$$

Since [B] is a linear combination of $[B_1]$ and an identity matrix, the transformation that diagonalizes $[B_1]$ will also diagonalize [B]. The diagonalization of $[B_1]$ depends on a recurrence relation for Tchebyshev polynomials of the first kind, which are given by

$$T_n(x) = \cos(n\theta), \qquad \cos\theta = x, \qquad n = 0, 1, \ldots. \tag{3.3.6}$$

The relevant three-term recurrence relation is (see *Special Functions for Physics and Chemistry*, Sneddon [M-11]):

$$\tfrac{1}{2}T_{n-1}(x) + \tfrac{1}{2}T_{n+1}(x) = xT_n(x). \tag{3.3.7}$$

If the first N polynomials are considered at the discrete values of x given by

$$x_k = \cos\left\{\frac{k\pi}{(N-1)}\right\}, \qquad k = 0, 1, \ldots, N-1, \tag{3.3.8}$$

then

$$\tfrac{1}{2}T_{n-1}(x_k) + \tfrac{1}{2}T_{n+1}(x_k) = x_k T_n(x_k), \qquad n = 1, 2, \ldots, N-2,$$
$$T_1(x_k) = x_k T_0(x_k),$$

and

$$T_{N-2}(x_k) = x_k T_{N-1}(x_k). \tag{3.3.9}$$

[Note: (3.3.9) can easily be verified by the compound angle expansion of cosine functions.] Equation (3.3.9) can be represented in matrix form if we define an N-

dimensional vector given by

$$\mathbf{v}_k = \left\{ \frac{T_0(x_k)}{\sqrt{2}}, T_1(x_k), \ldots, \frac{T_{N-1}(x_k)}{\sqrt{2}} \right\}^T. \tag{3.3.10}$$

Then

$$[B_1]\mathbf{v}_k = x_k \mathbf{v}_k, \qquad k = 0, 1, \ldots, N-1. \tag{3.3.11}$$

Equation (3.3.11) states that x_k and $\mathbf{v}_k$ are, respectively, the eigenvalue and eigenvector of the matrix $[B_1]$. For the matrix $[B]$ in (3.3.5), therefore,

$$[B]\mathbf{v}_k = w_k \mathbf{v}_k, \qquad k = 0, 1, \ldots, N-1, \tag{3.3.12}$$

where the eigenvalues w_k's are given by

$$w_k = (1 - \rho^2)^{-1} \left\{ 1 + \rho^2 - 2\rho \cos\left[\frac{k\pi}{(N-1)} \right] \right\}. \tag{3.3.13}$$

In (3.3.4) we note that as $N \to \infty$, $[B]$ should asymptotically approach $[A]^{-1}$. This is valid, at least in the transform domain, as we show in the following:

Let

$$[V] = [\hat{\mathbf{v}}_0, \hat{\mathbf{v}}_1, \ldots, \hat{\mathbf{v}}_{N-1}]^T, \tag{3.3.14}$$

where $\hat{\mathbf{v}}_k$ is the normalized eigenvector in (3.3.12). Since $\mathbf{v}_k$'s are real, the matrix in (3.3.14) is real and unitary. Thus $[V][V]^T = [I_N]$. Applying the similarity transformation to (3.3.4) gives

$$\begin{aligned}[V]^T[A]^{-1}[V] &= [V]^T[B][V] + [V]^T[R][V] \\ &= \operatorname{diag}(w_0, w_1, \ldots, w_{N-1}) + [V]^T[R][V]. \end{aligned} \tag{3.3.15}$$

By direct matrix multiplication, it can be shown that the nm-element of $[V]^T[R][V]$ is given by

$$\begin{aligned}([V]^T[R][V])_{nm} &= 0 \qquad \text{for } n + m = \text{odd} \\ &= k_n k_m 2\rho \frac{\left\{ -\rho + (2 - \sqrt{2})\left[\cos\left(\frac{m\pi}{N-1} \right) + \cos\left(\frac{n\pi}{N-1} \right) \right] \right\}}{[(N-1)(1-\rho^2)]} \\ &\qquad \text{for } n + m = \text{even}, \end{aligned} \tag{3.3.16}$$

where

$$\begin{aligned} k_m &= \frac{1}{\sqrt{2}} \qquad \text{if } m = 0 \text{ or } N-1 \\ &= 1 \qquad \text{otherwise.} \end{aligned}$$

For $\rho \neq 1$, (3.3.16) vanishes as $N \to \infty$. Therefore, using the matrix property of (3.3.1) and (3.3.2), we obtain asymptotically as $N \to \infty$ that

$$[V]^T[A][V] \approx \operatorname{diag}[\mu_0, \mu_1, \ldots, \mu_{N-1}],$$

where

$$\mu_k = \frac{1}{w_k}$$

and

$$[V]_{nm} = k_n k_m \left[\frac{2}{(N-1)}\right]^{1/2} \cos\left[\frac{nm\pi}{(N-1)}\right], \qquad n, m = 0, 1, \ldots, N-1. \tag{3.3.17}$$

Equation (3.3.17) is identical to the definition of DCT-I given in (2.3.6) if $N-1$ is replaced by N. This, then, is an example of asymptotic equivalence to the KLT as $N \to \infty$. We note that the equivalence is valid for $\rho \neq 1$. In fact, as $\rho \to 1$. $([V]^T[R][V])_{nm}$ in (3.3.16) becomes indeterminate even as $N \to \infty$. This predicts that the decorrelation power of DCT-I decreases as ρ increases to 1, even when N is very large.

It is interesting to note that the eigenvalues in (3.3.17) and those of the KLT in (3.2.15) are identical in form. In fact, if $w_n = n\pi/(N-1)$ is used in (3.2.15), the basis functions (3.3.17) are obtained (Problem 3.8). (Note: (3.2.15) is valid only for even N, and therefore the size of the covariance matrix must be even, making $(N-1)$ an odd integer). The implication here is that instead of solving (3.2.16) exactly for w_n, we have simply approximated it by $n\pi/(N-1)$. However, such a derivation is similar to the one used in Section 2.3, in which the asymptotic behavior of the transform is completely masked.

It should also be pointed out that, given a Markov-1 process, the eigenvectors as given by KLT are predetermined. However, its complex form has made any attempt in deriving fast algorithms quite futile, while the DCT-I and others that are asymptotically equivalent to it (either as ρ tends to a limit or as N tends to infinity) are amenable to fast algorithms, making real-time implementations feasible. This, in the final analysis, is the only justification for the two levels of approximation involved. [The first level is the assumption that the signal statistic is Markov-1, and the second level is the replacement of (3.2.15) by other simpler basis functions.]

3.3.2 *DCT-II*

We have seen in the case of DCT-I that the asymptotic equivalence as N tends to infinity was used to construct the eigenvalues and eigenvectors. A more direct approach can be taken here in considering the limit values for the correlation coefficient ρ for any given N. Ahmed and Flickner [G-4] provided just such a derivation for the DCT-II. For convenience, we repeat (3.2.15) and (3.2.16) here in the case of the KLT for stationary Markov-1 signals:

$$[\varnothing]_{nm} = \left\{\frac{2}{(N+\mu_m)}\right\}^{1/2} \sin\left\{w_m\left[(n+1) - \frac{(N+1)}{2}\right] + (m+1)\frac{\pi}{2}\right\},$$
$$m, n = 0, 1, \ldots, N-1, \tag{3.2.15}$$

where

$$\mu_m = \frac{(1-\rho^2)}{[1 - 2\rho\cos(w_m) + \rho^2]}$$

are the eigenvalues in which the w_m are the positive real roots of

$$\tan(Nw) = -\frac{(1-\rho^2)\sin(w)}{\{\cos(w) - 2\rho + \rho^2\cos(w)\}}. \tag{3.2.16}$$

Without resorting to the approximation for the auto-covariance matrix directly, one can examine what happens when ρ approaches the limit values. As $\rho \to 1$, we have for the transcendental equation (3.2.16) that

$$\tan(Nw) = 0,$$

thus giving

$$w_k = \frac{k\pi}{N}, \qquad \text{for } k = 0, 1, \ldots, N-1. \tag{3.3.18}$$

The eigenvalues μ_m vanish when w_m are nonzero. For μ_m when $m = 0$, the expression will seem to say that it should approach infinity. However, recalling the invariance of the trace of a matrix under similarity transformation, we have

$$\sum_{m=0}^{N-1} [A]_{mm} = \sum_{m=0}^{N-1} \mu_m,$$

and since the diagonal elements of $[A]$ in a Markov-1 signal are all 1's, we have immediately that $\mu_0 = N$. Combining these and substituting into (3.2.15), we obtain

$$[\varnothing]_{n0} = \frac{1}{\sqrt{N}}$$

and

$$[\varnothing]_{nm} = \sqrt{\left(\frac{2}{N}\right)}\sin\left[m(n+\tfrac{1}{2})\frac{\pi}{N} + \frac{\pi}{2}\right]$$

$$= \sqrt{\left(\frac{2}{N}\right)}\cos\left[m(n+\tfrac{1}{2})\frac{\pi}{N}\right], \qquad m \neq 0.$$

These, upon simplification, give

$$[\varnothing]_{nm} = \sqrt{\left(\frac{2}{N}\right)}\, k_m \cos\left[m(n+\tfrac{1}{2})\frac{\pi}{N}\right], \qquad m, n = 0, 1, \ldots, N-1 \tag{3.3.19}$$

and

$$k_m = \frac{1}{\sqrt{2}} \qquad \text{for } m = 0,$$

$$= 1 \qquad \text{otherwise.}$$

Equation (3.3.19) is exactly the definition given in (2.3.6) for DCT-II. Thus, it can be seen that DCT-II is asymptotically equivalent to KLT for Markov-1 signals as $\rho \to 1$. We note that this is independent of N. Also of interest is the fact that the only nonzero eigenvalue is μ_0, which is equal to N, and all other eigenvalues are zero as ρ goes to 1. In terms of the mean square error (MSE) representation discussed in the previous section, all the errors are packed into one single DCT-II eigenvalue when $\rho = 1$. As ρ deviates from this value, the MSE will be spread gradually among the other eigenvalues as well. This explains the empirical results in which DCT-II seems to have the best variance distribution compared with the other non-KLT discrete transforms (see Chapter 6). Since this asymptotic behavior is independent of N, one can understand why DCT-II performs so well for ρ near 1. Thus for any value of N, when $\rho = 1$ the DCT-II basis functions are the eigenvectors of a singular matrix whose elements are all 1's (Problem 3.9).

Since DCT-III is simply a transpose of DCT-II, it also diagonalizes the same singular matrix when $\rho = 1$ and has exactly the same asymptotic behavior as DCT-II.

When $\rho \to 0$, (3.2.16) is reduced to

$$\tan(Nw) = -\tan w,$$

giving

$$w_n = \frac{(n+1)\pi}{(N+1)} \qquad n = 0, 1, \ldots$$

and all $\mu_n = 1$.

It can be shown (Problem 3.10) that, in this case, the corresponding basis functions are those of one of the discrete sine transforms (see Jain [M-2]). One must, however, bear in mind that as ρ tends to zero, the covariance matrix is also approaching $[I_N]$, which is already diagonal.

DCT-IV is the only DCT we have not commented on in regard to its asymptotic behavior. It is not difficult to show that it is asymptotic to KLT in (2.3.6) as N tends to infinity, and one can also investigate the form of the covariance matrix that is diagonalized by the DCT-IV basis functions (Problem 3.11).

3.4 Asymptotic Equivalence and the Generation of Discrete Unitary Transforms

In our discussion of asymptotic equivalence between matrices, we have relied heavily on our intuitive understanding of the term *asymptotic equivalence*. The idea can be formalized, in particular with regard to the treatment of matrices. A well-known example of this formulation is the study of asymptotic distribution of eigenvalues of Toeplitz forms (see Szego's theorem [M-15]). In this regard,

Gray [M-17] has found that circulant matrices and Toeplitz matrices have similar asymptotic eigenvalue distributions. Pearl [M-16] undertook a formalization of the so-called asymptotic equivalence of spectral representations in various unitary transform domains. Yemeni and Pearl [M-7] extended this formalism to include a procedure for generating discrete unitary transforms based on the idea of asymptotic equivalence.

The derivation of DCT-II was based on its asymptotic behavior as the correlation coefficient ρ tends to 1. Does it also possess the necessary asymptotic behavior as N tends to infinity for a given ρ? The answer can be sought using a procedure similar to the one used in Section 3.3.1 for the derivation of DCT-I. However, there is much economy of thought and elegance of argument in a more general formulation of equivalent classes of matrices and their orthonormal representations. Such a formulation will admit not only the Markov-1 auto-covariance matrix that we have seen, but also a much more general class of signal covariance matrices. The equivalence as N tends to infinity of the DCT-II to such a class of covariance matrices demonstrates the reason behind its excellent performance under various criteria. In addition, the ready extension of such a formulation to the actual generation of discrete unitary transforms is sufficient reason for us to devote some space here to its consideration.

3.4.1 *The Hilbert–Schmidt Norms of a Matrix*

The formal discussion of equivalence rests on the definition of some norm. The vanishing of such a norm as $N \to \infty$ is usually taken as the necessary and/or sufficient condition for the asymptotic equivalence. Since signal covariance matrices are the primary concern here, we shall consider only real symmetric $N \times N$ matrices. Let $[A_N]$ be such a matrix whose eigenvalues are given by $\{\mu_i\}$, $i = 0, 1, 2, \ldots, N-1$. The *weak norm* of $[A_N]$ is defined by

$$|[A_N]|^2 \equiv N^{-1} \sum_{i=0}^{N-1} \sum_{j=0}^{N-1} (a_{ij})^2 = N^{-1} \sum_{i=0}^{N-1} (\mu_i)^2. \tag{3.4.1}$$

The *strong norm* of $[A_N]$ is defined by

$$\begin{aligned} \|[A_N]\|^2 &\equiv \sup\{\langle \mathbf{u}, [A_N]\mathbf{u}\rangle : \langle \mathbf{u}, \mathbf{u}\rangle = 1\} \\ &= \max_i \{(\mu_i)^2\}, \end{aligned} \tag{3.4.2}$$

where $\langle \cdot, \cdot \rangle$ denotes the inner product between two N-dimensional vectors. Both norms are invariant under a unitary transformation, i.e., if $[T_N]$ is an $N \times N$ unitary matrix, then

$$|[A_N]| = |[T_N]^H[A_N][T_N]| \quad \text{and} \quad \|[A_N]\| = \|[T_N]^H[A_N][T_N]\|, \tag{3.4.3}$$

where the superscript H denotes Hermitian conjugation. We note also that

$$\|[A_N]\| \geqslant |[A_N]|. \tag{3.4.4}$$

3.4.2 *Nets, Classes, and Sections*

For ease of discussion in treating sequences of matrices of increasing size N, we define the following terms (The treatment here follows closely that of Yemini and Pearl [M-7]):

(1) *Net:* A strongly bounded sequence of matrices $[A_N]$, $N = 1, 1, \ldots, \infty$, denoted by $\alpha = [A_N]_{N=0}^{\infty}$ (i.e., $\|[A_N]\| < M$, for some finite M).

(2) *Class:* A collection of nets with some common structural property, denoted by A. For example, $\cap$ is a diagonal class of matrices, if it contains nets $\delta = [\cap_N]_{N=0}^{\infty}$ such that $[\cap_N]_{ij} = d_i \delta_{ij}$.

(3) N-section: The collection of $N \times N$ matrices that belong to the nets in a class, denoted by A_N, e.g., $\cap_N$ is the N-section of the class $\cap$, i.e., the collection of all $N \times N$ diagonal matrices.

Thus, for example, the sequence of strongly bounded diagonal matrices $[\cap_N]$ is denoted by the net δ, the collection of δ's is denoted by the class $\cap$, and in the class $\cap$, the $N \times N$ diagonal matrices are denoted by the N-section $\cap_N$.

The motivation behind the definitions of these terms is to generalize the concept of asymptotic equivalence from a matrix-to-matrix level to a net-to-net level and, hopefully, also to a class-to-class level. Such a generalization provides the necessary framework for determining the equivalence between, for example, the class of circulant matrices and the class of Toeplitz matrices. So, instead of treating the equivalence of one particular matrix to another, as in the derivation of DCT-I in Section 3.3.1, a whole class of auto-covariance matrices with a common structural property may be examined.

We define *net* equivalence as follows:

Two nets, $\alpha = [A_N]_{N=0}^{\infty}$ and $\beta = [B_N]_{N=0}^{\infty}$ are asymptotically equivalent if the weak norm of the sequence of difference matrices $[A_N] - [B_N]$ vanishes as N tends to infinity, i.e.,

$$\alpha \approx \beta, \qquad \text{if } \lim_{N \to \infty} |[A_N] - [B_N]| = 0, \tag{3.4.5}$$

where $\approx$ is used to denote asymptotic equivalence. Note that definition (3.4.5) is reflexive, *Class* equivalence is more difficult to establish. In its place, the idea of "covering" may be introduced. If $\mathbf{A}$ and $\mathbf{B}$ are matrix classes, then $\mathbf{A}$ is said to asymptotically cover $\mathbf{B}$ if for any $\beta \in \mathbf{B}$, there exists a net $\alpha \in \mathbf{A}$ such that $\alpha \approx \beta$. This is given in the following definition:

$$\mathbf{A} \subset \mathbf{B} \text{ if for any } \beta \in \mathbf{B}, \text{ there exists } \alpha \in \mathbf{A} \text{ such that } \alpha \approx \beta. \tag{3.4.6}$$

Here we have introduced the symbol $\subset$ to denote the asymptotic covering. Since the definition is not reflexive, class equivalence $\mathbf{A} \approx \mathbf{B}$ occurs only if $\mathbf{A} \subset \mathbf{B}$ and $\mathbf{B} \subset \mathbf{A}$, i.e., only if $\mathbf{A}$ covers $\mathbf{B}$ and $\mathbf{B}$ covers $\mathbf{A}$ asymptotically. For example, it has been shown (see Pearl [M-12]) that the class of circulant matrices is asymptotically equivalent to the class of Toeplitz matrices.

The problem of diagonalization of a given signal covariance matrix can now be formulated more succinctly. Let τ be a given unitary transform net, and **S**, the class of signal covariance matrices, and let the transformed signal covariance class be denoted by $\tau^H \mathbf{S} \tau$. The question to ask now is whether the diagonal class $\cap$ covers the transformed class $\tau^H \mathbf{S} \tau$ asymptotically. Because of the invariance of the weak norm under unitary transformation, it can be seen that

$$\cap \subset \tau^H \mathbf{S} \tau \qquad \text{iff } \tau \cap \tau^H \subset \mathbf{S}. \tag{3.4.7}$$

Equation (3.4.7) indicates that the relevant question regarding the transform net τ is whether the transformed diagonal class $\tau \cap \tau^H$ covers the signal class asymptotically. An example will be discussed in Section 3.4.4.

3.4.3 *Spectral Representations and Asymptotic Equivalence*

In (3.4.7), $\tau^H \mathbf{S} \tau$ is said to be the τ-spectral representation of the class **S**. If the transform results in a diagonal class, **S** is said to be class diagonal in the transform net τ. For example, the class diagonal in the Fourier transform net is the class of circulant matrices (see, for example, Davis [M-9]). It is possible to examine the equivalence of two classes through their respective orthogonal spectral representations. Let **U** be the class diagonal with respect to the unitary transform net τ and let **V** be class diagonal with respect to the unitary transform net δ. Then, if $[U] \in \mathbf{U}_N$, $[V] \in \mathbf{V}_N$, $[T_N] \in \tau$, and $[D_N] \in \delta$,

$$[T_N]^H [U][T_N] = [\cap_\tau] \qquad \text{and } [D_N]^H [V][D_N] = [\cap_\delta], \tag{3.4.8}$$

where $[\cap_\tau] = \text{diag}[\mu_{\tau 0}, \mu_{\tau 1}, \ldots, \mu_{\tau(N-1)}]$ and $[T_N]$, $[D_N]$ are $N \times N$ unitary transform matrices. Let $\{\mathbf{t}\}$ and $\{\mathbf{d}\}$ be the sets of characteristic vectors making up the matrices $[T_N]$ and $[D_N]$, respectively. Then $\{\mathbf{t}\}$ completely characterizes $\mathbf{U}_N$ and $\{\mathbf{d}\}$, $\mathbf{V}_N$. The equivalence of the classes **U** and **V** can be examined through $\{\mathbf{t}\}$ and $\{\mathbf{d}\}$ in an asymptotic fashion. Consider the δ-spectral representation of $[U]$, i.e.,

$$[D_N]^H [U][D_N] = [D_N]^H [T_N][\cap_\tau][T_N]^H [D_N]. \tag{3.4.9}$$

The projection of $[U]$ on the space spanned by $\{\mathbf{d}\}$ is defined as

$$P_d[U] = [T_N][U'][T_N]^H, \tag{3.4.10}$$

where U' is a diagonal matrix made up of the diagonal elements of the matrix defined in (3.4.9), i.e.,

$$[U'] = \text{diag}[([D_N]^H [U][D_N])_{ii}]. \tag{3.4.11}$$

A measure of the asymptotic equivalence of $[U]$ and $[V]$, and thus of **U** and **V** can be established through the weak norm of the difference matrix given by

$$|[U] - P_d[U]|^2 = \left(\frac{1}{N}\right) \sum_{i=0}^{N-1} \{\mu_{\tau i}^2 - [U']_{ii}^2\}. \tag{3.4.12}$$

Define a vector using the eigenvalues $\mu_{\tau i}$, i.e.,

$$\boldsymbol{\mu}_\tau = [\mu_{\tau 0}, \mu_{\tau 1}, \ldots, \mu_{\tau(N-1)}]^T. \tag{3.4.13}$$

Then

$$\sum_{i=0}^{N-1} \mu_{\tau i}^2 = \boldsymbol{\mu}_\tau^T \boldsymbol{\mu}_\tau, \tag{3.4.14}$$

and

$$\sum_{i=0}^{N-1} [U']_{ii}^2 = \boldsymbol{\mu}_\tau^T [A^T][A] \boldsymbol{\mu}_\tau, \tag{3.4.15}$$

where

$$[A]_{ik} = \langle \mathbf{d}_i, \mathbf{t}_k \rangle^2.$$

Substitution of (3.4.14) and (3.4.15) in (3.4.12) leads to

$$|[U] - P_d[U]|^2 = \left(\frac{1}{N}\right) \boldsymbol{\mu}_\tau^T([I_N] - [A]^T[A])\boldsymbol{\mu}_\tau. \tag{3.4.16}$$

If as $N \to \infty$, the quantity in (3.4.16) vanishes, then the classes of matrices characterized by $\{\mathbf{t}\}$ and $\{\mathbf{d}\}$, respectively, are said to be asymptotically equivalent. Thus

$$\lim_{N\to\infty} \left(\frac{1}{N}\right) \{\boldsymbol{\mu}_\tau^T([I_N] - [A]^T[A])\boldsymbol{\mu}_\tau\} = 0 \tag{3.4.17}$$

is used as a sufficient condition for the asymptotic equivalence of $\{\mathbf{t}\}$ and $\{\mathbf{d}\}$ (also of classes **U** and **V**). When (3.4.17) is true for every bounded $\boldsymbol{\mu}_\tau$, a more compact form of the sufficient condition results:

$$\lim_{N\to\infty} \left\{1 - \left(\frac{1}{N}\right) \operatorname{tr}([A]^T[A])\right\} = 0. \tag{3.4.18}$$

It is perhaps useful to reflect on this particular discussion in terms of the actual problem in the class of signal covariance matrices. Let $[U]$ in (3.4.8) belong to the class **U** of auto-covariance matrices. Since it is class diagonal with respect to the net τ, τ contains the appropriate KLT matrix $[T]$. If one wants to examine another transform net δ as applied to the class **U**, (3.4.9) indicates that the diagonal elements of the δ-spectral representation of **U** should be considered. These diagonal elements are used to construct the projection $P_d[U]$ that belongs to a class diagonal with respect to τ. The asymptotic vanishing of the weak norm $|[U] - P_d[U]|^2$ is used as a sufficient condition for the τ-spectral representation (i.e., in terms of the KLT matrices $[T]$) and the δ-spectral representation (i.e., in terms of the transform matrices $[D]$) of $[U]$ to be equivalent. Put more simply, (3.4.18) ensures that the matrices $[D]$ will also diagonalize the signal auto-covariance matrices as $N \to \infty$. In (3.4.16), the weak norm of the difference between $[U]$ and its projection can also be used to derive a criterion of performance for discrete finite transforms (see Chapter 6).

3.4.4 *Gaussian Quadrature and Generation of Transforms*

The introduction of the concept of classes, nets, and sections in Section 3.4.2 enables us to consider the class of signal covariance matrices and its orthogonal spectral representations in the limit. This limit of $N \to \infty$, coupled with a finite support (assuming that the signal lasts for a finite duration), naturally evokes the transition from the discrete domain to the continuous domain. For example, the inner product of two N-dimensional vectors $\mathbf{u}$ and $\mathbf{v}$ is defined as

$$\langle \mathbf{u}, \mathbf{v} \rangle = \sum_{i=0}^{N-1} u_i^* v_i. \tag{3.4.19}$$

If $\mathbf{u}$ and $\mathbf{v}$ are vectors of finite duration Ω and N is allowed to increase indefinitely, it is not difficult to see that (3.4.19) in fact approaches the inner product of the continuous function $u(t)$ and $v(t)$, given by

$$\langle u, v \rangle = \int_{\Omega} u^*(t) v(t)\, dt. \tag{3.4.20}$$

If $\mathbf{u}_i$ and $\mathbf{v}_j$ are the ith and jth vector components of the matrices $[U_N]$ and $[V_N]$, respectively, then (3.4.19) and (3.4.20) represent the termwise asymptotic behavior of the infinite matrix product given by

$$([U_N]^H [V_N])_{ij} \xrightarrow[N \to \infty]{} \int_{\Omega} u_i^*(t) v_j(t)\, dt. \tag{3.4.21}$$

More generally, instead of dt, a measure given by $dn(t)$, where $n(t)$ is a given distribution, can be used.

It is now possible to formulate the asymptotic equivalence of two unitary transform nets $\tau = [T_N]_{N=0}^{\infty}$ and $\delta = [D_N]_{N=0}^{\infty}$, making use of the above concept. Let $\mathbf{U}$ be a class of signal covariance matrices diagonal in the net τ. Then $\tau^H \mathbf{U} \tau$ is the diagonal class $\cap$. It is therefore desired to find out if the δ-representation of the class $\cap$ asymptotically covers the signal class $\mathbf{U}$. To do this, the termwise behavior of the ki-element of the matrix $[D_N]^H [\cap_N][D_N]$ as $N \to \infty$ with support Ω and measure $dn(t)$ may be examined. Note that

$$([D_N]^H [\cap_N][D_N])_{ki} \xrightarrow[N \to \infty]{} \int_{\Omega} d_k^*(t) \mu(t) d_i(t) dn(t), \tag{3.4.22}$$

where $d_i(t)$ is the continuous limit of the ith vector component of the matrix $[D_N]$ and $\mu(t)$ is the continuous limit of the vector representing the diagonal of $[\cap_N]$. We have used * to denote complex conjugation. If the right-hand side of (3.4.22) is, or covers, the signal covariance matrix class $\mathbf{U}$, then $\mathbf{U}$ is asymptotically class diagonal with respect to the net δ. In other words, the matrix

$[D_N]^H[U_N][D_N]$ is diagonal as N tends to infinity. Thus the nets τ and δ are asymptotically equivalent.

Two aspects of (3.4.22) are of interest. First of all, when the integral represents an element of a matrix belonging to a class $\mathbf{U}$, it should have the structure common to the members of that class. For the signal covariance class, this is the Toeplitz structure. Secondly, the termwise convergence can be achieved with finite N, if the discrete left-hand side is interpreted as a quadrature approximation of the integral. It is well known that the integral of a polynomial of degree less than $2N - 1$ can be exactly reproduced by a numerical quadrature of order N (see, for example, Krylov [M-13]). It should be noted that for a given support Ω and a function $\mu(t)$, the measure $dn(t)$ and the orthonormal functions $d_i(t)$ can be chosen so that the integral in (3.4.22) will have the required Toeplitz structure. For example, for $\Omega = [-\pi, \pi]$, $dn(t) = dt$, and $d_i = \exp\{-jw_i t\}/(2\pi)$, we have

$$([D_N]^H[\cap][D_N])_{ik} \xrightarrow[N\to\infty]{} \left(\frac{1}{2\pi}\right) \int_{-\pi}^{\pi} \exp\{-j[w_k - w_i]t\}\mu(t)dt = \hat{\mu}(w_k - w_i), \tag{3.4.23}$$

where $\hat{\mu}(w)$ has been used to denote the Fourier transform of $\mu(t)$ and we note that it has a Toeplitz structure as shown by the form of the argument.

The general procedure for generating transform nets that will asymptotically diagonalize the class of signal covariance matrices is now clear:

(1) Choose a set of orthonormal functions (polynomials) with respect to a support Ω and a measure $dn(t)$ such that the integral in (3.4.22) will have the Toeplitz structure required.

(2) By discretizing the orthonormal polynomials and using the proper quadrature weights, so that the termwise convergence holds in (3.4.22), obtain the transform net that will asymptotically diagonalize the signal covariance matrix.

The discrete transform obtained in such a procedure is called the Gauss—Jacobi transform (see Yemini and Peal [M-7]). The actual generation of a discrete transform is briefly described below.

Consider a measure $dn(t)$ over $\Omega = [-1, 1]$. Let $\{P_k(t)\}$ be a set of real orthonormal polynomials with respect to the measure $dn(t)$. From the theory of numerical quadrature it can be shown that there exists a set of positive real weights α_i, for $i = 0, 1, \ldots, N - 1$, corresponding to the N zeros of the polynomial $P_N(t)$, denoted by $\{t_i\}$ such that the quadrature of a polynomial $f(t)$,

$$Q_N(f) = \sum_{i=0}^{N-1} \alpha_i f(t_i), \tag{3.4.24}$$

gives the exact value for the integral,

$$\int_{\Omega} f(t)dn(t),$$

if $f(t)$ is of degree less than $2N - 1$.

The orthonormality condition of the set $\{P_k(t)\}$ with respect to the measure $dn(t)$ can be stated as follows:

$$\int_{\Omega} P_k(t)P_i(t)dn(t) = \delta_{ki}, \qquad k, i = 0, 1, \ldots, N-1, \tag{3.4.25}$$

and since the product polynomial $P_k(t)P_i(t)$ is of degree less than $2N - 1$, its quadrature approximation is exact. Therefore,

$$\sum_{m=0}^{N-1} \alpha_m P_k(t_m)P_i(t_m) = \delta_{ki}, \qquad k, i = 0, 1, \ldots, N-1. \tag{3.4.26}$$

From this expression, the transform $\delta = [D_N]_{N=0}^{\infty}$ can be constructed so that the ki-element of $[D_N]$ is given by

$$[D_N]_{ki} = \sqrt{\alpha_k} P_i(t_k). \tag{3.4.27}$$

Alternately, the ith vector component of $[D_N]$ is

$$\mathbf{d}_i = \{\sqrt{\alpha_k} P_i(t_k)\}, \qquad k = 0, 1, \ldots, N-1. \tag{3.4.28}$$

We shall now conclude this section by applying the formalism contained in Sections 3.4.1 through 3.4.4 to the case of the DCT-II.

Let $\Omega = [-1, 1]$, $dn(t) = (1 - t^2)^{1/2}dt$, and $P_i(t) = \cos(i\theta)$, with $\cos(\theta) = t$. It is not difficult to see that the $P_i(t)$ defined here is the Tchebyshev polynomial of the first kind $T_i(t)$. To construct the N-point transform, the zeros of $P_N(t) = T_N(t) = 0$ are chosen, giving

$$t_k = \cos\left[\frac{(2k+1)\pi}{2N}\right] \qquad k = 0, 1, \ldots, N-1 \tag{3.4.29}$$

(see Fike [M-14]).

The quadrature weights α_k can be obtained by recalling the orthogonality condition for the discrete Tchebyshev polynomials, namely,

$$\sum_{m=0}^{N-1} T_k(m)T_i(m) = \left(\frac{N}{2}\right)\delta_{ki}, \tag{3.4.30}$$

where we have used the following definitions for the Tchebyshev polynomials:

$$T_0(m) = \frac{1}{\sqrt{2}}, \qquad T_k(m) = \cos\left[\frac{(2m+1)k\pi}{2N}\right]. \tag{3.4.31}$$

Thus, substitution of (3.4.31) into (3.4.28) provides the ith vector component of the transform matrix,

$$\mathbf{d}_i = \left\{\sqrt{\left(\frac{2}{N}\right)}\, T_i(m)\right\}, \qquad m = 0, 1, \ldots, N-1. \tag{3.4.32}$$

We note that (3.4.32) provides the basis functions for DCT-II.

By substituting $\Omega = [-1, 1]$, $dn(t) = (1 - t^2)^{-1/2}dt$ and using the Tchebyshev polynomials in (3.4.22), the structure of the matrix class diagonal with respect to the basis in (3.4.32) can be examined. Thus

$$\begin{aligned}\int_{\Omega} d_k^*(t)\mu(t)d_i(t)dn(t) &= \int_{-1}^{1} \cos(k\theta)\cos(i\theta)(1-t^2)^{-1/2}\mu(t)dt \\ &= \int_{-\pi}^{\pi} -\cos(k\theta)\cos(i\theta)\mu(\cos^{-1}\theta)d\theta. \end{aligned}\tag{3.4.33}$$

The right-hand side can be further simplified to

$$\frac{1}{2}\int_{-\pi}^{\pi} \cos[(k-i)\theta]f(\theta)d\theta + \frac{1}{2}\int_{-\pi}^{\pi} \cos[(k+i)\theta]f(\theta)d\theta, \tag{3.4.34}$$

where $f(\theta) = -\mu(\cos^{-1}\theta)$.

The first term in (3.4.34) has Toeplitz structure, and the second term has Hankel structure (i.e., opposite diagonals contain constant elements). It can be shown that the weak norm of the Hankel form vanishes as N tends to infinity, when a certain smoothness condition[1] is satisfied (see Yemeni and Pearl [M-7]). Thus the DCT-II, obtained by using the Gaussian quadrature on Tchebyshev polynomials, will diagonalize the Toeplitz class of matrices asymptotically. Since this class includes Markov signal of all orders, DCT-II is asymptotic to the KLT for the Markov signal of all orders, a fact that has important consequences for DCT-II.

[1] Consider the Hahkel form:

$$\begin{bmatrix} \alpha_0 & \alpha_1 & \alpha_2 & \alpha_3 & \cdots \\ \alpha_1 & \alpha_2 & \alpha_3 & \alpha_4 & \cdots \\ \alpha_2 & \alpha_3 & \alpha_4 & \cdots & \cdots \\ \vdots & \vdots & \vdots & \vdots & \vdots \end{bmatrix}$$

Its weak norm is given by $(1/N)\,\{\alpha_0^2 + 2\alpha_1^2 + 3\alpha_2^2 + \cdots + N\alpha_{N-1}^2\}$, which vanishes as N tends to infinity if the following condition is satisfied:

$$\sum_{m=1}^{\infty} m\alpha_{m-1}^2 < \infty.$$

This is referred to as the smoothness condition.

By varying the choice of Ω, $P_k(t)$, and $dn(t)$ in the quadrature procedure, other discrete unitary transforms such as the discrete sine transform and the discrete Legendre transforms can be generated.

3.5 Summary

In this chapter, we have detailed the relationship between the DCT and the statistically optimal KLT. The asymptotic behavior of the DCT is clearly demonstrated both as the sequence length N increases and also as the adjacent correlation coefficient increases. Although derivations are presented only for DCT-I and DCT-II, the results are similar for DCT-III and DCT-IV. A natural follow-up of this is the efficient implementation of these transforms based on fast algorithms that are developed in the next chapter.

The final section deals with the more general problem of so-called equivalent classes of matrices and their orthonormal representations. The emphasis here is on the asymptotic equivalence of different types of correlation matrices. The outcome of this development leads to a rather general and interesting procedure for generating certain discrete unitary transforms for a given class of signal covariance matrices.

PROBLEMS

3.1. Derive (3.2.10).

3.2. Derive (3.2.13).

3.3. Derive (3.2.15) and (3.2.16).

3.4. Derive (3.3.3).

3.5. Derive (3.3.9).

3.6. Derive (3.3.13).

3.7. Derive (3.3.16).

3.8. Toward the end of Section 3.3.1 it is stated that if $w_n = (n\pi)/(N - 1)$ is used in (3.2.15), the basis functions (3.3.17) are obtained. Verify this statement.

3.9. In Section 3.3.2 it is stated that the basis functions of DCT-II are the eigenvectors of a singular matrix whose elements are all 1's when $\rho = 1$. Prove this.

3.10. In Section 3.3.2 it is also stated that as $\rho \to 0$, the basis functions (3.2.15) lead to one of the discrete sine transforms (see [M-2]). Prove this.

3.11. Prove that the DCT-IV is asymptotic to KLT as $N \to \infty$. Investigate the form of the covariance matrix diagonalized by the DCT-IV basis functions.

CHAPTER 4

FAST ALGORITHMS FOR DCT-II

4.1 Introduction

The development of efficient algorithms for the computation of DCT (more specifically DCT-II) began soon after Ahmed *et al.* [G-1] reported their work on DCT. It was natural for initial attempts to focus on the computation of DCT by using the fast Fourier transform (FFT) algorithms. Although DCT-II was developed not as a discretized version of the Fourier cosine transform (FCT), its relations to the discrete Fourier transform (DFT) were exploited in the initial developments of its computational algorithms. Haralick [FF-3] reported the computation of an N-point DCT using two N-point FFTs. Other work along the same line soon followed. Narashima and Peterson [FF-1], Tseng and Miller [FF-2], Makhoul [FF-5], Duhamel [FF-6], and Vetterli and Nussbaumer [FF-4] all contributed to the advances in this direction.

Fast computational algorithms can also be obtained by considering the factorization of the DCT matrix. When the component factors of this factorization are sparse, the decomposition represents a fast algorithm. Chen, Smith, and Fralick [FRA-1] reported such a development. Since matrix factorization is not unique, there exist other forms of fast algorithms. Corrington [FRA-6], Wang [FRA-2], Lee [FRA-3, FRA-4], and Suehiro and Hatori, [FRA-12] among others, contributed to this development. Some of the factorization schemes fall into the decimation-in-time (DIT) category. Others fall into the decimation-in-frequency (DIF) category (see Yip and Rao [FRA-8, FRA-9, FRA-14]). Fast DCT algorithms can also be obtained through the computation of other discrete transforms (see Hein and Ahmed [DW-2]; Jones, Hein, and Knauer [DW-3]; Venkataraman et al. [DW-1]; Malvar [FRA-16, FRA-19, FRA-21]; Nagesha [FRA-28]), through prime factor decomposition (see Yang and Narashima [FRA-10]]), through recursive computation (see Hou [FRA-13]), and by planar rotations [DP-22, DP-27, FRA-23, FRA-26]. Other algorithms include split-radix [FRA-29].

In this chapter, we shall detail some of the developments for fast algorithms. Section 4.2 deals with algorithms related to FFTs. Since there are many versions of FFTs (see, for example, Elliott and Rao [D-2]), the speed and efficiency of computation using these algorithms are closely linked to the speed and efficiency of the particular FFT algorithm used. Section 4.3 discusses DCT matrix factorizations. These may be regarded as a direct attack on the computational problem. DIT and DIF algorithms are presented in Section 4.4, where the DCT of type I is seen to be involved. Computation via other discrete transforms is discussed in Section 4.5. Section 4.6 deals with additional algorithms, and the chapter is concluded with a summary and comparison of computational complexity of different algorithms for implementing DCT-II in Section 4.7.

4.2 DCT via FFT

In Section 2.2, where properties of the Fourier cosine transform are recalled, (2.2.2) shows that the Fourier transform of an even function results in the definition for the Fourier cosine transform. This very simple property can be exploited for the computation of the DCT. Let $\{x(n)\}$, $n = 0, 1, \ldots, N-1$ be a given sequence. Then an extended sequence $\{y(n)\}$ symmetric about the $(2N-1)/2$ point can be constructed so that

$$\begin{aligned} y(n) &= x(n) & n &= 0, 1, \ldots, N-1, \\ &= x(2N-n-1) & n &= N, N+1, \ldots, 2N-1. \end{aligned} \tag{4.2.1}$$

If we use W_{2N} to denote $\exp(-j2\pi/2N)$, as in Chapter 2, it can be seen that the DFT of $\{y(n)\}$ given by

$$Y(m) = \sum_{n=0}^{2N-1} y(n) W_{2N}^{nm} \tag{4.2.2}$$

is easily reduced to

$$\begin{aligned} Y(m) &= \sum_{n=0}^{N-1} x(n) W_{2N}^{nm} + \sum_{n=N}^{2N-1} y(n) W_{2N}^{nm} \\ &= \sum_{n=0}^{N-1} x(n) W_{2N}^{nm} + \sum_{n=N}^{2N-1} x(2N-n-1) W_{2N}^{nm} \\ &= \sum_{n=0}^{N-1} x(n) W_{2N}^{nm} + \sum_{n=0}^{N-1} x(n) W_{2N}^{(2N-n-1)m} \\ &= \sum_{n=0}^{N-1} x(n) [W_{2N}^{nm} + W_{2N}^{-(n+1)m}], \qquad m = 0, 1, \ldots, 2N-1. \end{aligned} \tag{4.2.3}$$

If we now multiply both sides of (4.2.3) by a factor of $\frac{1}{2} W_{2N}^{m/2}$, we directly obtain

$$\frac{1}{2} W_{2N}^{m/2} Y(m) = \sum_{n=0}^{N-1} x(n) \cos\left[(2n+1)\frac{m\pi}{2N}\right], \qquad m = 0, 1, \ldots, N-1. \tag{4.2.4}$$

Comparing (4.2.4) with the DCT-II definition in (2.3.6), it is easily seen that, except for the required scale factors in (2.3.6), (4.2.4) is the DCT-II of the N-point sequence $\{x(n)\}$. Equation (4.2.2) shows that $\{Y(m)\}$ is the $2N$-point DFT of $\{y(n)\}$ and (4.2.4) shows that for $m = 0, 1, \ldots, N-1$, the transformed sequence $\{Y(m)\}$ properly scaled is the DCT-II of the N-point sequence $\{x(n)\}$. Very simply then, (4.2.4) provides a means of computing the DCT-II of $\{x(n)\}$ via a $2N$-point DFT of the extended sequence $\{y(n)\}$. This concept was utilized in building a feasibility model based on high-speed pipeline FFT processing techniques (26 wirewrap modules of 13 types—total of 2816 TTL ICs) for real-time 4:1 bandwidth compression of images at a clock rate of 10 MHz [DP-7].

When the sequence $\{x(n)\}$ is real, $\{y(n)\}$ is real and symmetric. In this case, Haralick [FF-3] showed that $\{Y(m)\}$ can be obtained via two N-point FFTs rather than by a single $2N$-point FFT. Since an N-point FFT requires $N \log_2 N$ complex operations in general, such an approach represents a savings of $2N$ complex operations. Tseng and Miller [FF-2], however, pointed out that when the input sequence $\{x(n)\}$ is real, the radix-2 DFT of the real, even, $2N$-point sequence $\{y(n)\}$ requires only $(N/2)\log_2 N + N/2$ complex operations. The $N/2$ term arises from the scaling required as in (4.2.4). This represents a significant reduction in the complexity from that of computing two N-point DFTs.

A different rearrangement of the input sequence was proposed by Narasimha and Peterson [FF-1]. This provides an even more efficient algorithm. Consider an N-point sequence $\{x(n)\}$ and its DCT (specifically DCT-II). If N is even, the sequence $\{x(n)\}$ can be relabelled as

$$y(n) = x(2n)$$

and

$$y(N-1-n) = x(2n+1), \qquad n = 0, 1, \ldots, \frac{N}{2} - 1. \tag{4.2.5}$$

The DCT-II of $\{x(n)\}$ can now be reduced to (using notations introduced in Chapter 2)

$$X^{C(2)}(m) = \sum_{n=0}^{N/2-1} y(n) \cos\left[\frac{(4n+1)m\pi}{2N}\right] + \sum_{n=0}^{N/2-1} y(N-n-1) \cos\left[\frac{(4n+3)m\pi}{2N}\right]$$

$$m = 0, 1, \ldots, N-1. \tag{4.2.6}$$

If the sum over n is replaced by a sum over $N-n-1$ in the second term, the two summations in (4.2.6) can be combined into one, giving

$$X^{C(2)}(m) = \sum_{n=0}^{N-1} y(n) \cos\left[\frac{(4n+1)m\pi}{2N}\right], \qquad m = 0, 1, \ldots, N-1. \tag{4.2.7a}$$

The right-hand side of this equation is the real part of a scaled N-point DFT of the sequence $\{y(n)\}$. Thus

$$X^{C(2)}(m) = \mathrm{Re}[Z(m)],$$

where

$$Z(m) = W_{4N}^{m} Y(m) = W_{4N}^{m} \sum_{n=0}^{N-1} y(n) W_{N}^{nm}, \qquad m = 0, 1, \ldots, N-1. \tag{4.2.7b}$$

In (4.2.7b), $\{Y(m)\}$ is the N-point DFT of $\{y(n)\}$. Also, it is easy to verify that

$$Z(N-m) = -jZ(m)^*, \tag{4.2.8}$$

provided $\{y(n)\}$ is a real sequence. Therefore,

$$X^{C(2)}(m) = \mathrm{Re}[Z(m)],$$

and

$$X^{C(2)}(N-m) = -\mathrm{Im}[Z(m)], \qquad m = 0, 1, \ldots, N/2. \tag{4.2.9}$$

Equation (4.2.9) summarizes the procedure. The DCT-II of the N-point real sequence $\{x(n)\}$ is obtained by evaluating the real and imaginary parts of the first $(N/2 + 1)$ components of the N-point DFT of the rearranged sequence $\{y(n)\}$. Using a radix-2 algorithm for the DFT, the resulting number of complex multiplications is $(N \log_2 N - N + 1)/4$, which is about twice as fast as the algorithm in (4.2.4), as proposed by Tseng and Miller [FF-2]. Although we have compared the number of complex operations based on radix-2 algorithms for the DFT, the formulations are general. For example, the Winograd algorithm (see Silverman [M-24]) may be used to facilitate the DFT computation when N is not a power of 2. The expansions are valid regardless of what algorithm is being used in the computations of the DFTs.

In the rearrangement of the data sequence, (4.2.5) indicates that N should be even. This restriction can in fact be removed, as pointed out by Makhoul [FF-5]. As a result, (4.2.5) groups all the even-numbered points of $\{x(n)\}$ in increasing order followed by the odd-numbered points in reversed order. In Figs. 4.1 and 4.2, the block diagrams for DCT-II representing the two algorithms discussed above are shown.

The relation between the DCT and DFT can be exploited in a different way, as demonstrated by Vetterli and Nussbaumer [FF-4]. The approach taken is a recursive one, whereby the dimension of the DCT is repeatedly reduced until only trivial ($N = 2$ or 4) transforms are required. In particular, when the input sequence $\{x(n)\}$ is real, the real and imaginary parts of its DFT, $\{X(m)\}$, can be

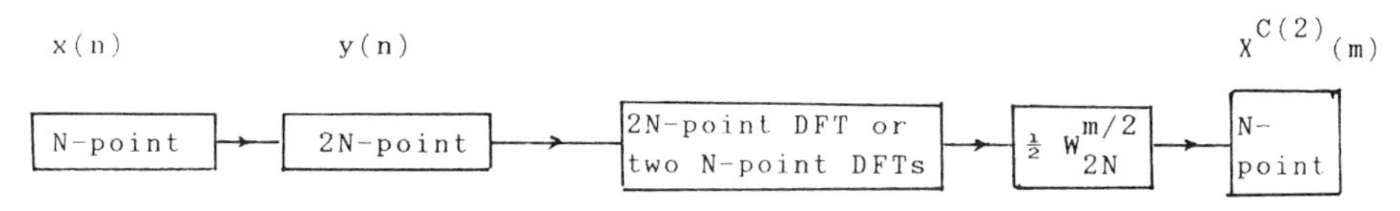

Fig. 4.1 N-point DCT via $2N$-point DFT.

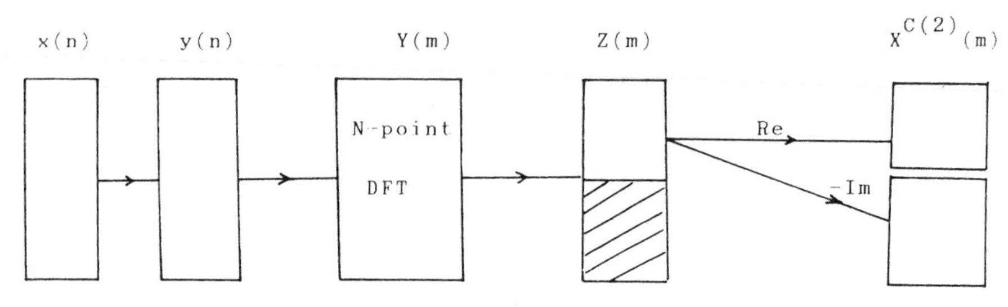

Fig. 4.2 N-point DCT via N-point DFT.

written as

$$\begin{aligned}\operatorname{Re}[X(m)] &= \sum_{n=0}^{N/2-1} x(2n)\cos\left[\frac{2nm\pi}{(N/2)}\right] \\ &\quad + \sum_{n=0}^{N/4-1} \{x(2n+1) + x(N-2n-1)\}\cos\left[\frac{2(2n+1)m\pi}{N}\right], \\ \operatorname{Im}[X(m)] &= \sum_{n=0}^{N/2-1} x(2n)\sin\left[\frac{2nm\pi}{(N/2)}\right] \\ &\quad + \sum_{n=0}^{N/4-1} \{x(2n+1) - x(N-2n-1)\}\sin\left[\frac{2(2n+1)m\pi}{N}\right], \\ &\quad m = 0, 1, \ldots, N-1. \end{aligned} \tag{4.2.10}$$

We note from (4.2.10) that for both the real and imaginary parts, the first terms on the right-hand side are essentially the real and imaginary parts of an $N/2$-point DFT of the even-numbered points in $\{x(n)\}$. The second term for the real part is a DCT-II of $N/4$ points, and the second term of the imaginary part can be further reduced to a $N/4$-point DCT-II by using the identity

$$\sin\left[\frac{2(2n+1)m\pi}{N}\right] = (-1)^n \cos\left[\frac{(2n+1)(N-4m)\pi}{2N}\right]. \tag{4.2.11}$$

It can also be verified that

$$X^{C(2)}(m) = \cos\left[\frac{m\pi}{2N}\right]\operatorname{Re}[Y(m)] - \sin\left[\frac{m\pi}{2N}\right]\operatorname{Im}[Y(m)], \tag{4.2.12}$$

where $\{Y(m)\}$ is the N-point DFT of the sequence $\{y(n)\}$ defined in (4.2.5). The computation of the DCT-II in (4.2.12) is summarized by the flow diagram shown in Fig. 4.3 (also see Problem 4.3).

This diagram shows the initial steps of recursive decomposition, excluding the actual scale factors. The rearrangement and recombinations of the input sequences are also seen. A continuation of this procedure will result in trivial transforms with $N = 2$ or 4. The resulting number of real multiplications

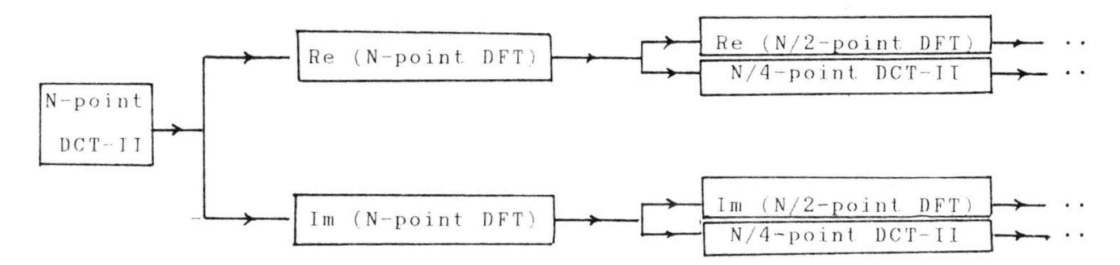

Fig. 4.3 Recursive computation of DCT-II. [FF-4].

required for an N-point DCT (radix-2) is given by

$$\left(\frac{N}{2}\right)\log_2 N. \tag{4.2.13}$$

The number of real multiplications described by (4.2.13) compares favorably with the number of complex multiplications required for the other algorithms discussed earlier in this section. A more comprehensive comparison will be made at the end of this chapter (see Section 4.7). It should be pointed out that the reduced number of multiplications is obtained at the cost of increased complexity in the permutation and combination of the input data samples and of the output transformed samples. This increased complexity in the topology of the overall transform is its major drawback when compared with the other algorithms.

4.3 Direct Computation by Sparse Matrix Factorizations

Consider the computation of the DCT-II of an input sequence $\{x(n)\}$ and let this sequence be represented by the column vector $\mathbf{x}$. The transformed sequence can be expressed in vector notations as follows:

$$\mathbf{X}^{C(2)} = \left(\frac{2}{N}\right)^{1/2}[A_N]\mathbf{x}, \tag{4.3.1}$$

where we have used the notations introduced in Chapter 2. Here, $[A_N]$ is an $(N \times N)$ matrix and is equal to $[C_N^{\mathrm{II}}]$, defined in (2.3.6) except for the normalization factor. When the matrix $[A_N]$ (or $[C_N^{\mathrm{II}}]$) is decomposed into sparse matrices, a fast algorithm will result. We show, in what follows, $[A_2]$ and $[A_4]$, and a sparse decomposition for $[A_4]$.

$$[A_2] = \begin{bmatrix} \frac{1}{\sqrt{2}} & \frac{1}{\sqrt{2}} \\ C_4^1 & C_4^3 \end{bmatrix}, \qquad [A_4] = \begin{bmatrix} \frac{1}{\sqrt{2}} & \frac{1}{\sqrt{2}} & \frac{1}{\sqrt{2}} & \frac{1}{\sqrt{2}} \\ C_8^1 & C_8^3 & C_8^5 & C_8^7 \\ C_8^2 & C_8^6 & C_8^6 & C_8^2 \\ C_8^3 & C_8^7 & C_8^1 & C_8^5 \end{bmatrix}, \tag{4.3.2}$$

where $C_k^i = \cos[i\pi/k]$ has been used to simplify the notations. The matrix $[A_4]$ can be decomposed as follows:

$$[A_4] = \begin{bmatrix} 1 & 0 & 0 & 0 \\ 0 & 0 & 0 & 1 \\ 0 & 1 & 0 & 0 \\ 0 & 0 & 1 & 0 \end{bmatrix} \begin{bmatrix} \frac{1}{\sqrt{2}} & \frac{1}{\sqrt{2}} & \frac{1}{\sqrt{2}} & \frac{1}{\sqrt{2}} \\ C_4^1 & C_4^3 & C_4^3 & C_4^1 \\ C_8^3 & -C_8^1 & C_8^1 & -C_8^3 \\ C_8^1 & C_8^3 & -C_8^3 & -C_8^1 \end{bmatrix}$$

$$= \begin{bmatrix} 1 & 0 & 0 & 0 \\ 0 & 0 & 0 & 1 \\ 0 & 1 & 0 & 0 \\ 0 & 0 & 1 & 0 \end{bmatrix} \begin{bmatrix} \frac{1}{\sqrt{2}} & \frac{1}{\sqrt{2}} & 0 & 0 \\ C_4^1 & C_4^3 & 0 & 0 \\ 0 & 0 & -C_8^1 & C_8^3 \\ 0 & 0 & C_8^3 & C_8^1 \end{bmatrix} \begin{bmatrix} 1 & 0 & 0 & 1 \\ 0 & 1 & 1 & 0 \\ 0 & 1 & -1 & 0 \\ 1 & 0 & 0 & -1 \end{bmatrix}. \tag{4.3.3}$$

Equation (4.3.3) can be expressed in more compact form as

$$[A_4] = [\bar{P}_4] \begin{bmatrix} [A_2] & 0 \\ 0 & [\bar{\bar{R}}_2] \end{bmatrix} [B_4], \tag{4.3.4}$$

where $[\bar{P}_4]$ is a permutation matrix, $[B_4]$ is a butterfly matrix, and $[\bar{\bar{R}}_2]$ is the remaining (2×2) block in the second factor matrix. The key is that $[A_4]$ is reduced in terms of $[A_2]$. In general for N, being a power of 2, (4.3.4) is in the form of

$$[A_N] = [\bar{P}_N] \begin{bmatrix} [A_{N/2}] & 0 \\ 0 & [\bar{\bar{R}}_{N/2}] \end{bmatrix} [B_N], \tag{4.3.5}$$

where $[\bar{P}_N]$ permutes the even rows in increasing order in the upper half and the odd rows in decreasing order in the lower half. The butterfly matrix $[B_N]$ can be expressed in terms of the identity matrix $[I_{N/2}]$ and the opposite identity matrix $[\tilde{I}_{N/2}]$ as follows:

$$[B_N] = \begin{bmatrix} [I_{N/2}] & [\tilde{I}_{N/2}] \\ [\tilde{I}_{N/2}] & -[I_{N/2}] \end{bmatrix}. \tag{4.3.6}$$

The matrix $[\bar{\bar{R}}_N]$ is obtained by reversing the orders of both the rows and columns of the matrix $[R_N]$ whose ik-element is given by

$$[R_N]_{ik} = \cos\left[\frac{(2i+1)(2k+1)\pi}{4N}\right], \qquad i, k = 0, 1, \ldots, N-1. \tag{4.3.7}$$

We note that (4.3.7) essentially defines the transform matrix for DCT-IV.

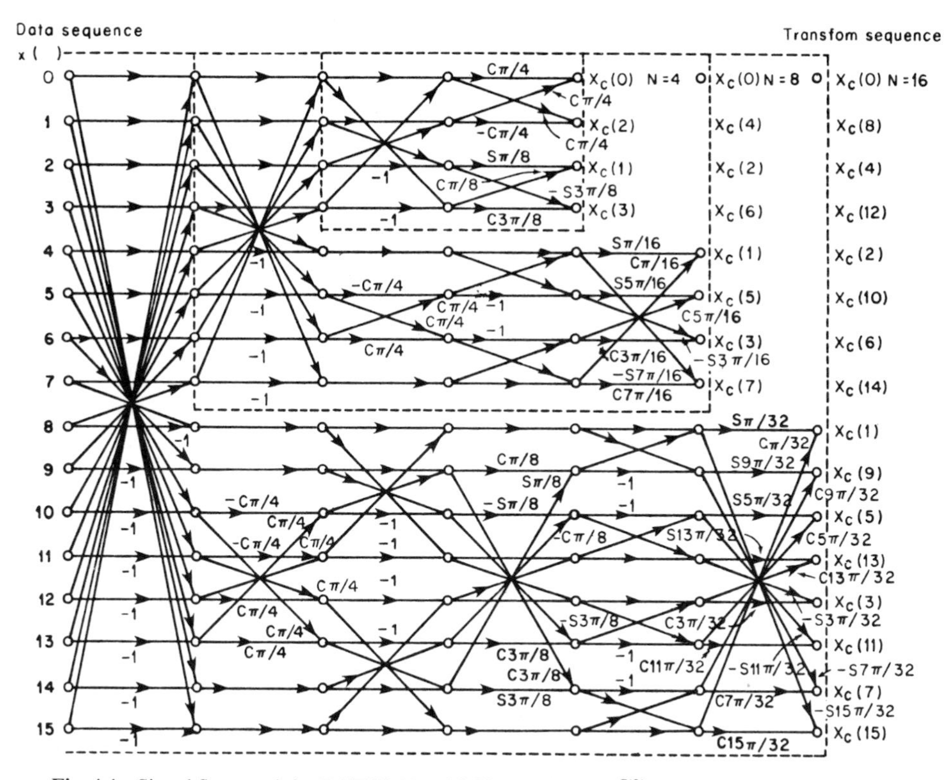

Fig. 4.4 Signal flowgraph for DCT-II, $N = 16$. Here $X_c(m) = X^{C(2)}(m)$ (FRA-1]. (© 1977 IEEE).

As can be seen in (4.3.5), the factorization is only partly recursive, since the matrix $[R_N]$ cannot be recursively factored. There is, however, regularity in its factorization, and it is found that it can be decomposed into five types of matrix factors, all of which have no more than two non-zero elements in each row. The interested reader is referred to the work by Wang [FRA-2] and Chen, Smith, and Fralick [FRA-1] for more detailed descriptions of the actual matrix factors for $[R_N]$. This algorithm requires

$$\frac{3N}{2}(\log_2 N - 1) + 2 \qquad \text{real additions}$$

and

$$N \log_2 N - \frac{3N}{2} + 4 \qquad \text{real multiplications.} \tag{4.3.8}$$

We conclude this section by displaying the signal flowgraph for the $N = 16$ case of the DCT-II in Fig. 4.4.

4.4 Decimation-in-Time (DIT) and Decimation-in-Frequency (DIF) Algorithms

Of all the algorithms described so far, only the scheme proposed by Vetterli and Nussbaumer [FF-4] is strictly recursive. The factorization presented in Section 4.3 is only partly recursive, since the factor $[R_N]$ cannot be so factored. Although its factorization is quite regular, it still lacks the desired property. The scheme proposed by Vetterli and Nussbaumer, on the other hand, involves both the DFT and DCT of reduced dimensions, and the topology of the index mapping is known to be complex.

In Section 4.2, a rearrangement of the input sequence $\{x(n)\}$ was described, as proposed by Narasimha and Peterson [FF-1]. Such a rearrangement [see (4.2.5)] makes possible the computation of DCT via an N-point DFT and is in fact typical of some of the fast algorithms for computing the DCT. Lee [FRA-3, FRA-4] studied just such an algorithm, where the key concept is to reduce an N-point DCT to an $N/2$-point DCT by permutation of the input sample points. The resulting algorithm contains the desirable recursive modularity of a fast decomposition. This arrangement can be accomplished either in the time domain or in the frequency domain. In the former case, it is generally referred to as decimation-in-time (DIT) and in the latter case, as decimation-in-frequency (DIF). Sparse matrix factors, flowgraphs, and computer programs for DCT-II and its inverse ($N = 8$ and 16) based on Lee's algorithms are provided in Appendix A.1. In this section, we describe in detail the DIT and DIF algorithms reported by Yip and Rao [FRA-8, FRA-14] for DCT-II.

4.4.1 *The DIT Algorithm*

For convenience, we repeat the definition for DCT-II from (2.3.6).

$$[C_N^{\mathrm{II}}]_{mn} = \sqrt{\left(\frac{2}{N}\right)}\, k_m \cos\left[\frac{(2n+1)m\pi}{2N}\right], \qquad m, n = 0, 1, \ldots, N-1, \tag{2.3.6}$$

where $k_m = 1/\sqrt{2}$ for $m = 0$ and $k_m = 1$ otherwise.

As in the previous sections, we shall leave out this scale factor for convenience. The transformed sequence $\{X^{C(2)}(m)\}$ is given by

$$X^{C(2)}(m) = \sum_{n=0}^{N-1} x(n) C_{2N}^{(2n+1)m}, \qquad m = 0, 1, \ldots, N-1, \tag{4.4.1}$$

where $C_k^i = \cos[i\pi/k]$ as before.

The first stage in the DIT algorithm consists of rearrangement of the input sample points. The second stage reduces the N-point transform to $N/2$-point transforms to establish the recursive aspect of the algorithm. These stages are described by the following equations:

Let

$$G(m) = \sum_{n=0}^{N/2} [x(2n) + x(2n-1)] C_{N/2}^{mn},$$

and

$$H(m) = \sum_{n=0}^{N/2-1} [x(2n) + x(2n+1)]C_N^{(2n+1)m}, \qquad m = 0, 1, \ldots, N/2-1, \tag{4.4.2}$$

with $x(-1) = x(N) = 0$.

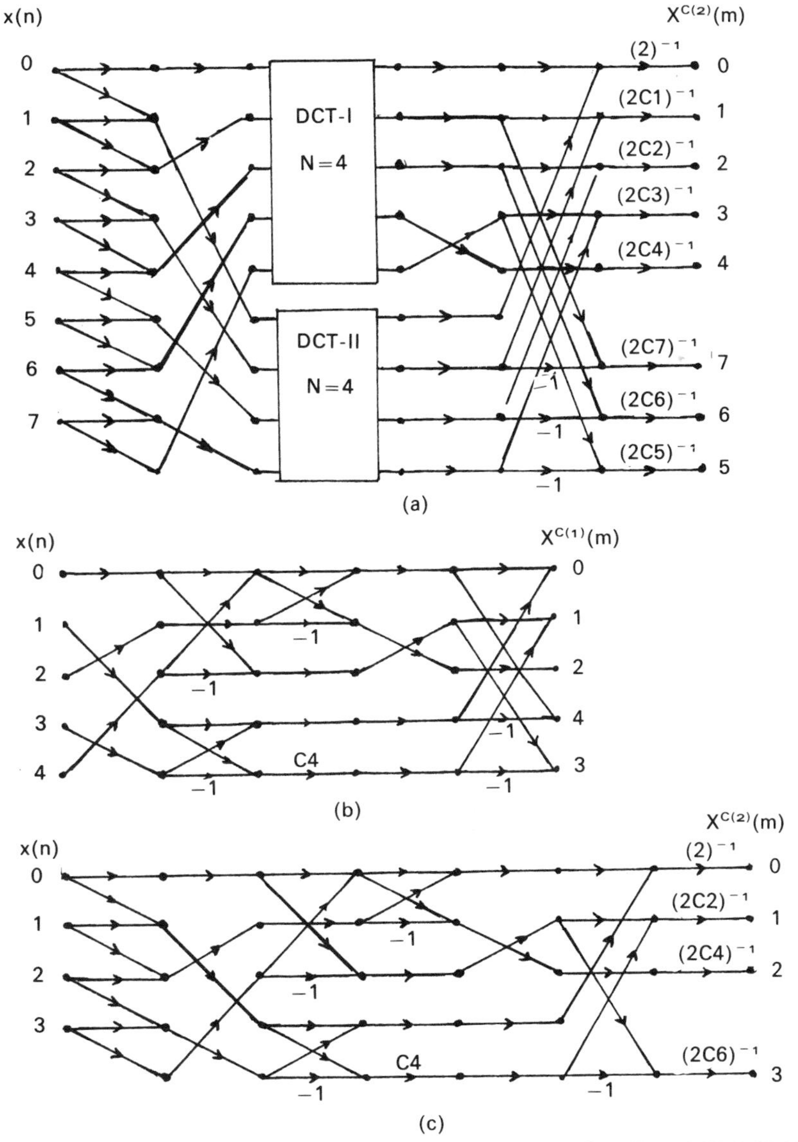

Fig. 4.5 Signal flowgraph for DIT DCT-II for $N = 8$. $Ck = \cos[k\pi/16]$. (a) DIT DCT-II for $N = 8$. (b) DIT DCT-I for $N = 4$. (c) DIT DCT-II for $N = 4$.

It is not difficult to show, using properties of the cosine functions, that (4.4.2) can be used in (4.4.1), resulting in

$$X^{C(2)}(m) = \frac{1}{2}\frac{[G(m) + H(m)]}{C_N^m},$$

and

$$X^{C(2)}(N - m - 1) = \frac{1}{2}\frac{[G(m+1) - H(m+1)]}{C_{2N}^{N-m-1}}, \qquad m = 0, 1, \ldots, N/2 - 1. \tag{4.4.3}$$

In (4.4.2), the sequence $\{H(m)\}$ is obtained as the DCT-II of $N/2$ sample points. However, $\{G(m)\}$ is obtained as a DCT-I of $(N/2 + 1)$ sample points (see Section 2.3). This part of the transform can be further reduced (see Yip and Rao [FRA-8]), leading to the desired recursive structure. Figure 4.5 shows the signal flowgraph for the DIT computation of an $N = 8$ DCT-II of the sequence $\{x(n)\}$. We note here that in [FRA-8], DCT-II has been incorrectly labeled as DCT-III.

The matrix factorization corresponding to the procedure expressed by (4.4.2) and (4.4.3) is given by

$$\begin{aligned}[C_N^{\mathrm{II}}] = {} & \mathrm{diag}\{[M_{N/2+1}^{\mathrm{I}}], [M_{N/2-1}^{\mathrm{IV}}]\}[A_N^{\mathrm{IV}}]\mathrm{diag}\{[I_{N/4+1}], [\tilde{I}_{N/4}], [I_{N/2}]\} \\ & \times \mathrm{diag}\{[C_{N/2+1}^{\mathrm{I}}], [C_{N/2}^{\mathrm{II}}]\}[P_{N+1}][B_{N+1}^{\mathrm{I}}],\end{aligned} \tag{4.4.4}$$

where $[C_{N+1}^{\mathrm{I}}]$ is the DCT-I transform matrix given by

$$\begin{aligned}[C_{N+1}^{\mathrm{I}}] = {} & [A_{N+1}^{\mathrm{I}}]\mathrm{diag}\{[I_{N/4+1}], [\tilde{I}_{N/4+1}], [M_{N/4+1}^{\mathrm{I}}], [M_{N/4-1}^{\mathrm{II}}] \\ & \times \mathrm{diag}\{[C_{N/2+1}^{\mathrm{I}}], [A_{N/2}^{\mathrm{II}}][C_{N/2+1}^{\mathrm{I}}][B_{N/2+1}^{\mathrm{I}}]\}[P_{N+1}].\end{aligned} \tag{4.4.5}$$

The salient feature of these factors is the presence of the DCT-I and DCT-II matrices of half the original size. The other factors, although complicated in appearance, are in fact quite simple. They are either butterfly matrices with +1 or −1 as elements, or are permutation matrices of some kind. The $[M]$ matrices are either diagonal or opposite diagonal, serving as scaling factor matrices. For completeness, we describe these other factor matrices in what follows:

$$[P_N] = \begin{bmatrix} 1 & 0 & 0 & 0 & \cdots & \cdots & \cdots \\ 0 & 0 & 1 & 0 & \cdots & \cdots & \cdots \\ \cdots & \cdots & \cdots & \cdots & \cdots & \cdots & \cdots \\ 0 & 1 & 0 & 0 & \cdots & \cdots & \cdots \\ 0 & 0 & 0 & 1 & \cdots & \cdots & \cdots \\ \cdots & \cdots & \cdots & \cdots & \cdots & \cdots & \cdots \end{bmatrix},$$

i.e., an $N \times N$ even-odd permutation matrix;

$$[B^{\mathrm{I}}_{N+1}] = \begin{bmatrix} 1 & 0 & 0 & 0 & \cdots & \cdots & \cdots \\ 1 & 1 & 0 & 0 & \cdots & \cdots & \cdots \\ 0 & 1 & 1 & 0 & \cdots & \cdots & \cdots \\ \cdots & \cdots & \cdots & \cdots & \cdots & \cdots & \cdots \\ \cdots & \cdots & \cdots & \cdots & \cdots & 1 & 1 \\ \cdots & \cdots & \cdots & \cdots & \cdots & 0 & 1 \end{bmatrix},$$

i.e., an $(N + 1) \times N$ matrix made up of an $N \times N$ lower bi-diag. matrix appended by a unit Nth row vector;

$$[A^{\mathrm{I}}_{N+1}] = \begin{bmatrix} [I_{N/2}] & \vdots & [I_{N/2}] \\ \cdots & 1 & \cdots \\ [I_{N/2}] & \vdots & [I_{N/2}] \end{bmatrix},$$

i.e., an $(N + 1) \times (N + 1)$ matrix with unit elements;

$$[A^{\mathrm{II}}_{N}] = \begin{bmatrix} & [I_{N-1}] & & 0 & 0 \\ & \vdots & & \vdots & \vdots \\ 0 & 0 & \cdots & 0 & 1 \end{bmatrix},$$

i.e., an $N \times (N + 1)$ matrix with unit elements;

$$[A^{\mathrm{IV}}_{N}] = \begin{bmatrix} & [I_{N/2}] & \vdots & & [I_{N/2}] \\ & \cdots & 1 & & \cdots \\ \vdots & [I_{N/2-1}] & \vdots & \vdots & -[I_{N/2-1}] \end{bmatrix},$$

i.e., an $N \times (N + 1)$ matrix with unit elements;

$$[M^{\mathrm{I}}_{N+1}] = \tfrac{1}{2}\,\mathrm{diag}\{1,\ 1/C^{1}_{4N},\ \ldots,\ 1/C^{N}_{4N}\};$$

$$[M^{\mathrm{II}}_{N-1}] = \tfrac{1}{2}\ \text{opposite diag}\{1/C^{N+1}_{4N},\ \ldots,\ 1/C^{2N-1}_{4N}\};$$

and

$$[M^{\mathrm{IV}}_{N-1}] = \tfrac{1}{2}\,\mathrm{diag}\{1/C^{2N-1}_{4N},\ 1/C^{2N-2}_{4N},\ \ldots,\ 1/C^{N+1}_{4N}\}. \tag{4.4.6}$$

The sparseness of these factor matrices is quite apparent, and simple difference equations can be established to delineate the computational complexity in terms of the numbers of additions and multiplications. Excluding scaling and normalization, it is found that for an N-point (radix-2) sequence $\{x(n)\}$, this DIT algorithm for DCT-II requires

$$\left(\frac{3N}{2} - 1\right)\log_2 N + \frac{N}{4} + 1 \qquad \text{real additions}$$

and

$$\left(\frac{N}{2}\right) \log_2 N + \frac{N}{4} \qquad \text{real multiplications.}$$

These numbers compare favorably with those of other algorithms.

4.4.2 *The DIF Algorithm*

When the rearrangement of the sample points results in the transformed sequence being grouped into even- and odd-frequency indexed portions, the decomposition is said to constitute a DIF algorithm. In the case of DCT-II, we shall see a very obvious recursive structure. We proceed by rearranging the transformed sequence $\{X^{C(2)}(m)\}$ into even and odd portions so that

$$X^{C(2)}(m) = G(m),$$

and

$$X^{C(2)}(2m+1) = H(m) + H(m+1), \qquad m = 0, 1, \ldots, \frac{N}{2} - 1, \tag{4.4.7}$$

where the sequences $\{G(m)\}$ and $\{H(m)\}$ are DCT-IIs of two rearranged and recombined sequences, as shown by the following equations.

$$G(m) = \sum_{n=0}^{N/2-1} [x(n) + x(N-n-1)] C_N^{(2n+1)m},$$

and

$$H(m) = \sum_{n=0}^{N/2-1} \tfrac{1}{2}[x(n) - x(N-n-1)] \frac{C_N^{(2n+1)m}}{C_{2N}^{(2n+1)}}. \tag{4.4.8}$$

Comparing with (4.4.1), it is easy to see that both $\{G(m)\}$ and $\{H(m)\}$ are DCT-IIs of $N/2$ points. Figure 4.6 shows the signal flowgraph for $N = 16$.

The matrix factorization corresponding to the above procedure is given by

$$\begin{aligned}[C_N^{\text{II}}] = [P_N]^T \operatorname{diag}\{[I_{N/2}], [B_{N/2}^{\text{I}}]^T\} \operatorname{diag}\{[C_{N/2}^{\text{II}}], [C_{N/2}^{\text{II}}]\} \\ \times \operatorname{diag}\{[I_{N/2}], [M_{N/2}^{\text{III}}]\}[A_N^{\text{III}}] \operatorname{diag}\{[I_{N/2}], [\tilde{I}_{N/2}]\}.\end{aligned} \tag{4.4.9}$$

All factor matrices have been defined in (4.4.6) except for $[M_N^{\text{III}}]$ and $[A_N^{\text{III}}]$, which are given by

$$[M_N^{\text{III}}] = \tfrac{1}{2} \operatorname{diag}\{1/C_{4N}^1, 1/C_{4N}^3, \ldots, 1/C_{4N}^{2N-1}\}$$

and

$$[A_N^{\text{III}}] = \begin{bmatrix} [I_{N/2}] & [I_{N/2}] \\ [I_{N/2}] & -[I_{N/2}] \end{bmatrix}. \tag{4.4.10}$$

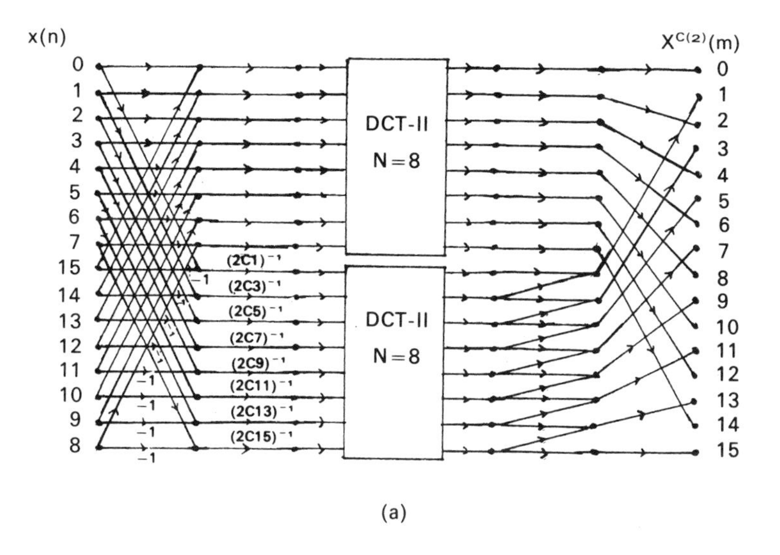

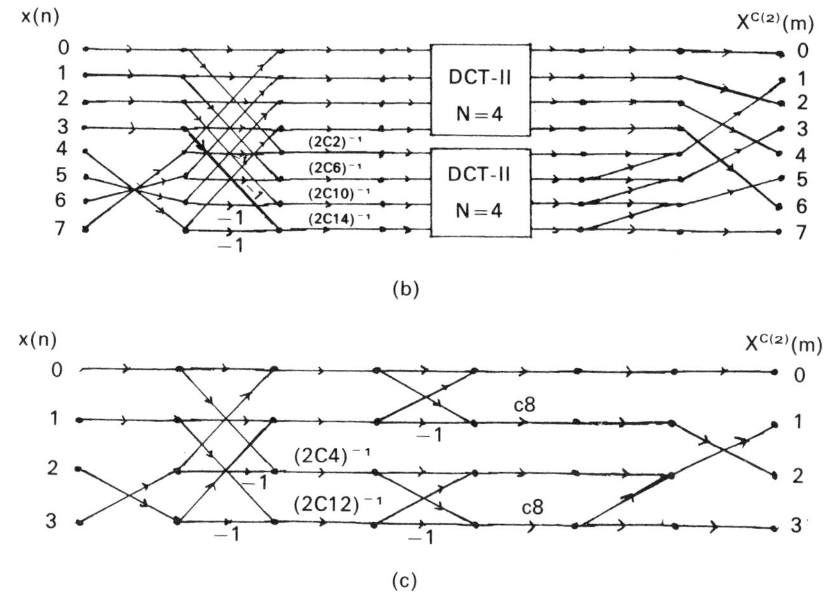

Fig. 4.6 Flowgraph for DIF computation of 16-point DCT-II. $Ck = \cos[k\pi/32]$. (a) DIF DCT-II for $N = 16$. (b) DIF DCT-II for $N = 8$. (c) DIF DCT-II for $N = 4$.

Similar considerations given to computational complexity produce, for an N-point (radix-2) sequence,

$$\left(\frac{3N}{2}\right)\log_2 N - N + 1 \qquad \text{real additions,}$$

and

$$\left(\frac{N}{2}\right)\log_2 N \qquad \text{real multiplications.}$$

4.5 DCT via Other Discrete Transforms

In Section 4.2, we have seen the development of some DCT algorithms via the computation of DFT. In the same spirit, the DCT computation may also be examined through other transforms. In this section, we consider two such algorithms, one via the Walsh–Hadamard transform (WHT) [DW-1, DW-2, DW-3] and the other via the discrete Hartley transform (DHT) [FRA-16, FRA-19, FRA-21, FRA-22]. The WHT is known to be fast since the computation involves no multiplications. Thus an algorithm for DCT via WHT may well utilize this advantage. The DHT, on the other hand, is very similar to DFT and has some computational advantages over the DFT if only real input sequences are to be processed. For these reasons we consider these two algorithms in some detail.

4.5.1 *DCT via WHT*

Let $\mathbf{x}$ be an $(N \times 1)$ column vector representing the input data and $\mathbf{X}^{C(2)}$ and $\mathbf{X}_W$ be $(N \times 1)$ column vectors representing the DCT-II and WHT of $\mathbf{x}$, respectively. Then

$$\mathbf{X}^{C(2)} = [C_N^{II}]\mathbf{x} \qquad \text{and } \mathbf{X}_W = [H_W]\mathbf{x}, \tag{4.5.1}$$

where $[C_N^{II}]$ and $[H_W]$ are the $(N \times N)$ transform matrices for the DCT and WHT (Walsh ordered) [B-1, B-2, B-3], respectively. Since $[H_W]$ is real and orthonormal, we see that

$$\begin{aligned}\mathbf{X}^{C(2)} &= [C_N^{II}][H_W]^T[H_W]\mathbf{x},\\ &= [T_N]\mathbf{X}_W,\end{aligned} \tag{4.5.2a}$$

where $[T_N] = [C_N^{II}][H_W]^T$ is the conversion matrix that takes the Walsh domain vector $\mathbf{X}_W$ and converts it to the DCT domain vector $\mathbf{X}^{C(2)}$. This conversion matrix has a block diagonal structure if the transformed vectors $\mathbf{X}^{C(2)}$ and $\mathbf{X}_W$ are arranged in bit-reversed order (BRO) (see Hein and Ahmed [DW-2]). Using the subscript BRO to denote such bit-reversed ordering, (4.5.2a) can be rewritten as

$$\begin{aligned}\mathbf{X}_{\mathrm{BRO}}^{C(2)} &= [C_N^{II}]_{\mathrm{BRO}}[H_W]_{\mathrm{BRO}}^T[H_W]_{\mathrm{BRO}}\mathbf{x},\\ &= [T_N]_{\mathrm{BRO}}[H_W]_{\mathrm{BRO}}\mathbf{x}.\end{aligned} \tag{4.5.2b}$$

For example, when $N = 8$, the conversion matrix is given explicitly as

$$[T_8]_{\mathrm{BRO}} = \begin{bmatrix} [1.0] & & & \\ & [1.0] & & 0 \\ & & \begin{bmatrix} 0.923 & 0.383 \\ -0.383 & 0.923 \end{bmatrix} & \\ & 0 & & \begin{bmatrix} 0.907 & -0.075 & 0.375 & 0.180 \\ 0.214 & 0.768 & -0.513 & 0.318 \\ -0.318 & 0.513 & 0.768 & 0.214 \\ -0.180 & -0.375 & -0.075 & 0.907 \end{bmatrix} \end{bmatrix}. \tag{4.5.3}$$

Since the WHT requires no multiplications, (4.5.3) accounts for all the multiplications required to compute $X_{\mathrm{BRO}}^{\mathrm{C(2)}}$. For $N = 8$, (4.5.3) indicates 20 real multiplications, as compared with 24 real multiplications needed for its computation using FFT (see Section 4.2). This computational advantage quickly disappears as N increases beyond 16. However, when many small transforms have to be computed, such as in the case of two-dimensional image processing, the accumulated saving may be substantial. In addition, the BRO as well as the block diagonal structure of the conversion matrix completely separates the even-frequency components from the odd-frequency components in the transformed vector, providing a possible means for data compression in the frequency domain. It may be noted that Venkataraman *et al.* [DW-1] have extended the conversion matrix up to $N = 32$ (see Problems 4.9 and 4.10).

4.5.2 *DCT via DHT*

The Hartley transform [M-26] has been revived in recent years in the literature [M-27, M-28, M-29] because of its close relationship to the Fourier transform and its apparent advantage in handling real data.

Recall that the kernel for the DFT of N points is given by

$$W_N^{mn} = \exp\left(-\frac{j2mn\pi}{N}\right) = \cos\left[\frac{2mn\pi}{N}\right] - j\sin\left[\frac{2mn\pi}{N}\right],$$

$$m, n = 0, 1, \ldots, N-1. \tag{4.5.4}$$

The kernel of an N-point DHT is given by

$$H_N^{mn} \equiv \mathrm{cas}\left(\frac{2mn\pi}{N}\right)^1 = \cos\left[\frac{2mn\pi}{N}\right] + \sin\left[\frac{2mn\pi}{N}\right], \qquad m, n = 0, 1, \ldots, N-1. \tag{4.5.5}$$

[1]Compare this notation with cis(·), sometimes used for exp(j·).

If $\mathbf{X}_F$ and $\mathbf{X}_H$ are used to denote the DFT and the DHT of $\mathbf{x}$, respectively, then it is not difficult to establish the relation

$$\mathbf{X}_H = \mathrm{Re}\{\mathbf{X}_F\} - \mathrm{Im}\{\mathbf{X}_F\}, \tag{4.5.6}$$

provided $\mathbf{x}$ is real. Equation (4.5.6) indicates that for any fast DFT algorithm, there should be a corresponding fast algorithm for DHT, as was indeed demonstrated by Sorensen *et al.* [M-29]. Therefore, there exists a relation between DCT and DHT analogous to the relation between DCT and DFT discussed in Section 4.2.

The following discussion (see [FRA-16, FRA-19, FRA-22, FRA-28]) delineates this relation and provides an algorithm to compute DCT via DHT. A similar relationship has also been developed by Hou [FRA-22]. The recursive structure of the fast Hartley transform (FHT) is exploited.

Let the DCT-II transformed sequence $\{X^{C(2)}(m)\}$ be given as in (4.4.1) by

$$X^{C(2)}(m) = \sum_{n=0}^{N-1} x(n) C_{2N}^{(2n+1)m}, \qquad m = 0, 1, \ldots, N-1. \tag{4.4.1}$$

We consider the DHT of a reordered sequence of $\{x(n)\}$ and its relation to $\{X^{C(2)}(m)\}$ in (4.4.1). First, we construct the sequence $\{y(n)\}$ from $\{x(n)\}$ by letting

$$\begin{aligned} y(n) &= x(2n), & n &= 0, 1, \ldots, N/2 - 1, \\ &= x(2N - 2n - 1), & n &= N/2, \ldots, N-1. \end{aligned} \tag{4.5.7}$$

The DHT of $\{y(n)\}$ is given by

$$Y_H(m) = \sum_{n=0}^{N-1} y(n) \operatorname{cas}\left(\frac{2mn\pi}{N}\right) = \sum_{n=0}^{N-1} y(n) H_N^{mn}. \tag{4.5.8}$$

Now, using (4.5.7) we obtain

$$Y_H(m) = \sum_{\text{even } n} x(n) \operatorname{cas}\left(\frac{mn\pi}{N}\right) + \sum_{\text{odd } n} x(n) \operatorname{cas}\left[-\frac{(n+1)m\pi}{N}\right]. \tag{4.5.9}$$

It is easy to show that $\operatorname{cas}(\alpha)\operatorname{cas}(\beta) = \operatorname{cas}(\alpha - \beta)$, and this identity can be used to get

$$X^{C(2)}(m) = \frac{1}{2}\left\{Y_H(m) \operatorname{cas}\left(-\frac{m\pi}{2N}\right) + Y_H(N-m) \operatorname{cas}\left(\frac{m\pi}{2N}\right)\right\}. \tag{4.5.10}$$

Using H_N^{mn} to denote $\operatorname{cas}(2mn\pi/N)$, (4.5.10) can be written more concisely,

$$\begin{bmatrix} X^{C(2)}(m) \\ X^{C(2)}(N-m) \end{bmatrix} = \frac{1}{2} \begin{bmatrix} H_{4N}^{-m} & H_{4N}^{m} \\ H_{4N}^{(N-m)} & -H_{4N}^{-(N-m)} \end{bmatrix} \begin{bmatrix} Y_H(m) \\ Y_H(N-m) \end{bmatrix},$$

$$m = 1, 2, \ldots, N/2 - 1. \tag{4.5.11a}$$

For $m = 0$ and $N/2$, the transformed points are determined by

$$X^{C(2)}(0) = Y_H(0) \qquad \text{and } X^{C(2)}\left(\frac{N}{2}\right) = \frac{Y_H\left(\frac{N}{2}\right)}{\sqrt{2}}. \tag{4.5.11b}$$

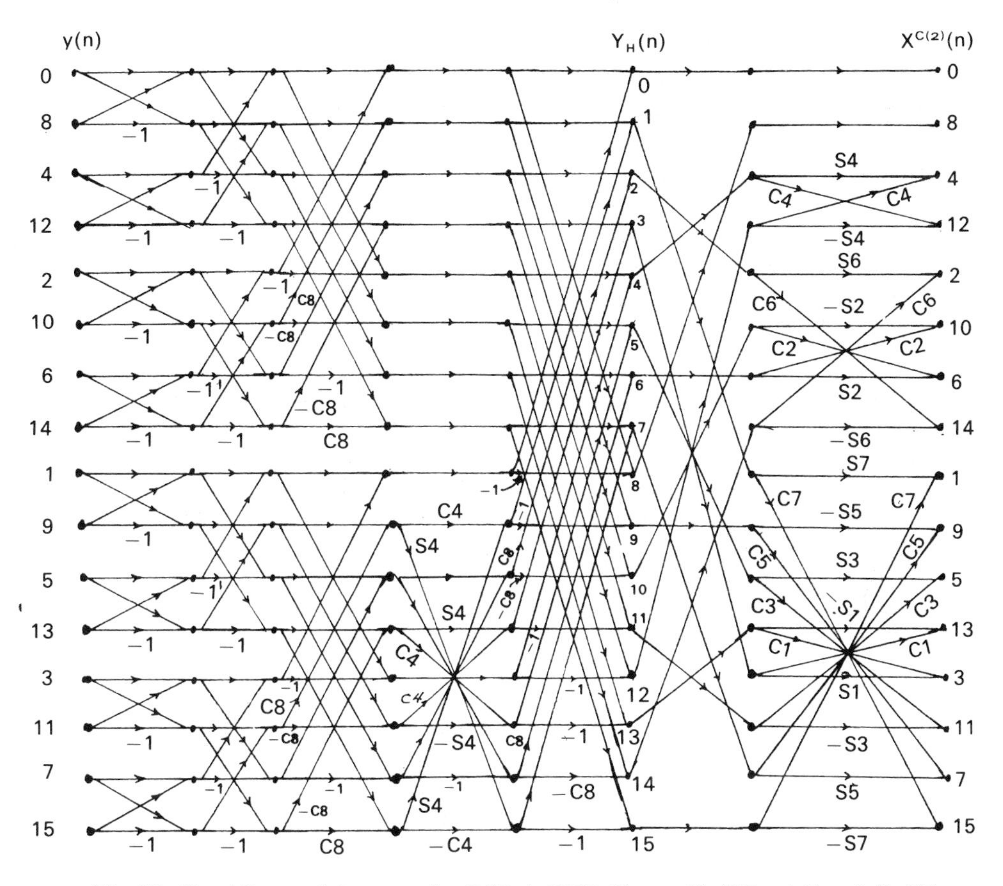

Fig. 4.7 Signal flowgraph for computing DCT via DHT. $Ck = \cos[k\pi/32]$ and $Sk = \sin\{k\pi/32)$, $N = 16$.

The amount of computation necessary for an N-point sequence compares favorably with that of other algorithms. We conclude this section by presenting a signal flowgraph for the computation of an $N = 16$ DCT-II via the DHT in Fig. 4.7.

4.6 Other Algorithms

In addition to the algorithms described in the previous sections, there are many more. Some are developed for specific applications. Others are developed for specific numbers of input sample points. Radix-2 algorithms, described in previous sections, may not be the most appropriate. A case in point is the implementation of transmultiplexer for the conversion between time-division multiplexing (TDM) and frequency-division multiplexing (FDM) in voice transmission over a telecommunication channel [T-1, T-2, T-3]. The required sampling rates dictate that the DCT-II used to implement the transmultiplexers have lengths of 14 and 72 points. Another situation arises in the consideration of systolic arrays implementation, where fast computation of DCT-II may be accomplished through the use of the Winograd Fourier transform algorithm (WFTA) [M-24, M-25]. In this case, radix-2 factorization is a special case. In Section 4.6.1, the prime factor algorithm (PFA) [FRA-10] is discussed as representative of the factorizations not based on radix-2.

Many of the algorithms discussed have been considered for VLSI implementation [DP-10 through DP-17, DP-20 through DP-24, DP-27 through DP-31, DP-33 through DP-35, DP-37 through DP-39, DP-41, DP-44 through DP-47, DP-49, DP-51 through DP-53]. (See Appendix B.1, where a partial list of DCT VLSI chip manufacturers is given.) In such an implementation, the orderly architecture and the numerical stability are important considerations, in addition to the recursive and modular nature of the algorithm. In Section 4.6.2, an algorithm that has these important ingredients is discussed. Unlike some of the others, this algorithm produces a 2^m-point DCT from two 2^{m-1}-point DCTs in much the same way as the Cooley–Tukey algorithm for the DFT [M-34].

Another aspect to be considered in the physical realization of the algorithm is the number of different basic processors required. In this regard, Ligtenberg *et al.* [DP-22, DP-27, FRA-23, FRA-26] proposed a procedure to compute the DCT using only planar rotations. This is described in Section 4.6.3.

4.6.1 *Prime Factor Algorithm (PFA)*

The prime factor algorithm (PFA) for DCT-II was developed (see [FRA-10]) based on the PFA for DFT [M-24, M-25]. Since DCT-II can be directly related to the DFT, as shown in Section 4.2, we describe first the PFA for DFT.

Consider an input sequence $\{y(n)\}$ of $N = N_1 N_2$ points. If N_1 and N_2 are relatively prime, the one-dimensional sequence $\{y(n)\}$ can be mapped to a two-dimensional array $\{y(n_1, n_2)\}$, where the indices are related by the equation,

$$n = n_1 r_1 N_2 + n_2 r_2 N_1 \text{ modulo } N. \tag{4.6.1}$$

Here, r_1 and r_2 are the inverses of N_2 modulo N_1 and N_1 modulo N_2, respectively, i.e.,

$$r_1 N_2 = 1 \text{ modulo } N_1 \qquad \text{and } r_2 N_1 = 1 \text{ modulo } N_2.$$

For example, when $N_1 = 3$ and $N_2 = 4$, $r_1 = 1$ and $r_2 = 3$. The input index mapping for n is given by Table 4.1.

$\{Y(m)\}$, the DFT of $\{y(n)\}$, can be expressed as a cascaded transform of the two-dimensional array $\{y(n_1, n_2)\}$ given by

$$\begin{aligned} Y(m) &= Y(m_1, m_2) \\ &= \sum_{n_1=0}^{N_1-1} \left[\sum_{n_2=0}^{N_2-1} y(n_1, n_2) W_{N_2}^{m_2 n_2} \right] W_{N_1}^{m_1 n_1}, \end{aligned} \tag{4.6.2a}$$

Table 4.1

Input index mapping for n.

	n_2	0	1	2	3
n_1			n		
0		0	9	6	3
1		4	1	10	7
2		8	5	2	11

where the output indices are mapped by the relation

$$m = m_1N_2 + m_2N_1 \text{ modulo } N. \tag{4.6.2b}$$

Figure 4.8 illustrates the essential features of this decomposition for $N_1 = 3$ and $N_2 = 4$.

Now, since the DCT-II is related to the DFT, we consider a scaled output sequence of $\{Y(m)\}$ given by

$$Z(m) = W_{4N}^{m}Y(m) = W_{4N}^{m}\sum_{n=0}^{N-1} y(n)W_N^{mn}. \tag{4.6.3}$$

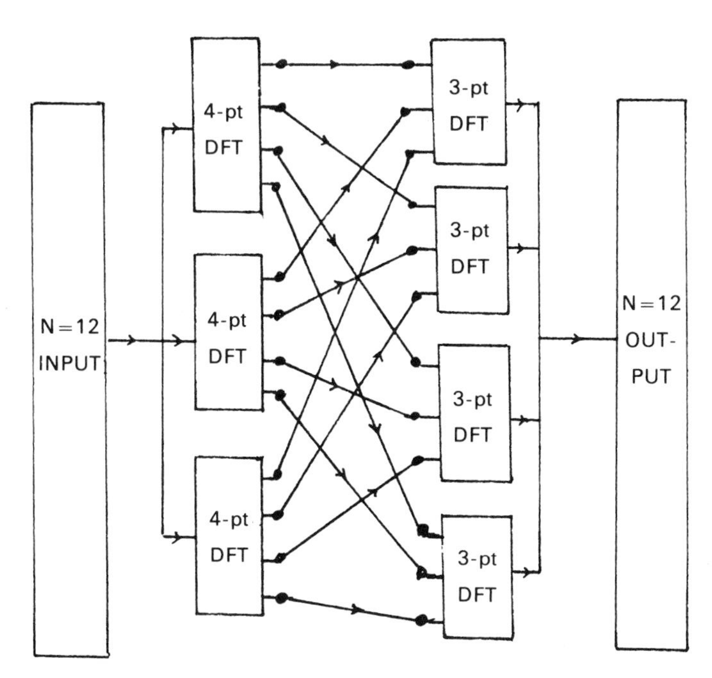

Fig. 4.8 Signal block diagram for the PFA of a 12-point DFT. pt = point.

Then, as is similarly shown in Section 4.2, the DCT-II of a sequence $\{x(n)\}$ is given by the real part of $\{Z(m)\}$, provided it is related to the sequence $\{y(n)\}$ by

$$y(n) = x(2n) \qquad \text{and } y(N - n - 1) = x(2n + 1), \qquad n = 0, 1, \ldots, N/2 - 1. \tag{4.6.4a}$$

In short, if $\{X^{C(2)}(m)\}$ is the DCT-II of $\{x(n)\}$, then

$$X^{C(2)}(m) = \text{Re}[Z(m)], \qquad m = 0, 1, \ldots, N - 1. \tag{4.6.4b}$$

Whereas (4.6.4a) is valid only for even N, a generalization is possible whereby the even-numbered input sample points are arranged in increasing order and the odd-numbered ones are arranged in decreasing order. Thus

$$\begin{aligned} y(n) &= x(2n) & n &= 0, 1, \ldots, (N - 1)/2, & &\text{odd } N, \\ & & &= 0, 1, \ldots, N/2 - 1, & &\text{even } N; \\ &= x(2N - 2n - 1) & n &= N - 1, \ldots, (N + 1)/2, & &\text{odd } N, \\ & & &= N - 1, \ldots, N/2, & &\text{even } N. \end{aligned} \tag{4.6.5}$$

It is easy to see that the DCT-II of $\{x(n)\}$ can be indirectly obtained by the PFA of DFT (or more correctly, the "phasor-modified" DFT). The drawback, as shown in (4.6.3), is that $\{Z(m)\}$ is a complex sequence, even though $\{y(n)\}$ (and in turn, $\{x(n)\}$) is a real sequence. Such a drawback may be overcome by directly examining the DCT-II of the two-dimensional sequence $\{y(n_1, n_2)\}$ whose indices are related to n, as indicated in (4.6.1). For a real input sequence, (4.2.9) ensures that

$$Z(0) = X^{C(2)}(0),$$

and

$$Z(m) = X^{C(2)}(m) - jX^{C(2)}(N - m), \qquad m = 1, 2, \ldots, N - 1. \tag{4.6.6}$$

Since $\{Z(m)\}$ is the phasor-modified DFT of $\{y(n)\}$, it is clear that $\{X^{C(2)}(m)\}$ is given by the DCT-II of $\{y(n_1, n_2)\}$. The idea of cascading the transform is now applied directly to compute the DCT-II of $\{y(n_1, n_2)\}$. First we identify a two-dimensional array $\{Z(m_1, m_2)\}$ with $\{Z(m)\}$ using the mapping in (4.6.2b). Then, we combine (4.6.2) and (4.6.3) to get

$$Z(m_1, m_2) = W_{4N}^{(m_1 N_2 + m_2 N_1 - hN)} Y(m_1, m_2), \tag{4.6.7a}$$

where

$$\begin{aligned} h &= 0 \qquad \text{if } m_1 N_2 + m_2 N_1 < N, \\ &= 1 \qquad \text{otherwise,} \end{aligned}$$

since the index is evaluated modulo N in (4.6.2b). The cascaded form of (4.6.7a) is now readily obtained,

$$Z(m_1, m_2) = W_4^{-h} Y_1(m_1, m_2), \tag{4.6.7b}$$

where

$$Y_1(m_1, m_2) = W_{4N_1}^{m_1} \left[\sum_{n_1=0}^{N_1-1} Y_2(n_1, m_2) W_{N_1}^{m_1 n_1} \right],$$

and

$$Y_2(n_1, m_2) = W_{4N_2}^{m_2} \left[\sum_{n_2=0}^{N_2-1} y(n_1, n_2) W_{N_2}^{m_2 n_2} \right].$$

We note that $\{Y_2(n_1, m_2)\}$ and $\{Y_1(m_1, m_2)\}$ are phasor-modified DFTs of $\{y(n_1, n_2)\}$ on n_2 and $\{Y_2(n_1, m_2)\}$ on n_1, respectively, and are in exactly the same form as (4.6.3). Therefore, for $\{Y_2(n_1, m_2)\}$ we have (since $\{y(n_1, n_2)\}$ is real)

$$Y_2(n_1, m_2) = Y_2^{C(2)}(n_1, m_2) - jY_2^{C(2)}(n_1, N_2 - m_2), \tag{4.6.8}$$

where $\{Y_2^{C(2)}(n_1, m_2)\}$ is the DCT-II of a sequence $\{y_2(n_1, n_2)\}$, which maps to the sequence $\{y(n_1, n_2)\}$ on n_2, as $\{x(n)\}$ maps to $\{y(n)\}$ in (4.6.5).

By repeating the same argument, a sequence $\{y_1(n_1, m_2)\}$ whose n_1 index maps it to $\{Y_1(n_1, m_2)\}$ as $\{x(n)\}$ maps to $\{y(n)\}$ in (4.6.5) has a DCT-II denoted by $\{Y_1^{C(2)}(m_1, m_2)\}$, related to $\{Y_1(m_1, m_2)\}$ as follows:

$$\begin{aligned} Y_1(m_1, m_2) = {} & Y_1^{C(2)}(m_1, m_2) - jY_1^{C(2)}(N_1 - m_1, m_2) - jY_1^{C(2)}(m_1, N_2 - m_2) \\ & - Y_1^{C(2)}(N_1 - m_1, N_2 - m_2). \end{aligned} \tag{4.6.9}$$

The result in (4.6.9), when combined with (4.6.7b) and (4.6.6), gives the relation between $\{X^{C(2)}(m)\}$ and $\{Y_1^{C(2)}(m_1, m_2)\}$. Thus

$$\begin{aligned} X^{C(2)}(m) &= Y_1^{C(2)}(m_1, m_2) \qquad \text{if } m_1 m_2 = 0; \\ &= -Y_1^{C(2)}(N_1 - m_1, m_2) - Y_1^{C(2)}(m_1, N_2 - m_2) \qquad \text{otherwise}, \\ & \qquad m = m_1 N_2 + m_2 N_1 \text{ modulo } N. \end{aligned} \tag{4.6.10}$$

It is useful to briefly summarize the procedure for the PFA of DCT-II, given an input sequence $\{x(n)\}$ of $N = N_1 N_2$ points:

(1) Map $\{x(n)\}$ to $\{y(n)\}$ using (4.6.5).
(2) Map $\{y(n)\}$ to $\{y(n_1, n_2)\}$ using (4.6.1).
(3) Compute the N_2-point DCT-II of $\{y(n_1, n_2)\}$ on the index n_2 to obtain $\{Y_2^{C(2)}(n_1, m_2)\}$.
(4) Compute the N_1-point DCT-II of $\{Y_2^{C(2)}(n_1, m_2)\}$ on n_1 to get $\{Y_1^{C(2)}(m_1, m_2)\}$.
(5) Obtain $\{X^{C(2)}(m)\}$ from (4.6.10).

Because the indices are rearranged each time a DCT-II is computed, the index mappings are not simple when one also includes the mappings required for cascading the transforms. Yang and Narasimha [FRA-10] have worked out the mappings, and we will state them here without proof.

A. *Input index mapping:* $n \leftrightarrow (n_1, n_2)$, for $N = N_1 N_2$,

$$n = \frac{(\bar{n} - 1)}{2} \quad \text{if } \bar{n} < 2N,$$
$$= \frac{(4n - \bar{n} - 1)}{2} \quad \text{if } \bar{n} > 2N, \tag{4.6.11}$$

where

$$\bar{n} = |(2n_1 + 1)r_1 N_2 - (2n_2 + 1)(N_2 - r_2)N_1| \bmod 4N, \quad n_1 + n_2 \text{ even},$$
$$= |(2n_1 + 1)r_1 N_2 + (2n_2 + 1)(N_2 - r_2)N_1| \bmod 4N, \quad n_1 + n_2 \text{ odd}, \tag{4.6.12}$$

where mod is the abbreviation for modulo.

Here, r_1 and r_2 are inverses of N_2 and N_1, modulo N_1 and N_2, respectively, [see (4.6.1)].

B. *Output index mapping:* $m \leftrightarrow (m_1, m_2)$, $N = N_1 N_2$,

$$m = m_1 N_2 + m_2 N_1 \quad \text{if } m_1 m_2 = 0. \tag{4.6.13}$$

For $m_1 m_2$ not equal to zero, a butterfly is required to determine $X^{C(2)}(m)$ from $Y_1^{C(2)}(m_1, m_2)$, as seen in (4.6.10). The butterfly is shown in Fig. 4.9.

Here,

$$q_1 = |m_1 N_2 - m_2 N_1|, \tag{4.6.14}$$
$$q_2 = |\bar{q}_2| \quad \text{if } \bar{q}_2 \leqslant N,$$
$$= |\bar{q}_2 - 2N| \quad \text{if } \bar{q}_2 > N,$$

where

$$\bar{q}_2 = (m_1 N_2 + m_2 N_1). \tag{4.6.15}$$

Figure 4.10 displays the detailed signal flowgraph of $N = 3 \cdot 4$ DCT-II using the PFA. The complete 12-point DCT-II requires 23 multiplications and 46 additions (see Problem 4.12).

More recently, Lee [FRA-18] has provided tables for mapping the indices for various sizes in the PFA of DCT-II.

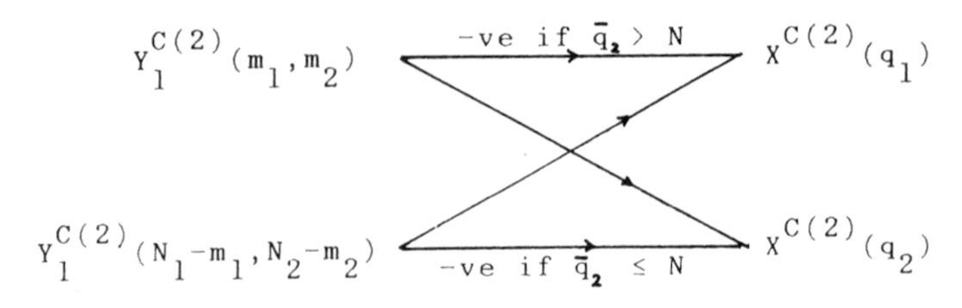

Fig. 4.9 Butterfly diagram for the last stage of PFA.

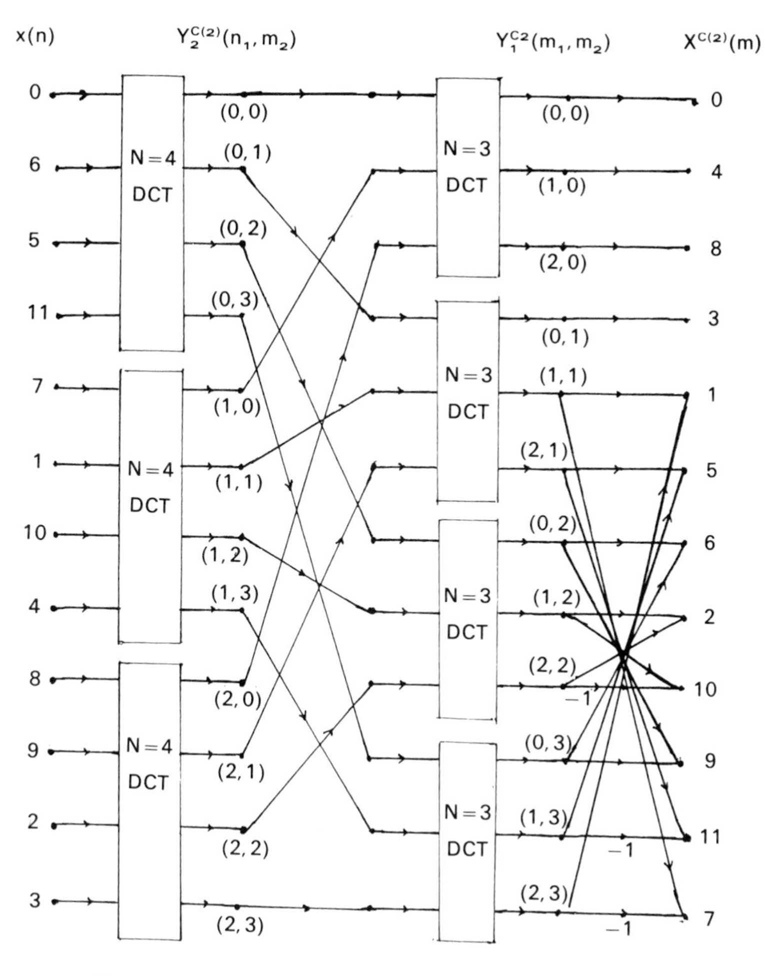

Fig. 4.10 Signal flowgraph for the PFA of DCT-II. $N = 3 \cdot 4$.

4.6.2 *A Fast Recursive Algorithm*

Recently, Hou [FRA-13] proposed a fast recursive algorithm for computing the DCT-II. The index permutation involved follows closely that of the Cooley–Tukey algorithm for the DFT. It is therefore not surprising that it finds its beginning at the same place as the Narasimha and Peterson algorithm [FF-1] discussed in Section 4.2. Recall that DCT-II can be obtained as the real part of a scaled DFT if the input sequence undergoes a reordering first, given by

$$y(n) = x(2n),$$

and

$$y(N - n - 1) = x(2n + 1), \qquad n = 0, 1, \ldots, N/2 - 1. \tag{4.2.5}$$

The transform sequence $\{X^{C(2)}(m)\}$ can then be written as

$$X^{C(2)}(m) = \sum_{n=0}^{N-1} y(n) \cos\left[\frac{(4n+1)m\pi}{2N}\right], \qquad m = 0, 1, \ldots, N-1. \tag{4.2.7a}$$

Instead of viewing (4.2.7a) as the real part of a scaled N-point DFT, Hou delves into the symmetry of the kernel in (4.2.7a) and partitions the transform matrix to obtain the recursive structure. The concept is best understood by examining the structures of the $N = 4$ and $N = 8$ transforms. For simplicity, we shall drop the C(2) superscript in the transformed sequence in what follows.

For $N = 4$, we have

$$\begin{bmatrix} X(0) \\ X(2) \\ X(1) \\ X(3) \end{bmatrix} = \begin{bmatrix} 1 & 1 & 1 & 1 \\ \alpha & -\alpha & \alpha & -\alpha \\ \beta & -\delta & -\beta & \delta \\ \delta & \beta & -\delta & -\beta \end{bmatrix} \begin{bmatrix} x(0) \\ x(2) \\ x(3) \\ x(1) \end{bmatrix}. \tag{4.6.16}$$

For $N = 8$, we have

$$\begin{bmatrix} X(0) \\ X(4) \\ X(2) \\ X(6) \\ X(1) \\ X(5) \\ X(3) \\ X(7) \end{bmatrix} = \begin{bmatrix} 1 & 1 & 1 & 1 & 1 & 1 & 1 & 1 \\ \alpha & -\alpha & \alpha & -\alpha & \alpha & -\alpha & \alpha & -\alpha \\ \beta & -\delta & -\beta & \delta & \beta & -\delta & -\beta & \delta \\ \delta & \beta & -\delta & -\beta & \delta & \beta & -\delta & -\beta \\ \sigma & \mu & -\tau & \varepsilon & -\sigma & -\mu & \tau & -\varepsilon \\ \mu & \tau & -\varepsilon & \sigma & -\mu & -\tau & \varepsilon & -\sigma \\ \varepsilon & -\sigma & \mu & \tau & -\varepsilon & \sigma & -\mu & -\tau \\ \tau & \varepsilon & \sigma & \mu & -\tau & -\varepsilon & -\sigma & -\mu \end{bmatrix} \begin{bmatrix} x(0) \\ x(2) \\ x(4) \\ x(6) \\ x(7) \\ x(5) \\ x(3) \\ x(1) \end{bmatrix}, \tag{4.6.17}$$

where $\alpha = 1/\sqrt{2}$, $\beta = C_8^1$, $\delta = S_8^1$, $\sigma = C_{16}^1$, $\varepsilon = C_{16}^3$, $\mu = S_{16}^3$, $\tau = S_{16}^1$, and $S_k^i = \sin(i\pi/k)$ and $C_k^i = \cos(i\pi/k)$ as usual.

In these input and output permutations, (4.6.16) and (4.6.17) suggest that if $[\hat{T}_N]$ is used to represent the permuted transform matrix for the DCT-II, then

$$[\hat{T}_N] = \begin{bmatrix} [\hat{T}_{N/2}] & [\hat{T}_{N/2}] \\ [\hat{D}_{N/2}] & -[\hat{D}_{N/2}] \end{bmatrix}. \tag{4.6.18}$$

If $[\hat{D}]$ can be further expressed in terms of $[\hat{T}]$ in a simple manner, the required recursive structure will have been established. We can demonstrate this relation by using the submatrices in (4.6.17) defined as

$$[\hat{T}_4] = \begin{bmatrix} 1 & 1 & 1 & 1 \\ \alpha & -\alpha & \alpha & -\alpha \\ \beta & -\delta & -\beta & \delta \\ \delta & \beta & -\delta & -\beta \end{bmatrix},$$

and (4.6.19a)

$$[\hat{D}_4] = \begin{bmatrix} \sigma & \mu & -\tau & \varepsilon \\ \mu & \tau & -\varepsilon & \sigma \\ \varepsilon & -\sigma & \mu & \tau \\ \tau & \varepsilon & \sigma & \mu \end{bmatrix}.$$

The regularity of these matrices can be further examined if we note that by defining $\phi_n = (4n + 1)\pi/2N$ for $N = 4$ we have

$$[\hat{T}_4] = \begin{bmatrix} 1 & 1 & 1 & 1 \\ \cos(4\phi_0) & \cos(4\phi_1) & \cos(4\phi_2) & \cos(4\phi_3) \\ \cos(2\phi_0) & \cos(2\phi_1) & \cos(2\phi_2) & \cos(2\phi_3) \\ \cos(6\phi_0) & \cos(6\phi_1) & \cos(6\phi_2) & \cos(6\phi_3) \end{bmatrix}$$

and (4.6.19b)

$$[\hat{D}_4] = \begin{bmatrix} \cos(\phi_0) & \cos(\phi_1) & \cos(\phi_2) & \cos(\phi_3) \\ \cos(5\phi_0) & \cos(5\phi_1) & \cos(5\phi_2) & \cos(5\phi_3) \\ \cos(3\phi_0) & \cos(3\phi_1) & \cos(3\phi_2) & \cos(3\phi_3) \\ \cos(7\phi_0) & \cos(7\phi_1) & \cos(7\phi_2) & \cos(7\phi_3) \end{bmatrix}.$$

Since $\cos\{(2k + 1)\phi\} = 2\cos(2k\phi)\cos(\phi) - \cos\{(2k - 1)\phi\}$, the elements in the matrix $[\hat{D}_4]$ can be rewritten as

$$[\hat{D}_4] = \begin{bmatrix} \cos(\phi_0), & \cdots \\ \cos(\phi_0)\{1 + 2\cos(4\phi_0) - 2\cos(2\phi_0)\}, & \cdots \\ \cos(\phi_0)\{-1 + 2\cos(2\phi_0)\}, & \cdots \\ \cos(\phi_0)\{-1 - 2\cos(4\phi_0) + 2\cos(2\phi_0) + 2\cos(6\phi_0)\}, & \cdots \end{bmatrix}, \tag{4.6.20}$$

where we have shown elements in the first column only. The remaining columns contain elements of the same form with ϕ_0 replaced by ϕ_1, ϕ_2, and ϕ_3, respectively. With this form, it is not difficult to verify the factorization for (4.6.20):

$$[\hat{D}_4] = [P'_4][L_4][P'_4][\hat{T}_4][Q_4], \tag{4.6.21}$$

where

$$[P'_4] = \begin{bmatrix} 1 & 0 & 0 & 0 \\ 0 & 0 & 1 & 0 \\ 0 & 1 & 0 & 0 \\ 0 & 0 & 0 & 1 \end{bmatrix}$$

is a 4×4 binary bit-reversal matrix,

$$[L_4] = \begin{bmatrix} 1 & 0 & 0 & 0 \\ -1 & 2 & 0 & 0 \\ 1 & -2 & 2 & 0 \\ -1 & 2 & -2 & 2 \end{bmatrix}$$

is a lower triangular matrix of the coefficients of the cosines in (4.6.20), and

$$[Q_4] = \mathrm{diag}\{\cos(\phi_0), \cos(\phi_1), \cos(\phi_2), \cos(\phi_3)\}.$$

Equation (4.6.21), which delineates the relation of $[\hat{D}]$ to $[\hat{T}]$ in the (4×4) case, is easily extended to the general $(N \times N)$ case, thus completing the recursive structure. The formal structure is made clear by the following equation:

$$\begin{bmatrix} (\mathbf{X}_e)_{\mathrm{BRO}} \\ (\mathbf{X}_o)_{\mathrm{BRO}} \end{bmatrix} = \begin{bmatrix} [\hat{T}_{N/2}] & [\hat{T}_{N/2}] \\ [K_{N/2}][\hat{T}_{N/2}][Q_{N/2}] & -[K_{N/2}][\hat{T}_{N/2}][Q_{N/2}] \end{bmatrix} \begin{bmatrix} (\mathbf{x}_e) \\ (\mathbf{x}_o)_{\mathrm{R}} \end{bmatrix}, \tag{4.6.22}$$

where $(\mathbf{X}_e)$ and $(\mathbf{X}_o)$ are the even- and odd-numbered transformed components, and $(\mathbf{x}_e)$ and $(\mathbf{x}_o)$ are the even and odd input points. BRO and R stand for bit-reversed order and reversed order, respectively. The matrix $[K]$ is used to denote the product $[P'][L][P']$ as shown in (4.6.21). Thus,

$$[K_{N/2}] = [P'_{N/2}][L_{N/2}][P'_{N/2}],$$

and

$$[Q_{N/2}] = \mathrm{diag}\{\cos(\phi_0), \ldots, \cos(\phi_{N/2-1})\}. \tag{4.6.23a}$$

Here, $[L_{N/2}]$ is the lower $(N/2) \times (N/2)$ triangular matrix given by

$$[L_{N/2}] = \begin{bmatrix} 1 & 0 & 0 & 0 & \cdots & 0 \\ -1 & 2 & 0 & 0 & \cdots & 0 \\ 1 & -2 & 2 & 0 & \cdots & 0 \\ -1 & 2 & -2 & 2 & \cdots & 0 \\ \vdots & \vdots & \vdots & \vdots & \cdots & 0 \\ -1 & 2 & -2 & 2 & \cdots & 2 \end{bmatrix}. \tag{4.6.23b}$$

It is clear from (4.6.22) that the factorization equation for the transform matrix is

$$[\hat{T}_N] = \begin{bmatrix} [\hat{T}_{N/2}] & [\hat{T}_{N/2}] \\ [K_{N/2}][\hat{T}_{N/2}][Q_{N/2}] & -[K_{N/2}][\hat{T}_{N/2}][Q_{N/2}] \end{bmatrix}. \tag{4.6.24}$$

It should be noted that in (4.6.22) the even- and odd-numbered components of the input and transformed vectors are separated, reversed, or bit reversed. It is therefore not strictly possible to classify this algorithm as being either decimation in time (DIT) or decimation in frequency (DIF).

The computational complexity of the algorithm can be delineated by considering the block matrix factorization of (4.6.24). This gives

$$[\hat{T}_N] = \begin{bmatrix} [I_{N/2}] & 0 \\ 0 & [K_{N/2}] \end{bmatrix} \begin{bmatrix} [\hat{T}_{N/2}] & 0 \\ 0 & [\hat{T}_{N/2}] \end{bmatrix} \times \begin{bmatrix} [I_{N/2}] & 0 \\ 0 & [Q_{N/2}] \end{bmatrix} \begin{bmatrix} [I_{N/2}] & [I_{N/2}] \\ [I_{N/2}] & -[I_{N/2}] \end{bmatrix}. \tag{4.6.25}$$

Now, $[Q_{N/2}]$ requires $(N/2)$ multiplications and $[K_{N/2}]$ can be implemented using shift-and-adds, which can be done using $N(N-2)/4$ additions. The last matrix factor can be implemented using N additions. Thus, the recurrence equations for the number of multiplications, μ_N, and for the number of additions, α_N, are given by

$$\mu_N = 2\mu_{N/2} + \frac{N}{2},$$

and

$$\alpha_N = 2\alpha_{N/2} + \frac{N(N-2)}{4} + N, \qquad N \geqslant 4. \tag{4.6.26a}$$

These equations can be solved by using $\mu_2 = 1$ and $\alpha_2 = 1$, giving

$$\mu_N = \left(\frac{N}{2}\right) \log_2 N,$$

and

$$\alpha_N = \left(\frac{3N}{2}\right) \log_2 N + \frac{N^2}{2} - N, \qquad N \geqslant 4. \tag{4.6.26b}$$

It is noted [FRA-13] that the second block matrix factor in (4.6.25) contains identical blocks and that the computation can be implemented using multiplexing techniques. This reduces the number of multipliers and adders required in the computation, making this algorithm quite competitive with other algorithms discussed in this chapter. We conclude this section by presenting a block flow diagram (Fig. 4.11) for the computation of an N-point DCT-II using this algorithm with the aid of multiplexing (See also Problem 4.4).

4.6.3 *DCT-II Realization via Planar Rotations*

One of the important considerations in physically realizing a DCT fast algorithm is the variety of processing units required. In this regard Ligtenberg *et al.* [DP-22, DP-27] and Loeffler *et al.* [FRA-23, FRA-26] have proposed an algorithm to compute the DCT that requires only rotations. This single type of processing unit replaces the adders, multipliers, and rotators required in the

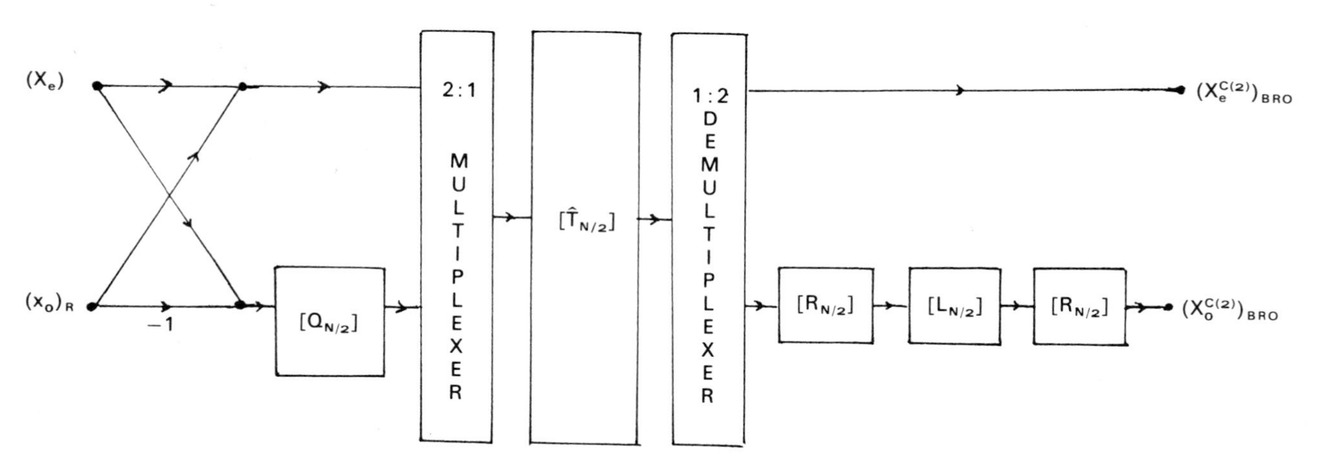

Fig. 4.11 Block diagram for N-point DCT with multiplexing [FRA-13]. (© 1987 IEEE).

implementation of other algorithms. To help us understand the basis of this algorithm, we briefly outline the theory.

It is well known that any matrix $[A]$ can be factored so that

$$[A] = [Q][R], \tag{4.6.27}$$

where $[Q]$ is an orthogonal matrix and $[R]$ is an upper triangular matrix. If, in addition, $[A]$ is an orthogonal matrix, then $[R]$ will also be orthogonal. Since an orthogonal upper triangular matrix must be diagonal, we have

$$[R]^T[R] = [I]. \tag{4.6.28}$$

Therefore, the diagonal elements of $[R]$ can only be $+1$ or -1. From (4.6.27) this implies that the columns of $[A]$ and $[Q]$ are identical, except along the ith column, wherever $[R]_{ii} = -1$, in which case they differ by a minus sign. The factorization in (4.6.27) is called a QR-decomposition and can be accomplished by using the so-called Givens rotations. The essential feature of this procedure is to null the lower off-diagonal elements of $[A]$ by means of a rotation of the proper angle. It can best be illustrated by an example in the two-dimensional case.

Consider a two-dimensional vector $[x_0, x_1]^T$. A (2×2) rotation matrix $[T(\theta)]$ can be constructed based on the values of x_0 and x_1 so that

$$[T(\theta)]\begin{bmatrix} x_0 \\ x_1 \end{bmatrix} = \begin{bmatrix} \hat{x}_0 \\ 0 \end{bmatrix}. \tag{4.6.29}$$

It is easy to see that if

$$[T(\theta)] = \begin{bmatrix} \cos(\theta) & \sin(\theta) \\ -\sin(\theta) & \cos(\theta) \end{bmatrix}, \tag{4.6.30}$$

then (4.6.29) is obtained when

$$\cos(\theta) = \frac{x_0}{\sqrt{(x_0^2 + x_1^2)}} \qquad \text{and } \sin(\theta) = \frac{x_1}{\sqrt{(x_0^2 + x_1^2)}}. \tag{4.6.31}$$

The nulling of the element x_1 can be applied, in general, to null the ij-element of the matrix $[A]$. The required matrix is seen to be $[T_{ij}(\theta)]$, where the angle of rotation θ is determined by the elements $[A]_{ij}$ and $[A]_{i-1,j}$ such that

$$\cos(\theta) = \frac{[A]_{i-1,j}}{D}, \qquad \text{and } \sin(\theta) = \frac{[A]_{ij}}{D}, \tag{4.6.32}$$

where

$$D = \{[A]_{i-1,j}^2 + [A]_{ij}^2\}^{1/2}.$$

If $[A]$ is an $(N \times N)$ orthogonal matrix, $[T_{ij}(\theta)]$ is seen to be an $(N \times N)$ identity matrix, except at the $(i-1, i-1)$, $(i-1, i)$, $(i, i-1)$, and (i, i) positions, where $\cos(\theta)$, $\sin(\theta)$, $-\sin(\theta)$, and $\cos(\theta)$ are located. We also note that in nulling the ij-element of $[A]$, the angle of rotation can be determined using $[A]_{ij}$ and any other element located above it in the same column. We have chosen to follow the more conventional procedure of using $[A]_{i-1,j}$ specifically. A systematic application of this type of rotations will eventually reduce the matrix $[A]$ to an upper triangular form and, in particular, when $[A]$ is orthogonal to begin with, a diagonal matrix with $+1$ or -1 as diagonal elements. This sequence of rotations on the matrix $[A]$ can be represented by

$$\{\pi_{j=1}^{N}\pi_{i=j+1}^{N}[T_{ij}(\theta)]\}[A] = [R]. \tag{4.6.33}$$

Thus,

$$\begin{aligned}[A] &= \{\pi_{j=1}^{N}\pi_{i=j+1}^{N}[T_{ij}(\theta)]\}^{T}[R], \\ &= \{\pi_{i=1}^{N}\pi_{j=i+1}^{N}[T_{ij}(\theta)]\}[R].\end{aligned} \tag{4.6.34}$$

Compared with (4.6.27), it is evident that the orthogonal matrix $[Q]$ is given by

$$[Q] = \pi_{i=1}^{N}\pi_{j=i+1}^{N}[T_{ij}(\theta)]. \tag{4.6.35}$$

Since $[A]$ and $[Q]$ have the same columns (up to a minus sign), (4.6.35) also represents a decomposition of $[A]$ into factor matrices, each of which is a planar rotation. Now, the DCT-II matrix is a real orthogonal matrix. It is therefore clear that the DCT-II matrix, $[C^{\text{II}}]$, can be factored into a sequence of planar rotations also. For an $(N \times N)$ matrix, there are $N(N-1)/2$ elements to be nulled, and the number of rotations required is thus of order N^2. We illustrate this using the DCT-II matrix for $N = 4$.

In this case,

$$[C^{\text{II}}] = \begin{bmatrix} 0.50 & 0.50 & 0.50 & 0.50 \\ 0.65 & 0.27 & -0.27 & -0.65 \\ 0.50 & -0.50 & -0.50 & 0.50 \\ 0.27 & -0.65 & 0.65 & -0.27 \end{bmatrix}. \tag{4.6.36}$$

The sequence of rotation matrices is given by

$$[T_{41}] = \begin{bmatrix} 1 & 0 & 0 & 0 \\ 0 & 1 & 0 & 0 \\ 0 & 0 & 0.88 & 0.48 \\ 0 & 0 & -0.48 & 0.88 \end{bmatrix}; \quad [T_{31}] = \begin{bmatrix} 1 & 0 & 0 & 0 \\ 0 & 0.75 & 0.66 & 0 \\ 0 & -0.66 & 0.75 & 0 \\ 0 & 0 & 0 & 1 \end{bmatrix};$$

$$[T_{21}] = \begin{bmatrix} 0.50 & 0.87 & 0 & 0 \\ -0.87 & 0.50 & 0 & 0 \\ 0 & 0 & 1 & 0 \\ 0 & 0 & 0 & 1 \end{bmatrix}; \quad [T_{42}] = \begin{bmatrix} 1 & 0 & 0 & 0 \\ 0 & 1 & 0 & 0 \\ 0 & 0 & 0.91 & 0.41 \\ 0 & 0 & -0.41 & 0.91 \end{bmatrix};$$

$$[T_{32}] = \begin{bmatrix} 1 & 0 & 0 & 0 \\ 0 & 0.58 & 0.82 & 0 \\ 0 & -0.82 & 0.58 & 0 \\ 0 & 0 & 0 & 1 \end{bmatrix}; \quad [T_{43}] = \begin{bmatrix} 1 & 0 & 0 & 0 \\ 0 & 1 & 0 & 0 \\ 0 & 0 & 0.71 & 0.71 \\ 0 & 0 & -0.71 & 0.71 \end{bmatrix}. \tag{4.6.37}$$

The product of these matrices gives the orthogonal matrix $[Q]$ in the QR-decomposition of the matrix $[C^{\mathrm{II}}]$, which is

$$[Q] = \begin{bmatrix} 0.50 & -0.50 & 0.50 & -0.50 \\ 0.65 & -0.27 & -0.27 & 0.65 \\ 0.50 & 0.50 & -0.50 & -0.50 \\ 0.27 & 0.65 & 0.65 & 0.27 \end{bmatrix}. \tag{4.6.38}$$

The resulting $[R]$, as expected, is diagonal and is given by

$$[R] = \mathrm{diag}\{1, -1, 1, -1\}. \tag{4.6.39}$$

Thus, for $N = 4$, six rotations are required to complete the transform. In general, when N is an integer power of 2, it is possible to first convert the transform matrix to a block diagonal one before applying the rotations. This approach will reduce the number of rotations to the order of $N \log_2 N$. Each rotation can be implemented by either four multiplications and two additions, or three multiplications and three additions, depending on the architecture one desires.

It is also possible to implement the QR-decomposition by another type of matrix operation. For the two-dimensional case, the operation is given by

$$[T'(\theta)] \begin{bmatrix} x_0 \\ x_1 \end{bmatrix} = \begin{bmatrix} x'_0 \\ 0 \end{bmatrix}, \tag{4.6.40}$$

where

$$[T'(\theta)] = \begin{bmatrix} \sin(\theta) & \cos(\theta) \\ \cos(\theta) & -\sin(\theta) \end{bmatrix}$$

and the angle θ is determined by (4.6.31) as before. We note here that $[T'(\theta)]$ is a symmetric matrix and its determinant is -1. Therefore $[T'(\theta)]$ is really a reflection of some sort. Vetterli and Ligtenberg [DP-15] have combined this type of operation with the recursive algorithm of Vetterli and Nussbaumer [FF-4] to propose a layout for an actual transform chip (see also Section 4.2). For example, the computation of an $N = 8$ DCT-II is accomplished via two $N = 2$ DCT and one $N = 4$ DFT, which are in turn implemented using $[T'(\theta)]$ operations. We note that in their work, they have referred to these operations as rotations also. Figure 4.12 shows the block diagram of the stages of the $N = 8$ DCT.

The signal flowgraph for the forward DCT-II, $N = 8$ is shown in Fig. 4.13 and that for IDCT-II is shown in Fig. 4.14. We note that in both cases, the

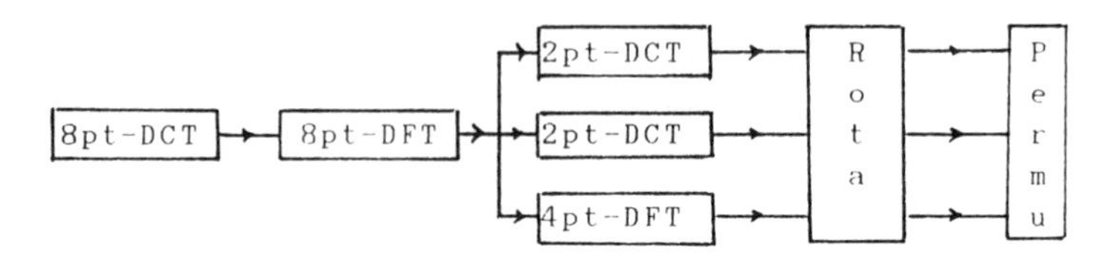

Fig. 4.12 Stages of $N = 8$ DCT based on [FF-4] and rotations. pt = point; Rota = rotation; Permu = permutation.

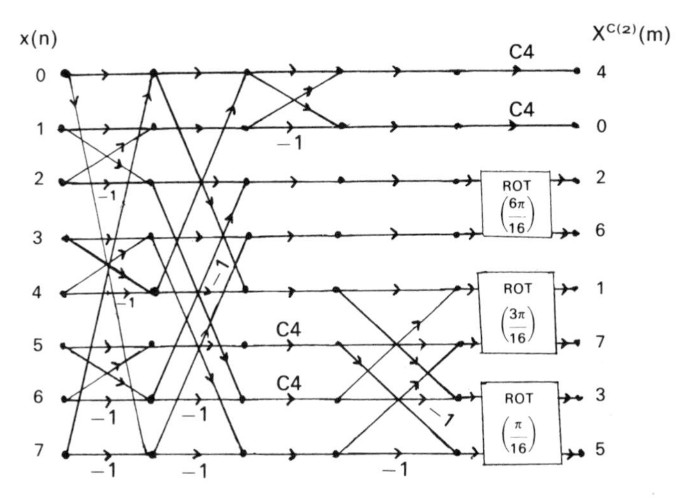

Fig. 4.13 Signal flowgraph for $N = 8$ DCT-II. The rotator here is defined in (4.6.40). $Ck = \cos(k\pi/16)$.

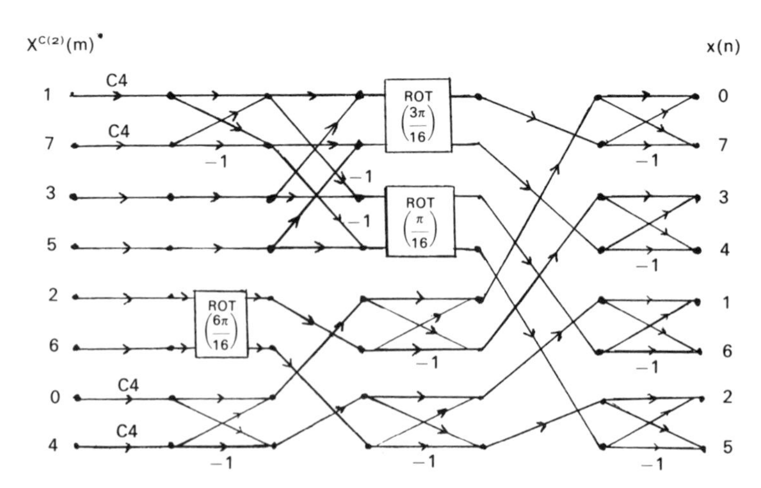

Fig. 4.14 Signal flowgraph for $N = 8$ IDCT-II. The rotator here is defined in (4.6.40). $Ck = \cos(k\pi/16)$.

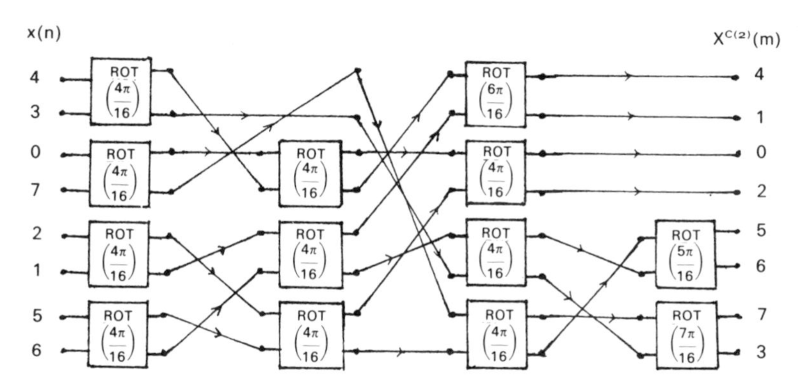

Fig. 4.15 Signal flowgraph of DCT-II. $N = 8$ using only rotations. The rotator here is as defined in (4.6.30).

implementation requires adders and multipliers in addition to the rotators. Figure 4.15 shows the signal flowgraph for DCT-II, $N = 8$ using rotators of the type in (4.6.30) only.

The rotator here is given by

$$\begin{matrix} x_i \rightarrow \\ x_j \rightarrow \end{matrix} \begin{bmatrix} \mathrm{Rot} \\ (\theta) \end{bmatrix} \begin{matrix} \rightarrow y_i \\ \rightarrow y_j \end{matrix} \quad \text{where} \begin{bmatrix} y_i \\ y_j \end{bmatrix} = \begin{bmatrix} \cos(\theta) & \sin(\theta) \\ -\sin(\theta) & \cos(\theta) \end{bmatrix} \begin{bmatrix} x_i \\ x_j \end{bmatrix}.$$

Note that the rotation matrix here is not exactly a Given's rotation that will null the second component of the two-dimensional vector $\mathbf{x}$. It may be considered as a combination of several Given's rotations.

For $N = 8$, we see that only 13 rotations are required.

4.7 Summary

In this chapter we have examined various approaches to the computation of the one-dimensional DCT-II. These are loosely classified as DCT via FFT (Section 4.2), sparse matrix factorization (Section 4.3), DIT and DIF algorithms (Section 4.4), DCT via other transforms (Section 4.5), and others (Section 4.6). Although the approaches and the resulting algorithms are quite different, the prime purpose of achieving speed and accuracy is the common goal. It is not easy to select, out of this myriad of methods, one that may be considered superior to all others. Aside from the simple consideration of the numbers of multiplications and additions (complex or real), the structure of the flowgraph, the mapping of input-to-output indices, and the recursivity are important but as yet nonquantifiable criteria that must somehow be taken into account.

In the following table we attempt to briefly summarize the computational complexity of the various algorithms and comment on their advantages and disadvantages as we perceive them. The one-dimensional DCT algorithms developed in this chapter can be utilized for implementing multidimensional

DCT, since it is a separable transform. Other algorithms for direct implementation of the multidimensional DCT have also been developed. These are covered in the following chapter.

Table 4.2

Comparison of Different Algorithms for DCT-II.

Algorithm	Complexity	Advantages	Disadvantages
Via N-point FFT [FF-1] (Section 4.2)	$\frac{1}{4}(N \log N - N + 2)$ complex multiplications	easy to implement using standard FFT routines	slow
Recur. FFT [FF-1, 2, 4] (Section 4.2, 4.6)	$\mu = \frac{1}{2}(N \log N)$ $\alpha = (3 \log N - N + 1)/2$	fast and recursive	complex index mapping
Sparse factors [FRA-1, 2] (Section 4.3)	$\mu = N \log N - 3N/2 + 4$ $\alpha = 3N(\log N - 1)/2 + 2$	reasonably fast	complex index mapping, nonrecursive
DIT [FRA-8] (Section 4.4)	$\mu = (N \log N)/2 + N/4$ $\alpha = (3N/2 - 1) \log N + N/4 - 1$	fast, recursive decimation in time	moderately complex index mapping
DIF [FRA-9] [FRA-14] (Section 4.4)	$\mu = (N \log N)/2$ $\alpha = (3N/2) \log N - N + 1$	fast, recursive decimation in frequency	moderately complex index mapping
Via WHT [DW-1] (Section 4.5)	dependent on WHT algorithm used	readily implementable, simple index mapping	conversion matrix needed, slow for $N \geqslant 16$
Via DHT [FRA-16] [FRA-19, 21] (Section 4.5)	same as using FFT	implementable using standard FFT routines	slow
PFA [FRA-10] (Section 4.6)	not applicable	not restricted to radix-2	very complex index mapping
Multiplexed [FRA-13] (Section 4.6)	$\mu = (N-1)$* $\alpha = (N^2 + N - 7/3)$*	fast, recursive simple index map	slower without multiplexing, shift required
Rotators [DP-22] (Section 4.6)	$(N \log N)/2$ rotations	single processor unit, good structure, simple index map	slow

μ = number of real multiplications, α = number of real additions; log is base 2.
*numbers without multiplexing are $\mu = (N \log N)/2$ and $\alpha = (3N/2) \log N + N^2/2 - N$.

PROBLEMS

4.1. Kou and Fjallbrant [FRA-17] have developed a fast algorithm for computing the DCT-II coefficients of a signal block composed of adjacent blocks. The composite block consists of the second half of the data of one block and the first half of the data of the following block. Generalize this technique to other cases. For example, a composite block is formed from the last quarter (or last third) of one block and the first three quarters (or first two thirds) of the following block. Compare the advantages of this technique with the traditional method.

4.2. Wang [DDD-4] has developed a technique for implementing DST-II via DCT-II. Develop a similar technique for implementing DCT-II via DST-II. Show all steps in detail.

4.3. The recursive computation of an N-point DCT-II (radix-2) has been developed by Vetterli and Nussbaumer [FF-4] (see Fig. 4.3). Verify this algorithm. For $N = 8$ and 16, develop flowgraphs and sparse matrix factors for DCT-II and its inverse. Write computer programs based on these flowgraphs and verify the same using random data. (Hint: DCT-II of a sequence followed by its inverse recovers the original sequence subject to computational errors—roundoff, truncation, etc.)

4.4. Hou [FRA-13] has developed a fast recursive algorithm for DCT-II (see Section 4.6.2). Repeat Problem 4.3 for this algorithm.

4.5. In Section 4.4, the DIT and DIF algorithms for DCT-II are presented [FRA-8, FRA-9, FRA-14]. Repeat Problem 4.3 for these algorithms.

4.6. The signal flowgraph for DCT-II, $N = 16$ is shown in Fig. 4.4 [FRA-1]. Write down the sparse matrix factors of the DCT-II matrix from an inspection of this flowgraph. Knowing that the DCT-II matrix is orthogonal (see 2.4.11), write down the sparse matrix factors for IDCT-II, $N = 16$ and then develop the flowgraph. Develop computer programs for these algorithms and check the same using random data (see Problem 4.3).

4.7. Ligtenberg and O'Neill [DP-22] and Ligtenberg, Wright, and O'Neill [DP-27] have developed a DCT-II algorithm based on planar rotations only. This has led to a VLSI chip for real-time image coding. Extend this algorithm for $N = 16$. Draw the flowgraphs for DCT-II and its inverse for $N = 16$. Write computer programs for $N = 8$ and 16, and verify the same (see Problem 4.3). See also [FRA-23, FRA-26] and Section 4.6.3.

4.8. In Fig. 4.7, the signal flowgraph for DCT-II via DHT is shown [FRA-16, FRA-19, FRA-21, FRA-28]. Repeat Problem 4.6 for this algorithm.

4.9. Derive (4.5.3). Sketch the flowgraph for implementing the DCT-II via WHT for $N = 8$ [DW-2]. Extend this to $N = 16$ [DW-1].

4.10. Sketch the flowgraphs for implementing the IDCT-II via WHT for $N = 8$ and 16 (see Prob. 4.9).

4.11. Similar to radix-2 FFT [B-2], Lee [FRA-3, FRA-4] has developed an algorithm for IDCT and has illustrated the algorithm for $N = 8$. Develop a similar algorithm for the DCT. Verify the validity of this algorithm by a flowgraph for $N = 8$. (Hint: Use the orthogonality property of the DCT matrix).

4.12. In Section 4.6.1 the PFA for DCT-II [FRA-10] is discussed. This algorithm is illustrated by a flowgraph/block diagram format for $N = 12$ in Fig. 4.10. Develop the flowgraph in detail and obtain the sparse matrix factors. Using the orthogonality property of the DCT-II, obtain the sparse matrix factors and then the flowgraph for IDCT, $N = 12$.

4.13. Duhamel [M-30] has developed the split-radix FFT algorithm. As DIT and DIF radix-2 algorithms [FRA-8, FRA-9, FRA-14] for the DCT have been developed, investigate whether split-radix DCT-II algorithms (both DIT and DIF) can be developed (see [FRA-29]). Compare the efficiency of these algorithms with the radix-2 algorithms.

4.14. In designing transmultiplexers (Section 7.5), Narasimha and Peterson [T-3] have developed a fast 14-point DCT-II algorithm. The flowgraph for the 14-point IDCT is shown in Fig. 10 of [T-3]. Obtain the matrix factors based on this flowgraph and obtain the number of multiplications and additions required for this algorithm. See also [T-2].

4.15. From Problem 4.14, transpose the matrix factors and draw the flowgraph based on the transposition. Show that this flowgraph yields the 14-point DCT-II as defined by (28) of [T-3]. Show that this flowgraph is a mirror image of the flowgraph shown in Fig. 10 of [T-3].

4.16. In realizing a 60-channel transmultiplexer, Narasimha *et al.* [T-1] utilized a 72-point DCT-II, which was implemented by the PFA [FRA-10]. Obtain the flowgraph for this DCT-II and comment on the computational complexity of this algorithm.

4.17. Repeat Problem 4.16 for a 72-point IDCT-II.

4.18. In [FRA-23] an algorithm tailored for custom DSP chips for 8-point DCT-II has been developed. Develop a similar algorithm for 8-point IDCT-II leading to the flowgraph similar to Fig. 9 of [FRA-23]. Extend this to a 16-point DCT-II and its inverse.

4.19. In [DP-13], a symmetric DCT-II and its inverse are, respectively, defined as follows:

$$X^{\mathrm{SC}(2)}(m) = \left[\frac{2}{(N-1)}\right]^{1/2} \sum_{n=0}^{N-1} C_n x(n) \cos\left[\frac{nm\pi}{(N-1)}\right], \qquad m = 0, 1, \ldots, N-1,$$

and

$$x(n) = \left[\frac{2}{(N-1)}\right]^{1/2} \sum_{m=0}^{N-1} C_m X^{\mathrm{SC}(2)}(m) \cos\left[\frac{nm\pi}{(N-1)}\right], \qquad n = 0, 1, \ldots, N-1,$$

where

$$C_m = \frac{1}{\sqrt{2}} \qquad \text{for } m = 0 \text{ or } N-1$$

$$= 1 \qquad \text{otherwise.}$$

The forward and inverse transform matrices are identical and are symmetric. These properties were utilized in developing 8-point and 16-point DCT VLSI chips. Determine if this transform is orthogonal. (Hint: Check for $N = 4$ and 8 specifically). Also determine if this transform matrix has sparse matrix factors. If so, develop the corresponding flowgraphs for $N = 4$, 8, and 16.

4.20. Develop the sparse matrix factors and the flowgraphs for the DCT and its inverse, for $N = 16$ based on Lee's algorithm [FRA-3, FRA-4]. List the number of additions and the number of multiplications required for implementing this algorithm ($N = 16$). Since this algorithm is recursive, identify parts of the flowgraphs that correspond to the DCT and its inverse for $N = 8$. Compare this with Problem 4.11.

4.21. In [DP-42, DP-43], an integer cosine transform (ICT) approximating the DCT-II for $N = 8$ has been developed. Of several possibilities, simple integers were chosen for implementing the ICT by LSI. Extend this approximation to $N = 16$. Similar to $N = 8$, focus on the selection of simple integers. Develop functional block diagrams and circuits for one-dimensional and two-dimensional ICTs similar to those illustrated for $N = 8$.

4.22. An algorithm-architecture approach using rotators that is appropriate for VLSI has been developed for an 8-point DCT (see Fig. 9 of [FRA-23]). Develop similar flowgraphs for 8-point IDCT and 16-point DCT/IDCT. List the number of additions and the number of multiplications required for these algorithms.

4.23. In Appendix A.1 the flowgraphs for DCT and IDCT ($N = 8$ and 16) are shown in Figs. A.1.1 through A.1.4. These are based on Lee's algorithm [FRA-3, FRA-4]. The sparse matrix factors for IDCT-II, $N = 8$ and 16 are described by (A.1.5) and (A.1.6), respectively. From the structure of these factors, obtain the corresponding matrix factors for DCT and IDCT ($N = 32$). Draw the corresponding flowgraphs. Identify parts of these flowgraphs that relate to DCT and IDCT ($N = 16$). List the number of additions and the number of multiplications required for the DCT and IDCT ($N = 8$, 16, and 32). Generalize these parameters for N being an integer power of two.

4.24. Repeat Problem 4.23 for the algorithms developed by Wang [FD-1] and Suehiro and Hatori [FRA-12]. See Appendix A.2.

4.25. The IDCT flowgraph for $N = 8$ based on Wang [FD-1] and Suehiro and Hatori [FRA-12] is shown in Fig. 4.16. Write down the matrix factors that correspond to this flowgraph. List the number of multiplications and the number of additions required for this algorithm.

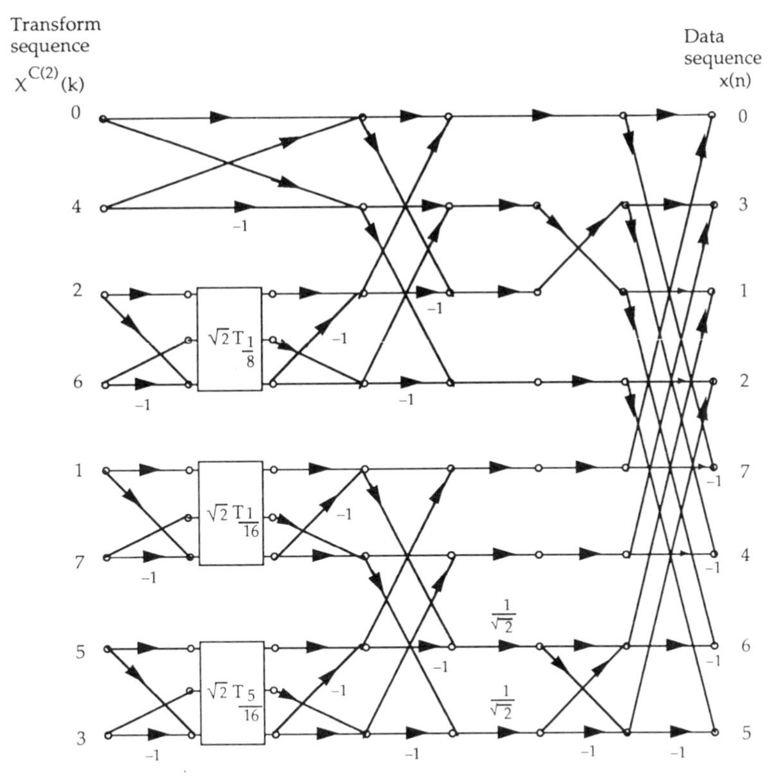

Fig. 4.16 Flow graph for fast implementation of IDCT, $N = 8$ based on [FD-1, FRA-12]. See Fig. 4.17.

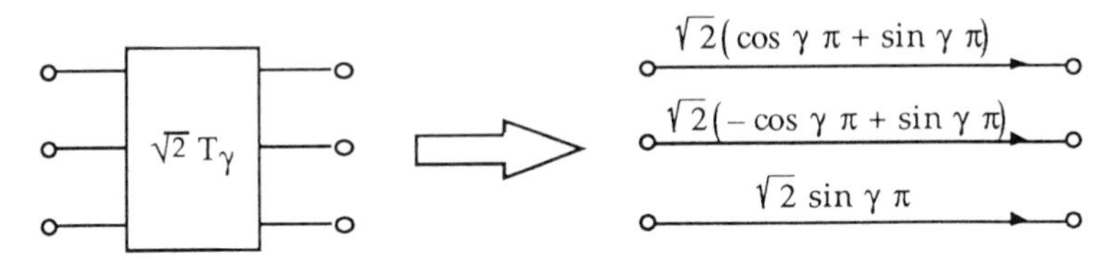

Fig. 4.17 Description of notation shown in Fig. 4.16.

4.26. As the DCT is an orthogonal transform, the matrix factors for the DCT can be developed from Problem 4.25. Using these matrix factors, draw the flowgraph for the DCT $N = 8$. Show that (including the normalization factor $2/N$): (matrix factors for DCT) (matrix factors for IDCT) = identity matrix.

4.27. In Fig. 4.13 the signal flowgraph for $N = 8$ DCT-II based on planar rotations [DP-22, DP-27] is shown. Develop the sparse matrix factors from this flowgraph. Carry out a matrix multiplication of these factors and obtain the DCT-II matrix.

4.28. Repeat Problem 4.27 based on Fig. 4.14.

4.29. A new discrete transform called the real-valued FFT (RVFFT) has been developed recently [PC-16]. This transform, which has been applied to image coding, is defined as follows:

$$X(k) = \sum_{n=0}^{N-1} x_p(n) \cos\left[\frac{2\pi nk}{N}\right], \qquad k = 0, 1, \ldots, N/2 - 1$$

and (P4.29.1)

$$X(k) = \sum_{n=0}^{N-1} x_p(n) \sin\left[\frac{2\pi nk}{N}\right], \qquad k = N/2, \ldots, N - 1,$$

where $\{x_p(n)\}$ is the permuted input sequence. The signal flowgraph for $N = 8$ is shown in Fig. 4.18. Write down the sparse matrix factors from this flowgraph and

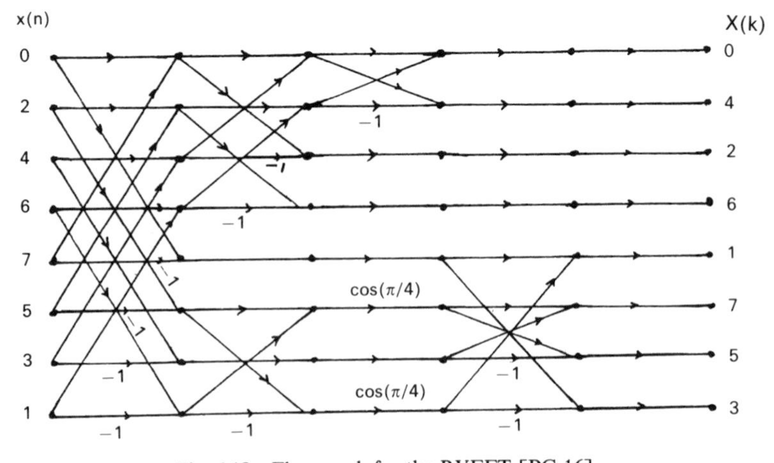

Fig. 4.18 Flowgraph for the RVFFT [PC-16].

then obtain the transform matrix. Express (P4.29.1) in vector-matrix form and show that for $N = 8$, the transform matrix is the same as that obtained from Fig. 4.18.

4.30. Obtain the sparse matrix factors for $N = 16$ RVFFT [PC-16]. Draw the corresponding signal flowgraph and obtain the transform matrix from the matrix factors.

4.31. Obtain the sparse matrix factors for $N = 8$ and 16 inverse RVFFT [PC-16]. Draw the corresponding signal flowgraphs. Obtain the inverse transform matrices from these matrix factors.

4.32. PFA for DCT has been developed by Yang and Narasimha [FRA-10]. Draw a signal flowgraph for 15-point DCT-II based on this algorithm. Obtain the sparse matrix factors from this flowgraph.

4.33. Repeat Problem 4.32 for 15-point IDCT-II.

4.34. An 8-point DCT algorithm that requires 11 multiplications and 29 additions only has been developed by Loeffler, Ligtenberg, and Moschytz [FRA-26]. Obtain the sparse matrix factors from Fig. 1 of [FRA-26] and show that this results in the 8×8 DCT-II matrix.

4.35. Draw the signal flowgraph for 8-point IDCT-II corresponding to Fig. 1 of [FRA-26]. Hint: Use the results of Problem 4.34. Also DCT-II is real and orthogonal.

4.36. Repeat Problems 4.34 and 4.35 for 16-point DCT-II. See Fig. 9 of [FRA-26].

4.37. In the conclusion of [FRA-26], it is stated "that an 8-point DCT can be calculated with 12 multiplications in parallel, i.e., with no signal path having more than one multiplication in cascade." Investigate if this concept can be extended to 16-point DCT-II.

CHAPTER 5

TWO-DIMENSIONAL DCT ALGORITHMS

5.1 Introduction

Digital picture processing has a very prominent position in the general discipline of digital signal processing (DSP). The importance of human visual perception has encouraged tremendous interest and advances in the art and science of digital picture processing. The major aspects of this field include image coding, image restoration, and image feature selection. Image coding represents the attempts to transmit pictures over digital communication channels in an efficient manner, making use of as few bits as possible to minimize the bandwidth required, while at the same time, maintaining distortions within certain limits. This concept is called data compression or bandwidth reduction, and is discussed in detail in Chapter 7. Practical applications include teleconferencing, remote medical consultation, facsimile transmission, satellite imaging, and picture phones, just to mention a few.

Image restoration represents efforts to recover the "true" image of the object. The coded image being transmitted over the communication channel may have been polluted by noise. The source of degradation may also have arisen originally in creating the image from the object. Operations such as focus correction, motion compensation, and aberration correction are considered as restoration techniques. In most cases, these techniques depend on some a priori knowledge as to how the degradation has arisen.

Feature selection refers to the selection of certain attributes of the picture. Such attributes may be required in the recognition, classification, and decision in a wider context. Scene classification, image matching, and robotic vision are only a few practical examples.

Basic to the analysis in all these aspects of digital picture processing is the ability to represent the pictures in a "space" in which the attributes of the pictures are not correlated. This is in no way different from the representation of a sampled sinusoid (see Section 3.2), which can be considered either as a sequence of data points (correlated) or as a collection of uncorrelated attributes (amplitude, phase, frequency, etc.). Assuming separability (see H. C. Andrews [B-15]), the representation of a two-dimensional picture in a space of uncorrelated coordinates is equivalent to the singular-value decomposition (SVD) of the two-dimensional signal. The SVD simply determines the "singular vectors" or eigenvectors for $[g][g]^T$ and $[g]^T[g]$, where $[g]$ is the matrix representing the sampled two-dimensional picture. It minimizes the least square error given by $\|[g] - [g_k]\|^2$, where $[g_k]$ is the reconstructed image using k singular vectors.

If, on the other hand, the expectation value of the square errors is to be minimized, i.e.,

$$E\{\|[g] - [g_k]\|^2\},$$

the resulting decomposition or transform is the Karhunen–Loeve transform (KLT) (see Section 3.2).

However, both the SVD and the KLT are dependent on the transmission of the two-dimensional signal. The SVD depends on it exactly, and the KLT depends on its correlation (both row and column) statistics. As in the case of one-dimensional signals, the DCT again is asymptotically equivalent to the KLT and is therefore the next best in the hierarchy of transforms to decompose the two-dimensional signal into uncorrelated "signal vectors." Thus in the schemes of two-dimensional digital signal analysis, DCT has as prominent a position as in the case of one-dimensional digital signal processing.

In this chapter, we shall discuss briefly a number of approaches in the implementation of two-dimensional DCT. Section 5.2 discusses the implementation of two-dimensional DCT by reducing it to a lexicographically ordered one-dimensional transform [FID-1]. This approach can be improved upon by considering the block matrix decomposition of the transform matrices and by aiming for a recursive structure for the algorithm. Such an approach was reported by Haque [FID-6] and is discussed in Section 5.3. As in the one-dimensional case, two-dimensional DFT can also be used as a means of obtaining the two-dimensional DCT. The works reported by Nasrabadi and King [FID-4] and Vetterli [FID-3] belong to this category and are briefly outlined in Section 5.4. Section 5.5 discusses the computation of two-dimensional DCT by using precomputed conversion matrices applied to the two-dimensional Walsh–Hadamard transform of the digital picture. Kashef and Habibi [DW-4] showed that computational advantages can be gained in processing pictures of small-to-moderate size using this approach. In Section 5.6, we report some recent attempts in the hardware implementation of two-dimensional DCT [DP-26, DP-28, DP-29, DP-31, DP-33, DP-34, DP-37, DP-39, DP-44 through DP-49, DP-52, DP-53]. Most of these VLSI chip designs are aimed at image processing and utilize the separability property of the two-dimensional DCT. A summary in Section 5.7 concludes this chapter.

5.2 Two-Dimensional DCT by Reduction to One-Dimensional DCT

Let an $(M \times N)$ matrix $[g]$ represent a black-and-white (for simplicity) digital picture, where the matrix element g_{mn} may be interpreted as the grey level or intensity of the pixel (picture element) at the (m, n) location. $[G]$, the two-dimensional orthogonal transform of the matrix $[g]$, and its inverse are, respectively, defined as

$$[G] = [T_M][g][T_N]^T \tag{5.2.1a}$$

and

$$[g] = [T_M]^T[G][T_N], \tag{5.2.1b}$$

where $[T_M]$ and $[T_N]$ are the $(M \times M)$ and $(N \times N)$ real transformation matrices, respectively. The form (5.2.1) assumes that the two-dimensional transform can be implemented by a series of one-dimensional transforms. This property is valid for any separable transform such as DCT, WHT, DFT, ST, HT, etc. An important consequence of this condition is immediately apparent. Suppose the rows (or columns) of the matrix $[g]$ are simply concatenated to create a column vector $\mathbf{x}$ of MN components. The one-dimensional transformed vector $\mathbf{X}$ is given by

$$\mathbf{X} = [T_{MN}]\mathbf{x}, \tag{5.2.2}$$

where $[T_{MN}]$ is of dimension $(MN \times MN)$. It can be seen that (5.2.1) and (5.2.2) are quite different in that (5.2.2) involves $(MN \times MN)$ elements in the transformation, whereas (5.2.1) involves $(M \times M + N \times N)$ elements. Thus it is seen that a simple concatenation to reduce a two-dimensional transform to a one-dimensional one requires additional considerations. We examine the specific case for DCT.

Let $[g]$ be an $(M \times N)$ matrix representing the two-dimensional data and $[G]$ be the two-dimensional DCT-II of $[g]$. Then the uv-element of $[G]$ is given by

$$G_{uv} = \frac{2c(u)c(v)}{\sqrt{(MN)}} \sum_{m=0}^{M-1} \sum_{n=0}^{N-1} g_{mn} \cos\left[\frac{(2m+1)u\pi}{2M}\right] \cos\left[\frac{(2n+1)v\pi}{2N}\right], \tag{5.2.3a}$$

where $u = 0, \ldots, M-1$, $v = 0, \ldots, N-1$, and

$$c(k) = \frac{1}{\sqrt{2}} \quad \text{if } k = 0,$$
$$= 1 \quad \text{otherwise.}$$

Similarly, the mn-element of $[g]$ is given by the two-dimensional IDCT-II of $[G]$, defined as

$$g_{mn} = \frac{2}{\sqrt{(MN)}} \sum_{u=0}^{M-1} \sum_{v=0}^{N-1} c(u)c(v)G_{uv} \cos\left[\frac{(2m+1)u\pi}{2M}\right] \cos\left[\frac{(2n+1)v\pi}{2N}\right], \tag{5.2.3b}$$

where $m = 0, \ldots, M-1$ and $n = 0, \ldots, N-1$.

The separability property of the two-dimensional DCT as indicated by (5.2.1) can be illustrated as follows:

Rewrite (5.2.3a) as

$$G_{uv} = \sqrt{\left(\frac{2}{M}\right)}\, c(u) \sum_{m=0}^{M-1} \left\{ \sqrt{\left(\frac{2}{N}\right)}\, c(v) \sum_{n=0}^{N-1} g_{mn} \cos\left[\frac{(2n+1)v\pi}{2N}\right] \right\} \times \cos\left[\frac{(2m+1)u\pi}{2M}\right], \tag{5.2.3c}$$

where $u = 0, \ldots, M-1$ and $v = 0, \ldots, N-1$.

The inner summation is an N-point one-dimensional DCT-II of the rows of $[g]$, whereas the outer summation represents the M-point one-dimensional DCT-II of the columns of the "semitransformed" matrix. This implies that a two-dimensional ($M \times N$) DCT can be implemented by M N-point DCTs along the rows of $[g]$, followed by N M-point DCTs along the columns of the matrix obtained after the row transformation (Fig. 5.1). The order in which the row transform and the column transform are done is theoretically immaterial. One can take the reverse order in the two sets of 1D DCTs, as shown in the following equation,

$$G_{uv} = \sqrt{\left(\frac{2}{N}\right)}\, c(v) \sum_{n=0}^{N-1} \left\{ \sqrt{\left(\frac{2}{M}\right)}\, c(u) \sum_{m=0}^{M-1} g_{mn} \cos\left[\frac{(2m+1)u\pi}{2M}\right] \right\} \times \cos\left[\frac{(2n+1)v\pi}{2N}\right]. \tag{5.2.3d}$$

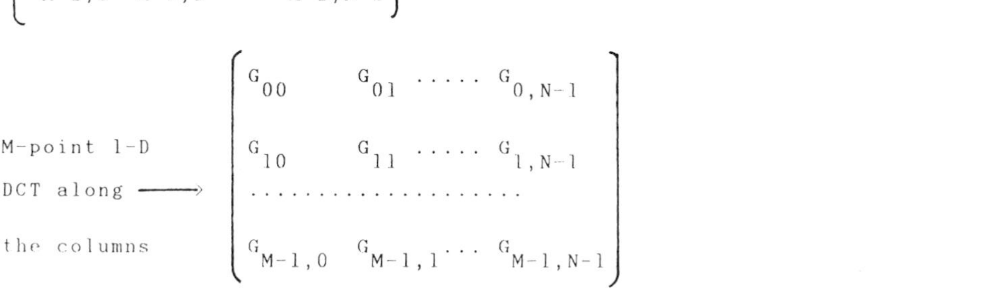

$$\begin{bmatrix} g_{00} & g_{01} & \cdots & g_{0,N-1} \\ g_{10} & g_{11} & \cdots & g_{1,N-1} \\ \cdots & \cdots & \cdots & \cdots \\ g_{M-1,0} & g_{M-1,1} & \cdots & g_{M-1,N-1} \end{bmatrix} \xrightarrow{\text{N-point 1-D DCT along the rows}} \begin{bmatrix} \text{Semi-transformed matrix} \end{bmatrix}$$

$$\xrightarrow{\text{M-point 1-D DCT along the columns}} \begin{bmatrix} G_{00} & G_{01} & \cdots & G_{0,N-1} \\ G_{10} & G_{11} & \cdots & G_{1,N-1} \\ \cdots & \cdots & \cdots & \cdots \\ G_{M-1,0} & G_{M-1,1} & \cdots & G_{M-1,N-1} \end{bmatrix}$$

Fig. 5.1 Implementation of $M \times N$ 2D DCT by 1D DCTs.

The above description of reducing a two-dimensional transform to one-dimensional ones can be summarized in Fig. 5.1.

Inspection of (5.2.3a) and (5.2.3b) shows that the separability property is equally valid for the 2D IDCT. In fact, this property can be extended to dimensions larger than two. As an example of a sequence of images (3D array), 3D DCT can be implemented by a series of 1D DCTs along each of the dimensions (Fig. 5.2). In summary, for a discrete orthogonal transform, such as the DCT, if the forward transform is separable, so also is the inverse. This has a number of advantages. If a forward 1D transform has a fast algorithm, so also does the forward multidimensional transform. For an orthogonal transform, this property is also valid for the inverse transformation. Indeed, if a particular architecture is utilized for implementing a 1D transform in hardware, the same concept can be extended to the multidimensional transform [DP-26, DP-28, DP-29, DP-31, DP-33, DP-34, DP-37, DP-39, DP-41]. Some of these concepts are discussed in Section 5.6.

In what follows, we again adopt the convention of leaving out the normalization constants $\sqrt{(MN)}$ and the scale factors $c(u)$ and $c(v)$ to simplify our discussions. Kamangar and Rao [FID-1] developed, from this property of separability, two versions of the equivalent one-dimensional DCT algorithms, one partially recursive and the other nonrecursive. We present only the partially recursive version here.

Let **x** and **X** represent the column vectors obtained by concatenating the rows of $[g]$ and $[G]$, respectively, so that

$$\mathbf{x} = [g_{0,0}; g_{0,1}; \ldots; g_{0,N-1}; g_{1,0}; \ldots; g_{M-1,N-1}]^T,$$

and

$$\mathbf{X} = [G_{0,0}; G_{0,1}; \ldots; G_{0,N-1}; G_{1,0}; \ldots; G_{M-1,N-1}]^T. \tag{5.2.4}$$

Thus, **x** and **X** are one-dimensional arrays obtained by lexicographically rearranging the two-dimensional data and the two-dimensional transform.

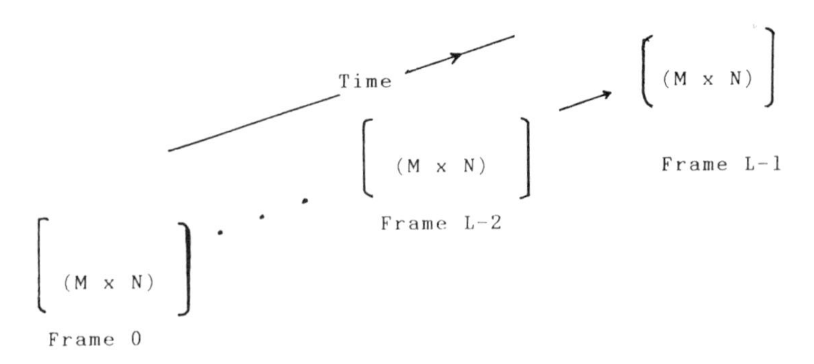

Fig. 5.2 Implementation of 3D DCT of a sequence of L frames MN L-point 1D transforms along the temporal direction following the 2D transforms of the frames.

Equation (5.2.2) can now be written as

$$\mathbf{X} = [D(M, N)]\mathbf{x}, \tag{5.2.5}$$

where the matrix $[D(M, N)]$ is the Kronecker or direct product of the two matrices $[C_M^{\mathrm{II}}]$ and $[C_N^{\mathrm{II}}]$, given by

$$[D(M, N)] = [C_M^{\mathrm{II}}] \otimes [C_N^{\mathrm{II}}]. \tag{5.2.6}$$

Here, $[C_M^{\mathrm{II}}]$ is the DCT-II transform matrix of dimension $(M \times M)$. (5.2.6) can be easily verified (see Problem 5.1). It is noted here that (5.2.6), where the transform matrix $[D]$ is written as a direct product of two DCT-II transform matrices, indicates that the two-dimensional transform on an $(M \times N)$ data matrix is not equivalent to a simple one-dimensional MN-point transform, a point that we emphasized earlier.

The partially recursive algorithm is developed by decomposing the matrix $[D(M, N)]$ into factor matrices given by

$$[D(M, N)] = [P_1(M, N)][H(M, N)][P_2(M, N)], \tag{5.2.7a}$$

where $[P_1(M, N)]$ and $[P_2(M, N)]$ are permutation matrices to be defined later, and $[H(M, N)]$ is further factored as

$$[H(M, N)] = \begin{bmatrix} \left[H\left(\frac{M}{2}, N\right)\right] & 0 \\ 0 & \left[Q\left(\frac{M}{2}, N\right)\right] \end{bmatrix} [A(MN)], \tag{5.2.7b}$$

where

$$\left[H\left(\frac{M}{2}, N\right)\right] = \begin{bmatrix} \left[H\left(\frac{M}{2}, \frac{N}{2}\right)\right] & 0 \\ 0 & \left[Q\left(\frac{M}{2}, \frac{N}{2}\right)\right] \end{bmatrix} \left[A\left(\frac{MN}{2}\right)\right]$$

and

$$[A(L)] = \begin{bmatrix} [I_{L/2}] & [\tilde{I}_{L/2}] \\ [\tilde{I}_{L/2}] & -[I_{L/2}] \end{bmatrix},$$

$[I]$ and $[\tilde{I}]$ being the unit and opposite unit matrices, respectively. The recursiveness of the algorithm lies in the factor matrix $[H]$. Both $[H]$ and $[Q]$ are made up of elements of cosine functions. Using the convention of $C_m^k = \cos(k\pi/m)$, we delineate the $[H]$ matrix for $M = 4$ and $N = 4$, as an example, in what follows.

Example for M = N = 4:

$$[H(4, 4)] = \begin{bmatrix} [H(2, 4)] & 0 \\ 0 & [Q(2, 4)] \end{bmatrix} [A(16)],$$

where

$$[H(2, 4)] = \begin{bmatrix} [H(2, 2)] & 0 \\ 0 & [Q(2, 2)] \end{bmatrix} [A(8)],$$

and

$$[H(2, 2)] = \begin{bmatrix} [H(1, 2)] & 0 \\ 0 & [Q(1, 2)] \end{bmatrix} [A(4)]. \tag{5.2.8}$$

In terms of the cosine functions, we have

$$[H(1, 2)] = \begin{bmatrix} C_4^1 C_4^1 & C_4^1 C_4^1 \\ C_4^1 C_4^1 & -C_4^1 C_4^1 \end{bmatrix},$$

$$[Q(1, 2)] = \begin{bmatrix} -C_4^1 C_4^1 & C_4^1 C_4^1 \\ C_4^1 C_4^1 & C_4^1 C_4^1 \end{bmatrix},$$

$$[Q(2, 2)] = \begin{bmatrix} C_4^1 C_8^1 & C_4^1 C_8^3 & & 0 \\ C_4^1 C_8^3 & -C_4^1 C_8^1 & & \\ & 0 & -C_4^1 C_8^1 & -C_4^1 C_8^3 \\ & & -C_4^1 C_8^3 & C_4^1 C_8^1 \end{bmatrix} [A(4)],$$

and

$$[Q(2, 4)] = \begin{bmatrix} [Q_1(2, 2)] & 0 \\ 0 & [Q_2(2, 2)] \end{bmatrix} [A(8)],$$

where

$$[Q_1(2, 2)] = \begin{bmatrix} C_8^1 C_4^1 & 0 & C_4^1 C_8^3 & 0 \\ 0 & C_4^1 C_8^3 & 0 & C_8^1 C_4^1 \\ C_4^1 C_8^3 & 0 & -C_8^1 C_4^1 & 0 \\ 0 & C_8^1 C_4^1 & 0 & -C_4^1 C_8^3 \end{bmatrix} \begin{bmatrix} 1 & 1 & 0 & 0 \\ 1 & -1 & 0 & 0 \\ 0 & 0 & 1 & 1 \\ 0 & 0 & 1 & -1 \end{bmatrix}$$

and

$$[Q_2(2, 2)] = (2)^{-1} \begin{bmatrix} 1 & 0 & 1 & 0 \\ 0 & -1 & 0 & -1 \\ 1 & 0 & -1 & 0 \\ 0 & -1 & 0 & 1 \end{bmatrix} \begin{bmatrix} C_4^1 & 0 & 0 & C_4^1 \\ 0 & 1 & 0 & 0 \\ 0 & 0 & 1 & 0 \\ C_4^1 & 0 & 0 & -C_4^1 \end{bmatrix}$$

$$\times \begin{bmatrix} 1 & 0 & 1 & 0 \\ 0 & 1 & 0 & 1 \\ 1 & 0 & -1 & 0 \\ 0 & 1 & 0 & -1 \end{bmatrix}. \tag{5.2.9}$$

As for the permutation matrices, we present examples for $[P_1(4, 4)]$ and $[P_2(4, 4)]$ and refer the reader to [FID-1] for details.

$$[P_2(4,4)] = \begin{bmatrix} \begin{bmatrix} [I_2] & 0 & 0 & 0 \\ 0 & 0 & [\tilde{I}_2] & 0 \\ 0 & 0 & 0 & [\tilde{I}_2] \\ 0 & [I_2] & 0 & 0 \end{bmatrix} & 0 \\ 0 & \begin{bmatrix} 0 & 0 & 0 & [\tilde{I}_2] \\ 0 & [I_2] & 0 & 0 \\ [I_2] & 0 & 0 & 0 \\ 0 & 0 & [\tilde{I}_2] & 0 \end{bmatrix} \end{bmatrix}$$

and

$$[P_1(4, 4)] = [\mathbf{u}_0, \mathbf{u}_2, \mathbf{u}_{10}, \mathbf{u}_8, \mathbf{u}_1, \mathbf{u}_3, \mathbf{u}_{11}, \mathbf{u}_9, \mathbf{u}_4, \mathbf{u}_{14}, \mathbf{u}_{12}, \mathbf{u}_6, \mathbf{u}_{13}, \mathbf{u}_5, \mathbf{u}_7, \mathbf{u}_{15}],$$

where $\mathbf{u}_i$ is a column vector with zeros, except a one in the ith row. The signal flowgraph for the 4×4 case is shown in Fig. 5.3.

For this case, the computation requires 24 real multiplications and 68 real additions. Based on the algorithm of Chen, Smith, and Fralick [FRA-1] (see Section 4.3), 48 multiplications and 64 additions are required when the

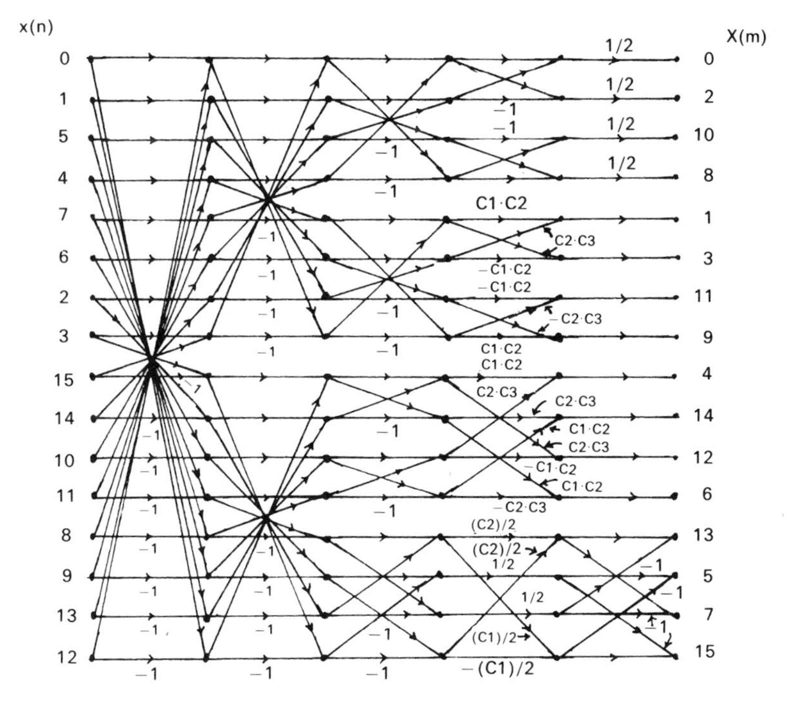

Fig. 5.3 2D-DCT signal flowgraph for a 4×4 block using lexicographical reordering [FID-1]. Here Ck denotes $\cos(k\pi/8)$. (© 1982 IEEE).

expression (5.2.3) is directly evaluated. Thus there is much to be gained in computational speed by this reduction to a one-dimensional transform. The minor drawbacks here are the complicated output mapping (i.e., the permutation matrix $[P_1(M, N)]$ and the nonrecursive nature of the factor block matrices $[Q(M, N)]$.

Instead of reducing the two-dimensional array lexicographically to a one-dimensional array, it is possible to operate directly on the two-dimensional array to obtain a fast two-dimensional algorithm, as reported by Haque [FID-6]. This we discuss in some detail now.

5.3 Block Matrix Decomposed Algorithm

Our discussion in Section 5.2 is based on concatenating columns (or rows) of the two-dimensional input data matrix, to reduce it to a one-dimensional array. The transform matrix is then factorized in block form. Unfortunately, only part of the factorization is really recursive. This makes the algorithm less than optimal in efficiency and not quite modular in structure. Haque [FID-6] reported a two-dimensional fast DCT algorithm based on a rearrangement of the elements of the two-dimensional input matrix into a block matrix form. Each block of the matrix is then put through a "half-size" two-dimensional DCT. The structure is thus quite modular and the algorithm is recursive.

The rearrangement of the elements in the original input matrix follows the decimation-in-frequency (DIF) scheme (see Section 4.4.2), thus combining the two frequency indices in even-even, even-odd, odd-even, and odd-odd groups. The derivation is more easily carried out for the inverse DCT-II (or the forward DCT-III), and we shall follow this approach. In addition, the normalization and scale factors are ignored, as in the one-dimensional case. We simply note here that for an $M \times N$ input matrix the normalization constant (adopting the symmetric definition, as in the one-dimensional case) is $2/\sqrt{(MN)}$, and in the transformed matrix the scaling factor for the first column and row is $1/\sqrt{2}$, except for the first element, which has a scale factor of $\frac{1}{2}$ [see (5.2.3)].

In order to clearly track the location of the signal element, we present the development of the algorithm in an element-by-element fashion before introducing the block matrix results. The principle, however, is the same as that used by Haque [FID-6].

Let $[\tilde{g}]$ and $[\tilde{G}]$ be the properly scaled and normalized two-dimensional input data and its DCT-II, respectively. Then the mn-element of $[\tilde{g}]$ is given by the inverse two-dimensional transform defined by

$$\tilde{g}_{mn} = \sum_{u=0}^{M-1} \sum_{v=0}^{N-1} \tilde{G}_{uv} C_{2M}^{(2m+1)u} C_{2N}^{(2n+1)v}, \tag{5.3.1}$$

for $m = 0, \ldots, M-1$ and $n = 0, \ldots, N-1$.

The double summation in (5.3.1) can be decomposed so that the expression becomes

$$\tilde{g}_{mn} = \tilde{g}_{mn}^{ee} + \tilde{g}_{mn}^{eo} + \tilde{g}_{mn}^{oe} + \tilde{g}_{mn}^{oo}, \tag{5.3.2}$$

where

$$\tilde{g}_{mn}^{ee} = \sum_{u=0}^{M/2-1} \sum_{v=0}^{N/2-1} \tilde{G}_{2u,2v} C_{2M}^{(2m+1)(2u)} C_{2N}^{(2n+1)(2v)},$$

$$\tilde{g}_{mn}^{eo} = \sum_{u=0}^{M/2-1} \sum_{v=0}^{N/2-1} \tilde{G}_{2u,2v+1} C_{2M}^{(2m+1)(2u)} C_{2N}^{(2n+1)(2v+1)},$$

$$\tilde{g}_{mn}^{oe} = \sum_{u=0}^{M/2-1} \sum_{v=0}^{N/2-1} \tilde{G}_{2u+1,2v} C_{2M}^{(2m+1)(2u+1)} C_{2N}^{(2n+1)(2v)},$$

and

$$\tilde{g}_{mn}^{oo} = \sum_{u=0}^{M/2-1} \sum_{v=0}^{N/2-1} \tilde{G}_{2u+1,2v+1} C_{2M}^{(2m+1)(2u+1)} C_{2N}^{(2n+1)(2v+1)}. \tag{5.3.3}$$

These are, respectively, the even-even, even-odd, odd-even, and odd-odd transforms of size $(M/2) \times (N/2)$.

We note that $\tilde{g}_{mn}^{oo}$ represents the IDCT-II, while the remaining ones can be reduced as follows:

$$\begin{aligned}\tilde{g}_{mn}^{eo} &= [2C_{2N}^{(2n+1)}]^{-1} \sum_{u=0}^{M/2-1} \sum_{v=0}^{N/2-1} \tilde{G}_{2u,2v+1} C_{M}^{(2m+1)u} [C_{N}^{(2n+1)(v+1)} + C_{N}^{(2n+1)v}] \\ &= [2C_{2N}^{(2n+1)}]^{-1} \sum_{u=0}^{M/2-1} \sum_{v=0}^{N/2-1} [\tilde{G}_{2u,2v-1} + \tilde{G}_{2u,2v+1}] C_{M}^{(2m+1)u} C_{N}^{(2n+1)v}.\end{aligned} \tag{5.3.4}$$

Equation (5.3.4) clearly shows this even-odd transform to be an IDCT-II of size $(M/2) \times (N/2)$. We have used the identity

$$2C_{2N}^{(2n+1)(2v+1)} C_{2N}^{(2n+1)} = C_{N}^{(2n+1)(v+1)} + C_{N}^{(2n+1)v}$$

and have assumed that $\tilde{G}_{2u,-1} = 0$.

The remaining terms in (5.3.2) can be decomposed in a similar fashion, giving

$$\tilde{g}_{mn}^{oe} = [2C_{2M}^{(2m+1)}]^{-1} \sum_{u=0}^{M/2-1} \sum_{v=0}^{N/2-1} [\tilde{G}_{2u-1,2v} + \tilde{G}_{2u+1,2v}] C_{M}^{(2m+1)u} C_{N}^{(2n+1)v},$$

and

$$\begin{aligned}\tilde{g}_{mn}^{oo} = [4C_{2M}^{(2m+1)} C_{2N}^{(2n+1)}]^{-1} \sum_{u=0}^{M/2-1} \sum_{v=0}^{N/2-1} [\tilde{G}_{2u-1,2v-1} + \tilde{G}_{2u-1,2v+1} \\ + \tilde{G}_{2u+1,2v-1} + \tilde{G}_{2u+1,2v+1}] C_{M}^{(2m+1)u} C_{N}^{(2n+1)v}.\end{aligned} \tag{5.3.5}$$

In the above formulae, all elements of $[\tilde{G}]$ outside the range of dimension of $(M \times N)$ are assumed to be zero. It is evident that (5.3.4) and (5.3.5) represent

$(M/2)\times(N/2)$ two-dimensional IDCT-II of the combined elements shown inside the square brackets. Thus, a two-dimensional $(M \times N)$ matrix to be inverse transformed has been reduced to four $(M/2)\times(N/2)$ inverse transforms of rearranged elements in $[\tilde{G}]$. Further note that

$$C_{2N}^{[2(N-n-1)+1](2v)} = C_{2N}^{(2n+1)(2v)}$$

and

$$C_{2N}^{[2(N-n-1)+1](2v+1)} = -C_{2N}^{(2n+1)(2v+1)} \tag{5.3.6}$$

Thus, the elements $\tilde{g}_{mn}$ making up the input matrix can be partitioned according to (5.3.2), giving

$$\begin{aligned}\tilde{g}_{mn} &= \tilde{g}_{mn}^{ee} + \tilde{g}_{mn}^{eo} + \tilde{g}_{mn}^{oe} + \tilde{g}_{mn}^{oo},\\ \tilde{g}_{m,N-n-1} &= \tilde{g}_{mn}^{ee} - \tilde{g}_{mn}^{eo} + \tilde{g}_{mn}^{oe} - \tilde{g}_{mn}^{oo},\\ \tilde{g}_{M-m-1,n} &= \tilde{g}_{mn}^{ee} + \tilde{g}_{mn}^{eo} - \tilde{g}_{mn}^{oe} - \tilde{g}_{mn}^{oo},\end{aligned}$$

and

$$\tilde{g}_{M-m-1,N-n-1} = \tilde{g}_{mn}^{ee} - \tilde{g}_{mn}^{eo} - \tilde{g}_{mn}^{oe} + \tilde{g}_{mn}^{oo}, \tag{5.3.7a}$$

where $m = 0, 1, \ldots, (M/2 - 1)$ and $n = 0, 1, \ldots, (N/2 - 1)$.

The four equations in (5.3.7a) can be combined into a block matrix equation by identifying them, respectively, as elements in the block matrices $[p]$, $[q]$, $[r]$, and $[s]$, i.e.,

$$p_{mn} = \tilde{g}_{mn},\ q_{mn} = \tilde{g}_{m,N-n-1},\ r_{mn} = \tilde{g}_{M-m-1,n} \quad \text{and } s_{mn} = \tilde{g}_{M-m-1,N-n-1}. \tag{5.3.7b}$$

The block matrix equation is then

$$\begin{bmatrix}[p]\\ [q]\\ [r]\\ [s]\end{bmatrix} = \begin{bmatrix} I & I & I & I\\ I & -I & I & -I\\ I & I & -I & -I\\ I & -I & -I & I\end{bmatrix}\begin{bmatrix}[\tilde{g}^{ee}]\\ [\tilde{g}^{eo}]\\ [\tilde{g}^{oe}]\\ [\tilde{g}^{oo}]\end{bmatrix}. \tag{5.3.8}$$

The block matrices $[\tilde{g}^{ee}]$, $[\tilde{g}^{eo}]$, $[\tilde{g}^{oe}]$, and $[\tilde{g}^{oo}]$ are, in turn, given by

$$[\tilde{g}^{ee}] = [C_{M/2}^{II}]^T[\tilde{G}^{ee}][C_{N/2}^{II}]; \qquad [\tilde{g}^{eo}] = [C_{M/2}^{II}]^T[\tilde{G}^{eo}][C_{N/2}^{II}][W_{N/2}];$$

$$[\tilde{g}^{oe}] = [W_{M/2}][C_{M/2}^{II}]^T[\tilde{G}^{oe}][C_{N/2}^{II}];$$

and

$$[\tilde{g}^{oo}] = [W_{M/2}][C_{M/2}^{II}]^T[\tilde{G}^{oo}][C_{N/2}^{II}][W_{N/2}], \tag{5.3.9}$$

where $[W_N] = 1/2\ \text{diag}[1/C_{2N}^1, 1/C_{2N}^3, \ldots, 1/C_{2N}^{(2N-1)}]$, and

$$\tilde{G}_{u,v}^{ee} = \tilde{G}_{2u,2v}, \qquad \tilde{G}_{u,v}^{eo} = \tilde{G}_{2u,2v-1} + \tilde{G}_{2u,2v+1},$$

$$\tilde{G}_{u,v}^{oe} = \tilde{G}_{2u-1,2v} + \tilde{G}_{2u+1,2v},$$

and

$$\tilde{G}^{oo}_{u,v} = \tilde{G}_{2u-1,2v-1} + \tilde{G}_{2u-1,2v+1} + \tilde{G}_{2u+1,2v-1} + \tilde{G}_{2u+1,2v+1}, \tag{5.3.10}$$

where $u = 0, 1, \ldots, M/2 - 1$ and $v = 0, 1, \ldots, N/2 - 1$.

Note that in (5.3.9) all the two-dimensional transforms are of size $(M/2) \times (N/2)$. It is evident that this procedure of reduction can be applied repeatedly until only 2×2, $2 \times 4, \ldots$, i.e., nonreducible two-dimensional transforms are left, thus rendering the overall algorithm a recursive one. We also note that the matrices $[\tilde{g}^{ee}]$, $[\tilde{g}^{eo}]$, etc. are multiplied by a 4×4 Walsh–Hadamard block matrix in (5.3.8).

We summarize the procedure for computing the inverse DCT-II of an $M \times N$ data array.

5.3.1 *Inverse* M × N *DCT-II Block Decomposed Algorithm*

(1) Decompose the matrix $[\tilde{G}]$ into four $(M/2) \times (N/2)$ sub-blocks $[\tilde{G}^{ee}]$, $[\tilde{G}^{eo}]$, $[\tilde{G}^{oe}]$, and $[\tilde{G}^{oo}]$, consisting of the even-even, even-odd, odd-even, and odd-odd elements in the original data matrix according to (5.3.10). All elements outside the $M \times N$ range are set equal to zero.

(2) Compute the $(M/2) \times (N/2)$ two-dimensional IDCT for each of the four sub-blocks and scale the results according to (5.3.9) using the diagonal matrices $[W_{N/2}]$ and $[W_{M/2}]$ to obtain the subblocks $[\tilde{g}^{ee}]$, etc.

(3) Premultiply the column vector consisting of the sub-blocks $[\tilde{g}^{ee}]$, etc. by a 4×4 Walsh–Hadamard matrix to obtain the column of sub-blocks $[p]$, $[q]$, $[r]$, and $[s]$.

(4) These matrices $[p]$, $[q]$, $[r]$, and $[s]$ are sub-blocks of the matrix $[\tilde{g}]$ determined by (5.3.7).

Fig. 5.4 shows the flow graph for a 4×4 case.

In step (2), the size of the two-dimensional IDCT-II can be further reduced by using step (1), thus resulting in the algorithm being recursive. We note that the above procedure is the IDCT-II. The forward DCT-II algorithm is obtained by reversing each step. We describe the procedure below.

5.3.2 *Forward* M × N *DCT-II Block Decomposed Algorithm*

(1) Decompose the $M \times N$ data matrix $[\tilde{g}]$ into four $(M/2) \times (N/2)$ sub-blocks, $[p]$, $[q]$, $[r]$, and $[s]$ according to (5.3.7b).

(2) Arrange the sub-blocks as a block column vector and premultiply this by $(1/4)[H]$, where $[H]$ is the 4×4 block Walsh–Hadamard matrix. The resulting block column vector consists of $[\tilde{g}^{ee}]$, etc. as in (5.3.8).

(3) Scale these sub-blocks by computing $[\tilde{g}^{ee}]$, $[\tilde{g}^{eo}][W_{N/2}]^{-1}$, $[W_{M/2}]^{-1}[\tilde{g}^{oe}]$, and $[W_{M/2}]^{-1}[\tilde{g}^{oo}][W_{N/2}]^{-1}$.

(4) Compute the four $(M/2) \times (N/2)$ two-dimensional DCT-II for the scaled sub-blocks in step (3) to obtain the matrices $[\tilde{G}^{ee}]$, $[\tilde{G}^{eo}]$, $[\tilde{G}^{oe}]$, and $[\tilde{G}^{oo}]$.

(5) Denote the even-even, even-odd, odd-even, and odd-odd elements of the

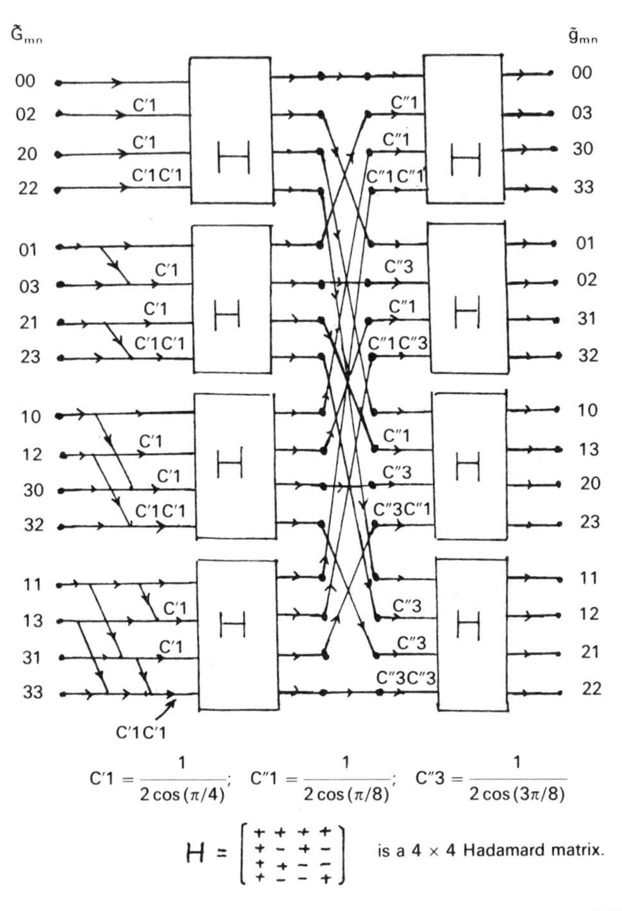

Fig. 5.4 Signal flowgraph for the IDCT-II of a 4×4 case [FID-6]. (© 1985 IEEE).

transformed matrix $[\tilde{G}]$ by $P_{u,v}$, $Q_{u,v}$, $R_{u,v}$, and $S_{u,v}$ such that

$$P_{u,v} = \tilde{G}_{2u,2v}, \qquad Q_{u,v} = \tilde{G}_{2u,2v+1}, \qquad R_{u,v} = \tilde{G}_{2u+1,2v},$$

and

$$S_{u,v} = \tilde{G}_{2u+1,2v+1}, \qquad u = 0, 1, \ldots, \frac{M}{2} - 1; \; v = 0, 1, \ldots, N/2 - 1. \tag{5.3.11}$$

(6) Invert (5.3.10) to obtain

$$[P] = [\tilde{G}^{ee}], \qquad [Q] = [L_{M/2}][\tilde{G}^{eo}], \qquad [R] = [\tilde{G}^{oe}][L_{N/2}]^T,$$

and

$$[S] = [L_{M/2}][\tilde{G}^{oo}][L_{N/2}]^T, \tag{5.3.12}$$

where $[L_M]$ is an $M \times M$ lower triangular matrix of the form

$$[L_M] = \begin{bmatrix} 1 & 0 & 0 & 0 & 0 & \cdots & \cdots & \cdots \\ -1 & 1 & 0 & 0 & 0 & \cdots & \cdots & \cdots \\ 1 & -1 & 1 & 0 & 0 & \cdots & \cdots & \cdots \\ \cdots & \cdots & \cdots & \cdots & \cdots & \cdots & \cdots & \cdots \\ \pm 1 & \cdots & \cdots & \cdots & \cdots & \cdots & -1 & 1 \end{bmatrix}. \tag{5.3.13}$$

Note that step 4 here contains the recursivity of the algorithm where the sizes involved may be reduced repeatedly. We conclude this section by combining all the steps into a single block-decomposed matrix equation for the forward DCT-II, using $[p]$, $[q]$, $[r]$, and $[s]$ as input blocks and $[P]$, $[Q]$, $[R]$, and $[S]$ as transformed blocks, as defined by (5.3.7b) and (5.3.11), respectively.

$$\begin{bmatrix} [P] \\ [Q] \\ [R] \\ [S] \end{bmatrix} = \begin{bmatrix} [I_{M/2}] & & & 0 \\ & [L_{M/2}] & & \\ & & [I_{M/2}] & \\ 0 & & & [L_{M/2}] \end{bmatrix} [I_4] \otimes [C^{\mathrm{II}}_{M/2}]$$

$$\times \begin{bmatrix} [I_{M/2}] & & & 0 \\ & [I_{M/2}] & & \\ & & [W_{M/2}]^{-1} & \\ 0 & & & [W_{M/2}]^{-1} \end{bmatrix} [H_4] \times [I_{M/2}]$$

$$\times (1/4) \begin{bmatrix} [p] \\ [q] \\ [r] \\ [s] \end{bmatrix} \begin{bmatrix} [I_{N/2}] & & & 0 \\ & [W_{N/2}]^{-1} & & \\ & & [I_{N/2}] & \\ 0 & & & [W_{N/2}]^{-1} \end{bmatrix} [I_4] \otimes [C^{\mathrm{II}}_{N/2}]^T$$

$$\times \begin{bmatrix} [I_{N/2}] & & & 0 \\ & [I_{N/2}] & & \\ & & [L_{N/2}]^T & \\ 0 & & & [L_{N/2}]^T \end{bmatrix}. \tag{5.3.14}$$

We have assumed in the foregoing discussion that M and N are integer powers of two. All component matrices have been defined previously. For a square $N \times N$ input data matrix the computation of DCT-II based on Haque's algorithm [FID-6] requires $(3/4)N^2 \log_2 N$ multiplications and $3N^2 \log_2 N - 2N^2 + 2N$ additions.

5.4 Computation via Two-Dimensional DFT

The close relationship between DCT-II and DFT can also be exploited in the two-dimensional case. What Narasimha and Peterson [FF-1] did for the one-dimensional DCT can be extended to the two-dimensional data matrix. Nasrabadi and King [FID-4] and Vetterli [FID-3] provided the necessary derivations. We briefly describe the procedure here.

Recall that the two-dimensional DCT-II of an input data matrix $[\tilde{g}]$ is given by

$$\tilde{G}_{u,v} = \sum_{m=0}^{M-1} \sum_{n=0}^{N-1} \tilde{g}_{m,n} C_{2M}^{(2m+1)u} C_{2N}^{(2n+1)v}, \tag{5.4.1}$$

where $u = 0, 1, \ldots, M-1$ and $v = 0, 1, \ldots, N-1$, and where M and N are both integer powers of two.

As shown by Nasrabadi and King [FID-4], a rearrangement of the input matrix elements easily leads to expressions involving evaluation of two-dimensional DFTs. The reordering is given by

$$\begin{aligned}
\tilde{y}_{m,n} &= \tilde{g}_{2m,2n}, && m = 0, 1, \ldots, \frac{M}{2} - 1; \quad n = 0, 1, \ldots, \frac{N}{2} - 1; \\
&= \tilde{g}_{2M-2m-1,2n}, && m = \frac{M}{2}, \ldots, M - 1; \quad n = 0, 1, \ldots, \frac{N}{2} - 1; \\
&= \tilde{g}_{2m,2N-2n-1}, && m = 0, 1, \ldots, \frac{M}{2} - 1; \quad n = \frac{N}{2}, \ldots, N - 1; \\
&= \tilde{g}_{2M-2m-1,2N-2n-1}, && m = \frac{M}{2}, \ldots, M - 1; \quad n = \frac{N}{2}, \ldots, N - 1.
\end{aligned} \tag{5.4.2}$$

Let $\tilde{Y}_{u,v}$ be the two-dimensional DFT of $\tilde{y}_{m,n}$ given in (5.4.2) such that

$$\tilde{Y}_{u,v} = \sum_{m=0}^{M-1} \sum_{n=0}^{N-1} \tilde{y}_{m,n} W_M^{mu} W_N^{nv}, \tag{5.4.3}$$

where $W_M^k = \exp[-j2\pi k/M]$.

Then, using simple compound angle formulae for the cosine functions in (5.4.1), it is possible to rewrite (5.4.1) so that

$$\tilde{G}_{u,v} = \left(\frac{1}{2}\right) \mathrm{Re}\{W_{4M}^u [W_{4N}^v \tilde{Y}_{u,v} + W_{4N}^{-v} \tilde{Y}_{u,N-v}]\}. \tag{5.4.4}$$

Equation (5.4.4) is sometimes referred as representing "phasor-modified" DFT components. The terms W_{4M}^u, W_{4N}^v, and W_{4N}^{-v} are referred to as phasors. Equation (5.4.4) may be further reduced to

$$\tilde{G}_{u,v} = \left(\frac{1}{2}\right) \mathrm{Re}\{\tilde{A}_{u,v} + j\tilde{A}_{u,N-v}\},$$

where

$$\tilde{A}_{u,v} = W_{4M}^{u} W_{4N}^{v} \tilde{Y}_{u,v}, \tag{5.4.5}$$

which clearly illustrates the idea of "phasor-modification".

Since $\tilde{Y}_{u,v}$ is the two-dimensional DFT, its implementation can be realized using any of the available two-dimensional algorithms. One of the most efficient methods is to use the idea of Nussbaumer [FID-2] and to compute the two-dimensional real DFT by means of the polynomial transforms. We briefly describe the procedure here.

Consider the evaluation of $\tilde{A}_{u,v}$ in (5.4.5), where we explicitly include the definition for $\tilde{Y}_{u,v}$ from (5.4.3). Thus,

$$\tilde{A}_{u,v} = W_{4M}^{u} \sum_{m=0}^{M-1} \sum_{n=0}^{N-1} \tilde{y}_{m,n} W_{4M}^{4mu} W_{4N}^{(4n+1)v}. \tag{5.4.6a}$$

Since $(4n + 1)$ is odd and M is a power of two, it is valid to permute m by $(4n + 1)$ modulo M. This modifies (5.4.6a) to

$$\tilde{A}_{u,v} = W_{4M}^{u} \sum_{m=0}^{M-1} \left\{ \sum_{n=0}^{N-1} \tilde{y}_{m(4n+1),n} W_{4N}^{(4n+1)v} \right\} W_{4M}^{4m(4n+1)u}. \tag{5.4.6b}$$

If we define the M odd DFTs [of N points, as described in the inner summation of (5.4.6b)] as

$$\tilde{Y}_{m,v} = \sum_{n=0}^{N-1} \tilde{y}_{m(4n+1),n} W_{4N}^{(4n+1)v}, \tag{5.4.7}$$

then these DFT outputs can be organized into M polynomials of N terms with

$$\tilde{Y}_{m}(z) = \sum_{v=0}^{N-1} \tilde{Y}_{m,v} z^{v} = \sum_{v=0}^{N-1} \left\{ \sum_{n=0}^{N-1} \tilde{y}_{m(4n+1),n} W_{4N}^{(4n+1)v} \right\} z^{v}, \tag{5.4.8}$$

where z is a complex variable. The polynomials in (5.4.8) are unchanged when they are multiplied by $z^{d} W_{4N}^{(4n+1)d}$ modulo $(z^{N} - j)$, where $d < N$ is a constant. (For details, see Nussbaumer [FID-2].) Thus we note that a multiplication of the polynomials by $W_{4N}^{(4n+1)d}$ is equivalent to multiplication by z^{-d} modulo $(z^{N} - j)$. Now, returning to (5.4.6b) for the computation of $\tilde{A}_{u,v}$, we note that the polynomial coefficients are being multiplied by the factor $W_{4M}^{4m(4n+1)u}$. Now, for $N \geqslant M/4$, we have

$$W_{4M}^{4m(4n+1)u} = W_{4N}^{4N(4n+1)mu/M} = z^{-4Nmu/M} \bmod (z^{N} - j). \tag{5.4.9}$$

Thus, using the polynomials $\tilde{Y}_{m}(z)$, we can construct the new polynomials,

$$\tilde{Y}_{u}(z) = \sum_{m=0}^{M-1} \tilde{Y}_{m}(z)\; z^{-4Nmu/M} \bmod (z^{N} - j). \tag{5.4.10}$$

Comparing this with (5.4.6b) and using (5.4.8) and (5.4.9) shows that $\tilde{A}_{u,v}$ are the coefficients in the polynomials in (5.4.10) given by

$$\tilde{Y}_{u}(z) = \sum_{v=0}^{N-1} \tilde{A}_{u,v} W_{4M}^{-u} z^{v}. \tag{5.4.11}$$

The main point of this discussion is that the second DFT required by (5.4.6b) along the index m can be replaced by the polynomial transforms and complex multiplication given by (5.4.10) and (5.4.11). This approach necessitates the following steps:

(1) Reorder the input data as in (5.4.2).
(2) Permute the input order and compute the M N-point DFTs in (5.4.7).
(3) Multiply the DFT output by W_{4N}^{v} as in (5.4.8). [Note that this factor is a root of $(z^N - j) = 0$.]
(4) Compute the inverse polynomial transform of (5.4.10).
(5) Multiply the inverse polynomial transform output by W_{4M}^{u} to obtain $\tilde{A}_{u,v}$ in (5.4.11).
(6) Substitute in (5.4.5) to get $G_{u,v}$.

The reduction in computational complexity is obtained by mapping the DFT on the index m to polynomial transform. Overall, an $M \times N$ point DCT is mapped onto M DFTs of lengths N, plus 2 MN complex multiplications and 3 MN complex additions. Further reduction is possible in the special case of a square input data matrix when $M = N$ (see [FID-3]). In this case a real $N \times N$ DCT requires

$$\left(\frac{N^2}{2}\right) \log_2 N + \frac{N^2}{3} - 2N + \frac{8}{3} \text{ multiplications}$$

and

$$\left(\frac{5N^2}{2}\right) \log_2 N + \frac{N^2}{3} - 6N + \frac{62}{3} \text{ additions.} \qquad (N > 4)$$

5.5 Two-Dimensional DCT via WHT

As discussed in Section 4.5.1, the computation of the DCT coefficients for a one-dimensional data vector can be accomplished through the use of a conversion matrix applied to the Walsh–Hadamard transform (WHT) [B-1, B-2] of the data vector. In the same spirit, a two-dimensional data matrix may be processed by first subjecting it to a two-dimensional WHT. The resulting transformed output matrix may then be converted using some precomputed conversion matrix. This particular approach exploits the efficiency of the WHT, which requires no multiplications, and the fact that the conversion matrix can be precomputed and stored. One additional advantage is that the $(2^M \times 2^M)$ DCT can be computed from the $(2^{M-1} \times 2^{M-1})$ DCT, providing further reduction in the computational complexity. We briefly outline the idea in what follows.

Let $[\tilde{g}_N]$ and $[\tilde{G}_N]$ denote the properly scaled and normalized two-dimensional input data matrix and its DCT of size $N \times N$, respectively. Then

$$[\tilde{G}_N] = [C_N^{\text{II}}][\tilde{g}_N][C_N^{\text{II}}]^T, \qquad (5.5.1)$$

where the subscript N is used explicitly to denote that the matrices involved are square and N is assumed to be an integer power of two.

If we now define the two-dimensional WHT of $[\tilde{g}_N]$ as $[\tilde{Y}_N]$, given by

$$[\tilde{Y}_N] = [H_N][\tilde{g}_N][H_N]^T, \tag{5.5.2}$$

where $[H_N]$ is an $N \times N$ Walsh–Hadamard matrix, then $[\tilde{G}_N]$ and $[\tilde{Y}_N]$ are related by a conversion matrix $[CM_N]$ such that

$$[\tilde{G}_N] = [CM_N][\tilde{Y}_N][CM_N]^T. \tag{5.5.3}$$

It is then obvious that $[CM_N]$ is a matrix given by

$$[CM_N] = [C_N^{\text{II}}][H_N]^T. \tag{5.5.4}$$

Note that (5.5.3) is the two-dimensional version of (4.5.2) and $[CM_N]$ is defined in exactly the same way. As noted in Section 4.5.1, the conversion matrix $[CM_N]$ has block diagonal structure if the rows and columns in $[\tilde{G}_N]$ and $[\tilde{Y}_N]$ are arranged in bit-reversed order (BRO). Let us use the subscript BRO to indicate bit reversal in both the rows and columns of a matrix and the circumflex ^ to denote bit reversal of rows only. Then

$$[\tilde{G}_N]_{\text{BRO}} = [\hat{C}_N^{\text{II}}][\hat{H}_N]^T[\hat{H}_N][\tilde{g}_N][\hat{H}_N]^T[\hat{H}_N][\hat{C}_N^{\text{II}}]^T$$

or

$$[\hat{C}M_N] = [\hat{C}_N^{\text{II}}][\hat{H}_N]^T. \tag{5.5.5}$$

Note the difference in notations here as compared with the one-dimensional case in Section 4.5.1. Equation (5.5.5) defines exactly the same conversion matrix given in (4.5.2) with the subscript BRO. We refer the reader to Section 4.5.1 for an explicit example of $[\hat{C}M_8]$.

Using the BRO, $[\hat{C}M_N]$ has the block diagonal structure given by

$$[\hat{C}M_N] = \begin{bmatrix} [\hat{C}M_{N/2}] & 0 \\ 0 & [\text{RLH}_{N/2}] \end{bmatrix}, \tag{5.5.6}$$

where $[\text{RLH}_{N/2}]$ is used to denote the entire right lower half of $[\hat{C}M_N]$.

Since the Walsh–Hadamard matrix $[\hat{H}_N]$ is recursive, such that

$$[\hat{H}_N] = \begin{bmatrix} [\hat{H}_{N/2}] & [\hat{H}_{N/2}] \\ [\hat{H}_{N/2}] & -[\hat{H}_{N/2}] \end{bmatrix}, \tag{5.5.7}$$

it is apparent that if one partitions the input data matrix $[\tilde{g}_N]$ into four submatrices of size $N/2 \times N/2$, the resulting equations will represent transforms of order $N/2 \times N/2$ instead of $N \times N$. This provides the possibility of reducing the computational complexity as well as preserving the recursivity for the algorithm.

Let $[\tilde{g}_N]$ be partitioned so that

$$[\tilde{g}_N] = \begin{bmatrix} [\tilde{g}_{11}] & [\tilde{g}_{12}] \\ [\tilde{g}_{21}] & [\tilde{g}_{22}] \end{bmatrix}, \tag{5.5.8}$$

where $[\tilde{g}_{ij}]$ are $N/2 \times N/2$ square data matrices. Then the WHT of $[\tilde{g}_N]$ may also be partitioned so that

$$[\tilde{Y}_N]_{\text{BRO}} = \begin{bmatrix} [\tilde{Y}_{11}] & [\tilde{Y}_{12}] \\ [\tilde{Y}_{21}] & [\tilde{Y}_{22}] \end{bmatrix}, \tag{5.5.9a}$$

where

$$[\tilde{Y}_{11}] = [\tilde{g}_{11}]_{\text{W}} + [\tilde{g}_{12}]_{\text{W}} + [\tilde{g}_{21}]_{\text{W}} + [\tilde{g}_{22}]_{\text{W}},$$

$$[\tilde{Y}_{12}] = [\tilde{g}_{11}]_{\text{W}} - [\tilde{g}_{12}]_{\text{W}} + [\tilde{g}_{21}]_{\text{W}} - [\tilde{g}_{22}]_{\text{W}},$$

$$[\tilde{Y}_{21}] = [\tilde{g}_{11}]_{\text{W}} + [\tilde{g}_{12}]_{\text{W}} - [\tilde{g}_{21}]_{\text{W}} - [\tilde{g}_{22}]_{\text{W}},$$

and

$$[\tilde{Y}_{22}] = [\tilde{g}_{11}]_{\text{W}} - [\tilde{g}_{12}]_{\text{W}} - [\tilde{g}_{21}]_{\text{W}} + [\tilde{g}_{22}]_{\text{W}}. \tag{5.5.9b}$$

Here, the subscript W denotes the $N/2 \times N/2$ two-dimensional WHT of the data matrix arranged in BRO, for example,

$$[\tilde{g}_{ij}]_{\text{W}} = [\hat{H}_{N/2}][\tilde{g}_{ij}][\hat{H}_{N/2}]^T.$$

Finally, if the DCT of $[\tilde{g}_N]$, arranged in BRO, i.e., $[\tilde{G}_N]_{\text{BRO}}$, is partitioned also, so that

$$[\tilde{G}_N]_{\text{BRO}} = \begin{bmatrix} [\tilde{G}_{11}] & [\tilde{G}_{12}] \\ [\tilde{G}_{21}] & [\tilde{G}_{22}] \end{bmatrix}, \tag{5.5.10}$$

then using (5.5.5) and (5.5.6), we obtain the relations between $[\tilde{G}_{ij}]$ and $[\tilde{Y}_{ij}]$ as

$$[\tilde{G}_{11}] = [\hat{C}M_{N/2}][\tilde{Y}_{11}][\hat{C}M_{N/2}]^T,$$

$$[\tilde{G}_{12}] = [\hat{C}M_{N/2}][\tilde{Y}_{12}][\text{RLH}_{N/2}]^T,$$

$$[\tilde{G}_{21}] = [\text{RLH}_{N/2}][\tilde{Y}_{21}][\hat{C}M_{N/2}]^T,$$

and

$$[\tilde{G}_{22}] = [\text{RLH}_{N/2}][\tilde{Y}_{22}][\text{RLH}_{N/2}]^T. \tag{5.5.11}$$

Note that in (5.5.9) only $(N/2 \times (N/2)$ two-dimensional WHTs are required. Thus (5.5.11) represents the computationally intensive part of this approach. Kashef and Habibi [DW-4] have shown that this approach has an advantage over the conventional direct two-dimensional DCT up to $N = 32$.

5.6 Hardware Implementation of DCT Processor

Recently there has been a flurry of activities around the world focusing on the development of VLSI and chip implementation of the two-dimensional DCT (see Appendix B.1 for a list of DCT VLSI chip manufacturers). Its relations to the KLT, its energy compaction ability, and its fast algorithms have resulted in almost universal recognition. International standards of image processing in both TV and still-image transmission (see Chapter 7) have included two-dimensional DCT as a standard processing component in many applications. This has provided the impetus for laboratories around the world to develop hardware for its implementation.

As distinct from our discussions on two-dimensional algorithms for DCT in the foregoing sections, most, if not all, of the efforts of hardware development has been concentrated on basic block sizes of 8×8 or 16×16. These have been found (see Chapter 7) to provide sufficient details and localized activities of the picture to enable reasonable adaptive processing of the images. Because of the small sizes, much of the chip development work has not included the mapping of fast, two-dimensional (or even one-dimensional) algorithms onto silicon. Instead, regularity and layout seem to be the prime concern, together with a realistic throughput rate for real-time implementations. There have, however, been attempts to map Lee's algorithm (see Chapter 4 and [FRA-3, FRA-4]) onto silicon. As well, chips based on a single processing element (PE) of rotation (see [DP-22, DP-27, FRA-23]) are also being developed.

In this section we shall briefly describe some recent work in this area. It is not meant to be either exhaustive or complete. The purpose is to provide some highlights in this area of development and to give credence to the claim that DCT is indeed an important processing tool in many applications. For more detailed information, the reader is directed to the papers listed under DCT processors (DP) in the references. Table B.1.1 in Appendix B.1 also contains a more exhaustive list of DCT chip manufacturers.

Before beginning this brief description, it is appropriate to note the one common approach of all the chip developments. All the two-dimensional DCT processors developed so far have made use of the separability property of the two-dimensional DCT. None have attempted to directly map a specific two-dimensional DCT algorithm to silicon. Except for the actual means of implementation, Fig. 5.5 shows a block diagram summarizing this "row-column" transform approach to realize the two-dimensional DCT.

Variations on this basic block structure are many. Some use a single, one-dimensional DCT processor to perform both row and column transformations. Others use special devices for the transposition operation. We shall not be

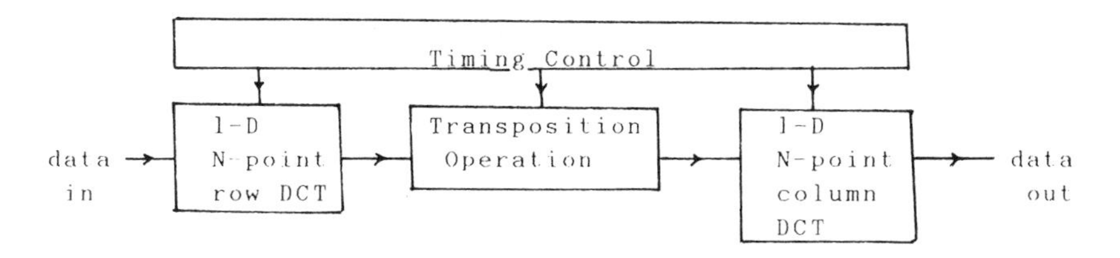

Fig. 5.5 Block diagram of the row-column approach for 2D DCT.

concerned with these details. In the hardware implementations described below, we shall briefly touch on the highlights of the processor chips. These chips are chosen to be representative of the efforts in industry and university research laboratories around the world. They are identified by their institution and location, but are not arranged in any particular order.

5.6.1 *Linkoping University, Linkoping, Sweden* [*DP-18, DP-38*]

Afghahi *et al.* [DP-18] and Sikstrom *et al.* [DP-38] reported the development of an array processor for the two-dimensional modified symmetric discrete cosine transform (MSDCT) (or modified DCT-I according to the definition in Chapter 2). This two-dimensional MSDCT is defined by

$$[G]_{uv} = \sum_{m=0}^{N-1} \sum_{n=0}^{N-1} c_m c_n [g]_{mn} \cos\left[\frac{mu\pi}{(N-1)}\right] \cos\left[\frac{nv\pi}{(N-1)}\right],$$

$$u, v = 0, 1, \ldots, N-1; \tag{5.6.1}$$

where $[g]$ and $[G]$ are the $N \times N$ data image and transformed image, respectively, and c_k are scale factors such that

$$c_k = \tfrac{1}{2} \quad \text{for } k = 0 \text{ or } N-1,$$
$$= 1 \quad \text{otherwise.}$$

This particular transform is chosen because of its symmetry. The inverse and forward transforms are identical. However, unlike the other DCTs, it is only "approximately" orthogonal. Distributed arithmetic is used in the processing elements (PE). Coefficients for the 16-point transform are contained in a look-up table stored in ROM. Thus, the processor consists of only shift-accumulators. Figure 5.6 shows the block diagram for an N-point one-dimensional DCT. We

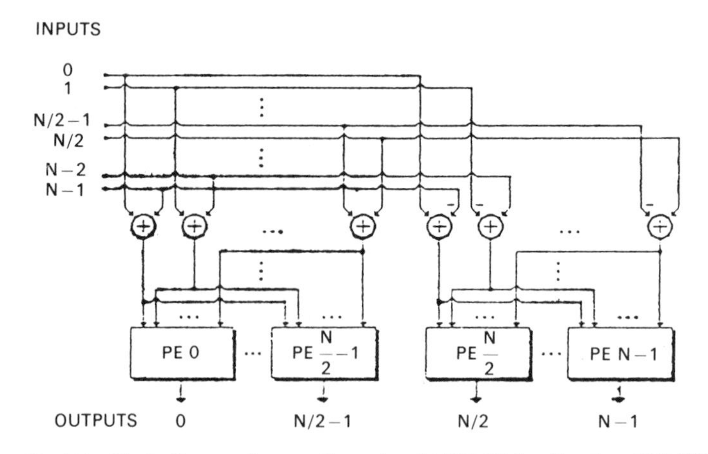

Fig. 5.6 Block diagram for one-dimensional MSDCT for N points [DP-38].

note that this represents computation of inner products and is not a mapping of any special fast algorithm.

Table 5.1 summarizes some technical specifications.

Table 5.1

Specifications for a 16×16 DCT chip [DP-38].

Technology	3μ CMOS bimetallic
Die area	5.4 mm × 6.4 mm
No. of devices	42,000
Clock rate	25 MHz

It is claimed that for a TV image of 512×768 pels, the chip is capable of processing 30 images per second in real time.

5.6.2 *SGS–Thomson Microelectronics, Grenoble, France* [DP-28, DP-33, DP-41]

Jutand *et al.* [DP-28] and Artieri *et al.* [DP-33] reported a multiformat VLSI chip for real-time two-dimensional DCT (more specifically, DCT-II). Again the row-column approach is taken and the algorithm used for the one-dimensional DCT is the fast recursive one reported by Lee [FRA-3, FRA-4]. The implementation exploits the recursive property of this algorithm so that DCT computation of various lengths can be handled by the same hardware. The different sizes of image blocks that can be processed by this chip are 4×4, 4×8, 8×4, 8×8, 8×16, 16×8, and 16×16. The chip is said to be capable of 420 million additions per second using 16-bit accuracy.

In Fig. 5.7, the detailed architecture of the two-dimensional DCT processor chip is shown. The serial operative part consists of a direct mapping of Lee's algorithm, and the transposition memory stores and transposes the DCT coefficients for the row-column approach of the two-dimensional DCT. Table 5.2 contains the technical specifications for this 16×16 chip.

Table 5.2

Specifications for the multiformat VLSI chip for 2D DCT [DP-33].

Technology	1.25 μ CMOS bimetallic
Die area	5.39 mm × 7.53 mm
No. of devices	114,300
Max. speed	13.5 MHz

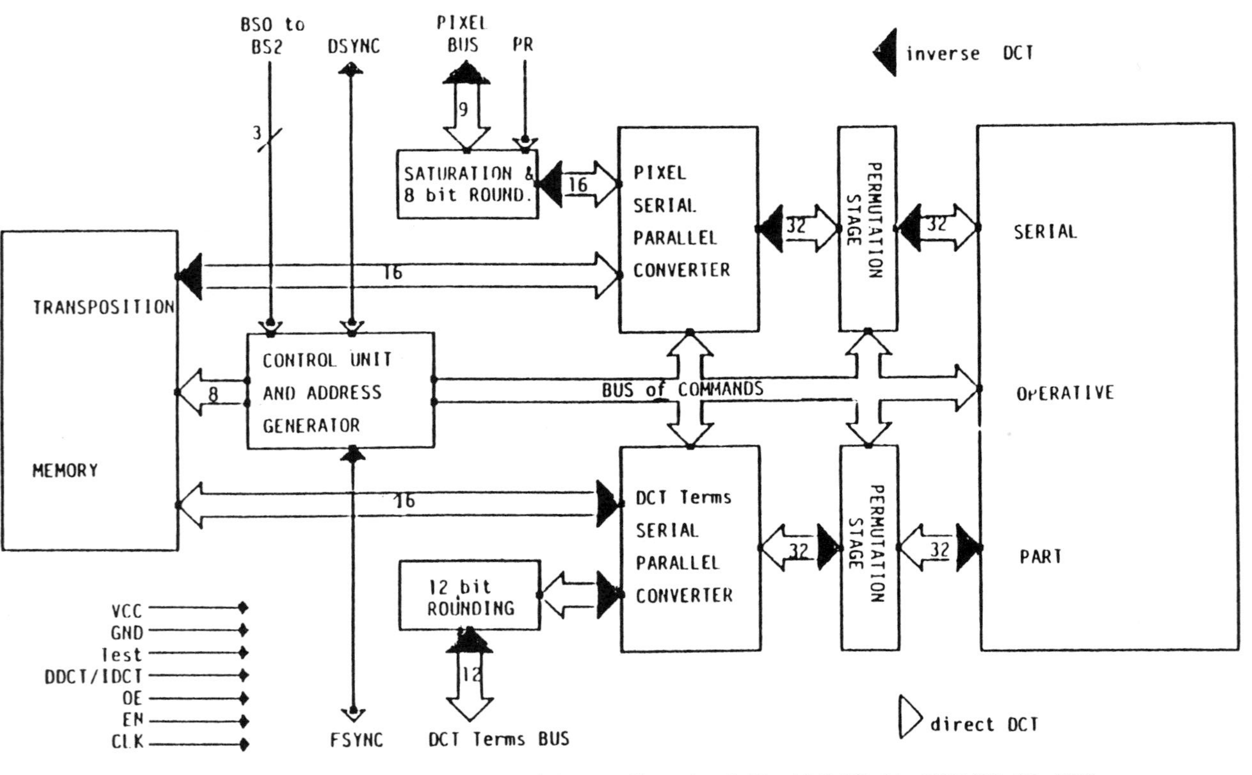

Fig. 5.7 Detailed architecture of the two-dimensional 16×16 DCT chip [DP-33]. (© 1988 IEEE).

5.6.3 *Bell Communications Research Inc., New Jersey, U.S.A.* [DP-29, DP-34, DP-37, DP-39, DP-49, DP-56]

Sun *et al.* [DP-29, DP-39, DP-56], Chen *et al.* [DP-34], and Gottlieb *et al.* [DP-49] have reported, at various times, a concurrent architecture for VLSI implementation of DCT. As in Section 5.6.1, distributed arithmetic is used in the PEs, together with bit-serial and bit-parallel data structures, to implement vector inner products concurrently. The regularity provided by this approach has resulted in an implementation that is a direct matrix-vector multiplication. No fast algorithm of any type is mapped. The circuit requires only memories, adders, and shift registers. Precalculated look-up tables for specific sizes of transforms are stored in ROMs. The ROM and accumulator (RAC) performs the matrix-vector multiplications required in the two-dimensional DCT. Figure 5.8a shows the concurrent architecture for a 16×1 DCT. Here $x_{km}^{(j)}$ is the jth bit of the km-element of the data matrix and y_{km} is the km-element of the transform output. Figure 5.8b shows the block structure of a RAC for a 16×1 DCT. The mask layout of the VLSI chip is shown in Fig. 5.8c.

Table 5.3

Specifications for a concurrent VLSI chip for 2D DCT [DP-29].

Technology	2 μ CMOS bimetallic
Die area	8.3 mm × 8.1 mm
No. of devices	73,000
Max. speed	15.1 MHz

5.6.4 *Centre Commun d'Etudes de Telediffusion et de Telecommunications, Cesson-Sevigne, France* [DP-44]

Recently, Carlach *et al.* [DP-44] reported a two-dimensional DCT chip for an 8×8 data block. The row-column approach is again used, and the one-dimensional 8-point DCT is based on the algorithm of Duhamel and H'Mida [DP-21]. The general formulation is related to cyclic convolution. However, for the $N = 8$ case, the cyclic convolution is reduced to a block matrix factorization given by the following equation:

$$\begin{bmatrix} u(0) \\ u(4) \\ u(2) \\ u(6) \\ u(1) \\ u(5) \\ u(3) \\ u(7) \end{bmatrix} = \begin{bmatrix} a & a & a & a & & & & \\ a & -a & -a & a & & 0 & & \\ f & -g & g & -f & & & & \\ g & f & -f & -g & & & & \\ & & & & b & d & c & e \\ & & & & d & e & -b & c \\ & & 0 & & c & -b & -e & -d \\ & & & & e & c & -d & -b \end{bmatrix} \begin{bmatrix} x(0)+x(7) \\ x(2)+x(5) \\ x(1)+x(6) \\ x(3)+x(4) \\ x(0)-x(7) \\ x(2)-x(5) \\ x(1)-x(6) \\ x(3)-x(4) \end{bmatrix}, \tag{5.6.2}$$

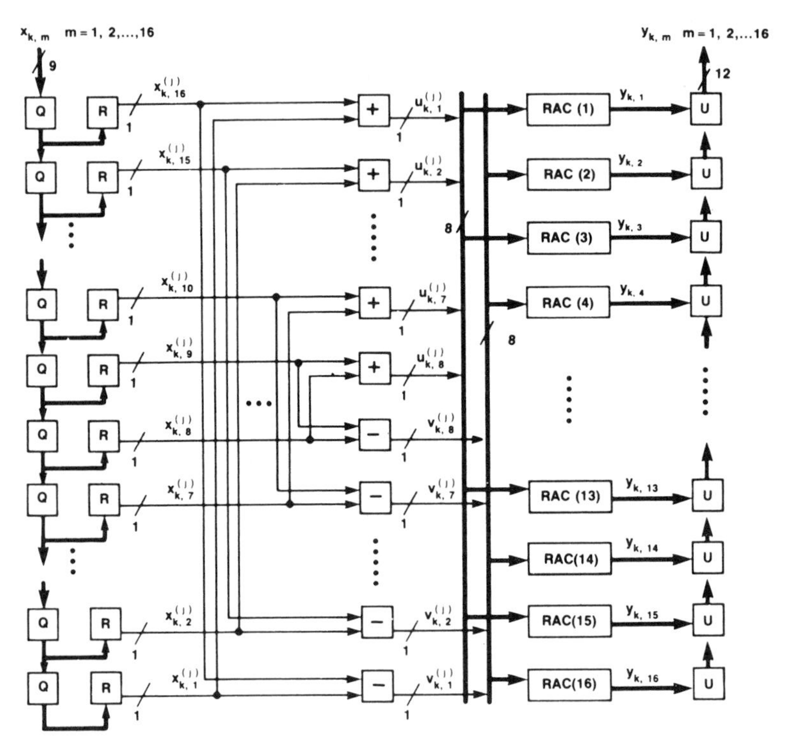

Fig. 5.8a Concurrent architecture of a 16 × 1 DCT [DP-29].

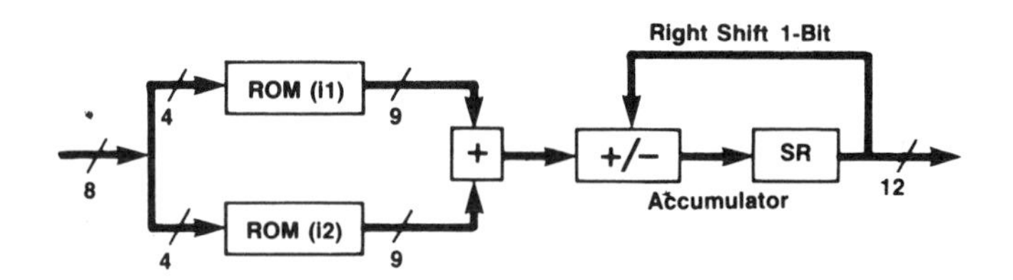

Fig. 5.8b Block structure of an RAC [DP-29].

where $x(m)$ and $u(m)$ are the mth components of the input and output vectors of the transform, respectively, and $a, b, \ldots, f, g$ are the DCT coefficients. This one-dimensional transform is realized based on bit-serial computing. The coefficients are stored in ROM, and the multiplication is implemented with ROM and an accumulator similar to the one described in Section 5.6.3. Thus, the circuit consists of ROM, adders, and shift-registers only. Figure 5.9a shows the implementation of a one-dimensional 8-point DCT. The transposition operation required between the one-dimensional row DCT and the column DCT is

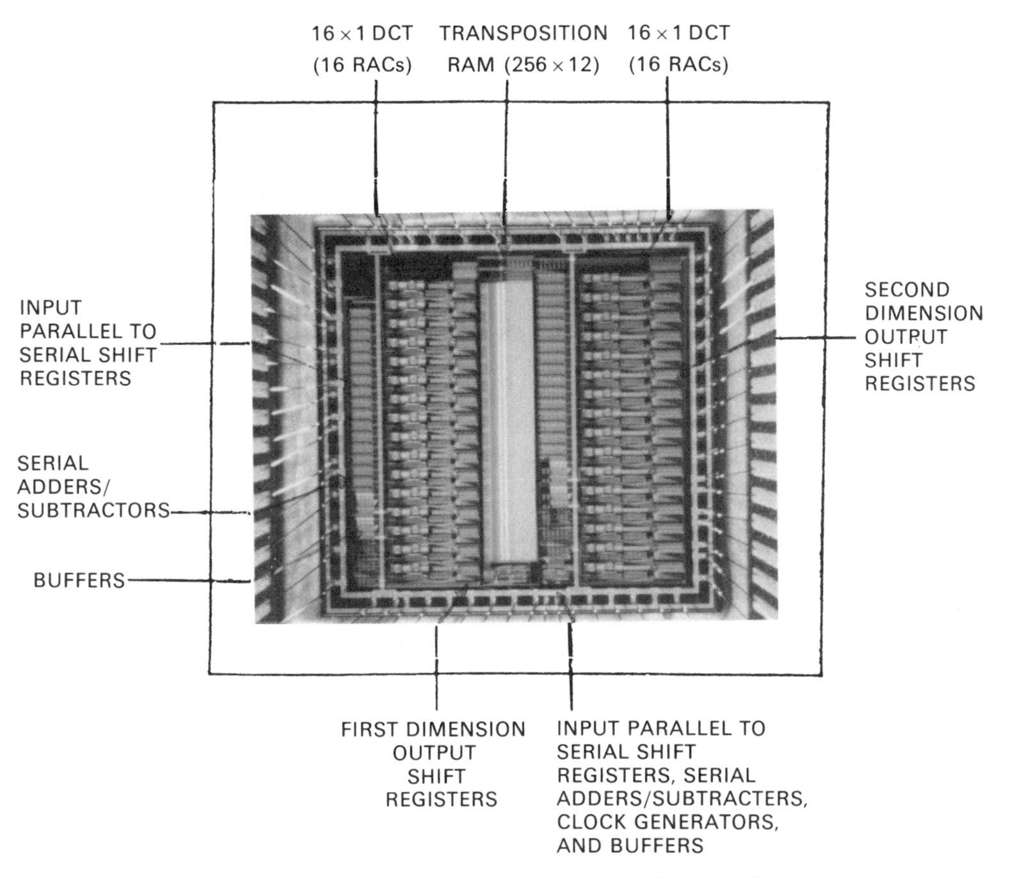

Fig. 5.8c Mask layout of the 16 × 16 DCT chip [DP-29].

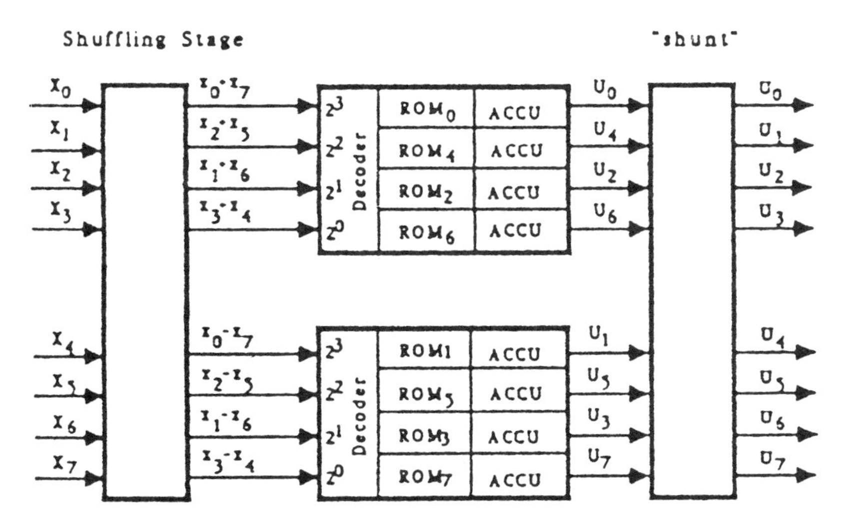

Fig. 5.9a Implementation of an 8-point DCT [DP-44] (© 1989 IEEE).

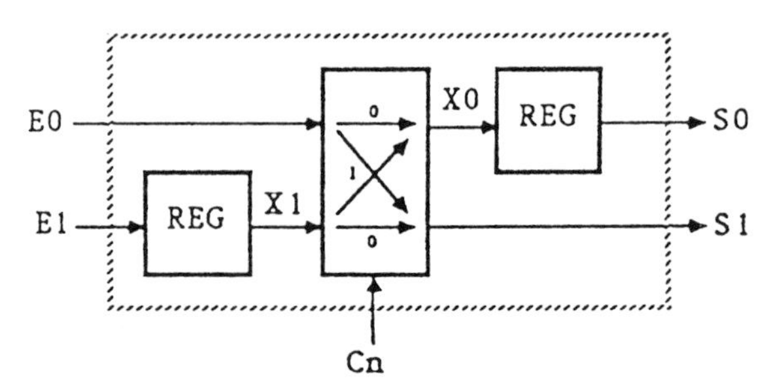

Fig. 5.9b An elementary transposition cell [DP-44] (© 1989 IEEE).

accomplished by using shift-registers. Figure 5.9b shows an elementary transposition cell for the 2×2 case.

By a simple modification of the upper 4×4 matrix in (5.6.2), IDCT can be computed using the same hardware, provided the "shunt" and "shuffle" stages in Fig. 5.9a are properly reversed. Table 5.4 contains the specifications for the 8×8 chip (referred to as the TCAD = circuit de *t*ransformee en *c*osines discrete en *a*rithmetique *d*istribuee).

Table 5.4

Specifications for 8×8 TCAD [DP-44].

Technology	1.2 μ CMOS
Die area	26 mm square
No. of devices	50,000
Clock rate	54 MHz (max. 27 MHz data rate)

5.6.5 *Philips Research Laboratories, Eindhoven, The Netherlands* [DP-46]

A different 8×8 two-dimensional DCT chip was reported by Matterne *et al.* [DP-46]. In this case, no specific fast algorithms or matrix factorizations are used. The transform is mapped on a very long instruction word (VLIW) multiprocessor architecture (see references [1] and [2] in [DP-44]). The resulting IC is claimed to be as efficient as a hard-wired datapath implementation while retaining the flexibility of a microprogrammable VLIW machine.

Figure 5.10 shows the IC architecture for the 8×8 two-dimensional DCT. Here, processor A and processor B perform the one-dimensional DCTs, and each contains four PEs: one multiplier, two adder/subtractors, and one rounder/limiter, all operating at a clock rate of 27 MHz.

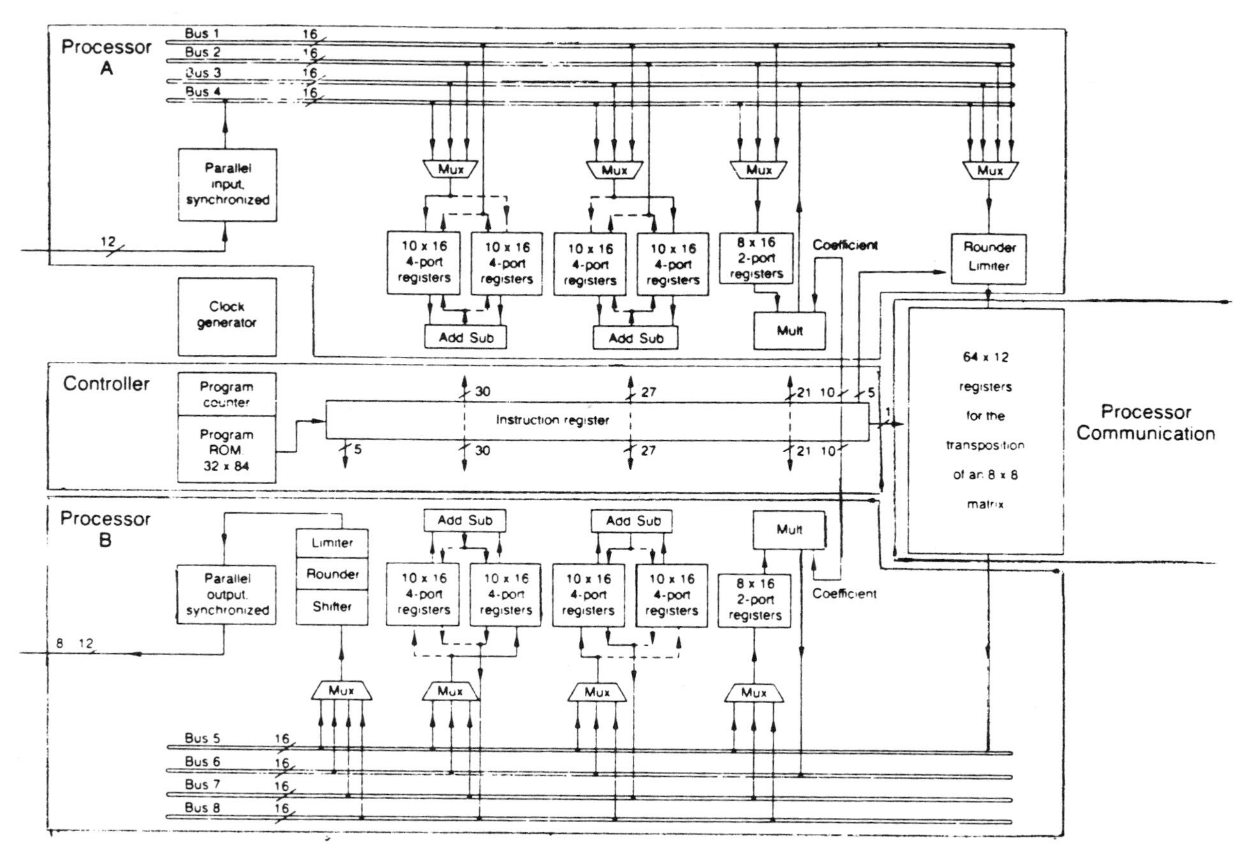

Fig. 5.10 IC architecture of the 8 × 8 DCT chip [DP-46]. (© 1989 IEEE).

Table 5.5

Specifications for an 8 × 8 VLIW chip [DP-46].

Technology	1.6 μ CMOS
Die area	7.5 mm × 7.8 mm
No. of devices	56,000
Clock rate	27 MHz

5.6.6 *Lincoln Laboratory, MIT, Massachusetts, U.S.A.* [DP-26, DP-31]

A novel analog two-dimensional DCT processor based on charge-coupled device (CCD) technology has been reported by Chiang [DP-26, DP-31]. The speed, power, weight, and throughput rate advantages of the CCD technology make it ideal for low-cost image transform codec. The transform is achieved via direct vector-matrix multiplication (VMP) by a CCD, while the DCT coefficients are stored as fixed weights in the device. A block diagram showing the CCD VMP device is shown in Figure 5.11a. Figure 5.11b shows the block diagram for the two-dimensional DCT processor. The device is claimed to have a 60-dB dynamic range and 40-dB harmonic distortion for a 16 × 16 data input.

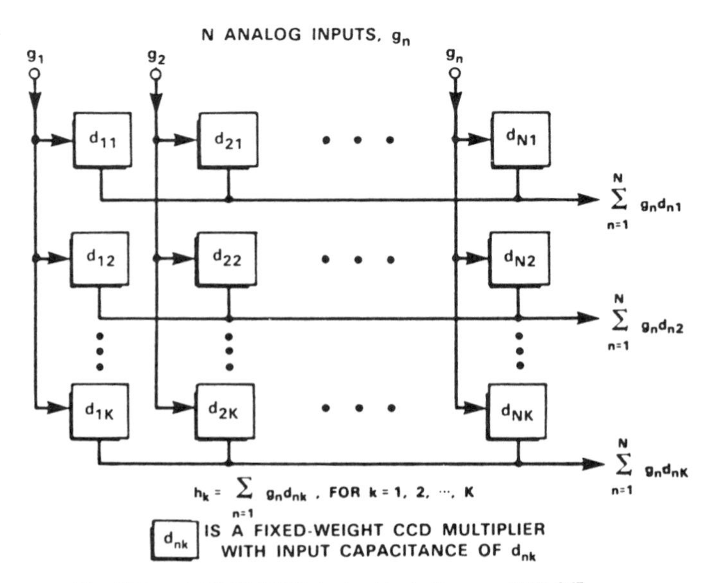

Fig. 5.11a A CCD VMP device block diagram [DP-26].

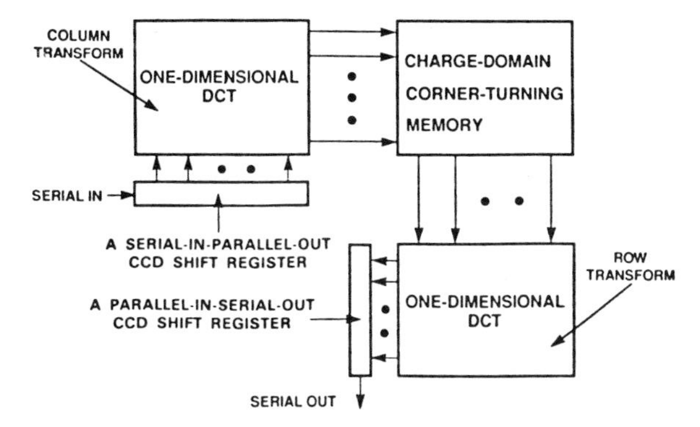

Fig. 5.11b Block diagram for the CCD 2D-DCT processor [DP-26]. Reprinted with permission of Lincoln Laboratory, Massachusetts Institute of Technology, Lexington, Massachusetts.

Table 5.6 summarizes the specifications for this device.

Table 5.6

Specifications for a CCD 2D DCT [DP-26].

Technology	CCD
Die area	8 mm × 8 mm
No. of devices	256 in 16 × 16 format
Clock rate	10 MHz

5.6.7 *Institut de Microtechniq‹, Universite de Neuchatel, Neuchatel, Switzerland* [DP-45]

Defilippis *et al.* [DP-45] at the University of Neuchatel, Switzerland reported a 16 × 16 DCT chip. The discrete cosine transform used is the MSDCT, identical to the one defined in (5.6.1). Distributed arithmetic is used for the computation of the matrix-vector product, and thus no fast algorithm is actually mapped

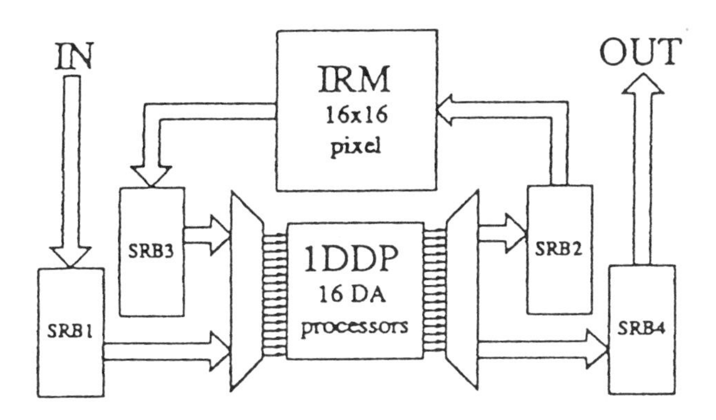

Fig. 5.12 Block diagram of the 2D-DCT 16 × 16 [DP-45].

onto silicon. The transform coefficients are stored in ROM, and the multiplications are achieved by adder/subtractor, accumulator, and shift-registers. These make up a distributed arithmetic (DA) processor and a 16-point one-dimensional DCT requires 16 such DAs. As shown in the block diagram (Fig. 5.12), the two-dimensional transform is accomplished by one-dimensional transforms in the row-column approach. In the diagram, SRBs are shift-register banks, IRM is the intermediate result memory, and the 1D DCT processor (1DDP) is the core of the chip. The transposition required is managed by the SRBs.

Table 5.7 gives the details of the chip.

Table 5.7

Specifications for a 16 × 16 DCT chip [DP-45].

Technology	2 μ CMOS bimetallic (CMN20a)
Core area	4.5 mm × 6.5 mm
No. of devices	95,000
Clock rate	16 MHz

5.6.8 *Universite Catholique, Louvian-la-Neuve, Belgium* [DP-48]

A multiprocessor chip based on a systolic array structure for matrix multiplication was reported by Denayer *et al.* [DP-48]. The chip is designed for general image processing, computing 8 × 8 matrix products. By cascading several such devices, two-dimensional DCT of arbitrary sizes can be computed. Figure 5.13 shows the systolic pipeline structure for computing matrix products.

The chip can be adapted to any matrix multiplication and is coefficient programmable. One distinct advantage of the systolic structure is the fact that no intermediate storage is required for the two successive matrix multiplications necessary in the computation of the two-dimensional DCT. Table 5.8 summarizes the specifications for the chip.

Table 5.8

Specifications for the systolic multiprocessor 8 × 8 2D DCT processor [DP-48].

Technology	2 μ CMOS bimetallic
Die area	9 mm × 8 mm
No. of devices	90,000
Clock rate	20 MHz

There are many other groups active in the development of hardware implementation of the two-dimensional DCT. The work of Rampa *et al.* [DP-47] at CNR Milano and Stanford University, Li *et al.* [DP-50] at Lehigh

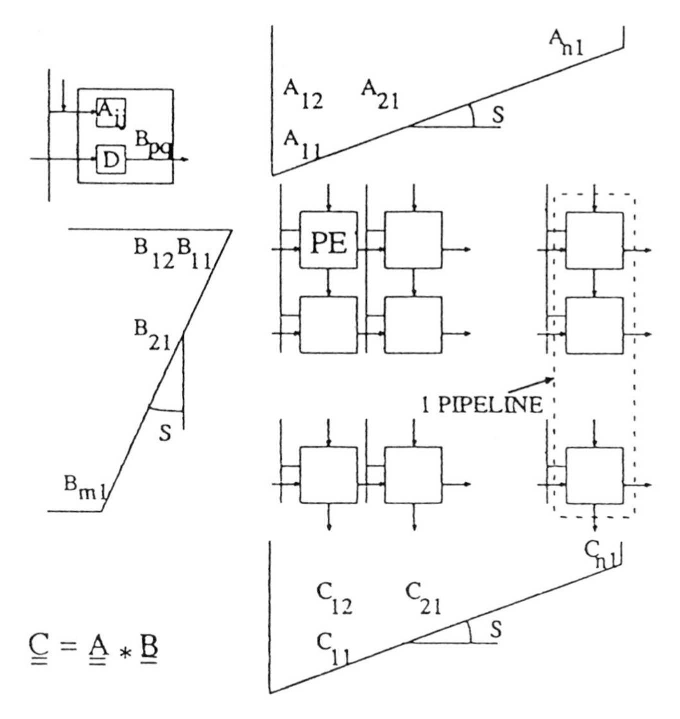

Fig. 5.13 Systolic pipeline structure for computing matrix products [DP-48]. (© 1988 IEEE).

University, Parkhurst *et al.* [DP-35] at University of California at Davis, and Ligtenberg *et al.* [DP-27] at AT&T Bell Labs, and Columbia University are worthy of mention. In addition, there may be other active groups of whom we are unaware.

5.7 Summary

In this chapter we have detailed some of the algorithms designed to implement the two-dimensional DCT. As is evident in our discussions, the most prominent property is the separability property, which is exploited both in the algorithms and in the chip designs.

It has been shown that the two-dimensional DCT can be implemented by reducing it to a lexicographically ordered one-dimensional transform that is partly recursive. More efficient algorithms result when the two-dimensional DCT is block decomposed and computations are carried out via the two-dimensional DFT. It is also possible to precompute the conversion matrix to obtain the DCT coefficients from the WHT coefficients. In Table 5.9, we briefly summarize the computational complexities of the different algorithms discussed in this chapter. As in Table 4.2, we also note the relative merits of the algorithms as we perceive them. It should be mentioned that these algorithms generally require a smaller number of multiplications compared with the direct application of cascaded one-dimensional DCT-II algorithms.

We have also briefly described some hardware implementations of the two-dimensional DCT. These are representative of what active research groups are doing in the chip design area. The list, however, is not meant to be either complete or exhaustive. It is safe to say that there will be great advances in this direction, given the technological changes in the area of digital devices. Also, it is perhaps worth mentioning the work of Guglielmo [FWE-3] in relation to the error behavior of a two-dimensional DCT, in which error propagation due to quantization and rounding is described. General references on finite word-length effects in hardware implementation of DCT are listed under [FWE-1] through [FWE-4]. Having described the fast algorithms for efficient implementation of one-dimensional (Chapter 4) and two-dimensional DCT, it is only logical to compare the performance of DCT with other discrete transforms based on some standard criteria. This aspect is described in Chapter 6.

Table 5.9

Computational complexity of two-dimensional DCT-II algorithms for an $N \times N$ block input.

Algorithm	Complexity	Advantages	Disadvantages
Lexicographical re-order 1D [FID-1]	$\alpha = 68$ $\mu = 24$ for 4×4	simple index mapping	only partially recursive
Block factors [FID-6]	$\alpha_N = 3N^2 \log N/4$ $\mu_N = 3N^2 \log N - 2N^2 + 2N$	recursive	moderately complex index map
Via 2D FFT [FID-3]	$\alpha_N = (5N^2/2) \log N + N^2/3 - 6N + 62/3$ $\frac{1}{2}N^2 \log N + N^2/3 - 2N + 8/3$	easy to use with any FFT	complex index map
Via WHT [DW-4]	depends on WHT used	easy to use with any WHT	slower for $N \geqslant 32$ conversion matrix needed

α represents the number of additions and μ represents the number of multiplications; log is to the base 2.

PROBLEMS

5.1. Verify (5.2.6).

5.2. Verify the identity following (5.3.4).

5.3. Prove (5.3.6).

5.4. Vector radix [M-36, M-39] and vector split radix algorithms [M-37, M-38] for the multidimensional DFT have been developed. Using the algorithms described in [FRA-3, FRA-4, FRA-8, FRA-9, FRA-14] develop the following for the 2D DCT-II:
(a) Vector radix-2 algorithms (both DIT and DIF),
(b) Vector split-radix algorithm (both DIT and DIF).

5.5. Obtain the matrix factors from the 2D-DCT signal flowgraph shown in Fig. 5.3. Show that these correspond to the 2D-DCT algorithm described in Section 5.2.

5.6. Develop a signal flowgraph for 2D-IDCT corresponding to that shown in Fig. 5.3. Obtain the corresponding matrix factors.

5.7. The signal flowgraph for the 2D-IDCT developed by Haque [FID-6] is shown in Fig. 5.4. Obtain the matrix factors from this flowgraph and show that these correspond to the 2D-DCT algorithm described in Section 5.3.

5.8. Develop a signal flowgraph for the 4×4 2D-DCT based on [FID-6]. Obtain the corresponding matrix factors.

CHAPTER 6

PERFORMANCE OF THE DCT

6.1 Introduction

In Chapter 3, the asymptotic equivalence of DCT-I and DCT-II to KLT (as $N \to \infty$ or as $\rho \to 1$, respectively) is shown. This equivalence property is also valid for DCT-III, as it is simply a transpose of DCT-II. As the KLT is optimal under a variety of criteria (see Section 3.2), it is not surprising that the DCTs perform well in the general area of data compression or bandwidth reduction. Apart from the DFT, DCT has been most widely used in digital signal and image processing (for details see Chapter 7). The object of this chapter is to compare the performance of the DCT with other well-known discrete transforms such as DST, slant transform (ST) [IC-57], DFT, WHT, discrete Legendre transform (DLT) [DW-5, BSI-7], C-matrix transform (CMT [IC-11], etc., based on variance distribution, energy packing efficiency, rate distortion, Wiener filtering, residual correlation, and maximum reducible bits (mrb). A similar comparison for some of these discrete transforms has been performed by Kekre and Solanki [PC-3].

6.2 Variance Distribution

In Section 3.2 it was shown that for the KLT, if only the first D of the N transform coefficients were utilized in reconstructing the original N-point sequence (assuming the eigenvectors are arranged such that their corresponding eigenvalues are in nonincreasing order, i.e., $(\mu_0 \geqslant \mu_1 \geqslant \mu_2 \geqslant \cdots \geqslant \mu_{N-1})$, then the MSE between the original signal $\mathbf{x}$ and its reconstructed signal $\hat{\mathbf{x}}$ is equal to the sum of the eigenvalues of the remaining $(N - D)$ eigenvectors [see (3.2.13)]. Since the trace of the auto-covariance matrix in the transform domain for any unitary transform is invariant, one can judge the performance of a discrete transform by its variance distribution for a random sequence governed by some specific probability distribution. It is therefore desirable to have a few transform coefficients with large variances (consequently the remaining transform coefficients will have small variances). The later transform coefficients can be discarded in the reconstruction process with negligible MSE between the original and reconstructed signals. This concept can be extended to multidimensional signals such as images, under the constraint that the statistics along each dimension can be independently defined (for purposes of mathematical tractability). This aspect will be described in detail in Chapter 7. Also, since the variances represent the energy or information content of the corresponding transform coefficients, the transform coefficients with large variances are candidates containing significant features in a pattern-recognition application. Further increase in the bit-rate reduction in a block quantization scheme can be achieved by not only discarding low variance transform coefficients, but also by resorting to a bit allocation scheme based on the variances or log of the variances of the remaining transform coefficients. These techniques have been extensively utilized in image coding. One standard comparison is based on the assumption that the random sequence $\mathbf{x}$ is governed by a first-order Markov process with the adjacent correlation coefficient ρ specified.

Following (3.2.9), the auto-covariance matrix in the transform domain is

$$\begin{aligned} A_{\mathrm{tr}} &= E[\mathbf{X}\mathbf{X}^T] \\ &= E([\phi(n)]\mathbf{x}\mathbf{x}^T[\phi(n)]^T) \\ &= [\phi(n)](E[\mathbf{x}\mathbf{x}^T])[\phi(n)]^T \\ &= [\phi(n)]A[\phi(n)]^T, \end{aligned} \tag{6.2.1}$$

where $[\phi(n)]$ is any $(2^n \times 2^n)$ orthogonal matrix. The diagonal elements of A_{tr} then are the variances of the transform coefficients $X_0, X_1, \ldots, X_{N-1}$, where $2^n = N$. The variance distribution based on (6.2.1) for various discrete transforms for a first-order Markov process when $\rho = 0.9$ for $N = 8$, 16, and 32 is shown in Figs. 6.1, 6.2, and 6.3, respectively. This is also listed in a tabular form in Tables 6.1 through 6.3. An inspection of these figures and tables reveals that the DCT-II variance distribution, in general, decreases most rapidly, indicating its effectiveness in data compression or feature selection. Also the variance distribution of the DCT-II is extremely close to that of the KLT, thus confirming its near optimality.

6.3 Energy Packing Efficiency (EPE)

Kitajima [M-19] first proposed the energy packing efficiency (EPE) for the WHT. This is defined as the energy portion contained in the first M of N transform coefficients, i.e.,

$$\mathrm{EPE}(M) = \frac{\sum_{p=0}^{M-1} E[X_p^2]}{\sum_{p=0}^{N-1} E[X_p^2]}. \tag{6.3.1}$$

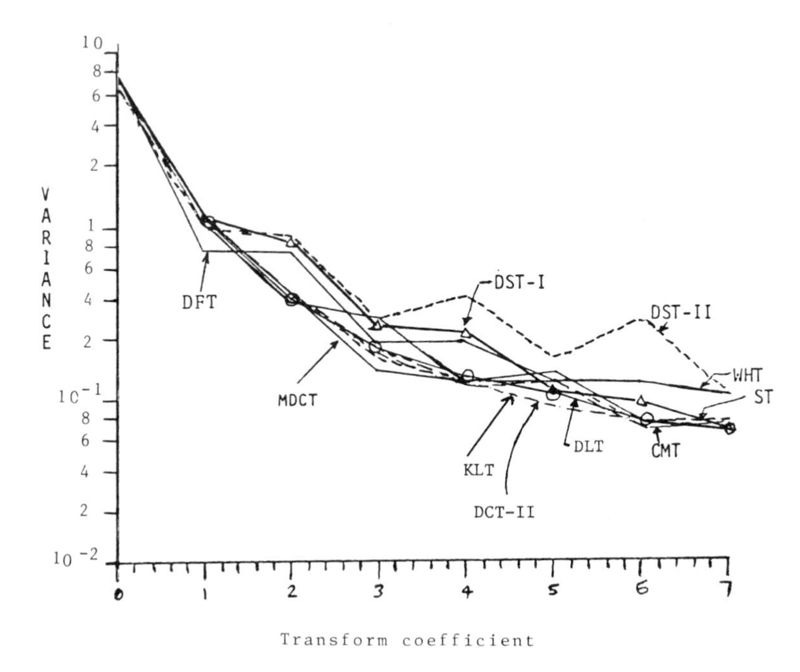

Fig. 6.1 The variance distribution for various discrete transforms for $N = 8$ and $\rho = 0.9$.

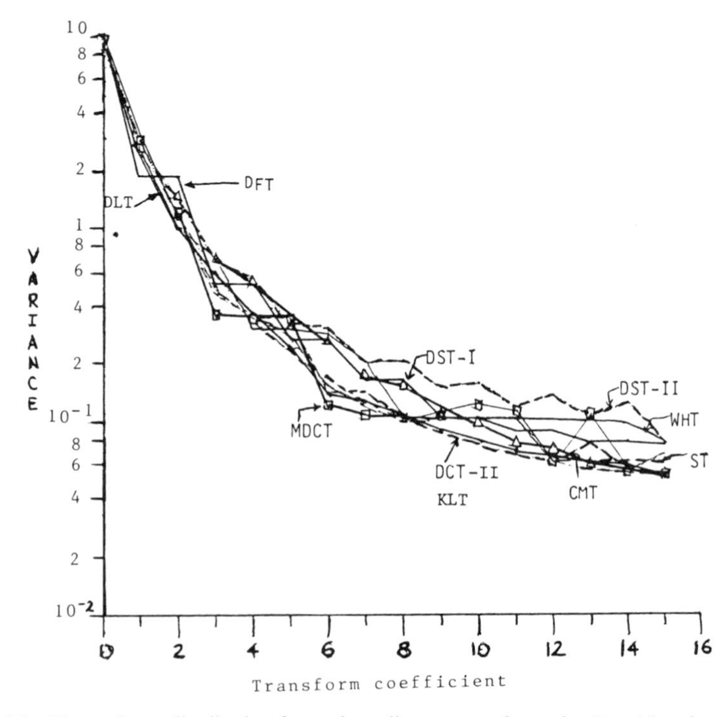

Fig. 6.2 The variance distribution for various discrete transforms for $N = 16$ and $\rho = 0.9$.

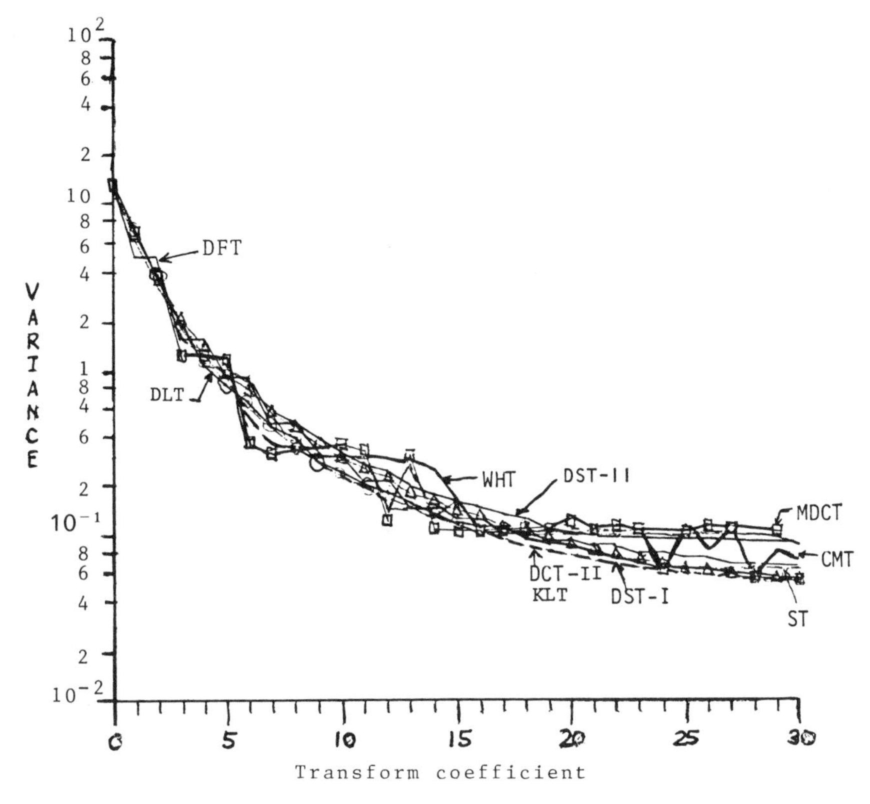

Fig. 6.3 The variance distribution for various discrete transforms for $N = 32$ and $\rho = 0.9$.

Table 6.1

Variance distribution for first-order Markov process defined by $\rho = 0.9$ and $N = 8$. i is the ith transform coefficient.

i	WHT	ST	DCT-II	DST-I	DST-II	CMT	DFT	DLT
0	6.185	6.185	6.185	5.76	5.384	6.185	6.185	6.185
1	0.863	0.989	1.006	0.931	0.823	0.992	0.585	0.989
2	0.305	0.345	0.346	0.661	0.74	0.346	0.585	0.332
3	0.246	0.146	0.166	0.222	0.246	0.153	0.175	0.173
4	0.105	0.105	0.105	0.197	0.333	0.100	0.175	0.113
5	0.104	0.104	0.076	0.093	0.145	0.105	0.103	0.083
6	0.103	0.063	0.062	0.08	0.240	0.057	0.103	0.067
7	0.088	0.063	0.055	0.056	0.088	0.062	0.088	0.057

Table 6.2

Variance distribution for first-order Markov process defined by $\rho = 0.9$ and $N = 16$. i is the ith transform coefficient.

i	HT	CHT	WHT	DCT-II	DFT	ST	(SHT)1	$(HHT)_1$	$(HHT)_2$	DST-I	DST-II	CMT	DLT
0	9.8346	9.8346	9.8346	9.8346	9.8346	9.8346	9.8346	9.8346	9.8346	9.218	8.914	9.835	9.8346
1	2.5464	2.5364	2.5360	2.9328	1.8342	2.8536	2.7765	2.5364	2.5364	2.64	2.458	2.885	2.8532
2	0.8638	0.8635	1.0200	1.2108	1.8342	1.1963	1.0208	1.0209	1.0209	1.468	1.433	1.196	1.1356
3	0.8638	0.8635	0.7060	0.5814	0.5189	0.4610	0.4670	0.7061	0.7061	0.709	0.689	0.489	0.5912
4	0.2755	0.2755	0.3070	0.3482	0.5189	0.3468	0.3092	0.2946	0.3066	0.531	0.561	0.348	0.3656
5	0.2755	0.2755	0.3030	0.2314	0.2502	0.3424	0.3031	0.2946	0.3031	0.314	0.332	0.283	0.2523
6	0.2755	0.2755	0.2830	0.1664	0.2502	0.1461	0.2837	0.2562	0.2864	0.263	0.312	0.157	0.1875
7	0.2755	0.2755	0.2060	0.1294	0.1553	0.1460	0.2059	0.2562	0.2059	0.174	0.206	0.125	0.1471
8	0.1000	0.1000	0.1050	0.1046	0.1553	0.1047	0.1042	0.1024	0.1038	0.153	0.211	0.101	0.1202
9	0.1000	0.1000	0.1050	0.0876	0.1126	0.1044	0.1042	0.1024	0.1038	0.110	0.149	0.105	0.1014
10	0.1000	0.1000	0.1040	0.0760	0.1126	0.1044	0.1034	0.1024	0.1034	0.099	0.162	0.105	0.0878
11	0.1000	0.1000	0.1040	0.0676	0.0913	0.0631	0.1034	0.1024	0.1034	0.078	0.121	0.105	0.0775
12	0.1000	0.1000	0.1030	0.0616	0.0913	0.0631	0.1010	0.0976	0.1013	0.071	0.138	0.062	0.0695
13	0.1000	0.1000	0.1020	0.0574	0.0811	0.0631	0.1010	0.0976	0.1013	0.061	0.107	0.079	0.0634
14	0.1000	0.1000	0.0980	0.0548	0.0811	0.0631	0.0913	0.0976	0.0913	0.057	0.128	0.057	0.0585
15	0.1000	0.1000	0.0780	0.0532	0.0780	0.0631	0.0913	0.0976	0.0913	0.054	0.073	0.07	0.0544

Table 6.3

Variance distribution for first-order Markov process defined by $\rho = 0.9$ and $N = 32$. i is the ith transform coefficient.

i	WHT	ST	DCT-II	DST-I	DST-II	CMT	DFT	DLT
0	13.568	13.569	13.569	13.244	13.08	13.568	13.568	13.568
1	6.101	6.59	6.847	6.352	6.157	6.611	4.938	6.590
2	3.128	3.658	3.720	3.460	3.352	3.659	4.938	3.423
3	1.965	1.561	2.023	2.048	1.983	1.602	1.613	1.986
4	1.046	1.23	1.244	1.386	1.355	1.229	1.613	1.273
5	0.995	1.163	0.832	0.960	0.943	0.916	0.770	0.881
6	0.867	0.464	0.596	0.729	0.726	0.511	0.770	0.646
7	0.545	0.46	0.448	0.5551	0.553	0.359	0.452	0.496
8	0.308	0.349	0.350	0.446	0.456	0.350	0.452	0.394
9	0.305	0.345	0.282	0.356	0.367	0.299	0.300	0.321
10	0.303	0.343	0.233	0.300	0.318	0.298	0.300	0.268
11	0.303	0.342	0.197	0.249	0.266	0.286	0.217	0.229
12	0.293	0.146	0.169	0.216	0.238	0.273	0.217	0.198
13	0.280	0.146	0.147	0.184	0.204	0.165	0.167	0.173
14	0.247	0.146	0.130	0.163	0.187	0.158	0.167	0.154
15	0.165	0.146	0.116	0.142	0.165	0.155	0.135	0.138
16	0.105	0.105	0.105	0.128	0.154	0.129	0.135	0.125
17	0.105	0.105	0.096	0.114	0.138	0.116	0.113	0.114
18	0.104	0.104	0.088	0.104	0.131	0.116	0.113	0.105
19	0.104	0.104	0.081	0.094	0.119	0.110	0.098	0.097
20	0.104	0.104	0.076	0.087	0.115	0.110	0.098	0.090
21	0.104	0.104	0.071	0.080	0.106	0.108	0.087	0.085
22	0.104	0.104	0.068	0.075	0.104	0.107	0.087	0.079
23	0.104	0.104	0.064	0.070	0.097	0.105	0.079	0.075
24	0.104	0.063	0.062	0.066	0.096	0.105	0.079	0.071
25	0.103	0.063	0.059	0.062	0.090	0.104	0.074	0.068
26	0.103	0.063	0.057	0.060	0.090	0.100	0.074	0.065
27	0.102	0.063	0.056	0.057	0.086	0.082	0.070	0.062
28	0.100	0.063	0.055	0.056	0.087	0.080	0.070	0.060
29	0.097	0.063	0.054	0.054	0.084	0.073	0.068	0.057
30	0.088	0.063	0.053	0.053	0.085	0.062	0.068	0.055
31	0.068	0.063	0.053	0.053	0.068	0.057	0.068	0.053

The EPE for other transforms was evaluated by Yip and Rao [M-20], who concluded that the EPE is invariant with respect to those discrete transforms that display block spectral structure. It may be shown that the EPE is directly equivalent to the ratio of the sum of the first M variances to the sum of all the N variances (Problem 6.1). Thus EPE is an alternative to the variance distribution in terms of the performance criteria.

6.4 Residual Correlation

Another performance measure of a discrete transform is how well it can decorrelate a given sequence. The KLT as shown in Section 3.2 is an optimal transform, as it completely decorrelates a Markov-1 sequence. Other discrete transforms, although suboptimal, can be judged by how well or how closely they approximate the KLT in decorrelating the sequence. Whereas the auto-covariance matrix in the KLT domain is diagonal, the covariance matrices for the other discrete transforms have nonzero off-diagonal elements. Hamidi and Pearl [PC-1] proposed a measure called fractional correlation (FC) left undone by a transform. Comparison of DCT-II with DFT [PC-1] and with DHT [PC-15] shows that the DCT is superior to both DFT and DHT based on the residual correlation for all N (sequence length) and ρ (adjacent correlation coefficient) for Markov-1 signals. The FC as defined by Hamidi and Pearl [PC-1] can be developed as follows. Let $[A_{\mathrm{tr}}]_{jj}$ be the diagonal matrix representing the diagonal elements of $[A_{\mathrm{tr}}]$ defined in (6.2.1). Then an auto-covariance matrix A' can be obtained by inverse transforming $[A_{\mathrm{tr}}]_{jj}$ such that

$$A' = [\phi(n)]^T [A_{\mathrm{tr}}]_{jj} [\phi(n)]. \tag{6.4.1}$$

The FC left undone by the transform is then defined as

$$\mathrm{FC} = \frac{|A - A'|^2}{|A - I_N|^2}, \tag{6.4.2}$$

where $| \ |^2$ denotes the Hilbert–Schmidt weak norm defined in (3.4.1). For KLT as A_{tr} is a diagonal matrix, FC is zero. The denominator of (6.4.2) is a measure of the cross correlation present in the original sequence. FC versus ρ (Markov-1 signals) is shown in Figs. 6.4 through 6.6 and is listed in Tables 6.4 through 6.6, ($N = 8$, 16, and 32) for different discrete transforms. In general the DCT-II (hence all the DCTs) is superior to all the other transforms based on the degree of decorrelation. For weakly correlated data, DST-I, however, performs better than the DCT-II. The break-even point for ρ depends on the sequence length N.

6.5 Rate Distortion and Maximum Reducible Bits

The rate distortion function yields the minimum information rate in bits per transform coefficient needed for coding when a certain maximum average distortion $D(\theta)$ is allowed for any source probability distribution. For Gaussian distribution, Davisson [M-21] has shown that the rate distortion can be expressed as

$$R(\theta, D) = \frac{1}{2N} \sum_{j=0}^{N-1} \max\left(0, \log_2 \frac{A_{\mathrm{tr}}(j, j)}{\theta}\right), \tag{6.5.1}$$

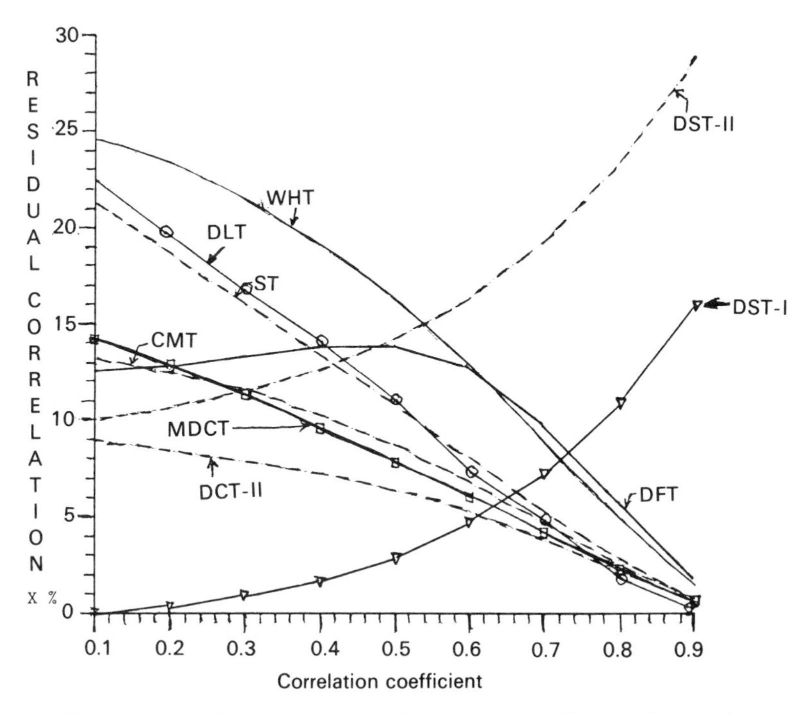

Fig. 6.4 Residual correlation versus the correlation coefficient ρ for $N = 8$.

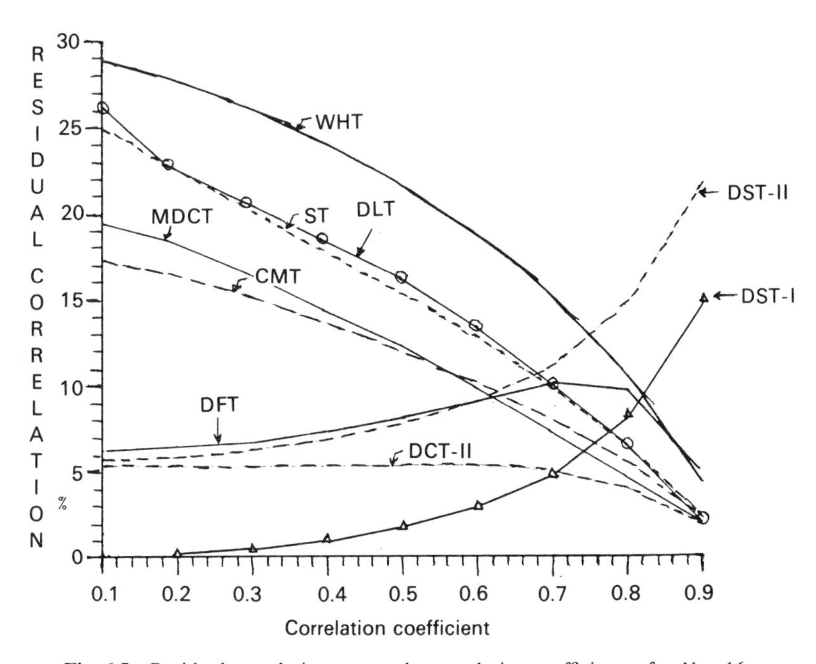

Fig. 6.5 Residual correlation versus the correlation coefficient ρ for $N = 16$.

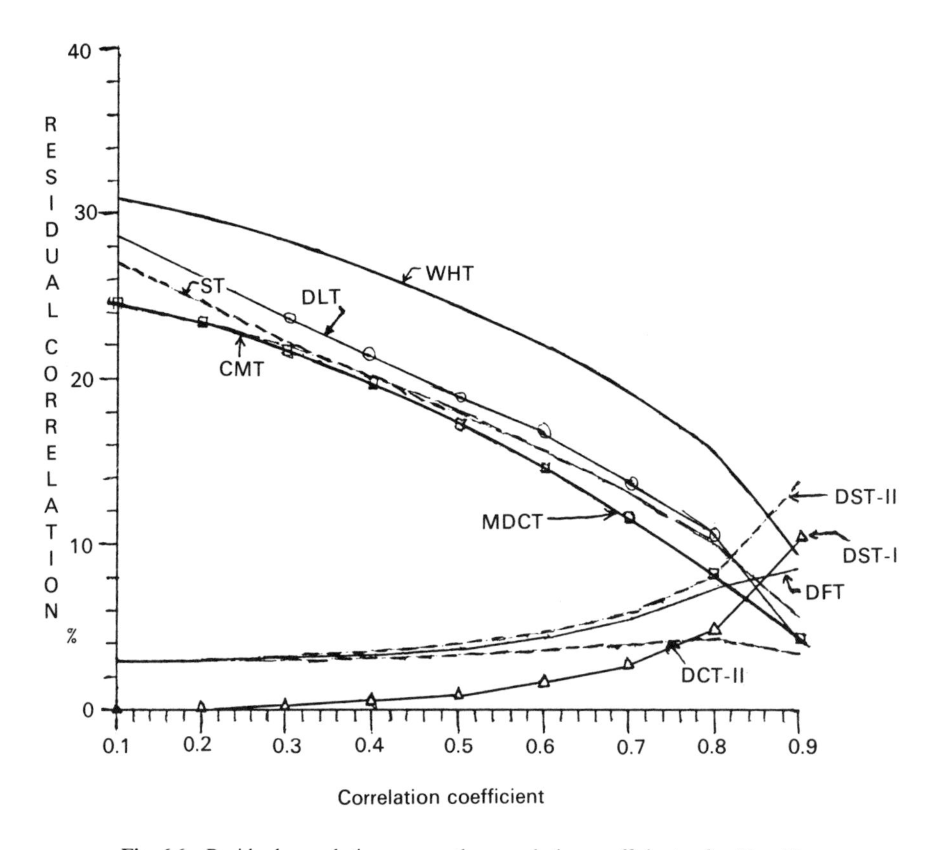

Fig. 6.6 Residual correlation versus the correlation coefficient ρ for $N = 32$.

Table 6.4

Residual correlation left undone for a I order Markov process $\rho = 0.9$ and $N = 8$.

ρ	WHT	DFT	ST	DCT-II	CMT	DST-I	DST-II	DLT
0.1	24.583	12.61	21.39	9.012	13.2	0.135	10.066	22.2
0.2	23.365	12.93	18.64	8.422	12.5	0.405	10.684	19.4
0.3	21.490	13.44	16.03	7.881	11.6	0.929	11.551	16.7
0.4	19.1	13.93	13.44	7.25	10.3	1.751	12.698	14.0
0.5	16.24	13.99	10.81	6.424	8.7	2.966	14.239	11.2
0.6	12.86	12.96	8.1	5.307	6.8	4.742	16.363	8.3
0.7	8.97	10.31	5.34	3.850	4.7	7.323	19.339	5.5
0.8	4.86	6.18	2.75	2.166	2.5	11.006	23.479	2.8
0.9	1.425	1.95	0.77	0.661	0.74	16.008	28.991	0.8

Table 6.5

Residual correlation left undone for a I order Markov process $\rho = 0.9$ and $N = 16$.

ρ	WHT	DFT	ST	DCT-II	DST-II	CMT	DST-I	DLT
0.1	28.8	6.31	25.1	5.392	5.68	17.2	0.136	26.19
0.2	27.7	6.49	22.54	5.321	5.05	16.4	0.251	23.77
0.3	26.04	6.82	20.08	5.315	6.35	15.1	0.550	21.27
0.4	23.97	7.34	17.67	5.347	6.93	13.6	1.038	18.77
0.5	21.54	8.13	15.25	5.377	7.78	11.9	1.787	16.20
0.6	18.69	9.24	12.69	5.328	9.06	10.1	2.954	13.45
0.7	15.20	10.37	9.84	5.006	11.11	8.0	4.865	10.34
0.8	10.56	9.96	6.4	3.990	14.76	5.5	8.280	6.63
0.9	4.30	5.23	2.41	1.83	21.84	2.2	14.962	2.45

Table 6.6

Residual correlation left undone for a I order Markov process $\rho = 0.9$ and $N = 32$.

ρ	WHT	DST-II	DCT-II	DFT	ST	CMT	DST-I	DLT
0.1	30.9	2.96	2.88	3.16	27.06	24.4	0.058	28.6
0.2	29.9	3.11	2.94	3.25	24.67	23.4	0.131	26.1
0.3	28.4	3.31	3.03	3.42	22.34	22.0	0.294	23.7
0.4	26.5	3.59	3.17	3.70	20.09	20.2	0.558	21.7
0.5	24.4	4.03	3.38	4.12	17.88	18.1	0.968	19.1
0.6	22.0	4.71	3.67	4.79	15.67	15.7	1.618	16.7
0.7	19.2	5.86	4.03	5.92	13.20	13.1	2.723	14.1
0.8	15.6	8.09	4.3	7.89	10.21	10.0	4.876	10.8
0.9	9.3	13.79	3.44	8.72	5.567	5.6	10.403	5.8

where $A_{tr}(j, j)$ is the jth diagonal element of A_{tr} and θ is chosen to give a predetermined value to D, i.e.,

$$D(\theta) = \frac{1}{N} \sum_{j=0}^{N-1} \min(\theta, A_{tr}(j, j). \tag{6.5.2}$$

This measures the extent to which the transform coefficients are decorrelated. The distortion $D(\theta)$ can be spread uniformly in the transform domain, thus minimizing the bit rate required for transmitting the information. It can be observed from Fig. 6.7 that DCT-II performs nearly as well as KLT based on the rate distortion criterion. If θ in (6.5.2) is chosen to be smaller than all $A_{tr}(j,j)$ then $D(\theta) = \theta$ and (6.5.1) reduces to

$$R(\theta, D) = \frac{1}{2N} \sum_{j=0}^{N-1} \log_2[A_{tr}(j, j)] - \tfrac{1}{2}\log_2[D(\theta)]. \tag{6.5.3}$$

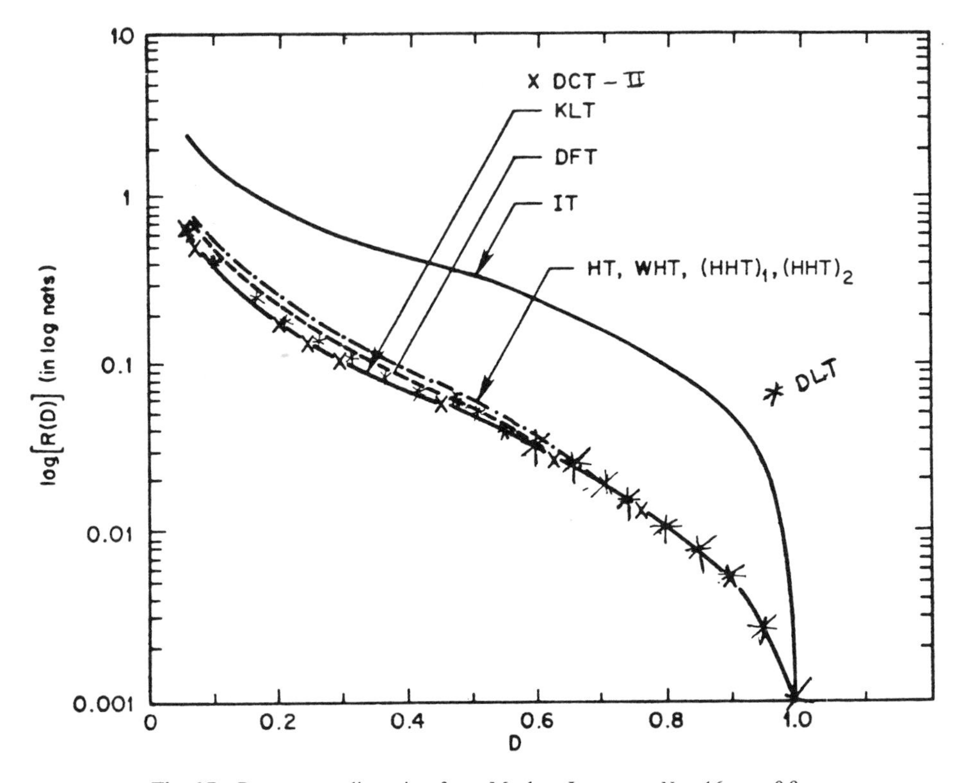

Fig. 6.7 Rate versus distortion for a Markov-I process. $N = 16$, $\rho = 0.9$.

Wang and Hunt [PC-4] have defined the first term in (6.5.3) as the negative maximum reducible bits (mrb) for each transform coefficient, i.e.,

$$\text{mrb} = -\frac{1}{2N} \sum_{j=0}^{N-1} \log_2[A_{\text{tr}}(j, j)]. \tag{6.5.4}$$

The larger the mrb is, the better is the transform performance in the sense of bit reduction. This figure of merit is listed in Table 6.7 for various discrete transforms. As expected DCT-II has the largest value.

6.6 Generalized Wiener Filtering

Pratt [M-22] extended the concept of Wiener filtering to discrete transforms. This problem can be formulated as follows. Given a random sequence $\mathbf{x}$ of length N corrupted by an additive noise $\mathbf{n}'$ (Both $\mathbf{x}$ and $\mathbf{n}'$ have zero mean and are uncorrelated with each other) the problem is to design a generalized Wiener filter $[G(n)]$ in the transform domain, such that the inverse transformation of the

Table 6.7

Maximum reducible bits for $N=16$ and $\rho=0.9$.

Transform	mrb
HT	0.9311
WHT	0.9374
DCT-II	1.1172
DFT	0.9485
ST	1.0744
DST-I	0.9752

filtered output $\hat{\mathbf{x}}$ yields the best estimate of $\mathbf{x}$ with the least mean square error (MSE) (Fig. 6.8). Pratt [M-22] has shown that the minimum MSE is independent of any discrete transform $\phi(n)$ utilized for Wiener filtering (Prob. 6.2). The computational requirements, however, for carrying out the filtering operation can be formidable. Thus by constraining the filter matrix $G(n)$ to be diagonal (this is true when the transform is KLT), one can minimize the computational effort, but the filter is no longer optimal. For this suboptimal filtering (also called scalar filtering), individual transform coefficients are appropriately weighted by the filter parameters. The optimum scalar filter and the corresponding MSE, $\varepsilon_{\text{scalar}}$ can be expressed as (Problem 6.3):

$$[G(n)]_{\text{scalar}} = \text{diag}\left[\frac{A_{\text{tr}}(j,j)}{A_{\text{tr}}(j,j)+B_{\text{tr}}(j,j)}\right] \tag{6.6.1}$$

and

$$\varepsilon_{\text{scalar}} = \sum_{j=0}^{N-1} A_{\text{tr}}(j,j) - \sum_{j=0}^{N-1}\left[\frac{A_{\text{tr}}^2(j,j)}{A_{\text{tr}}(j,j)+B_{\text{tr}}(j,j)}\right], \tag{6.6.2}$$

where $[B_{\text{tr}}(n)]$ is the transformed auto-covariance matrix of the noise and $B_{\text{tr}}(j,j)$ is its jth diagonal element. Assuming that $\mathbf{x}$ is a Markov-1 signal and that the noise auto-covariance matrix $[(B(n)]$ is diagonal with all the diagonal elements equal to K_0, the performance of various discrete transforms in scalar Wiener filtering is shown in Fig. 6.9 (see also Table 6.8) for various values of

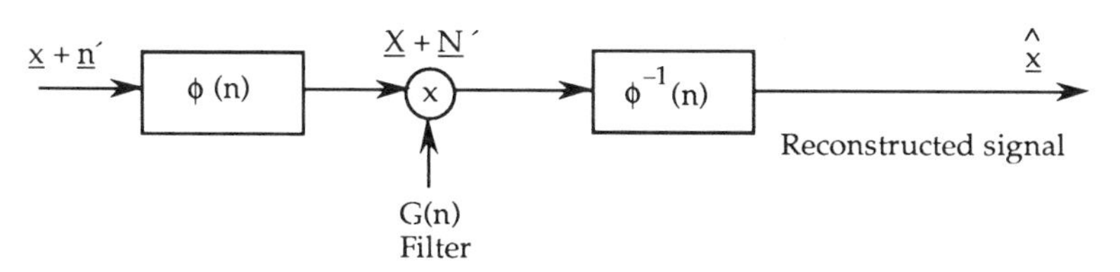

Fig. 6.8 Generalized Wiener filter [M-22]. (© 1972 IEEE).

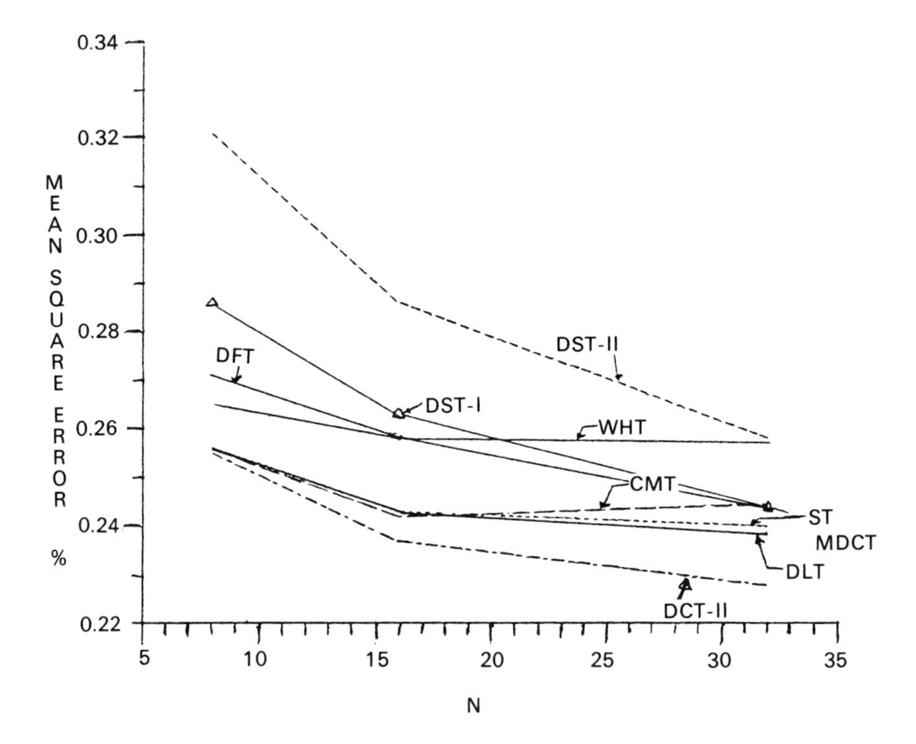

Fig. 6.9 Mean square error versus sequence length, N, for various discrete transforms for $\rho = 0.9$ and SNR, $K_0 = 1$.

sequence length N. This indicates the superiority of DCT-II compared with all the other discrete transforms. KLT, of course, is optimal since the corresponding filter matrix is diagonal.

6.7 Summary

The performance of the DCT-II is compared with other discrete transforms based on some standard performance criteria such as variance distribution,

Table 6.8

Mean square error for scalar Wiener filtering, $\rho = 0.9$.

N	WHT	ST	DCT	DST-I	DST-II	CMT	DFT	MDCT	DLT	DHART
8	0.265	0.256	0.255	0.286	0.321	0.256	0.271	0.256	0.256	0.267
16	0.258	0.243	0.237	0.263	0.286	0.242	0.258	0.243	0.243	0.425
32	0.257	0.240	0.228	0.244	0.258	0.245	0.244	0.240	0.239	

residual correlation, scalar Wiener filtering, and rate distortion. This comparison shows that the DCT, in general, is superior to all the other suboptimal discrete transforms (KLT being statistically optimal). This has led to its widespread application in signal and image processing, which forms the subject of the next chapter.

PROBLEMS

6.1. In Section 6.3 it is stated that the EPE is directly related to the variance distribution in the transform domain. Derive this relationship.

6.2. In the discussion on generalized Wiener filtering, it is stated that the minimum MSE is independent of any discrete orthogonal transform. Prove this. Obtain the optimal filter matrix $G(n)$ and the minimum MSE.

6.3. Derive (6.6.1) and (6.6.2) for the scalar Wiener filter.

6.4. Derive (6.5.1).

6.5. The variance distribution for DCT-II is shown in Figs. 6.1 through 6.3 and listed in Tables 6.1 through 6.3. Obtain this parameter for the remaining DCTs (see Section 2.3) and compare their performance with the other discrete transforms.

6.6. Repeat Problem 6.5 based on residual correlation. See Figs. 6.4 through 6.6 and Tables 6.4 through 6.6.

6.7. Repeat Problem 6.5 based on rate distortion and mrb. See Fig. 6.7 and Table 6.7.

6.8. Repeat Problem 6.5 based on scalar Wiener filtering. See Fig. 6.8 and Table 6.8.

6.9. In Section 6.2 it is stated that the trace of the auto-covariance matrix in the transform domain for any unitary transform is invariant. Prove this. (In Section 3.2 it is shown that, for the KLT, this matrix reduces to a diagonal matrix (3.2.12) with the diagonal elements as the eigenvalues of the auto-covariance matrix.) This implies that the energy of a signal is preserved under orthogonal transformation.

CHAPTER 7

APPLICATIONS OF THE DCT

7.1 Introduction

Ever since the DCT-II was discovered in 1974 [G-1], it has attracted the attention of the scientific, engineering, and research communities. Even a casual perusal of the technical literature indicates that after the DFT, DCT is most widely used. This is not surprising, since the DCT has the energy packing capabilities and also approaches the statistically optimal transform (i.e., KLT) in decorrelating a signal governed by a Markov process (see Chapters 3 and 6). The development of various fast algorithms for the efficient implementation of DCT, which involves real arithmetic only (see Chapters 4 and 5), further contributed to its popularity. The reduction in the computational complexity of these algorithms and the recursive structure resulted in simpler hardware. As mentioned earlier, the DCTs are orthogonal and separable. The former implies that the energy (information) of a signal is preserved under transformation (mapping into the DCT domain). The latter implies that a multidimensional DCT can be implemented by a series of one-dimensional (1D) transforms. The benefit of this property is that the fast algorithms developed for the 1D DCT can be directly extended to multidimensional transforms (Figs. 5.1 and 5.2). This is advantageous from a simulation viewpoint as well as for hardware realization. As described in Chapter 5, fast algorithms for multidimensional transforms have also been directly developed. It may be stated that other discrete transforms such as the DFT, slant, Haar, DHT, WHT, and DLT [B-1 through B-3, DW-5, BSI-7, FRA-16, FRA-19, FRA-22, FRA-23] also possess the separability and orthogonality properties. Some of the DCT fast algorithms have been specifically tailored for VLSI implementation [DP-12, DP-13, DP-15, DP-17, DP-22, DP-23, DP-26 through DP-29, DP-31, DP-33, DP-34, DP-37 through DP-39, DP-41, DP-44 through DP-49, DP-53, IC-67, IC-73]. In the early stages, DCT was implemented with discrete components at the board level. This was followed by its implementation using general purpose (DSP) chips [DP-9, DP-

19, DP-32, DP-50, LBR-7, LBR-11, LBR-15, LBR-16, LBR-18, IC-68, PIT-29, LBR-45 through LBR-47, DP-55]. Also, image compression boards and multiprocessor workstations (see Appendix B.2.) based on DCT have been developed by industry [PIT-29, DP-40, IC-68, LBR-47]. In general, these boards are designed to go into the expansion slots of the IBM PC, PC/XT, PC/AT, or compatibles, and can lead to simple and basic image compression systems [IC-34, DP-40, IC-68]. More recently, research labs and industry have been prominent in developing VLSI chips for real-time implementation of the DCT at video rates (15–80 MHz) (see Appendix B.1). The incentive for this has been the recommendations [LBR-23 through LBR-27, LBR-35 through LBR-41, LBR-44, PIT-14, PIT-17, PIT-18, PIT-21, PIT-26, PIT-27, PIT-30, PIT-33 through PIT-37] by the standards committees (CCITT, CCIR, CMTT, CEPT, ISO) to include the DCT as one of the major components in data compression for applications in teleconferencing, videophones, progressive image transmission, videotex, etc. Also, the coding scheme for the HIVITS (high-quality videotelephone and high-definition television system) project of RACE (Research in Advanced Communication Technologies for Europe) is based on the hybrid (predictive/DCT) technique. The objective of HIVITS is to offer multimedia communication services ranging from videophone (64 KBPS) to high-definition TV (HDTV) (70–140 MBPS) [PV-9]. Also, processing of the HDTV signal for transmission in the range of 70–140 MBPS using DCT as the main compression tool is being investigated [HDTV-1, HDTV-8 through HDTV-12, HDTV-18] (see Section 7.11). Telettra [IC-53, IC-67, IC-75, IC-76, IC-78] has developed DCT-based codecs for broadcast-quality NTSC TV at DS-3 level (44.736 MBPS) and PAL TV at 34.368 MBPS (European digital hierarchy). Also, coding of the 4:2:2 digital TV signal (CCIR rec. 601 [G-21]) (Table 7.1) at the H_2 level (32.768–44.736 MBPS) of the CCITT hierarchy (see Tables 7.2 and 7.3) [IC-60] and at 15 MBPS [IC-61] based on DCT has shown promising results. In addition DCT compressed storage of TV in home digital VTR is being developed [IC-59]. DCT was also the main compression tool in coding of Y, U, and V components sampled at 10.125 MHz and 3.35 MHz, respectively, resulting in picture quality corresponding to category 3 of CCIR 500-2 at 15 MBPS [IC-63]. While the data compression techniques for storage and retrieval of images in digital storage media (DSM) such as VTR, optical disc, CDROM, video disc, audio DAT, data DAT, Winchester drives, telecommunication networks etc. are currently being investigated, it is speculated that DCT may play a powerful role in this arena also [IC-59, IC-78]. The applications of DSM are random-access image data banks, multimedia interactive systems, and systems where video, audio, data, and computer programs are stored on the same media. An experts group on moving picture coding (MPEG) has been set up under WG8 of SC 2 of the Joint ISO/IEC Technical Committee with its mission to define a draft proposal of standards for coded representation of moving pictures for DSM having a maximum throughput of 1.5 MBPS [IC-80]. The coding algorithm has to be designed to maintain high audio and video

Table 7.1

Digital video standard specifications. (4:2:2 TV signal) CCIR recommendation 601 [IC-78].

Parameters	Systems	
Lines per frame	525	625
Fields per sec.	60	50
Coded signals	Y, C_R, C_B	
Total number of samples per line:		
luminance Y	858	864
chrominance C_B, C_R	429	432
Sampling structure	orthogonal, line, field and picture repetitive C_R and C_B samples co-sited with odd (1°, 3°, . . .) Y samples in each line	
Sampling frequency		
luminance Y	13.5 MHz	
chrominance C_B, C_R	6.75 MHz	
Form of coding	uniformly quantized PCM, 8 bits per sample, for the luminance signal and each color-difference signal	
Number of samples per digital active line		
luminance Y	720	
chrominance C_B, C_R	360	
Correspondence between video signal levels and quantization levels		
luminance	220 quantization levels: black = level 16 white = level 235	
chrominance C_B, C_R	225 quantization levels: zero = level 128	
Code work usage	0 and 255 for syncs, from 1 to 254 for video	

Table 7.2

Multiplexing hierarchy for digital transmission [IC-78] (rec. CCITT G.703).

Level	Bit rate (MBPS)		
1	1.544		2.048
2	6.312		8.448
3	32.064	44.736	34.368
4	97.728		139.264
Adopted in	Japan	USA	Europe

Note: The first level is given by synchronous multiplexing of a certain number of 64 KBPS channels, with additional data for switching and system control; 1.544 MBPS correspond to 24 channels at 64 KBPS, 2.048 MBPS to 30 channels. The higher levels are obtained by asynchronous multiplexing of the lower ones, with justification, to compensate light variations of nominal bit rate.

Table 7.3

Digital transmission hierarchy H levels for ISDN [IC-78].

Level	Bit rate (MBPS)	
H0	0.384	
H1	1.536	1.920
H2	43–44.736	32.768
H3	60–70	
H4	132–138.24	

Note: The final values of some levels are not yet determined, and ranges are indicated. The level H4 should be common to both hierarchies. Level H3 will be studied at lower priority. Level H21, at 32.768 MBPS, will be adopted in Europe; level H22, at about 44 MBPS, in the United States.

quality during various operations, such as random access, freeze frame, reverse playback, fast forward, etc. The spectrum of DCT applications includes filtering, quadrature mirror filters, transmultiplexers, multispectral scanner data [IC-55], speech coding, image coding (still frame, sequence in monochrome or color, image storage [IC-59]), pattern recognition, image enhancement, SAR image coding [SIC-1], surface texture analysis [STA-1], infrared image coding [IR-1], and printed image coding [IR-1]. It may be cautioned that the DCT is not a panacea to all the problems in digital signal processing. While it has established itself as a significant tool in specific application areas, other techniques such as vector quantization (VQ) (Section 7.9), sub-band coding, arithmetic coding, and pyramid coding have recently emerged as promising and powerful processes in DSP applications. While some of the concepts may themselves not be new, their impact has been profound. For example, techniques such as codebook design, codevector search, multistage, gain/shape, adaptive VQ, etc. have contributed to their practical implementation in video codecs. Using systolic processor architecture, VLSI chips for codevector selection based on full-search VQ (options include various suboptimal schemes) applicable to speech and image coding have been developed [DV-40]. The distortion measures include the mse (7.9.1) and the weighted mse (7.9.5). Similar modifications/improvements to the basic DCT operation, i.e., various scanning schemes, adaptive features, different quantizers, human visual system (HVS) weighting, progressive transmission of transform coefficients, and supplementing with other operations such as prediction, VQ, and sub-band coding, have enhanced the effectiveness of the DCT in image coding. The DCT has thus entrenched itself firmly in the arsenal of the DSP tools. The combination of various mutually complementing factors such as the introduction of ISDN, development of desk-top teleconferencing terminals, videophones, videotex work stations, availability of image compression boards, and design and development of application specific ICs (ASICs) such as those for DCT [DP-20 through DP-24, DP-26 through DP-31, DP-33 through DP-35, DP-41 through DP-49, DP-52, DP-53], DPCM [IC-65, IC-69 through IC-72], motion estimation [MC-30, through MC-33, MC-37 through MC-43], filtering [F-10 through F-14], error detection and correction [HDTV-9], etc. can lead to other, as yet unforeseen applications. The DFT, because of its fundamental frequency decomposition and also the simple convolution-multiplication property, has a prime place in the DSP arena. The DCT, on the other hand, because of its powerful bandwidth reduction capability (this is evident from the performance comparison of the discrete transforms described in Chapter 6), will continue to be a primary force in data compression devices. However, as mentioned earlier, other techniques such as VQ and sub-band coding may either complement the DCT (see Section 7.9 on DCT/VQ) or may eventually overtake the DCT in some specific applications. In fact, VQ-based video codecs have been designed, built, and marketed. The objective of this chapter is to outline the various DCT applications with a view that a novice can delve deeply into the details of certain fields by complementing it with the extensive literature (see the references listed at the end of this book). This chapter may spark the specialist to come up with additional applications. It may seduce the student into unexplored areas. In essence, the philosophy will be based on a block diagram format rather than an in-depth description of the multitude of applications.

7.2 Filtering

The convolution property for the DCT-II (Section 2.8) was originally developed by Chen and Fralick [IE-1]. This is, of course, not as simple as the corresponding property for the DFT. As is true for any separable transform, the convolution property (2.8.14) can be extended to multiple dimensions. Some examples of DCT filtering of images are illustrated in Figs. 7.1 through 7.3. By shaping the filter response, the desired filtered image can be obtained. By selecting a high-frequency emphasis filter in Fig. 7.1, Chen and Fralick [IE-1] obtained sharpened images (high-frequency details) with no edge-effect artifacts. The latter are characteristic of DFT filtering on image blocks, as the DFT

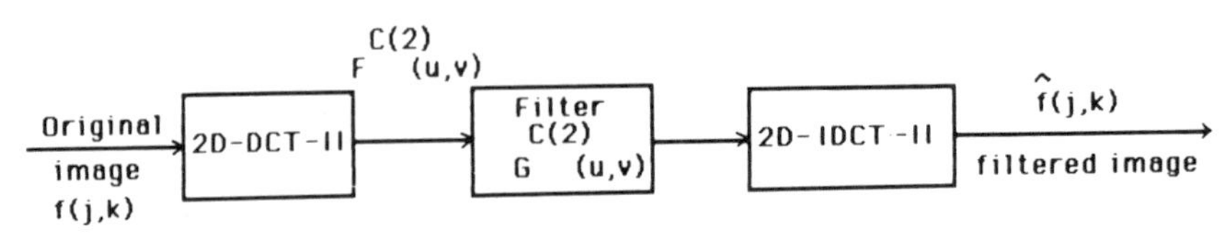

Fig. 7.1 DCT-II filtering of images [IE-1].

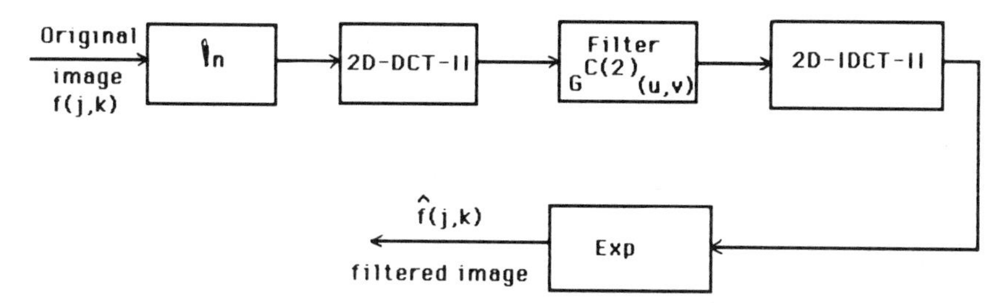

Fig. 7.2 Homomorphic/DCT-II filtering of images [IE-1].

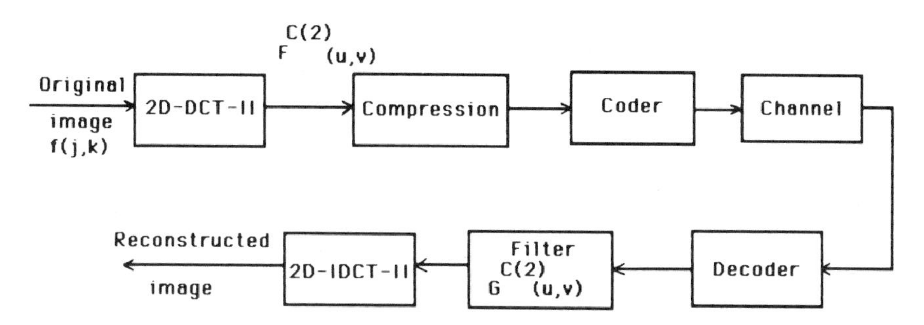

Fig. 7.3 Enhancement of DCT-II compressed images [IE-1].

assumes the periodicity of the signal/image. When extreme brightness transitions occur, the image dynamic range can be reduced by homomorphic filtering. An example of homomorphic/DCT-II filtering is shown in Fig. 7.2. Filtering of DCT compressed images can also be performed (Fig. 7.3) [IE-1].

By introducing a logarithmic preprocessor at the input and an exponential postprocessor at the output in Fig. 7.3, homomorphic filtering of compressed images can be achieved. Details of these operations and simulation results are discussed in [IE-1]. Low-pass filtering of images in the DCT domain (Fig. 7.4) has also been investigated by Ngan and Clarke [F-1] (Problem 7.2). Ngan, Leong, and Singh [HTC-5 through HTC-7] utilized the DCT convolution property [IE-1] in incorporating a human visual system (HVS) model in the DCT coding of images (Fig. 7.5). As mentioned in Section 2.8, when the impulse response was real and even, a simpler convolution-multiplication property for the DCT-II was developed in [F-7]. This property has been helpful in improving the quality of progressive image transmission based on the DCT coefficients whose transmission is controlled by the HVS [PIT-12].

7.3 Decimation and Interpolation

Both subsampling (decimation) and upsampling (interpolation) can also be performed in the DCT domain (Fig. 7.6a). This process efficiently combines the two operations, i.e., filtering and transform coding. Ngan [F-4] and Adant *et al.*

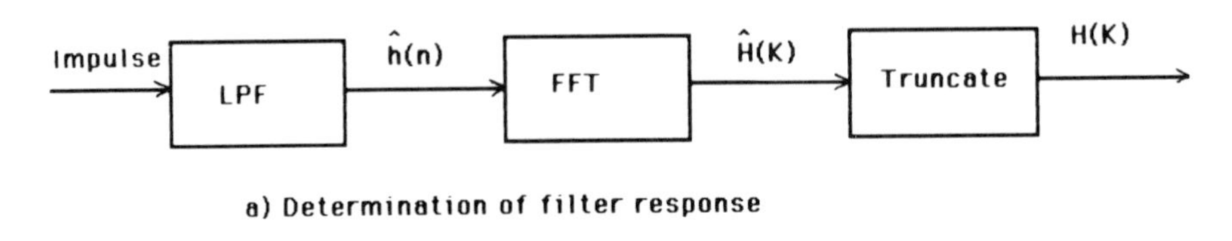

a) Determination of filter response

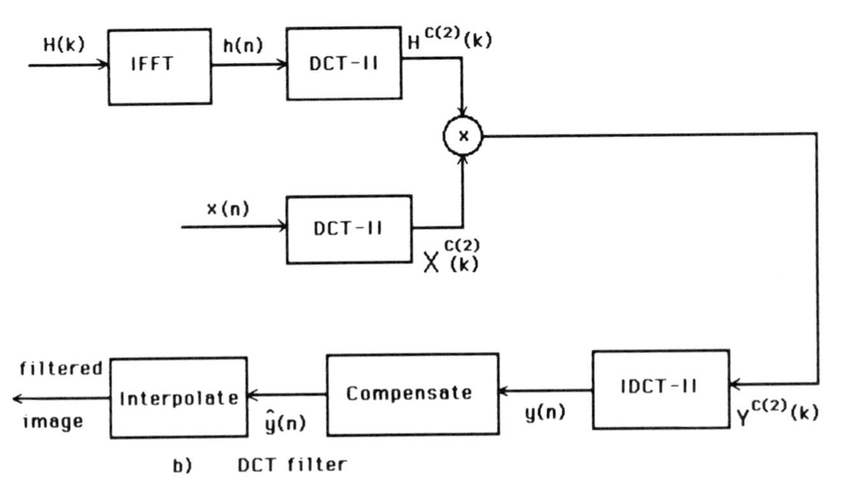

b) DCT filter

Fig. 7.4 Low-pass filtering in the DCT domain [F-1]. (© 1980 IEEE).

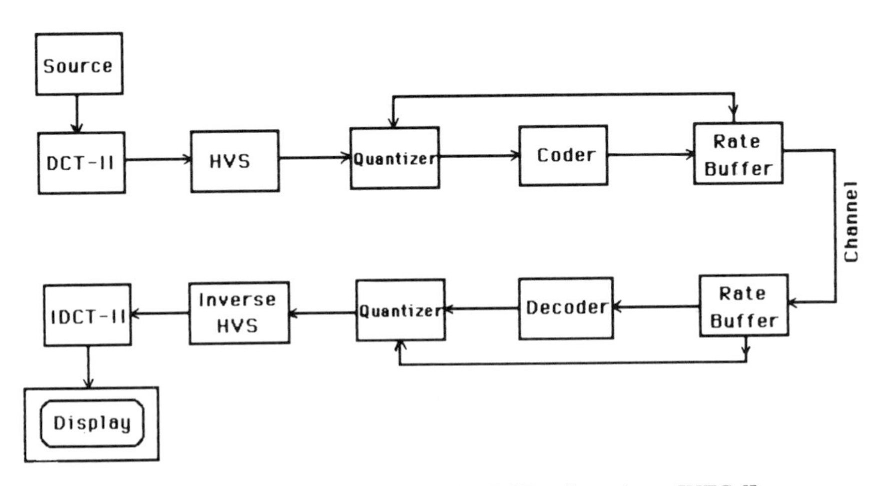

Fig. 7.5 Block diagram of the adaptive DCT coding scheme [HTC-5].

[F-6] have developed these techniques and have demonstrated their feasibility with applications to image coding. Using the DCT convolution-multiplication property, Ngan [F-4] has implemented the filtering and subsampling operations in the DCT domain and has applied the same to images based on 2:1 and 4:1 decimation. He has also compared this process with other transforms such as

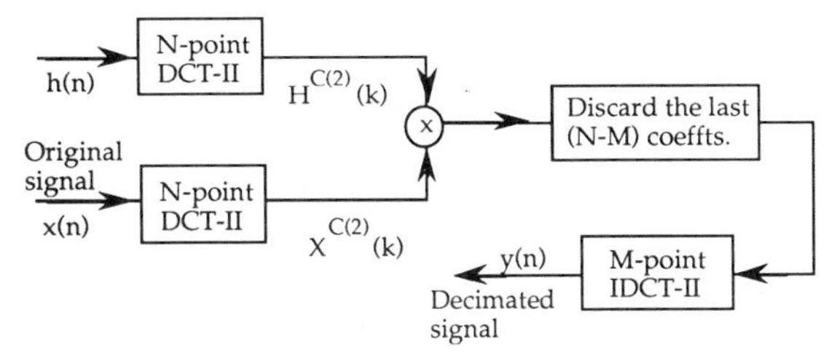

Fig. 7.6a Subsampling in the DCT domain.

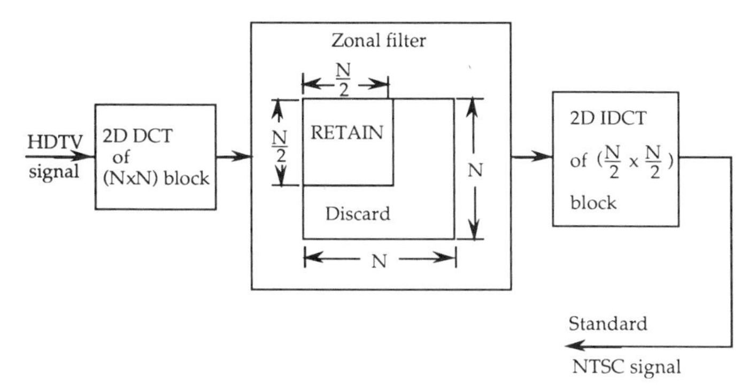

Fig. 7.6b Embedded HDTV coding scheme [HDTV-4, HDTV-8].

DFT and WHT. The combination of low-pass filtering and decimation in the DCT domain has an interesting application [HDTV-4, HDTV-8]. For converting a HDTV signal into standard NTSC signal for reception by a standard monitor, geometrical zonal filtering (see Fig. 7.29a) is applied to the 2D DCT of $N \times N$ blocks of the original HDTV image. The zonal filter retains only the $N/2 \times N/2$ low-frequency coefficients. 2D IDCT of $N/2 \times N/2$ blocks of the filter output yields the standard NTSC signal (Fig. 7.6b).

Adant *et al.* [F-6] have shown that the DSP operations such as filtering, down or up sampling, and orthogonal transformation can be combined into a single matrix multiplication and have demonstrated the utility of DCT transcoding by converting the color images from the 4:2:2 format (CCIR rec. 601 [G-21]) to the 3:1:1 format. This requires decimations of 4:3 and 2:1 for the luminance and color difference signals, respectively. They claim by applying the 8×8 DCT-II on the component signals, absolutely no perceptible impairments were observed in the subsampled images. Based on the modified symmetric DCT (MSDCT) [DP-13, DP-24], Van Caille *et al.* [F-8] have utilized a single matrix multiplication to obtain the standards conversion from 4:2:2 to the common source intermediate format (CSIF) [LBR-23 through LBR-26, LBR-35, LBR-37, LBR-38] followed by the DCT for further processing leading to a codec for teleconferencing or videophone (Problem 7.5).

7.4 LMS Filtering

Adaptive least-mean-square (LMS) filters both in the time domain and in the transform domain have been developed. The objective of the LMS algorithm is to minimize the mse between the output signal $y(n)$ and the desired signal $d(n)$ by adaptively adjusting the weights $a_{n0}, a_{n1}, \ldots, a_{n(N-1)}$ (Fig. 7.7) for the time domain filter or the weights $b_{n0}, b_{n1}, \ldots, b_{n(N-1)}$ (Fig. 7.8) for the transform domain filter. Only recently Narayan, Peterson, and Narasimha [LF-1] have shown that frequency domain filtering reduces the convergence rate (or increases the convergence speed) of the LMS algorithm. They have shown that for speech-related applications such as spectral analysis and echo cancellers, DCT domain filtering has a faster convergence rate compared with temporal

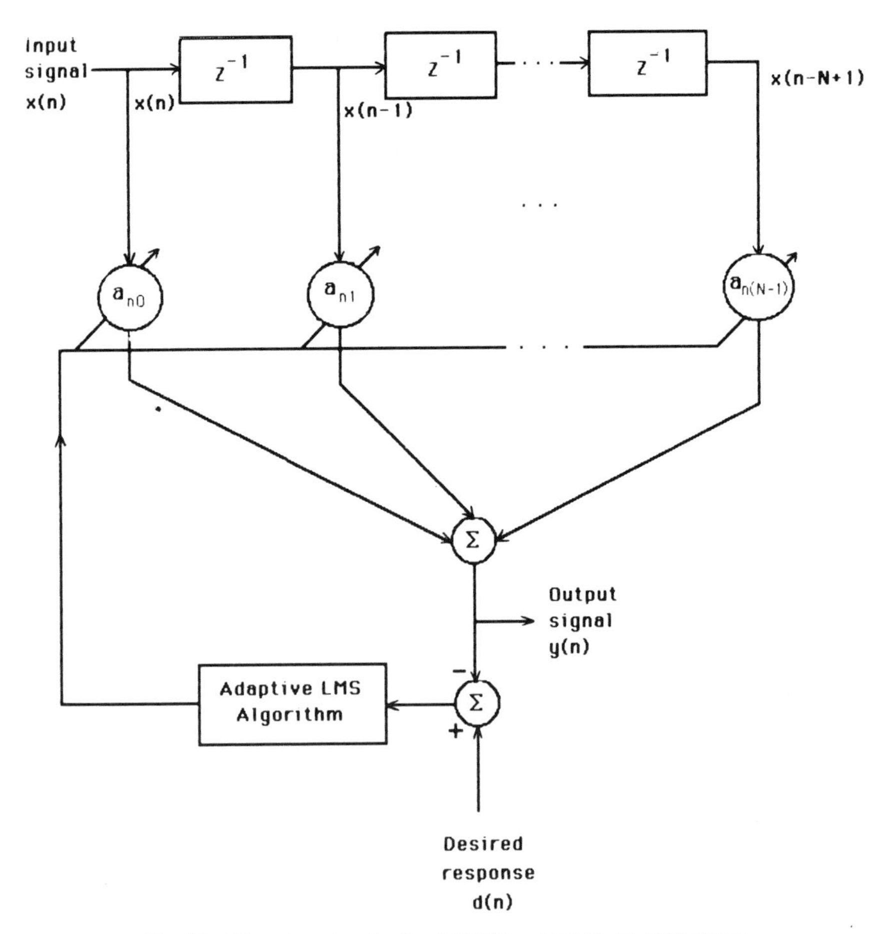

Fig. 7.7 Time domain adaptive LMS filter [LF-1]. (© 1983 IEEE).

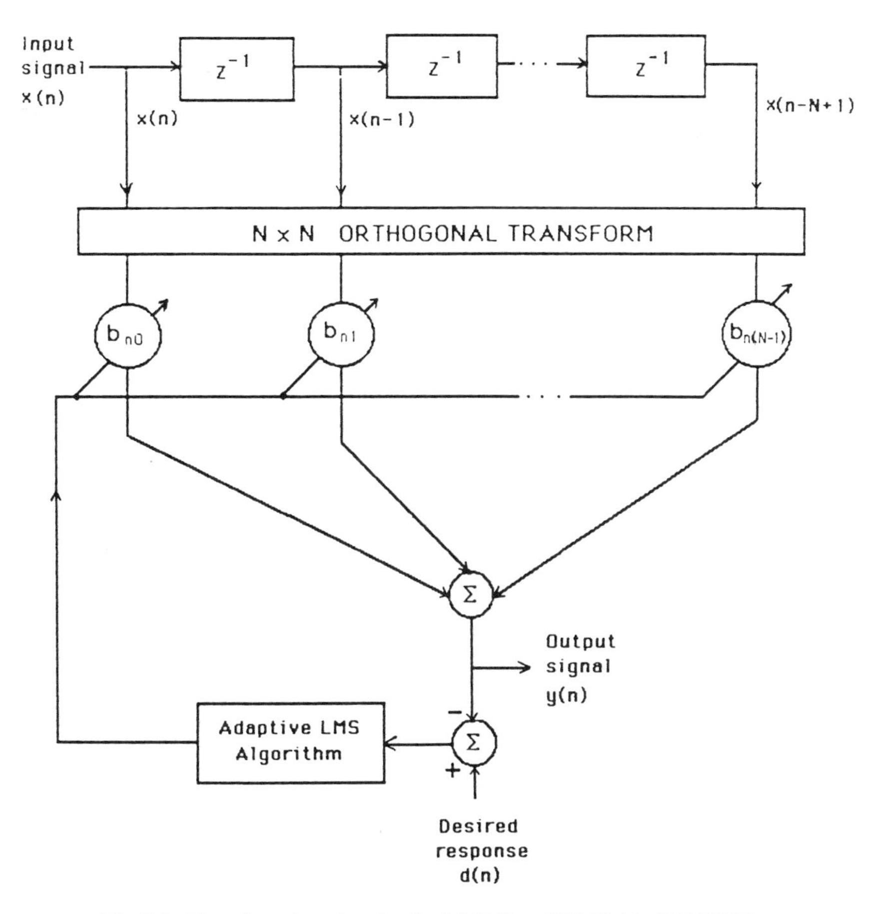

Fig. 7.8 Transform domain adaptive LMS filter [LF-1]. (© 1983 IEEE).

and DFT domain filtering. They have also applied the DCT-LMS approach to an adaptive line enhancer.

Lee and Un [LF-2] have extended the transform domain LMS filtering [LF-1] to other discrete transforms such as the DST, SCT, and KLT, and have conducted an in-depth investigation of temporal and transform domain filtering. They have obtained the optimum Wiener solution and the minimum MSE in the transform domain. They have also compared the performance of the LMS filtering for all these transforms based on computational complexity, convergence conditions, etc.

7.5 Transmultiplexers

Bandpass digital filter banks are used in transmultiplexers (TDM ⇔ FDM conversions) [T-1 through T-6] and in sub-band coding [B-4, SC-9, HDTV-8,

HDTV-12]. While there are several techniques for realizing such filter banks, Narasimha and Peterson [T-3] have shown that the filter bank can also be realized by a weighting network and the DCT-II. They have also developed an efficient algorithm for implementing a 14-point DCT-II and its inverse [T-3]. Narasimha *et al.* [T-1] have also utilized a 72-point DCT and a weighting network in realizing a 60-channel CCITT transmultiplexer. In both the 24-channel [T-2, T-3] and the 60-channel transmultiplexers, the DCT-II and its inverse are realized at the board (discrete components) level. Nussbaumer [T-6] has shown that the filter banks can also be realized by a set of real polyphase filters followed by the DCT-II. Marshall [T-4] has used the DCT-II and a polyphase network of 14 filters for realizing a 12-channel transmultiplexer based on DSP chips. Based on the DCT, Malvar [BSI-5, BSI-6, SC-29] has developed a new transform called a lapped orthogonal transform (LOT) for block signal processing. This has the basis functions overlapping the DCT, and it is effective in reducing the blocky effect (see Section 7.12) in low bit-rate speech and image coding. Malvar [T-7] has shown that the LOT is equivalent to an efficient quadrature mirror filter (QMF) bank in which the analysis and synthesis filters have identical FIR responses (Problem 7.31). Another filter bank interpretation of the DCT based on the concepts of short-time spectral analysis has been provided by Gunde [F-9].

7.6 Speech Coding

Two excellent review articles [SC-7, SC-13] address the general topic of speech coding. Speech coding can generally be classified into two categories: source coders or vocoders and waveform coders. In the latter category, another article [SC-9] discusses the frequency domain (sub-band and transform coding) coding of speech. Each of these, in turn, has many subcategories. For example, the spectrum of PCM, DPCM, DM, sub-band coding, adaptive transform coding, LPC, etc. fall into waveform coding. Zelinski and Noll [SC-1] have conducted an in-depth investigation into adaptive transform coding of speech based on various discrete transforms such as the KLT, DCT, DFT, WHT, and slant transforms [B-1, B-2] and have compared their performance with PCM and DPCM. The speech is low-pass filtered to 4 KHz (sometimes low-pass filtered to 3.4 KHz or bandpass filtered (300–3400 Hz)) and sampled at 8 KHz. Adaptive transform coding (ATC) is applied to speech segments (typically 16 ms long, 128 samples). The transform coefficients are adaptively quantized and coded (both the number of bits assigned and the step size of an uniform quantizer (see Fig. 7.46) are controlled by the side information such as the block speech spectrum). Both the coded transform coefficients and the side information are transmitted over a digital channel for inverse processing at the receiver (Fig. 7.9) [SC-3].

As indicated in Problem 7.10, the bit assignments for the transform coefficients are based on their variances. This bit assignment is adaptively controlled

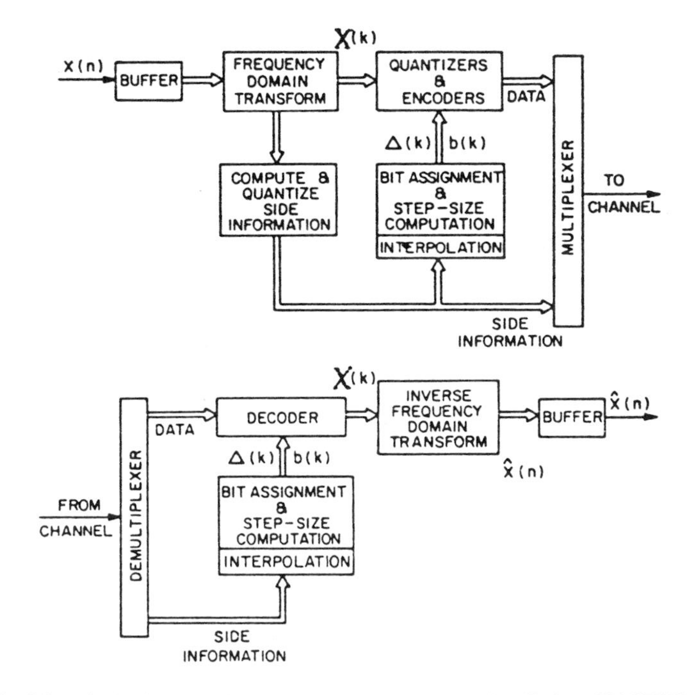

Fig. 7.9 Block diagram of adaptive transform coding [SC-3]. (© 1981 IEEE).

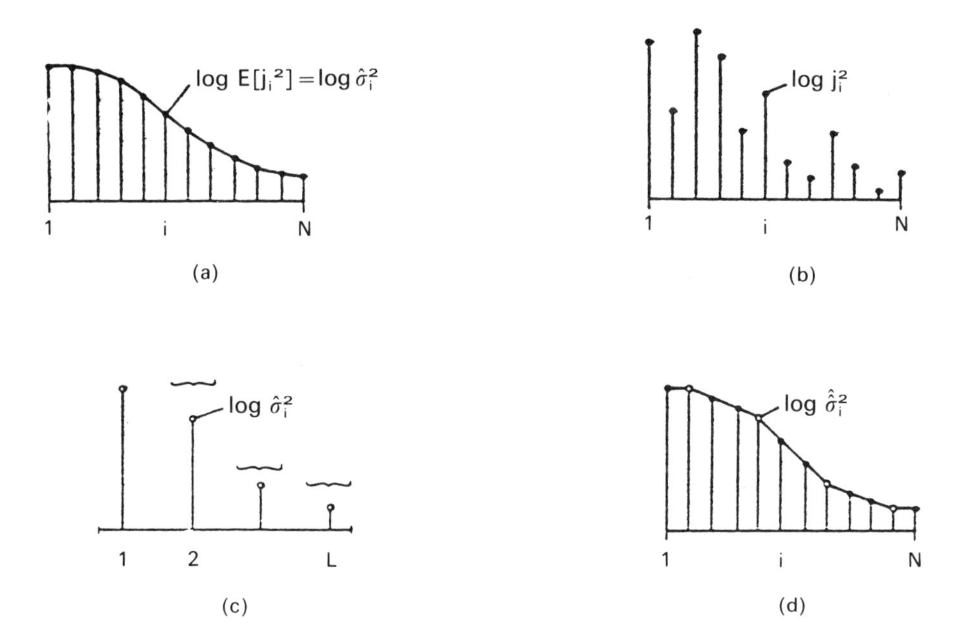

Fig. 7.10 Estimation of the basis spectrum (example: $N = 12$, $M = 3$) [SC-1]. (a) Basis spectrum of speech; (b) actual squared amplitudes of the transform coefficients; (c) averaged samples $\hat{\sigma}_i^2$ derived from b; (d); estimated basis spectrum $\hat{\hat{\sigma}}_i^2$ obtained by interpolation using the quantized version of the averaged samples of c. (© 1977 IEEE).

on a block-by-block basis. To reduce the overhead for conveying this side information to the receiver, the N transform coefficients are squared and averaged over M neighboring samples (Fig. 7.10). The N/M average variances are used to estimate the basis spectrum of the specific speech segment. The quantized versions of the averaged variances are transmitted to the receiver as side information. Both the transmitter and receiver use this side information to estimate the basis spectrum (linear interpolation in the logarithmic domain) and then compute the bit assignment. The estimated variances are also used to control the step sizes of the quantizers. Other types of adaptation such as the "vocoder-driven" ATC [SC-3, SC-9, SC-13] have also been proposed.

The bit assignment scheme (P7.10.1) is modified as follows:

$$R_i = \bar{R} + \frac{1}{2}\log_2 \hat{\hat{\sigma}}_i^2 - \frac{1}{2N}\sum_{j=1}^{N}\log_2 \hat{\hat{\sigma}}_j^2 \qquad i = 1, 2, \ldots, N. \tag{7.6.1}$$

R_i is the number of bits assigned to the ith transform coefficient and $\bar{R}$ is the average bit rate.

As shown in Fig. 7.10, $\hat{\hat{\sigma}}_i^2$ represents the estimated (quantized) power spectrum obtained from the side information (Fig. 7.10c). Zelinski and Noll [SC-1] have utilized 16-ms speech segments (128 samples) for transform coding with 16 averaged variances per segment (see Fig. 7.10c) as side information. Transform processing in blocks assumes no correlation between adjacent blocks. At low bit rates this can result in blocky structure in image coding and "click" and "burbling" noises at the block rate in speech coding [SC-9]. Several techniques to reduce or eliminate this artifact in image coding (Section 7.12) have been investigated. In speech coding, this boundary effect can be reduced by overlapping successive blocks. For example, 10% of the samples of a block are common to adjacent blocks, with the common samples weighted by a window such as a trapezoidal window (Fig. 7.11) [SC-9].

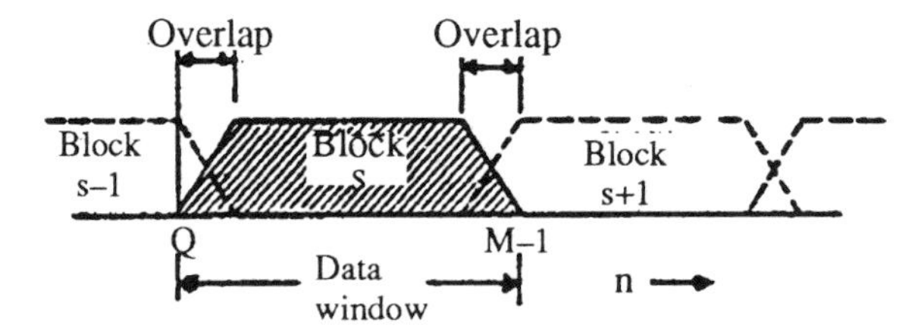

Fig. 7.11 Trapezoidal window for DCT analysis/synthesis [SC-9]. (© 1979 IEEE).

In the ATC scheme at bit rates below 12 KBPS, some of the high-frequency coefficients based on (7.6.1) are discarded resulting in "low-pass effects" [SC-2, SC-9]. To solve this problem, some modifications to (7.6.1) have been proposed. Also, techniques for reducing the data rates required for the side information have been proposed [SC-2]. An alternate scheme to simplify the coder involves

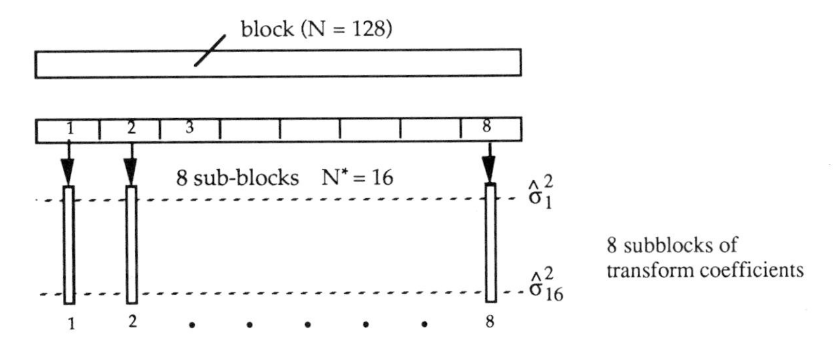

Fig. 7.12a ATC scheme with small blocks [SC-2]. (© 1979 IEEE).

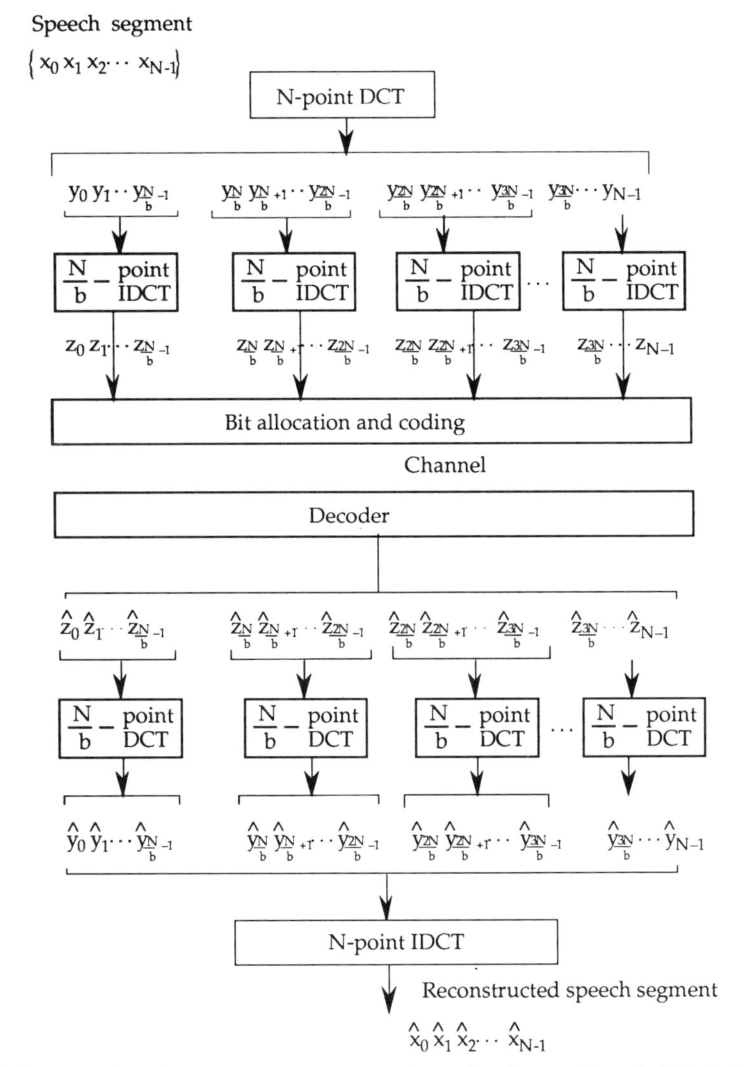

Fig. 7.12b Transform-based speech coder. The bands can be of unequal lengths [SC-22, SC-34].

dividing the 128 sample block into eight sub-blocks, each having 16 samples. The variances of the transform coefficients averaged over these eight sub-blocks form the side information (Fig. 7.12a). The quantized and coded version of the transform coefficients of each sub-block are sent over the digital channel to the receiver. A modification of the ATC scheme with small blocks (Fig. 7.12b) involves splitting the N transform coefficients of the speech signal into a number of bands and separately inverse transforming the coefficients in each of these bands [SC-22, SC-34]. This is followed by a adaptive bit allocation and quantization, and transmission of side information (energy in each band), which is similar to the ATC scheme [SC-1, SC-2]. Both the number of bands and the length of the speech segment N can be varied, as also the equal or unequal lengths of the bands. The performance of the transform-based split band codes (TSBC) is compared with adaptive DPCM (see Fig. 7.33) sub-band coding (see Fig. 7.14) and ATC for speech signals at 16 KBPS, and is shown to perform as well as SBC at a much lower complexity.

The ATC schemes of Zelinski and Noll [SC-1, SC-2] form the basis for comparing the performances of the C-matrix transform (CMT) [M-33], symmetric cosine transform (SCT) [S-1], and DCT-II in speech coding [SC-6]. In all these cases DCT-II proved to be superior to all the other deterministic transforms, both in the SNR and also in the perceptual quality of the reconstructed speech. Although an adaptive orthogonal transform based on the autoregressive model (AOT-AR) of the input speech waveform proved to be marginally superior to the DCT-II, the computational complexity of the AOT-AR precludes its practical application in speech coding [SC-8]. The performance of the ATC can be further improved by combining it with time-domain harmonic scaling (TDHS) [SC-4] (Fig. 7.13). Such a combined system is noted as ATC/HS. In [SC-4] the TDHS was also combined with the sub-band coding (SBC) (Fig. 7.14), called SBC/HS, which proved to be superior to ATC/HS at 9.6 KBPS from a hardware-implementation point of view. At 16 KBPS the 32-band sub-band coder proposed in [SC-5] is nearly comparable with the homomorphic side information-ATC scheme [SC-3]. An ATC whose bit assignment map is based on the autoregressive Gaussian hidden Markov model (AGHMM) has been proposed by Farvardian and Hussain [SC-16]. They have demonstrated an improved performance of this scheme over that of [SC-1].

DCT-II was also used in medium-band (4.8–9.6 KBPS) speech coding [SC-10, SC-11]. A combination of schemes, i.e., phase equalization, VQ of pulse pattern, VQ/SQ of spectral components, adaptive bit allocation, etc. [SC-10] (Fig. 7.15) resulted in a 6-bit log PCM coder. A modification of this technique is simulated in [SC-11] (Fig. 7.16a). In this modified scheme the LPC residue is mapped into the DCT domain and vector quantized (see Section 7.9 on DCT/VQ) based on a weighted distance measure. Simulation of the TC-WVQ (transform coder-weighted vector quantization) has shown that the speech quality based on this scheme at 7.2 KBPS is comparable with that of a 5.5-bit log PCM coder [SC-11]. A modification of the TC-WVQ speech coder is the introduction of two-channel conjugate VQ (see Fig. 7.65b), resulting in a

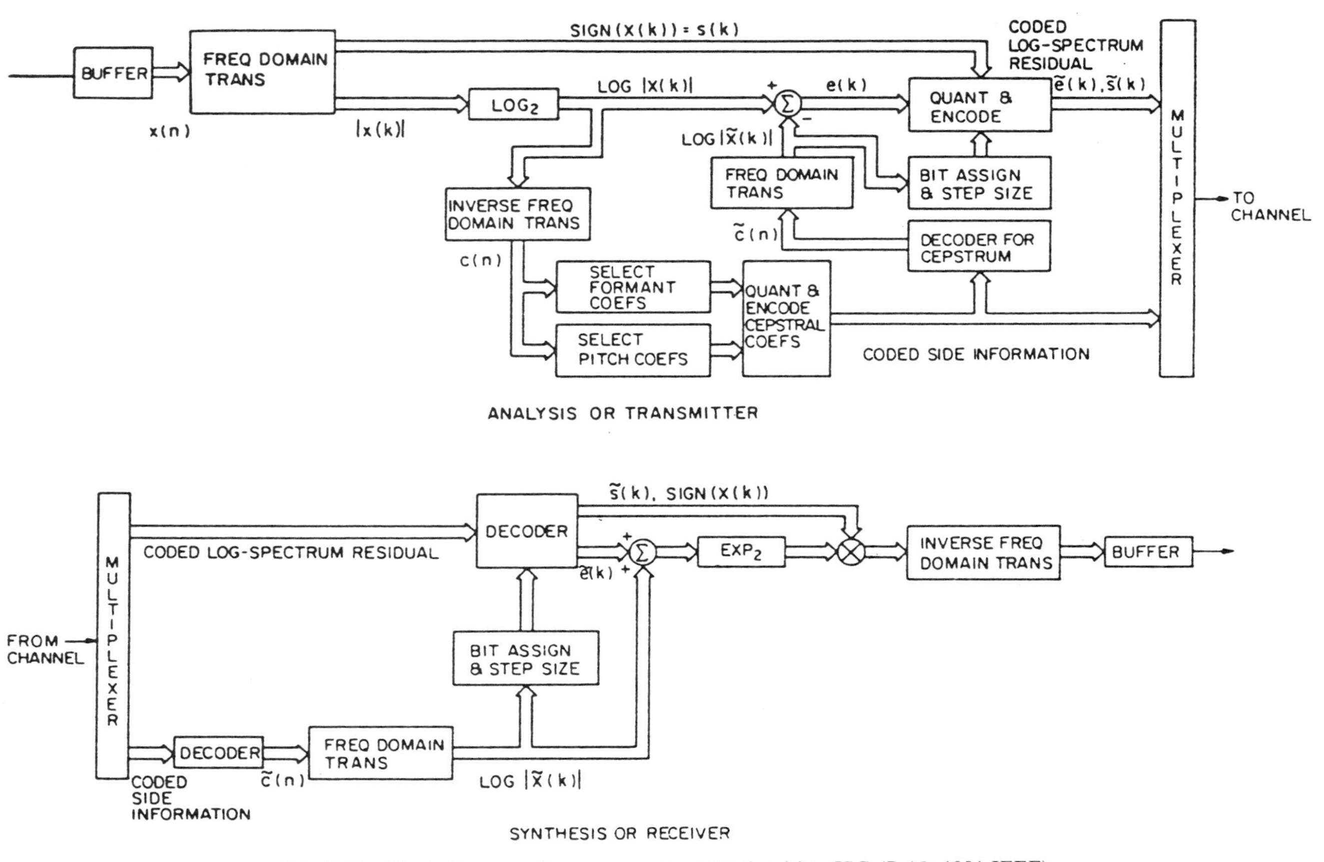

Fig. 7.13 Block diagram of homomorphic ATC algorithm [SC-4]. (© 1981 IEEE).

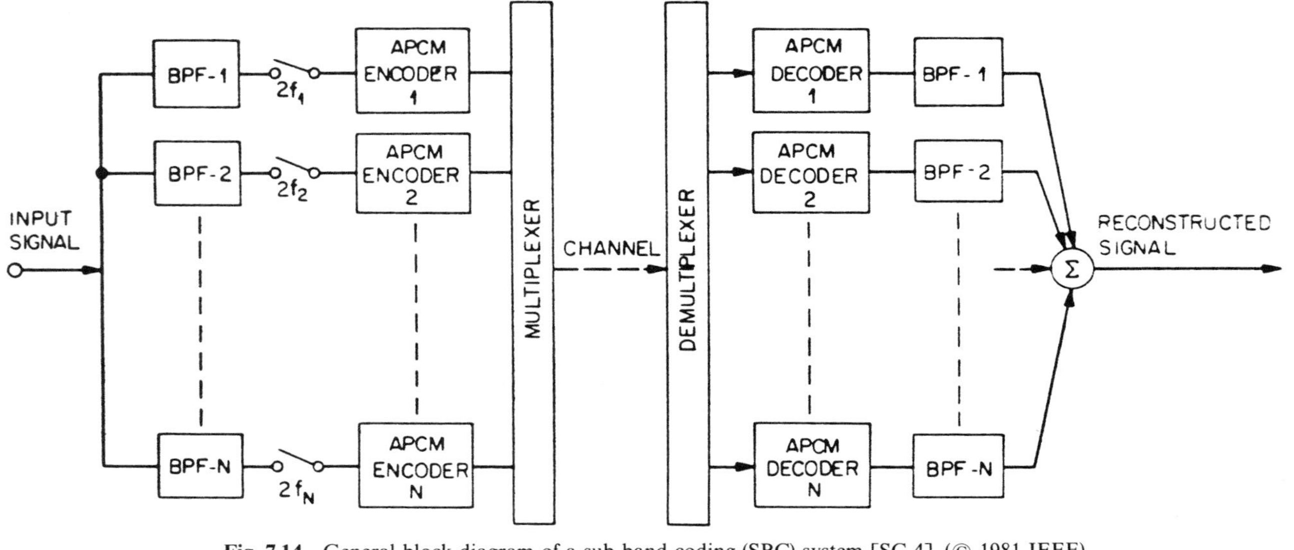

Fig. 7.14 General block diagram of a sub-band coding (SBC) system [SC-4]. (© 1981 IEEE).

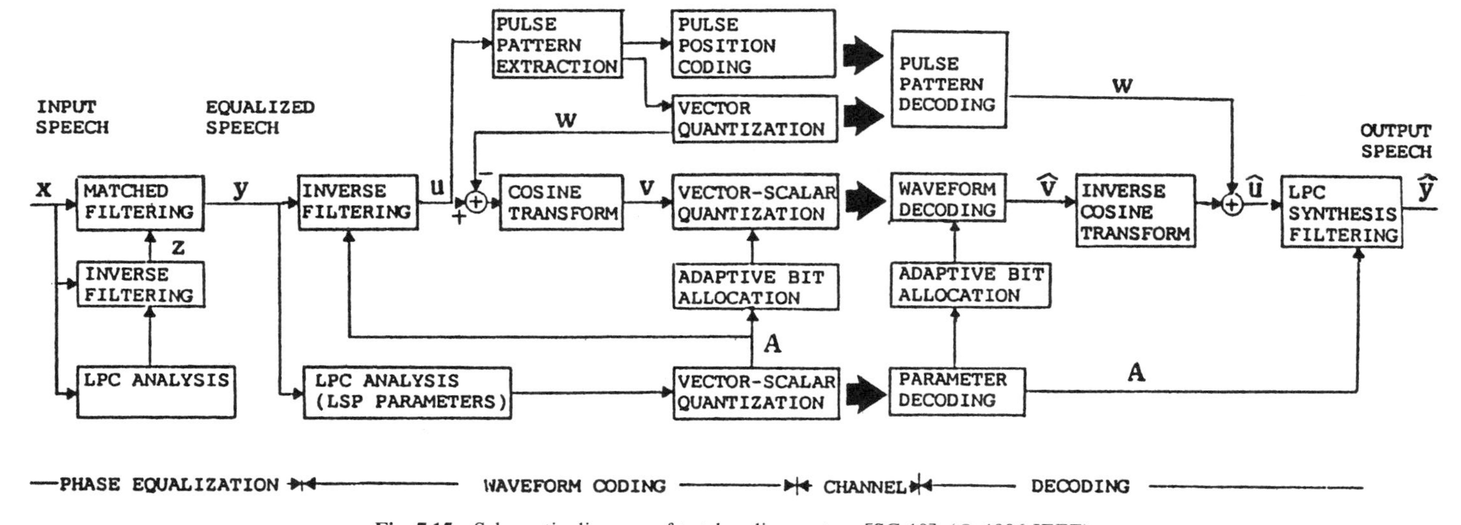

Fig. 7.15 Schematic diagram of total coding system [SC-10]. (© 1986 IEEE).

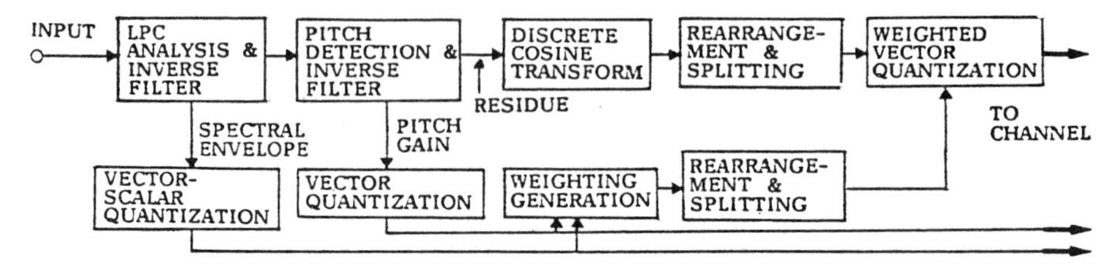

Fig. 7.16a Proposed speech coder [TC-WVQ) [SC-11]. (© 1988 IEEE).

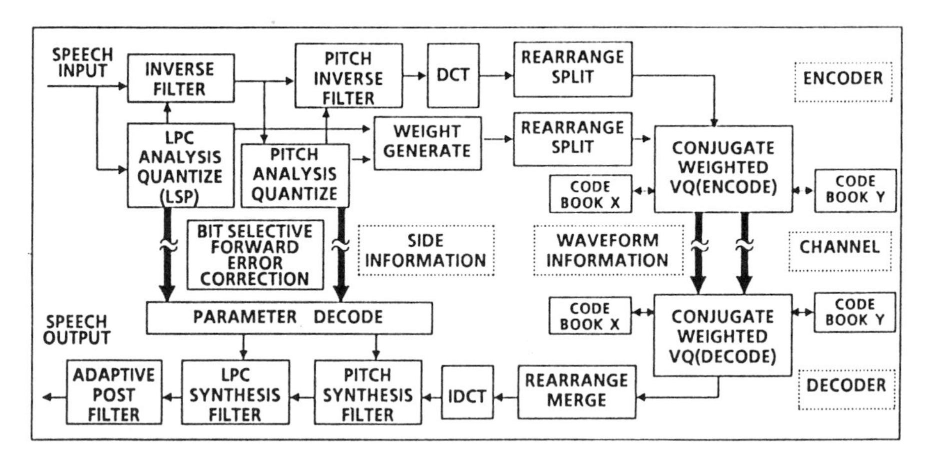

Fig. 7.16b TC-WVQ two-channel conjugate VQ speech coder [DV-48]. (© 1988 IEEE).

8 KBPS speech coder [DV-48] (Fig. 7.16b) that is less sensitive to channel errors. Additional advantages of the two-channel conjugate VQ are reduced computational complexity and memory requirements. Other schemes utilizing the DCT, called the transform trellis code (TTC), have also been proposed for speech coding [SC-12, SC-15] (Fig. 7.17).

7.6.1 *Wideband Transform Speech Coding*

So far transform speech coding has been based on the narrow-band approach, i.e., transform of long speech segments (128 or 256 samples corresponding to 16 or 32 ms of speech). One reason for this is the desire to have the most harmonics from the "block end effects" below the 300 Hz–3.4 KHz speech spectrum. Fjallbrant and others [SC-17 through SC-20, SC-23 through SC-26, SC-30, SC-31], however, have introduced a wide-band approach in which 8-point DCT of speech sampled at 8 KHz, corresponding to 1-ms segments, followed by adaptive vector quantization and dynamic bit allocation of the transform coefficients, coupled with other parameters such as magnitude and phase-derivative time trajectories in the frequency domain, was simulated. The vector-

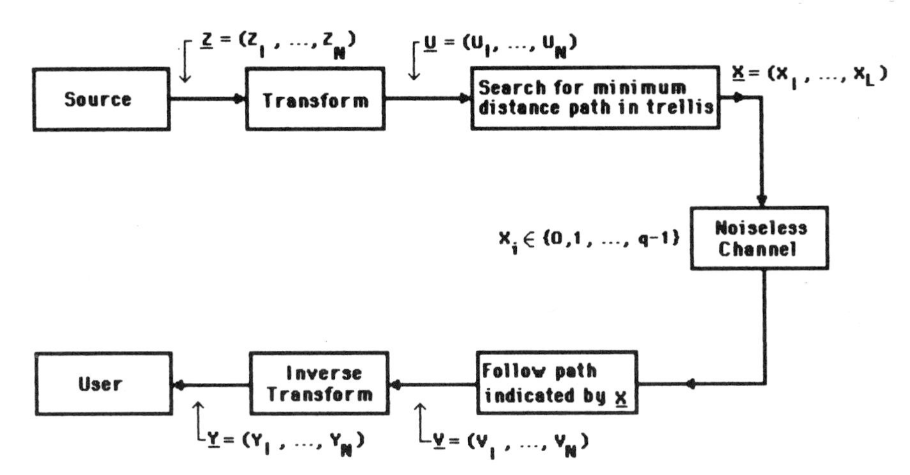

Fig. 7.17 Trellis source coding system [SC-15]. (© 1985 IEEE).

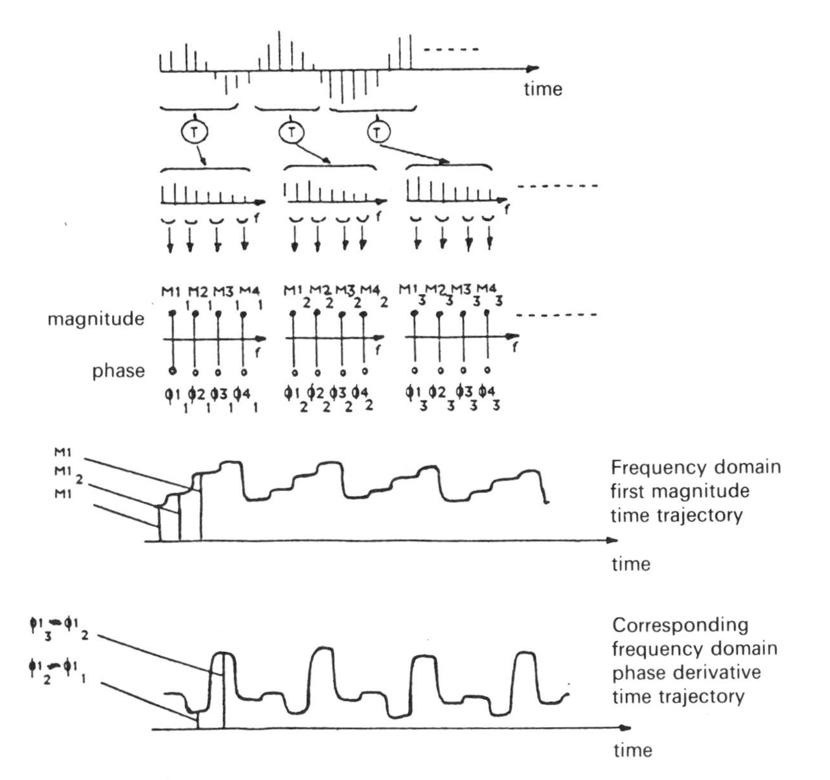

Fig. 7.18 Computation of frequency domain magnitude and phase derivative time trajectories [SC-18].

adaptive wide-band transform product code vector quantization (VAWTPCVQ) coder was implemented on a single TMS 32020 DSP chip, and good communication-quality speech was obtained at rates between 7 KBPS and 16 KBPS. This algorithm is characterized by robust performance in a noisy channel environment (bit errors affect only 1–8 ms speech segments); short coding delay, thus avoiding the necessity of echo cancellers; and real-time implementation on a DSP chip. In the time vector approach of the VAWTPCVQ coder [SC-19, SC-23], the DCT coefficients of consecutive 1-ms speech segments are utilized to obtain the magnitude, phase, and phase derivative time trajectories in the frequency domain (adjacent coefficients are combined), (Fig. 7.18). For an 8-point DCT, the magnitude and phase derivative time trajectories reflect the pitch periodicity in voiced speech because of the relationship between the speech segment (1 ms) and pitch periods (3–12 ms). The trajectories of the eight DCT coefficients are vector quantized based on different codebooks, codebook sizes, and vector lengths for each of these coefficients (Table 7.4). Adaptation is incorporated by dynamically varying the codebook sizes or the codebooks for the trajectories of the low-frequency coefficients, as these coefficients have the most energy and also contribute most to the formant and pitch structure, even for short pitch periods. This dynamic adaptation exploits the perception properties of the human ear. Adaptation has also been obtained in this time vector approach by incorporating gain adaptation of a medium-band speech coder whose performance at 9.6 KBPS was comparable to that of the time trajectories for the first four transform coefficients [SC-27]. In a frequency vector approach of the VAWTPCVQ coder, a gain adaptation strategy has been developed, including an adjustment-vector predictor, adjusting each transform coefficient during the partial distance search [SC-28].

Table 7.4

Bit allocation to transform coefficient time trajectories [SC-23].

Transform coefficient number	1	2	3	4	5	6	7	8
Codebook size in bits/vector	9	9	7	7	7	7	6	6
Vector dimension = vector length in ms (for 8-point transform, 8 KHz sampl. freq.)	4	4	4	4	8	8	8	8
Data rate Kbits/s	2.25	2.25	1.75	1.75	7/8	7/8	3/4	3/4
Total data rate				11.25 Kbits/s				

An alternative scheme for the wide-band approach to the ATC has been proposed by Fjallbrant and Mekuria [SC-18]. The DCT of primary blocks (eight sample speech segments) are followed by secondary blocks whose lengths are adaptively adjusted to approximately one pitch period (Fig. 7.19). The DCT coefficients of only every third secondary block are transmitted (after adaptive

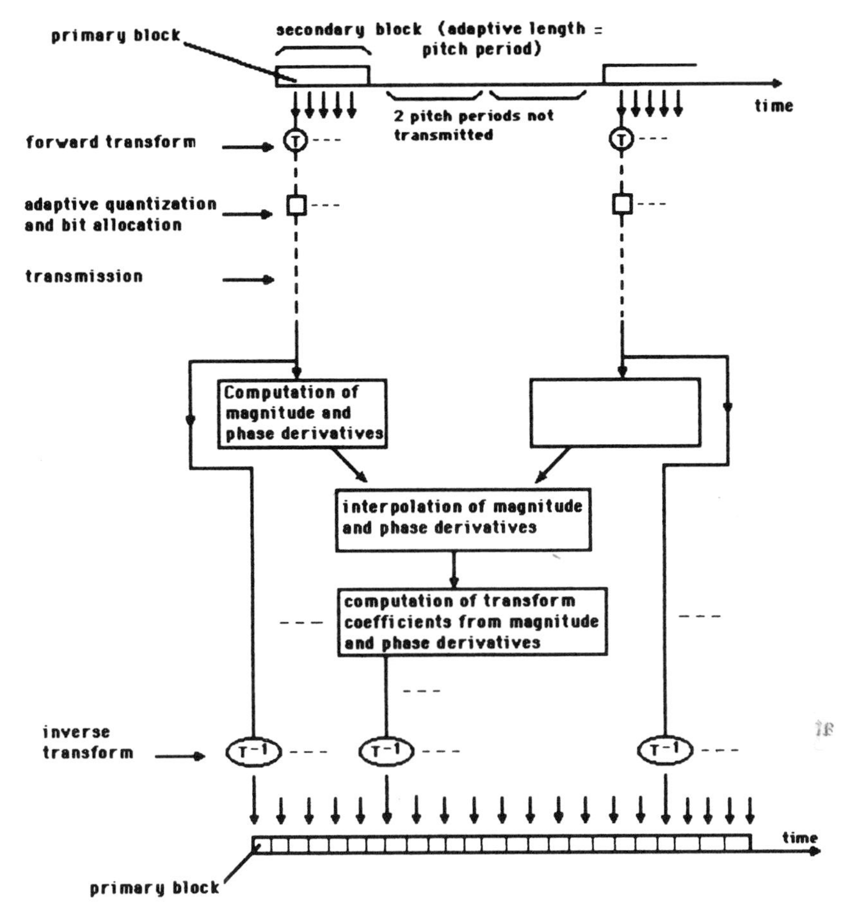

Fig. 7.19 A wideband ATC system including the use of frequency domain phase derivative time trajectories [SC-18].

quantization and bit allocation). Also, the frequency domain magnitude and phase-derivative time trajectories for this secondary block are developed from the transmitted DCT coefficients of this block. The corresponding time trajectories for the other two secondary blocks, whose DCT coefficients are not transmitted, are interpolated from the time trajectories of the blocks preceding and following the two secondary blocks. The DCT coefficients are obtained from these interpolated trajectories. Inverse DCT of all the primary blocks results in the reconstructed speech signal.

7.6.2 *Classified VQ Transform Speech Coder*

Fjallbrant, Mekuria, and Kou [SC-20] have implemented a speech coder on a TMS 32020 DSP chip based on product code VQ of DCT coefficients of 1-ms speech segments (sampled at 8 KHz) coupled with adaptive codebook selection

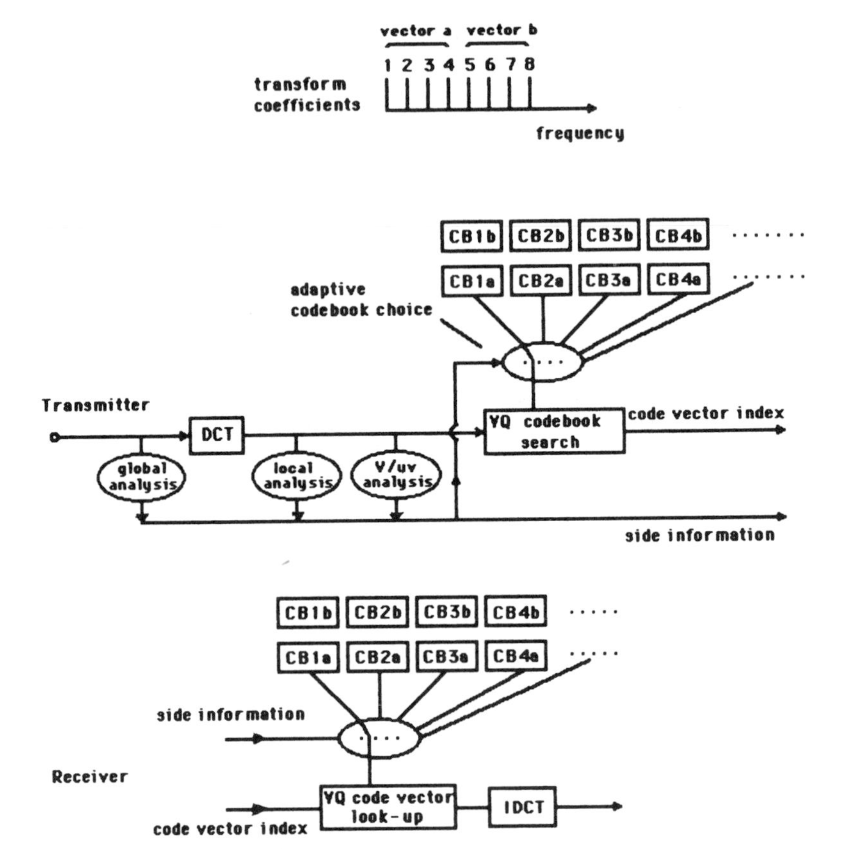

Fig. 7.20 Block diagram of transmitter and receiver [SC-20]. (© 1988 IEEE).

(Fig. 7.20). The product code VQ has two sub-blocks with a total of 256 codevectors, and the eight transform coefficients are divided into two vectors representing the low- and high-frequency regions. Adaptive selection of the 10 codebooks (nine for voiced speech and one for unvoiced speech) is based on the classification results from global and local gain levels for 12 ms of speech, whereas a frequency-domain local analysis leads to the gain levels for each pair of adjacent transform coefficients on the basis of 3 ms of speech. Overhead information (approximately 0.6 KBPS) representing the global, local, and voiced/unvoiced analyses (the latter is based on 36 ms of speech) is transmitted to the receiver along with the index for the selected codevector. Inverse DCT of the codevector based on the index and the side information completes the speech reconstruction process. This hybrid system has reproduced speech at 16 KBPS whose quality is comparable to other types of speech coders. The authors [SC-20] suggest that the time-domain global analysis and the voiced/unvoiced classification can be performed using shorter segments so that the need for echo cancellers does not arise. To reduce the bit rate further, the authors are investigating DCTs of 16 and 32 speech samples, using four to eight sub-blocks in the product code VQ and varying lengths of different codevectors and sizes of different codebooks (Problem 7.33).

7.6.3 *Adaptive Vector Transform Quantization for Speech Coding*

To improve the performance while preserving the complexity at a reasonable level, Cuperman [SC-33] has developed adaptive vector transform quantization (AVTQ) (see Section 7.9 on DCT-VQ) for speech coding (Fig. 7.21). In the AVTQ, implement the following operations:

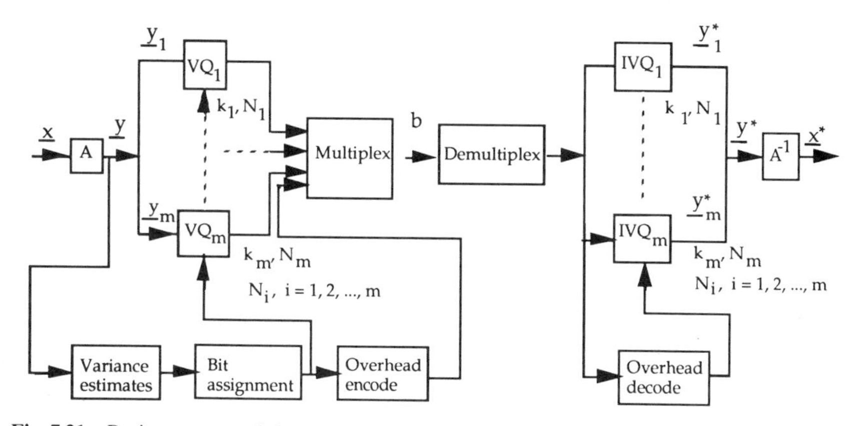

Fig. 7.21 Basic structure of the vector transform quantization system [SC-33]. (© 1989 IEEE).

1. Map each consecutive segment of M speech samples into the DCT domain, i.e.,

$$\mathbf{y} = [C_M^{\mathrm{II}}]\mathbf{x} \tag{7.6.2}$$

where $\mathbf{x} = [x_0, x_1, \ldots, x_{M-1}\}^T$ is the input vector and $\mathbf{y} = [y_0, y_1, \ldots, y_{M-1}\}^T$ is the transform vector.

2. Divide $\mathbf{y}$ into m subvectors $\mathbf{y}_1, \mathbf{y}_2, \ldots, \mathbf{y}_m$ of dimensions $k_1, k_2, \ldots, k_m$, respectively, such that

$$\sum_{i=1}^{m} k_i = M,$$

where $k_i r_i \leqslant \log_2 N_{\max}$, $i = 1, 2, \ldots, m$.

$k_i r_i$ is the number of bits allocated to $\mathbf{y}_i$, and $N_{\max}$ is the maximum codebook size.

$$N_i = 2^{k_i r_i} \qquad \text{and } N = \sum_{i=1}^{m} N_i,$$

where N_i is the codebook size for VQ_i, $i = 1, 2, \ldots, m$.

3. For the subvectors $\mathbf{y}_i$, $i = 1, 2, \ldots, m$, develop the corresponding codebooks C_i. The m vector quantizers VQ_i, $i = 1, 2, \ldots, m$, are equivalent to a VQ having a product codebook C

$$C = \mathop{X}_{i=1}^{m} C_i, \tag{7.6.3}$$

where X denotes the Cartesian product. The product codebook C is the collection of all possible concatenations of the m subvector codewords drawn successively from the m codebooks C_i.

4. For each block of n speech segments (each block has nM samples) develop an adaptive optimal bit assignment as follows:

$$R_i = \bar{R} + \beta_i + \tfrac{1}{2}\log_2\left[\frac{\left(\prod_{j=1}^{k_i} \hat{\sigma}_{ij}^2\right)^{1/k_i}}{\left(\prod_{i=1}^{m}\prod_{j=1}^{k_i} \hat{\sigma}_{ij}^2\right)^{1/M}}\right] \tag{7.6.4}$$

and

$$\hat{\sigma}_{ij}^2 = \frac{1}{n}\sum_{h=1}^{n} y_{ij(h)}^2, \tag{7.6.5}$$

where $y_{ij(h)}$ is the jth DCT coefficient of $\mathbf{y}_i$ in the hth segment (each of length M), $h = 1, 2, \ldots, n$.

β_i depends on quantization coefficients and on vector dimensions. Observe the similarity between (7.6.5) and (7.6.1).

5. Transmit the m bit assignments for each of the nM samples along with the codewords for $\mathbf{y}_i$, $i = 1, 2, \ldots, m$.

6. Inverse operations at the receiver result in the reconstructed speech.

The performance of the AVTQ system for a speech test sequence ($M = 32$, $n = 8$, $m = 5$) as a function of $N_{\max}$ is shown in Table 7.5. Corresponding performance as a function of M is shown in Table 7.6. Changing n, the number of segments from 1 through 8 in a block, has negligible effect on the results. Cuperman [SC-33] has shown that the ATVQ performs better than other techniques such as ATC, VQ, ADPCM, etc.

Table 7.5

Performance of the AVTQ system for different values of the maximum codebook size $N_{\max}$. The rate is 1 bit/sample for $N_{\max} = 1024$ [SC-33]. (© 1989 IEEE).

$N_{\max}$	8	32	128	512	1024
SNR(dB)	4.6	7.8	11.2	14.6	15.9

Table 7.6

Performance of the AVTQ system versus segment size M [SC-33]. (© 1989 IEEE).

M	32	64	128
k_i values	2*2, 3, 16, 9	4*2, 3, 4, 6, 8, 2*16, 3	11*2, 2*3, 5, 6, 2*7, 4*16, 11
SNR (dB)	15.9	17.1	18.6

4*2 means $k_i = 2$, $i = 1, 2, 3, 4$.

A hybrid technique, called waveform-parametric coding, resulting in perceptually good communication quality speech at 9.6 KBPS, has been developed by Shoham and Gersho [SC-21] (Fig. 7.22). Initially the speech is split into variable-length segments (50 to 256 samples for transitional segments versus 256 to 768 samples for steady-state segments; sampling rate = 8 KHz). Pitch extracted during these segments rearranges the segment samples into a variable-size, two-dimensional array before 2D DCT is applied. This is followed by three-level HVQ (see Fig. 7.91a). Side information representing the segment length and the pitch, along with the output of the HVQ is sent to the decoder, where inverse operations reconstruct the speech signal. A number of variables, such as weighted distortion measures, improved codebooks, full search versus tree search, etc., may further improve the speech quality at the same bit rate.

This section has presented a partial spectrum of speech coding techniques especially those involving the DCT. There are a number of other techniques, such as harmonic coding, LPC vocoders, predictive trellis coders, residual

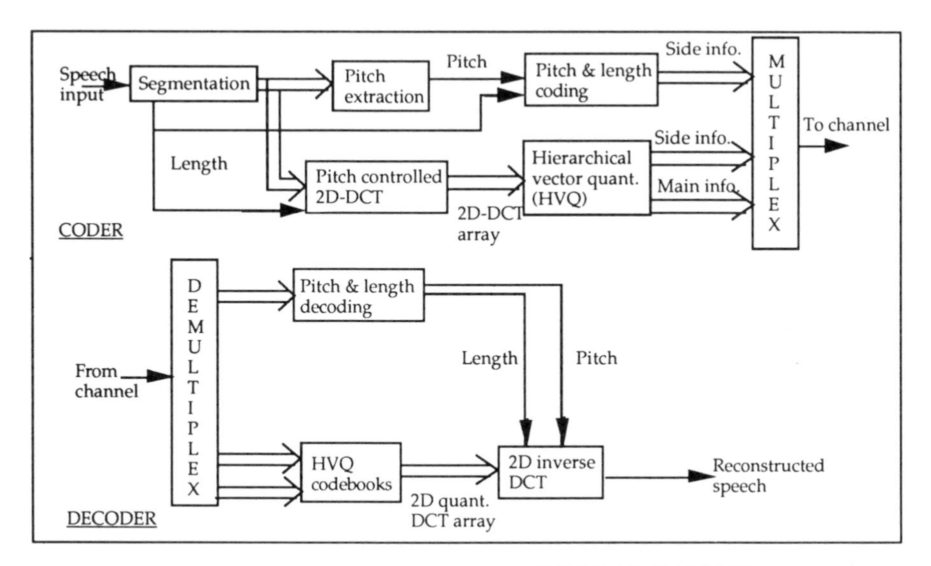

Fig. 7.22 Coder and decoder structures [SC-21]. (© 1984 IEEE).

excited LPC (RELP) vocoder, etc. Also, as outlined earlier, DPCM, VQ, SBC, ATC, TDHS, AR, AGHMM, etc. can be meaningfully combined in different ways to come up with additional avenues. In any case, the overall criterion in selecting a particular algorithm for the speech coder has to be based on speech quality, bit rate, and complexity of implementation. As the industry and research labs have developed the DCT chips [DP-20 through DP-24, DP-26 through DP-31, DP-33 through DP-35, DP-41 through DP-44], speech coders based on DCT-supported algorithms may be quite attractive for specific applications. Also, as the processing time is much slower compared with video coding, DSP chips can be readily used in DCT-driven speech coders [SC-17 through SC-20, SC-23 through SC-26].

7.7 Cepstral Analysis

Hassanein and Radko [CA-1] have applied a modified version of the DCT to the cepstral analysis of speech signals. By forcing a real sequence to be symmetrical, the DCT of the real, even sequence can be computed via the DFT (see Section 4.2). For example, define $X_F(k)$ as the N-point DFT of a M-point, real sequence $x(n)$, for $0 \leqslant n \leqslant M-1$ and zero elsewhere, as

$$X_F(k) = \sum_{n=0}^{N-1} x(n) W_N^{nk} \qquad k = 0, 1, \ldots, N-1, \tag{7.7.1}$$

where $N \geqslant 2M - 2$.

The N-point even sequence $s(n)$ is obtained from $x(n)$ as follows:

$$s(n) = \begin{cases} x(n) & 0 < n < N/2 \\ 2x(n) & n = 0, N/2 \\ x(N-n) & N/2 < n \leqslant N-1. \end{cases} \tag{7.7.2}$$

$S_F(k)$, the DFT of (7.7.2) reduces to (Problem 7.16)

$$S_F(k) = 2\left[x(0) + (-1)^k x\left(\frac{N}{2}\right) + \sum_{n=1}^{(N/2)-1} x(n) \cos\left(\frac{2\pi nk}{N}\right)\right], \tag{7.7.3}$$

which is a modified form of the DCT of $s(n)$.

Note that

$$S_F(k) = \mathrm{Re}[X_F(k)], \qquad k = 0, 1, 2, \ldots, N-1. \tag{7.7.4}$$

It is well known that the DFT of a real, even sequence such as $s(n)$ is real and even (Problem 7.17). The modified DCT (7.7.3) can be computed using an $N/4$-point complex input FFT [M-34]. The cosine transform real cepstrum (CTRC), i.e., the real cepstrum of $s(n)$ can be evaluated as shown in Fig. 7.23. In Fig. 7.23, FFTSYM corresponds to (7.7.3) and $(\text{FFTSYM})^{-1}$ is its inverse. Linear filtering of the cepstrum has been applied to speech, seismic, and EEG signals. Similar to CTRC, its counterpart, cosine transform complex cepstrum (CTCC), can also be

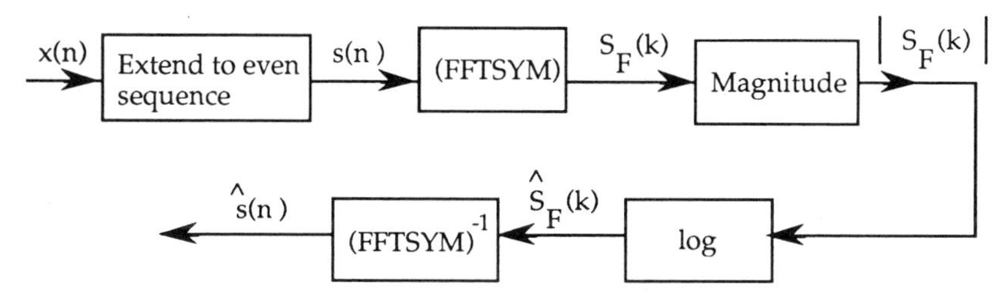

Fig. 7.23 Cosine transform real cepstrum (CTRC) of $s(n)$ [CA-1]. (© 1984 IEEE).

defined [CA-1]. It is shown [CA-1] that the CTRC can be computed with half the complexity of the traditional method and without seriously affecting the true real cepstrum. It is also shown [CA-1] that the CTCC may actually improve the quality of the encoded speech.

7.8 Image Coding

In both Sections 5.1 and 7.1, it was briefly stated that the DCT (so also the other discrete transforms) can be applied to image coding strictly from a bandwidth reduction or data compression viewpoint. The objective here is very simple, i.e., how an image (spatial domain) or sequence of images (spatial-temporal domains) can be mapped into the transform domain such that the bandwidth for transmission or the memory for storage can be reduced with subsequent recovery of the image or image sequence by inverse transformation with negligible distortion. Minimizing the distortion is meaningful in a subjective sense, even though many quantitative measures have been defined. The latter provide a preliminary but partial insight into the efficiency and effectiveness of transform coding. The human viewer, however, is the ultimate judge regarding the quality of the processed images. In transform coding of images, therefore, adaptive features based on psychovisual or perceptual aspects are, in general, introduced such that the image distortions are either not perceived or tolerable. The recovered images hopefully are a near replica of the original images. While other aspects of image processing, such as filtering, enhancement, classification, pattern recognition, and progressive transmission, are discussed elsewhere in this chapter, the objective of this section is to present the fundamentals of transform and/or hybrid (transform/DPCM) coding. Invariably, the comprehensive coding scheme involves many other operations such as quantizers, coders (both source coders and channel coders), buffer, buffer control, error detection and correction, encryption, multiplexers (multiplexing video, data, and audio), interfaces, various sync, control and alarm signals, etc. Also, both preprocessors and postprocessors such as A/D (analog-to-digital), D/A (digital-to-analog), filters, and composite ⇔ component transformations (when color images are coded in a component format) constitute integral parts of the codec (coder and decoder). Power, size, cost (hence marketable realization), versatility, modular expandability, and compatibility with other codecs are also significant factors.

7.8.1 *Quantitative Criteria*

As indicated earlier, various quantitative criteria, in general, are not true evaluators of the performance of a video codec. To obtain a comprehensive judgment regarding the codec efficiency, subjective testing (visual inspection of the processed images) is essential. Also, the codec is subjected to some standard test signals. In the literature, however, quantitative measures such as mean square error (MSE) and signal-to-noise ratio (SNR) are often used to compare different processors. In this context, we can define the following performance measures:

7.8.1.1 Mean Square Error (MSE)

$$E\{[x(m,n) - \hat{x}(m,n)]^2\} \approx \frac{1}{MN} \sum_{m=0}^{M-1} \sum_{n=0}^{N-1} [x(m,n) - \hat{x}(m,n)]^2, \quad (7.8.1)$$

where $x(m,n)$ and $\hat{x}(m,n)$ are the intensities of the original and reconstructed images in row m and column n, respectively. The image size is $M \times N$. The difference between the original and the reconstructed images is the reconstruction error. The approximation described by the right side of (7.8.1) improves as the MSE is computed over a large number of images. Henceforth, we will describe the remaining quantitative measures by their approximations.

7.8.1.2 Normalized MSE (NMSE)

(a)

$$\text{NMSE}_a = \frac{\dfrac{1}{MN} \displaystyle\sum_{m=0}^{M-1} \sum_{n=0}^{N-1} [x(m,n) - \hat{x}(m,n)]^2}{\dfrac{1}{MN} \displaystyle\sum_{m=1}^{M-1} \sum_{n=1}^{N-1} [x(m,n)]^2}. \quad (7.8.2)$$

(b)

$$\text{NMSE}_b = \frac{\dfrac{1}{MN} \displaystyle\sum_{m=0}^{M-1} \sum_{n=0}^{N-1} [x(m,n) - \hat{x}(m,n)]^2}{x_{pp}^2}. \quad (7.8.3)$$

In (7.8.2) and (7.8.3), the MSE is normalized by the image energy and the peak-to-peak image intensity x_{pp}, respectively. For an 8-bit PCM, x_{pp} is 255.

7.8.1.3 Mean Absolute Error (MAE)

$$\mathrm{MAE} = \frac{1}{MN} \sum_{m=0}^{M-1} \sum_{n=0}^{N-1} |x(m, n) - \hat{x}(m, n)|. \tag{7.8.4}$$

7.8.1.4 Normalized MAE (NMAE)

$$\mathrm{NMAE} = \frac{\mathrm{MAE}}{\dfrac{1}{MN} \displaystyle\sum_{m=0}^{M-1} \sum_{n=0}^{N-1} |x(m, n)|}. \tag{7.8.5}$$

The corresponding SNR in dB can be defined by expressing $-10\log_{10}(\mathrm{NMSE})_a$ and $-10\log_{10}(\mathrm{NMSE})_b$. For an image sequence or sequences, the errors are computed over the temporal domain also by expressing $-10\log_{10}(\mathrm{MSE})$, $-10\log_{10}(\mathrm{NMSE})_a$, and $-10\log_{10}(\mathrm{NMSE})_b$. For example, for the MSE in (7.8.1), the errors are computed over the temporal domain also, i.e., MSE in (7.8.1) changes to

$$\frac{1}{MNP} \sum_{m=0}^{M-1} \sum_{n=0}^{N-1} \sum_{p=0}^{P-1} [x(m, n, p) - \hat{x}(m, n, p)]^2, \tag{7.8.6}$$

where the third index p refers to the temporal domain. The remaining distortion measures can be similarly extended based on the image sequences.

7.8.1.5 Normalized Correlation Coefficient (ncc)

$$\mathrm{ncc} = \frac{\displaystyle\sum_{m=0}^{M-1} \sum_{n=0}^{N-1} x(m, n)\hat{x}(m, n)}{\left\{\displaystyle\sum_{m=0}^{M-1} \sum_{n=0}^{N-1} [x(m, n)]^2 \sum_{m=0}^{M-1} \sum_{n=0}^{N-1} [\hat{x}(m, n)]^2\right\}^{1/2}}. \tag{7.8.7}$$

For an ideal processor (reconstructed image is same as the original image) ncc is 1.

7.8.1.6 Essential Maximum (em) or Peak Differential Value This can be best illustrated by an example. If 99% em of a processor is 7, this implies that 99% of the |reconstruction errors| lie in the range 0 to 7. For this to be meaningful, the intensity resolution of the original image has to be defined. As an example, for an 8-bit PCM original image, the possible range of the |reconstruction errors| is 0 to 255. For a good processor the mse, nmse, mae, nmae, and em are to be as small as possible (the appropriate SNR should be large), whereas the normalized correlation coefficient (7.8.7) should be large.

7.8.2 *Subjective Evaluation*

For subjective evaluation, a group of observers (assuming they are image coding experts) view both the original and processed images at random under proper lighting conditions and viewing distances, and arrive at a mean opinion score (mos) based on some rating scale such as Table 7.7. The mos is the weighted average based on the rating scale.

Table 7.7

Subjective impairment scale.

Opinion	Points
Not perceptible	7
Barely perceptible	6
Definitely perceptible but only slight impairment to the image	5
Impairment to the image but not objectionable	4
Somewhat objectionable	3
Definitely objectionable	2
Extremely objectionable	1

While the quantitative and subjective (visual) criteria are the benchmarks for evaluating an algorithm, equally important are other factors such as implementation complexity (hardware realization) and various optional features (for example, split screen, text, documents, annotation, freeze frame, speaker insert, video and/or audio mute, encryption, and overlay on a document capabilities of a teleconferencing codec) pertinent to a specific application.

In Chapter 3, it was shown that the statistically optimal transform is the KLT, which maps a sequence into uncorrelated coefficients and also packs the most energy into the fewest coefficients. Because of the implementation problems (not to mention the complexity involved in generating the KLT), however, KLT is rarely used. As the DCT approaches the KLT in performance based on a number of criteria, as described in Chapter 6, it has been extensively applied in digital signal processing, including image coding. By discarding and/or coarsely quantizing some of the transform coefficients, bit-rate reduction is achieved. By introducing adaptive features in this truncation/coarse quantization process, the image fidelity can be improved.

In transform coding of images, an $N \times N$ image is generally divided into blocks, each of size $L \times L$. For simplicity the rows and columns of an image, and also of the block, are assumed to be of the same size. No generality is lost in this assumption.

In general, block sizes of 8×8 and 16×16 have been utilized in image coding. These small block sizes enable an engineer to introduce adaptive features based on activity or detail in the block. Hardware complexity (memory size and logic) is also reduced considerably compared with the 2D DCT of the entire frame. (A single frame may be of size 512×512.) Although there may be some correlation between neighboring blocks, this is too insignificant to warrant large block size transform beyond 16×16 [IC-20, LBR-6]. Also, based on combined source-channel coding of images, Modestino, Daut, and Vickers [IC-2] have concluded that a 16×16 block size is the best compromise, taking into account the statistical distribution of images, quantizers, and noisy channels. In fact, DCT-based videoconferencing codecs, videophones, and still-frame processors have been developed by the industry based on 8×8 or 16×16 blocks. In low bit-rate coding, block transform processing, however, can result in a blocky structure of the reconstructed image. Various techniques to eliminate or alleviate this artifact have been developed (see Section 7.12). Also 2D DCT of 8×8 blocks has been recommended both by the CCITT SG/XV [LBR-23, LBR-35, LBR-37 through LBR-41, LBR-44] for video teleconferencing at $n \times 384$ KBPS ($n = 1, 2, \ldots, 5$) (this has later changed to $p \times 64$ KBPS, $p = 1, 2, \ldots, 30$ [LBR-35]) and by the Joint Photographic Experts Group (JPEG) [PIT-14, PIT-17, PIT-18, PIT-20 through PIT-27, PIT-30, PIT-33, PIT-34] for progressively transmitting photovideotex, still-frame TV, etc.

Also, the basic still-picture coding algorithm (i.e., 2D DCT of 8×8 blocks) recommended by the JPEG has been extended to moving images (image sequences), with possible transmission over packet networks [PV-5]. The basic JPEG algorithm is applicable to still and moving images covering the spectrum from narrowband to broadband ISDN [G-21]. One of the proposals under consideration by the T1. Y1. 1 group for developing broadcast-quality NTSC codecs at the DS-3 level (44.736 MBPS) is based on 2D DCT of 8×8 pseudo-luminance and -chrominance components [IC-53, IC-67].

After mapping the $L \times L$ block into the transform domain (Fig. 7.24), the selector discards some of the coefficients either adaptively or in a fixed manner. A physical interpretation of this selection process can be observed from Fig. 7.25 in which the basis images (eigen images) of the 2D 8×8 DCT are shown. The top left block is of uniform intensity, representing the average of an image block. Progression from left to right represents an increasing number of vertical edges. Similarly, progression from top to bottom represents an increasing number of horizontal edges, with the bottom right block yielding the maximum mix of the vertical and horizontal edges (this block has a chess pattern). Discarding high-frequency coefficients in the 2D-DCT domain implies deletion of corresponding basis images from the original image, as the 2D DCT is a decomposition process mapping from the spatial domain into the DCT basis images. This structural decomposition can be utilized in adaptively selecting the transform coefficients on a block-by-block basis for further processing (see Figs. 7.29b and 7.161) by assigning each block to one of a finite number of classes based on the distribution of its coefficients. Even though overhead bits indicating the block classification are required, increased bit-rate reduction and improved subjective quality of the reconstructed images can be achieved. Subsequent operations include quantization and coding before transmitting the image information in a serial bit stream over a digital link (microwave, coaxial cable, optical fiber, or satellite) to the receiver for inverse operations. Although Fig. 7.24 shows DCT image coding, it is equally valid for image coding by any orthogonal transform.

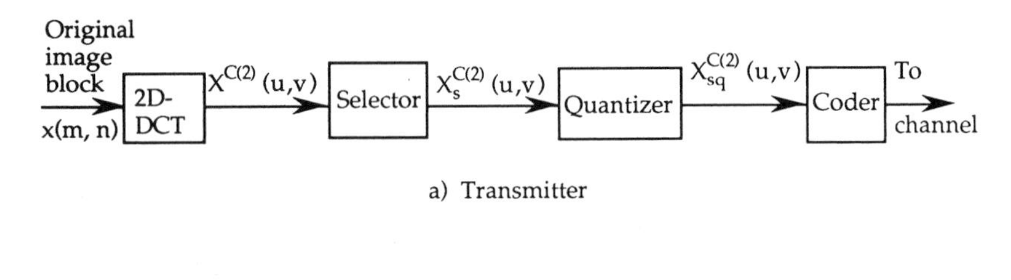

a) Transmitter

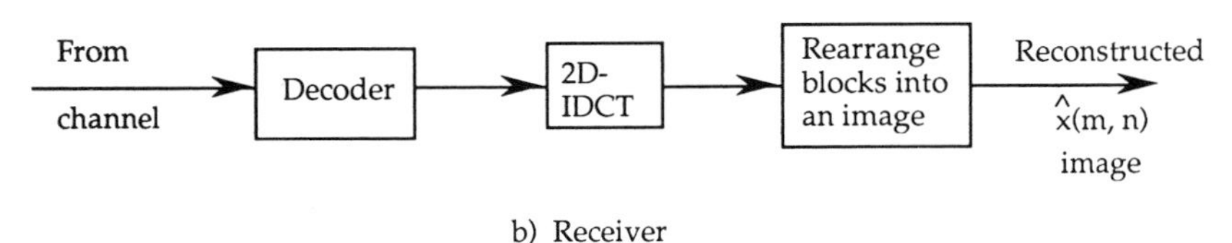

b) Receiver

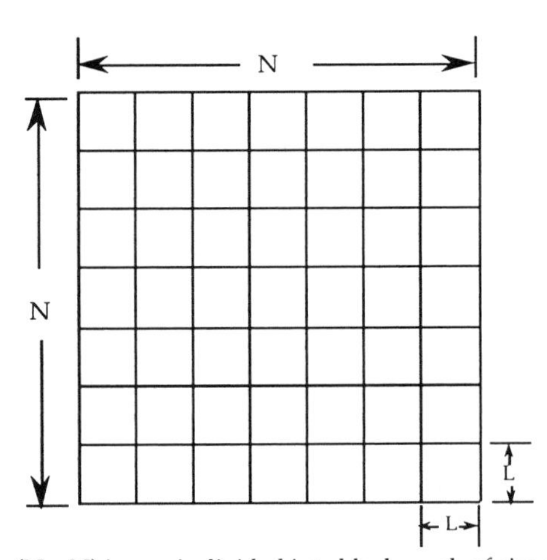

c) An (N x N) image is divided into blocks each of size (L x L)

Fig. 7.24 2D-DCT image coding.

In Fig. 7.24, $X_s^{C(2)}(u, v)$ is the DCT coefficient selected for further processing (see Fig. 7.29). $X_{sq}^{C(2)}(u, v)$ is the quantized value of $X_s^{C(2)}(u, v)$. This is also the output of the decoder, provided the channel is noise free. The error image $[x(m, n) - \hat{x}[m, n)]$ is the cumulative contribution of quantization noise, selector, any bit errors not corrected from the noisy channel, and finite word-length effects of DCT and its inverse. A comprehensive image coding system may include various other operations, such as buffer, buffer control, coding for error detection and correction, multiplexer, sync signals, network interfaces, etc. Even though the transform and its inverse can be any number of the family of discrete unitary transforms such as DFT, DCT, DST, WHT, ST, HT, DHT, DLT, and CMT [B-1 through B-3], we will necessarily limit our discussion to the DCT

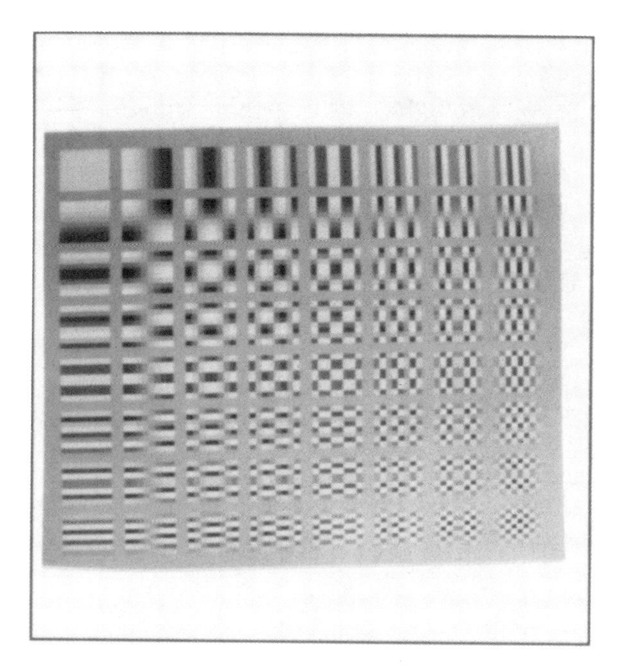

Fig. 7.25 8×8 basis images for the DCT.

and, in particular, to the DCT-II, as the latter is most widely used in image coding. The fundamentals, principles, and concepts in image coding are valid for all the discrete transforms (unless otherwise stated, it is implied that these transforms are unitary). Theoretically a different transform can be applied along each dimension (for example, DCT horizontally and WHT vertically). In practical applications, however, this has been a rarity [IC-51].

Inverse mapping naturally must correspond to the forward transformation. Also, as described in Chapter 5, although efficient algorithms have been developed for direct implementation of 2D DCT-II, hardware (discrete components in a board form or VLSI chip) has been designed and developed [DP-26 through DP-29, DP-31, DP-33, DP-34, DP-37, DP-39, DP-41 through DP-44, IC-67, IC-73, HDTV-9] taking advantage of the separability property of the DCT-II, (see Fig. 5.1). In quantitative terms the operations of transform image coding can be described as follows:

2D DCT-II

$$X^{C(2)}(u, v) = \frac{4C(u)C(v)}{L^2} \sum_{m=0}^{L-1} \sum_{n=0}^{L-1} x(m, n) \cos\left[\frac{(2m+1)u\pi}{2L}\right] \cos\left[\frac{(2n+1)v\pi}{2L}\right]$$

$$u, v, = 0, 1, \ldots, L-1, \qquad (7.8.8a)$$

where $C(u)$ and $C(v)$ are defined in (5.2.3).

2D IDCT-II

$$x(m, n) = \sum_{u=0}^{L-1} \sum_{v=0}^{L-1} C(u)C(v)X^{C(2)}(u, v) \cos\left[\frac{(2m+1)u\pi}{2L}\right] \cos\left[\frac{(2n+1)v\pi}{2L}\right]$$

$$m, n, = 0, 1, \ldots, L-1, \qquad (7.8.8b)$$

where $X^{C(2)}(u, v)$ is the 2D DCT-II of the image data $x(m, n)$. A careful reader may notice the difference between (5.2.3) and (7.8.8) regarding the normalization factor. Whereas in (5.2.3), the normalization factor $4/(MN)$ is split equally between the forward and inverse transforms, in (7.8.8) the corresponding factor $4/L^2$ appears entirely in the forward transform. Alternately, we can split this factor equally between the two transforms or move it entirely to the inverse transform. Theoretically any of these formats is immaterial, although in hardware realization, scaling, overflow, and other considerations may influence a particular choice.

As the image energy is preserved under an orthogonal transformation, the following is valid.

$$\sum_{m=0}^{L-1} \sum_{n=0}^{L-1} [x(m, n)]^2 = \sum_{u=0}^{L-1} \sum_{v=0}^{L-1} [X^{C(2)}(u, v)]^2. \qquad (7.8.9)$$

Whereas $X^{C(2)}(0, 0)$ represents the mean or average intensity of the image block, the remaining coefficients $X^{C(2)}(u, v)$ are components of increasing horizontal and vertical frequencies (Fig. 7.26). By applying zonal filtering (maximum variance zonal sampling, MVZS) in the 2D-DCT domain, only those coefficients with large variances (the remaining coefficients are set to zero) are processed further (quantized and coded). The variance distribution in the 2D-DCT-II domain can be developed based on a large number of test images. Because of the energy packing capability of the DCT-II (see Figs. 6.1–6.3 and Tables 6.1–6.3), most of the image energy is still retained in the few but relevant transform

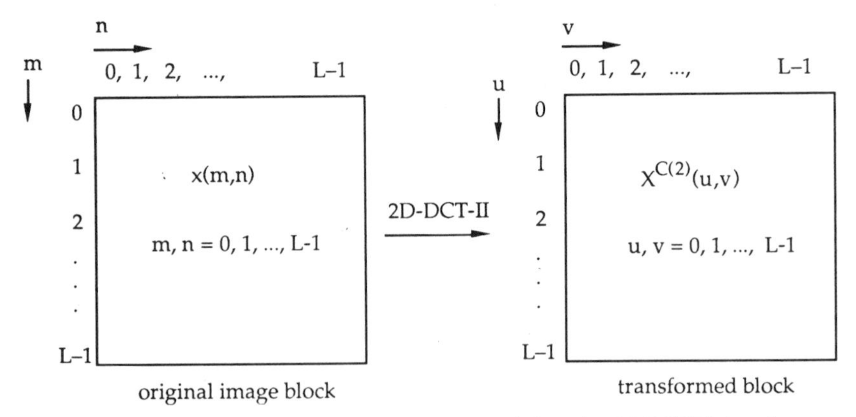

Fig. 7.26 Mapping an $L \times L$ original image block into the 2D-DCT-II domain.

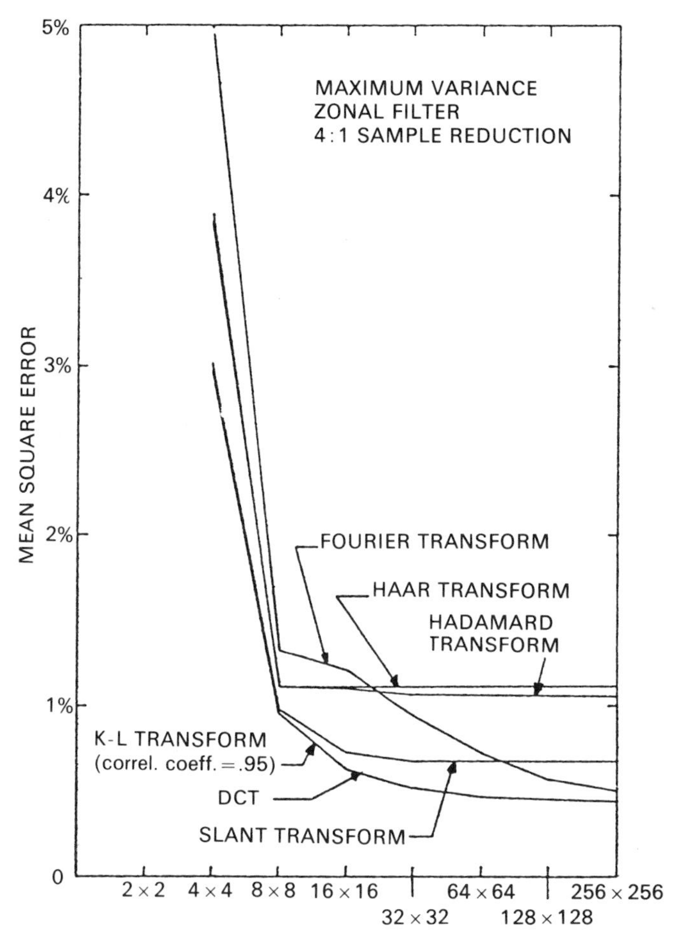

Fig. 7.27 MSE of image transforms as a function of block size. The image statistics along both rows and columns are assumed to be a first-order Markov process with a correlation coefficient of $\rho = 0.95$. No quantization and coding [IC-15]. (© 1974 IEEE).

coefficients. This can also be observed in Fig. 7.27, where the mse between the original and reconstructed images is shown for MVZS as a function of block size [IC-15]. For the various discrete transforms, only 25% of the transform coefficients having large variances are retained (the remaining coefficients are set to zero) for inverse transformation. This figure illustrates two properties: (a) The MSE for the DCT is almost the same as for the statistically optimal KLT. (b) There is no significant decrease in MSE in block processing of images beyond 16×16. At this stage, no overhead bits need be sent to the receiver, as the variance distribution can be set up as a look-up table in the receiver memory. In threshold sampling, however, large overhead is required, as the receiver has to know the locations of transform coefficients being coded. This type of sampling, is adaptive, i.e., only those coefficients whose magnitudes are above a threshold are transmitted. For example, the number of transform coefficients of "GIRL" below threshold versus the threshold level for various discrete transforms is shown in Fig. 7.28 [IC-57]. This indicates that a large number of transform coefficients have small magnitudes, confirming the energy packing property of these transforms. As in Fig. 7.27, DCT approaches the performance of the KLT. The image "GIRL" is of size 256 × 256, with each pel quantized to 8-bit PCM. Another type of selection is based on geometrical zones, i.e., only those coefficients within a specified zone are processed (Fig. 7.29a). If only low-frequency coefficients are selected, this process is tantamount to a low-pass filter. Some high-frequency information relating to edges or contours can be lost in this selection. The frequency distribution of these coefficients and the features of the block they represent are shown in Fig. 7.29b [HTC-8].

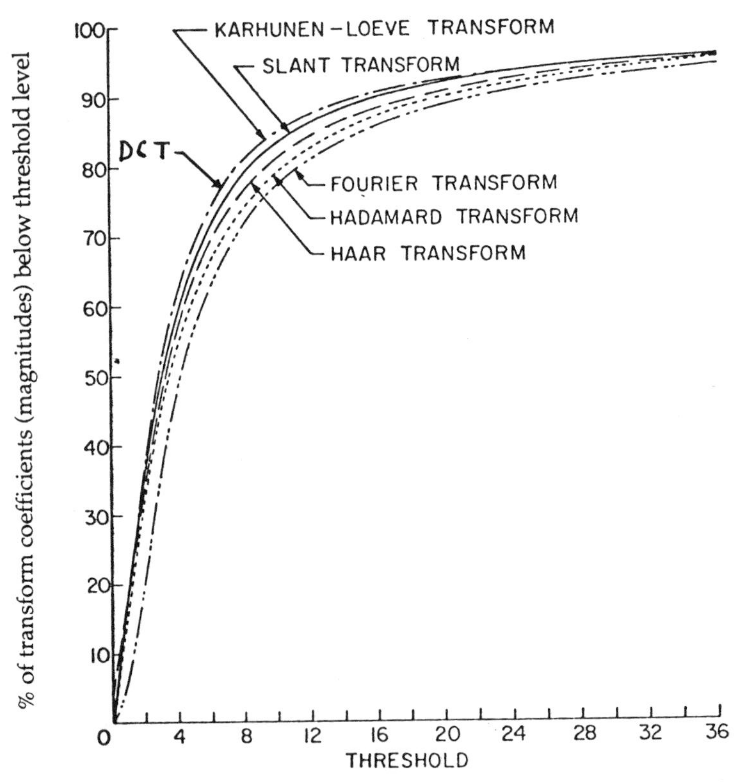

Fig. 7.28 Number of transform coefficients of GIRL below threshold versus threshold level. (Transform is performed in 16 × 16 blocks.) [IC-57].

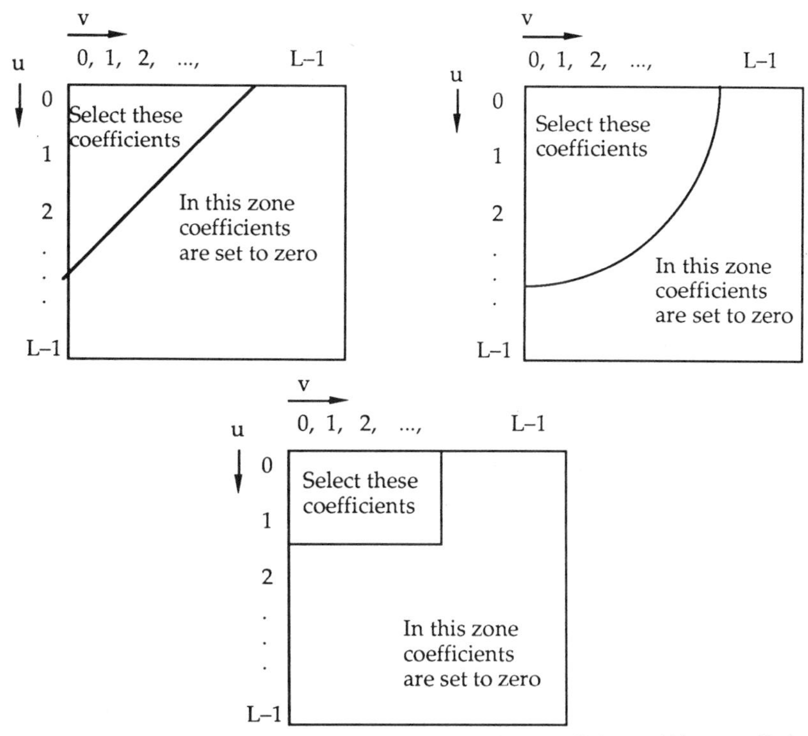

Fig. 7.29a Geometrical zonal sampling. Only those transform coefficients within a specified zone are processed further, with the remaining set to zero. This selection corresponds to frequency filtering.

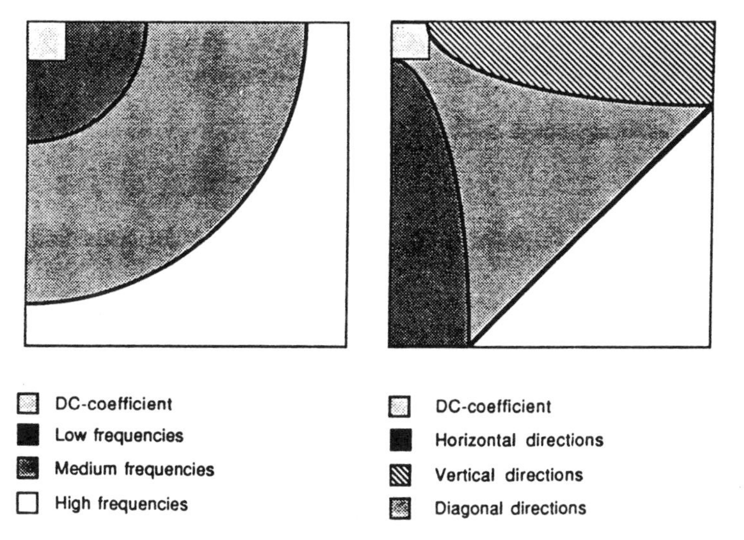

Fig. 7.29b Frequency distribution of the 2D-DCT coefficients and the block features they represent [HTC-8] (see Fig. 7.25).

A more popular selection mechanism is based on zig-zag scanning (see Fig. 7.174) of the coefficients and selecting those coefficients above a threshold. By designing an appropriate quantizer, some of the low-magnitude coefficients can be automatically set to zero. Tescher and Cox [IC-20] have shown that the DCT coefficient variances are highly correlated along the zig-zag scan. Also, Ngan [IC-4] has shown that the variances of the zig-zag scanned DCT coefficients can be approximated by a linear relationship in the logarithmic domain (Fig. 7.30). In this case only the slope and intercept need be transmitted. Also, these variances can be iteratively estimated [IC-19, IC-20, IC-39] as follows:

$$\hat{\sigma}_n^2 = A_1\hat{\sigma}_{n-1}^2 + (1 - A_1)[X_{sq}^{C(2)}(n-1)]^2, \tag{7.8.10}$$

where the weighting factor A_1 has been experimentally chosen as 0.75. $\hat{\sigma}_n^2$ is the variance estimate of the nth 2D-DCT coefficient, and $X_{sq}^{C(2)}(n-1)$ is the quantizer output (Fig. 7.24) both scanned along the zig-zag path.

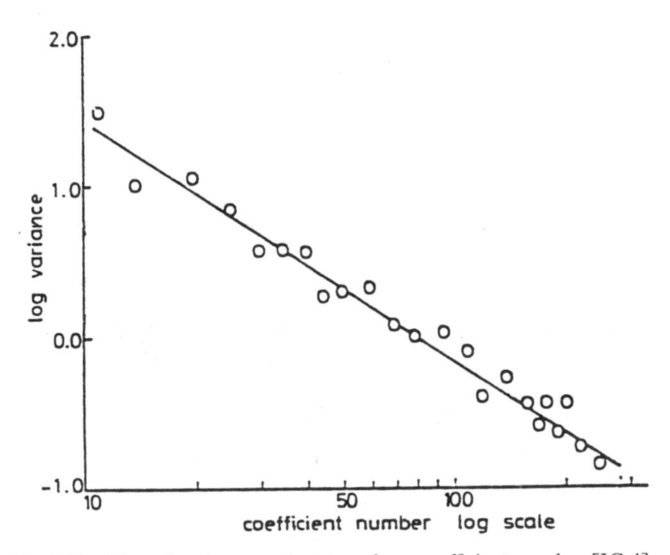

Fig. 7.30 Plot of variance against transform coefficient number [IC-4].

In HDTV image coding (Section 7.11) the zig-zag scanning is extended to the 3D DCT domain [HDTV-10]. After global motion estimation (Section 7.10), 3D DCT of $8 \times 8 \times 4$ blocks is performed (8 pels, 8 lines, and 4 frames). Prior to quantization and coding, the 3D DCT coefficients are scanned as follows:

7.8.2.1 Moving Areas Scan corresponding coefficients of successive frames from low to high spatial frequencies. For spatial scanning follow the zig-zag format (see Fig. 7.174).

7.8.2.2 Fixed Areas 3D-DCT coefficients are rearranged based on the zig-zag scanning of the first frame, followed by the zig-zag scanning of the successive frames.

3D-DCT coding provides any number of scanning mechanisms, the objective being to tailor these variations to detail, activity, motion, etc.

Intraframe image coding (Fig. 7.24) can be extended to interframe image coding, wherein the 3D DCT is applied to an image sequence (Fig. 5.2). Under this transformation the spatial and temporal domains are transformed to corresponding frequency domains. Though such a process has been simulated based on actual image sequences [IC-8, IC-18, IC-48, IC-54], 3D transform coding has been seldom used in practice because of the considerable computational complexity and extensive memory required to process and store the frames. For example, for a 3D DCT of $16 \times 16 \times 16$ pel cubes, 16 frames need to be stored in the codec memory. As indicated in Chapter 5, the 3D DCT, being a separable transform, can be implemented by a series of 1D DCTs along each dimension (Fig. 5.2). 3D DCT, however, has been applied to 6:1 compression of multispectral scanner data based on $4 \times 4 \times 4$ blocks composed of 4×4 blocks from each of the four bands [IC-55]. Another application is 3D DCT of $16 \times 16 \times 16$ blocks involving human heart slices at the same instant (3D image) and the same slice recorded at various instants (moving image sequence) in which, by adaptive threshold sampling and adaptive quantization, 32:1 data compression was achieved [IC-48]. Another application is 3D DCT of 3D blocks displaced by global motion estimation performed on each frame. These blocks are generated from noninterlaced HDTV frames [HDTV-10]. 3D DCT of $8 \times 8 \times 4$ blocks of TV sequences in component form was also simulated [IC-78]. The 3D blocks are composed of 8×8 blocks from four consecutive fields.

7.8.2.3 Variance Distribution In MVZS (Fig. 7.24) the variance distribution of the transform coefficients is useful for bit allocation among the transform coefficients. This is called block quantization, as the bit assignment scheme is based on the variances of the transform coefficients (Problem 7.10). A fast computational method for block quantization based on marginal analysis that achieves superior performance compared with the rate-distortion theoretic approach [Q-5] has been developed by Tzou [Q-6]. An example of a bit assignment map in the 2D-DCT domain based on variance distribution is shown in Fig. 7.31. This map indicates the number of bits allocated to the corresponding transform coefficients. Observe that most of the bits are allocated to the low-frequency zone, reflecting the energy compaction property of the DCT. The original 8-bit PCM signal is compressed to 1 BPP by the transform coding (Fig. 7.24). The variances can be expressed as

$$\tilde{\sigma}^2(u, v) = E[X^{C(2)}(u, v) - \bar{X}^{C(2)}_{(u,v)}]^2 \qquad u, v = 0, 1, \ldots, L-1, \tag{7.8.11}$$

$\rightarrow f_x$

	7	6	4	4	3	3	3	3	2	2	2	1	1	1	0	0
	6	4	4	4	3	3	3	2	2	2	1	1	1	0	0	0
	5	4	4	3	3	3	2	2	2	2	1	1	1	0	0	0
	4	4	3	3	3	2	2	2	2	1	1	1	0	0	0	0
↓	4	3	3	3	2	2	2	2	1	1	1	1	0	0	0	0
	3	3	3	2	2	2	2	1	1	1	1	0	0	0	0	0
f_y	3	2	2	2	2	2	1	1	1	1	0	0	0	0	0	0
	2	2	2	2	2	1	1	1	1	0	0	0	0	0	0	0
	2	2	2	1	1	1	1	1	0	0	0	0	0	0	0	0
	2	2	1	1	1	1	1	0	0	0	0	0	0	0	0	0
	2	1	1	1	1	1	0	0	0	0	0	0	0	0	0	0
	1	1	1	1	1	0	0	0	0	0	0	0	0	0	0	0
	1	1	1	1	0	0	0	0	0	0	0	0	0	0	0	0
	1	1	1	0	0	0	0	0	0	0	0	0	0	0	0	0
	1	0	0	0	0	0	0	0	0	0	0	0	0	0	0	0
	0	0	0	0	0	0	0	0	0	0	0	0	0	0	0	0

Fig. 7.31 Bit assignment map in the 2D-DCT domain based on variance distribution of the transform coefficients [IC-4].

where $\tilde{\sigma}^2(u, v)$ is the variance of $X^{C(2)}(u, v)$ and the overbar on the right side of (7.8.11) indicates the statistical mean. In practice, a large number of test images can be used to estimate these variances. If the image blocks are assumed to be governed by a zero mean separable Markov process described by

$$[\Sigma_r] = \sigma_r^2 \begin{bmatrix} 1 & \rho_r & \rho_r^2 & \cdots & \rho_r^{L-1} \\ \rho_r & 1 & \rho_r & \cdots & \rho_r^{L-2} \\ \cdots & \cdots & \cdots & \cdots & \cdots \\ \rho_r^{L-1} & \rho_r^{L-2} & \rho_r^{L-3} & \cdots & 1 \end{bmatrix} \tag{7.8.12a}$$

and

$$[\Sigma_c] = \sigma_c^2 \begin{bmatrix} 1 & \rho_c & \rho_c^2 & \cdots & \rho_c^{L-1} \\ \rho_c & 1 & \rho_c & \cdots & \rho_c^{L-2} \\ \cdots & \cdots & \cdots & \cdots & \cdots \\ \rho_c^{L-1} & \rho_c^{L-2} & & \cdots & 1 \end{bmatrix}, \tag{7.8.12b}$$

where $[\Sigma]$, σ^2, and ρ represent the correlation matrix, variance, and adjacent correlation coefficient (all in the spatial domain), with the subscripts r and c denoting the rows and columns, respectively, then the transform coefficient variances are [IC-8, IC-54]:

$$\tilde{\sigma}^2(u, v) = \tilde{\sigma}_r^2(u, u)\tilde{\sigma}_c^2(v, v) \qquad u, v = 0, 1, \ldots, L-1, \tag{7.8.13}$$

where $\tilde{\sigma}_r^2(u, u)$ and $\tilde{\sigma}_c^2(v, v)$ are the diagonal elements of the row and column variance matrices in the transform domain. For the DCT-II these matrices are:

$$[\tilde{\Sigma}_r] = [C_L^{II}][\Sigma_r][C_L^{II}]^T \tag{7.8.14}$$

and

$$[\tilde{\Sigma}_c] = [C_L^{II}][\Sigma_c][C_L^{II}]^T,$$

where $[\tilde{\Sigma}_r]$ and $[\tilde{\Sigma}_c]$ are 2D DCT of $[\Sigma_r]$ and $[\Sigma_c]$, respectively.

For an image sequence assuming separable zero-mean Markov process, the transform coefficient variances can be developed similar to (7.8.13).

In Fig. 7.24, if the image is reconstructed from the selector output, then the MSE between the original and reconstructed images is given by

$$\text{MSE} = \sum_{u=0}^{L-1} \sum_{v=0}^{L-1} E[X^{C(2)}(u, v)]^2[1 - S(u, v)], \tag{7.8.15}$$

where $S(u, v)$ is 1 or 0 depending on whether the transform coefficient is selected or not.

An example of DCT image coding applied to NTSC component signals, Y, I, and Q is shown in Fig. 7.32. As the bandwidths of the luminance Y and chrominance I and Q are 4.2 MHz, 1.5 MHz, and 0.5 MHz, respectively, the color components are subsampled.

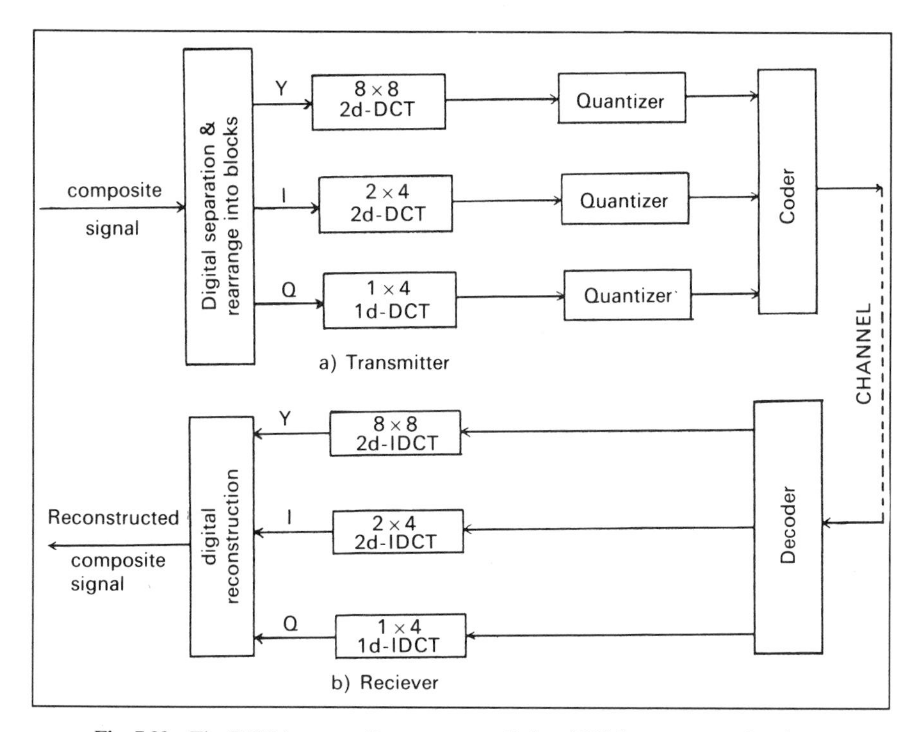

Fig. 7.32 The DCT image coding system applied to NTSC component signals.

7.8.2.4 Hybrid (Transform/DPCM) Coding Transform/DPCM, called hybrid coding, was first proposed by Habibi [IC-56]. (Hybrid is more general in that it implies the combination of two or more different schemes.) Transform coders are quite efficient for large-scale compression (say 6:1 or higher), more robust to channel noise because of the inherent averaging property, but suffer from computational complexity. Fast algorithms (see Chapters 4 and 5) and dedicated chips have minimized the latter handicap. DPCM (differential pulse code modulation) coders, on the other hand, are simple to implement, fairly effective for small bit-rate reduction (say 2:1 through 6:1), but are highly sensitive to channel noise. The compression ratio threshold of 6:1 between these two techniques is purely arbitrary. It may be noted that both these techniques have been used over a wide range of compression ratios, the specific application being the deciding factor. In any case, the philosophy behind the transform/DPCM coding is to take advantage of their strong points while minimizing their weaknesses. Before describing the different hybrid methods, an elementary introduction to DPCM is in order.

DPCM: Only the basic concepts of DPCM will be presented. The stress will be on various schemes for bandwidth compression using predictors and quantizers. For an in-depth discussion the reader is referred to standard books such as [B-4]. As signals, in general, are highly correlated, a reasonable estimation of the present sample $x(n)$ can be made based on the previous sample(s). A basic DPCM coder and decoder are shown in Fig. 7.33a. Assuming a lossless channel the following expressions can be developed:

$$e_p(n) = x(n) - x_p(n) \qquad \text{prediction error}$$

$$e_{p_q}(n) = e_p(n) - q(n) \qquad \text{quantized prediction error}$$
$$q(n) = e_p(n) - e_{p_q}(n) \qquad \text{quantization error}$$

$$\hat{x}(n) = x_p(n) + e_{p_q}(n) \qquad \text{reconstructed signal}$$

For a noise-free channel, the reconstruction error, $x(n) - \hat{x}(n) = q(n)$, is the quantization error. The predictors in the coder and decoder are identical and are synchronized. Also, prediction is based on previously predicted samples rather than previous samples, as the former are also available at the decoder. This is a simple memoryless prediction scheme. To improve the coding efficiency various adaptive predictors and/or quantizers can be incorporated. A simple example of adaptive prediction is predicting the present sample x based on its closest match with the nearest neighbors (Fig. 7.34) described as follows:

$$x_p(n) = A, \text{ or } B, \text{ or } C, \text{ or } D \text{ based on } \min(|x - A|, |x - B|, |x - C|, |x - D|). \qquad (7.8.16)$$

Overhead bits indicating which one among the pels (A, B, C, D) chosen for prediction need to be sent to the receiver. One way to avoid the overhead information is to apply the adaptive scheme to the present pel based on the previous pel prediction, which is available at the receiver. It is clear that the

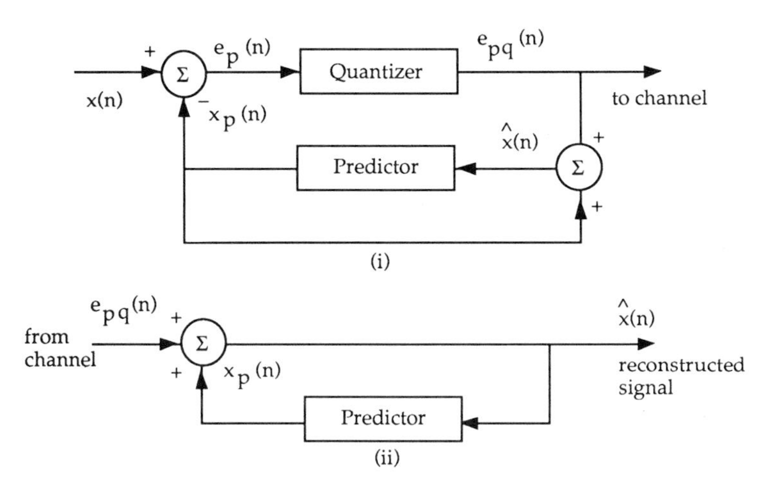

Fig. 7.33a DPCM coding: (i) coder, (ii) decoder. Adaptive DPCM involves adaptive predictor and/or adaptive quantizer.

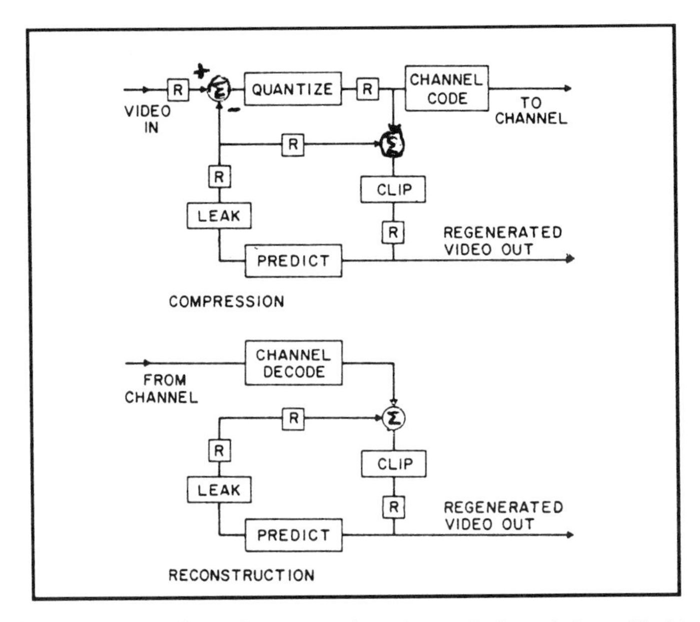

Fig. 7.33b Data compression and reconstruction using predictive techniques. The blocks labeled R are registers [IC-65]. (© 1987 IEEE).

adaptive prediction requires more logic compared with the fixed prediction. Another example is to evaluate a number of different predictors over a group of pels and to choose the predictor that yields the minimum of the absolute sum of the prediction errors for that group. In this case overhead bits indicating the

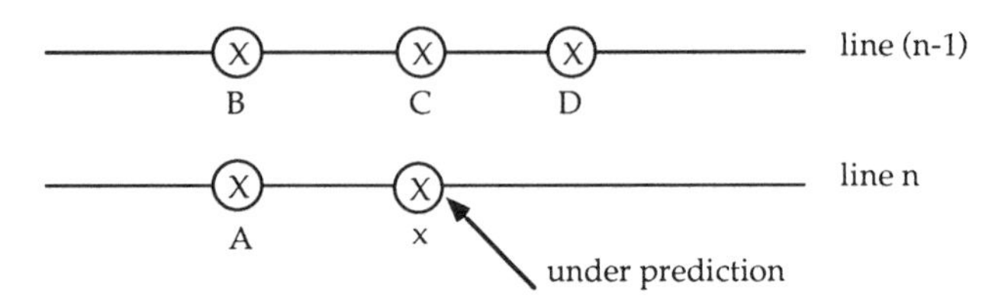

Fig. 7.34 Adaptive prediction of pel x based on its nearest neighbors, A, B, C, and D.

predictor selection have to be transmitted to the receiver. Alternately, one can choose the best among the intraline (1D), intrafield (2D), and interfield or interframe (3D) predictors. A predictor described by (7.8.17) is called an Nth order linear predictor.

$$x_p(n) = \sum_{m=1}^{N} a_m \hat{x}(n-m). \tag{7.8.17}$$

The coefficients a_m are the weights and they reflect the correlations between the present and previous samples. The predictor is classified as 1D, 2D, or 3D depending on $\hat{x}(n-m)$ in (7.8.17). For example, if prediction is based on previous samples from the present frame and also from the previous frame, then it is called a *3D predictor*. For a well-designed prediction scheme, the entropy of prediction error is less than that of the input signal and thus can be quantized to fewer levels. This is the key to the bandwidth reduction of the original signal. Further reduction can be achieved by variable word-length (VWL) coding of the quantizer tables. In general, the prediction error distribution tends to be Laplacian (see Figs. 7.44 and 7.45). Hence by assigning smaller bit size (word length) to inner levels having larger probabilities and vice versa, additional compression can be achieved. As VWL coding tends to be more sensitive to a noisy channel compared with the fixed bit-size coding, error detection and correction become more complex. Also in DPCM and in transform/DPCM coding, periodic refresh (or even demand refresh) is introduced in which the digitized input signal is transmitted periodically without any coding. Distortions of DPCM image coding are granular noise (grainy structure in uniform regions), contouring patterns, slope overload near edges (visible losses in edge and contrast rendition), and edge busyness (Fig. 7.35). Reducing granular noise requires fine quantizer levels, whereas attenuation of slope overload can be accomplished by large quantization steps. These conflicting requirements compel the quantizer to be designed based on art rather than on the basis of a mathematical model. Quantizers matched to the human visual sensitivity (visibility thresholds) have been designed [IC-64]. These quantizers mask the DPCM distortions and generate visually acceptable images. In smooth areas the quantizer reconstruction levels tend to oscillate around a uniform level, resulting in a grainy structure as a visual artifact. Edge busyness is defined as a noisy edge when the prediction scheme is unable to reconstruct the precise

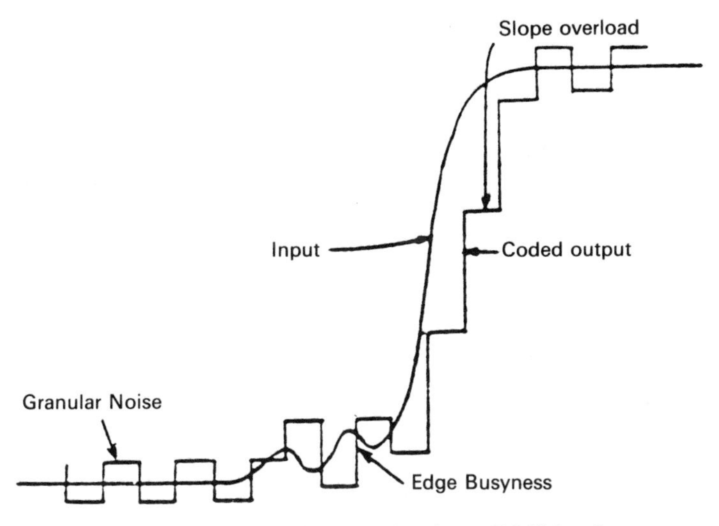

Fig. 7.35 Quantizing distortion due to DPCM coding.

continuity of an edge. In interframe or interfield coding, this artifact appears as a blurred moving edge. When the number of quantization levels are too small, the images have layered effects (both in intensity and color), generating contouring patterns. In predictive coding, a noisy channel can cause error propagation, resulting in visual degradation. To counteract this impairment, a combination of schemes, such as error detection and correction, periodic refresh (the input signal is transmitted in PCM format, periodically circumventing the DPCM loop), and introducing a leak factor into the predictor output, is necessary. The leak function, which is very close to 1, say 15/16 or 31/32, attenuates the error propagation and forces the predictor output to converge towards the middle of its dynamic range. To prevent the overflow or underflow that can be generated by the addition of predictor output and the quantized prediction error, a clip circuit is introduced before the predictor [IC-65]. The DPCM coder and decoder shown in Fig. 7.33a are modified in Fig. 7.33b, showing the leak and the clip circuits.

In both transform and transform/DPCM coding of images, because of quantization noise and zonal filtering (i.e., dropping of some transform coefficients), in the reconstructed image, distortion is distributed over the entire block, even when spatial activity is distributed locally in the block. This equal distortion distribution is visually perceptible around slowly moving heads and shoulders of speakers in a videoconferencing or a videophone environment. This visible artifact is called the mosquito effect.

7.8.2.5 One-Dimensional Transform/DPCM Also, a hybrid technique in which a block is processed based on 1D transform horizontally followed by the

DPCM vertically in the transform domain has been proposed by Habibi [IC-56] (Fig. 7.36). Theoretically 1D transform vertically and DPCM horizontally also can be implemented. A 1D-DCT/DPCM coder for the Army's AQUILA RPV has been designed and built by RCA [DP-5] (Fig. 7.37). The hybrid coder, together with spatial and temporal subsampling, achieves data transmission rates at 200, 400, 800, and 1600 KBPS. The hybrid technique has also been applied to NTSC composite video signal sampled at $3f_{SC}$ (f_{SC} is the subcarrier frequency). The process involves 1D DCT of 32 samples of the same subcarrier phase along a horizontal line followed by a first-order predictor along the vertical direction [IC-6].

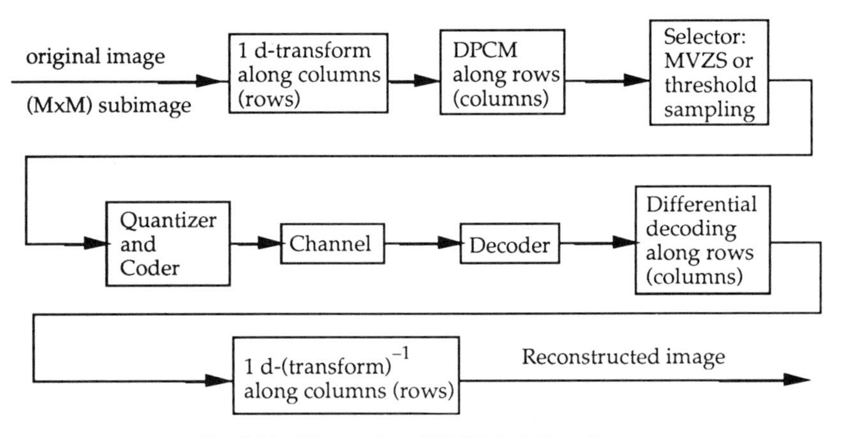

Fig. 7.36 1D-transform/DPCM hybrid coding.

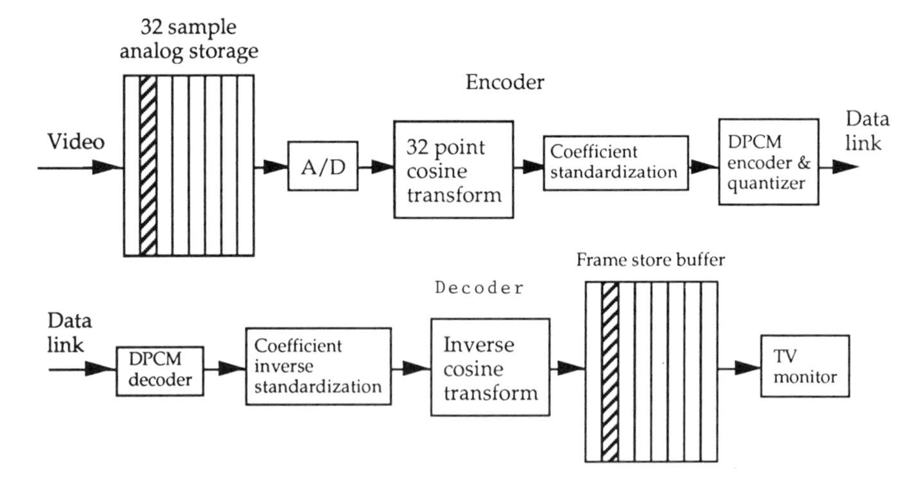

Fig. 7.37 Block diagram of DCT/DPCM TV encoding system [DP-5].

7.8.2.6 TWO-DIMENSIONAL TRANSFORM/DPCM Block transform processing neglects correlation among adjacent blocks. Habibi [IC-56] has proposed DPCM between adjacent blocks in the 2D transform domain to reduce or remove this correlation (Fig. 7.38). An adaptive interblock DPCM in which the 4×4 transformed block is predicted based on its closest match among the four nearest neighbors (west, northwest, north, and northeast) has been applied to the NTSC composite signal (Fig. 7.39) [IC-6]. The prediction errors of the 2D DCT coefficients are quantized to different numbers of levels. The block processing yielded subjectively better picture quality at lower bit rates compared with line processing. It may be noted that the transform/DPCM processing requires a bank of DPCM loops, as shown in Fig. 7.40 for the 2D transform/DPCM. In general, 2D transform/DPCM has been extensively applied to image sequences. The object here is to remove or reduce the spatial correlation by the transform and the temporal correlation by the predictor. Improved temporal prediction

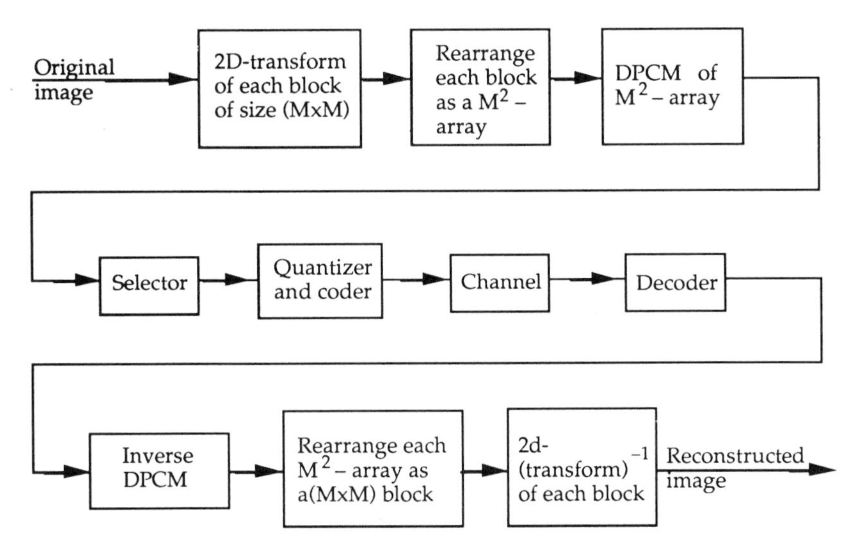

Fig. 7.38 2D-transform/DPCM image processing.

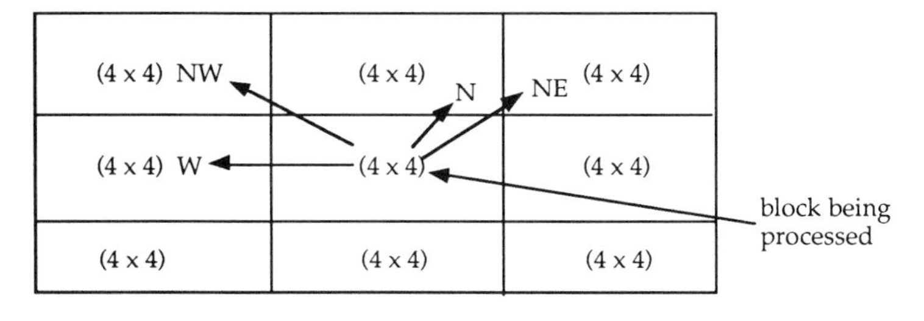

Fig. 7.39 Adaptive prediction of the center block (2D DCT/DPCM) [IC-6]. (© 1982 IEEE).

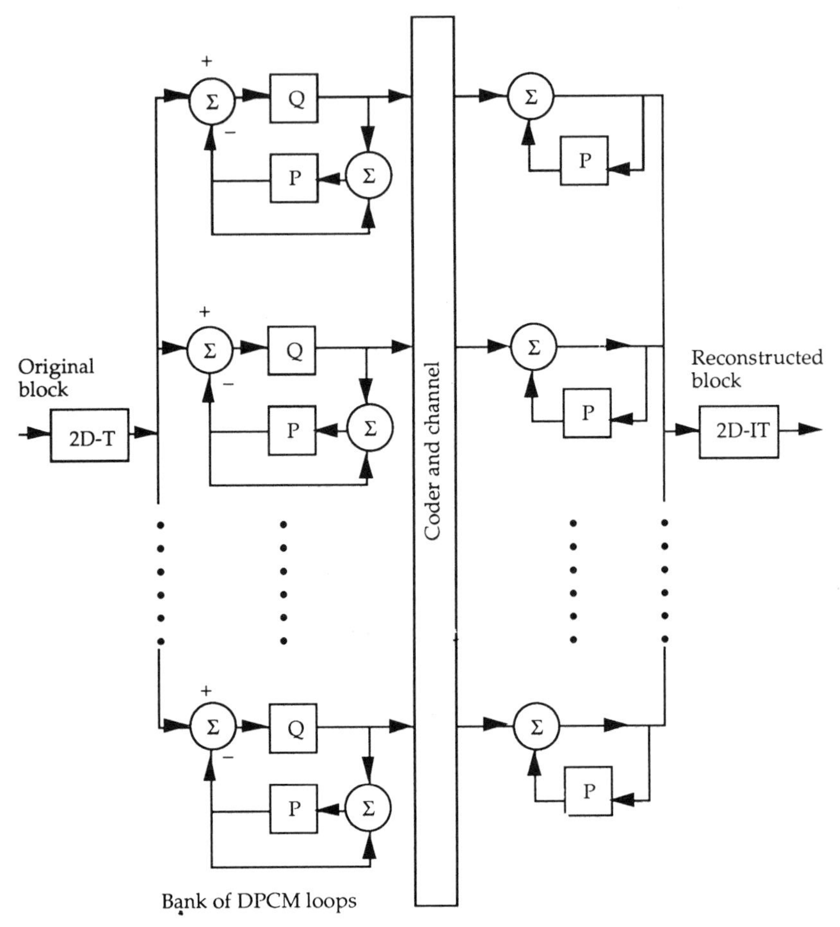

Fig. 7.40 2D-transform/DPCM coding system. P = predictor; Q = quantizer; 2D-T = 2D transform; 2D-IT = 2D inverse transform [IC-56]. (© 1974 IEEE).

can be obtained by motion estimation and compensation. In this scheme, the prediction is along the estimated motion trajectory. Various block motion estimation algorithms (see Figs. 7.108–7.113) have been developed, and some of them have been incorporated in teleconferencing and videophone codecs. Also, motion estimation VLSI chips are being designed and developed (see Appendix B.3). Motion compensated hybrid coding is discussed in Section 7.10.

The 2D DCT/DPCM is applied to color component signals Y, I, and Q or Y, U, and V (see Fig. 7.118). In the hybrid systems described so far (Fig. 7.36–Fig. 7.41), the prediction is in the transform domain (Fig. 7.42). Theoretically, however, the order can be interchanged, i.e., prediction in the spatial or temporal domain followed by transform of the prediction errors (Fig. 7.43). Ericsson [IC-26] has shown that the two systems (Figs. 7.42 and 7.43) are equivalent provided

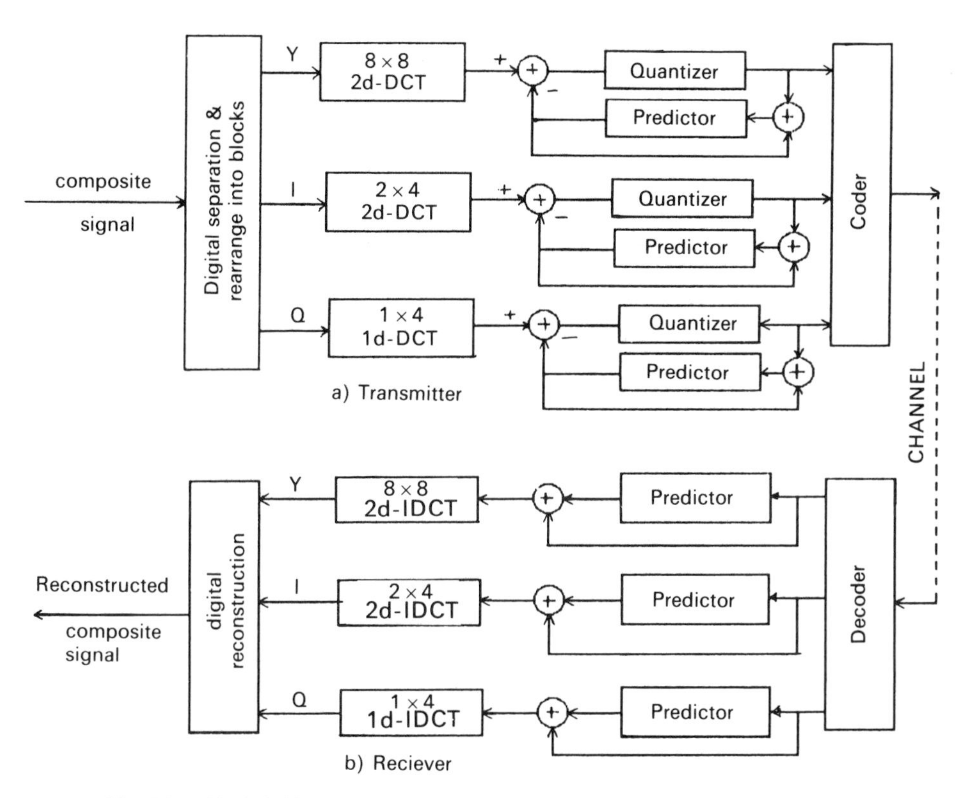

Fig. 7.41 The hybrid (DCT/DPCM) coding system applied to component signals.

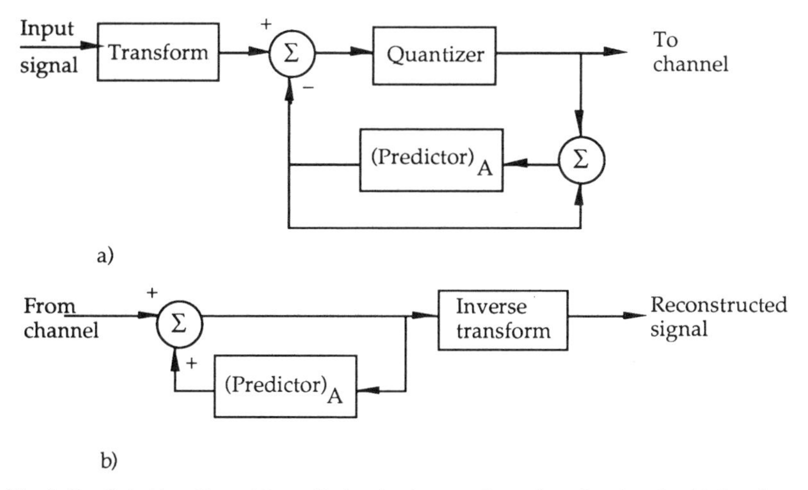

Fig. 7.42 Hybrid coding with prediction in the transform domain. a) coder; b) decoder.

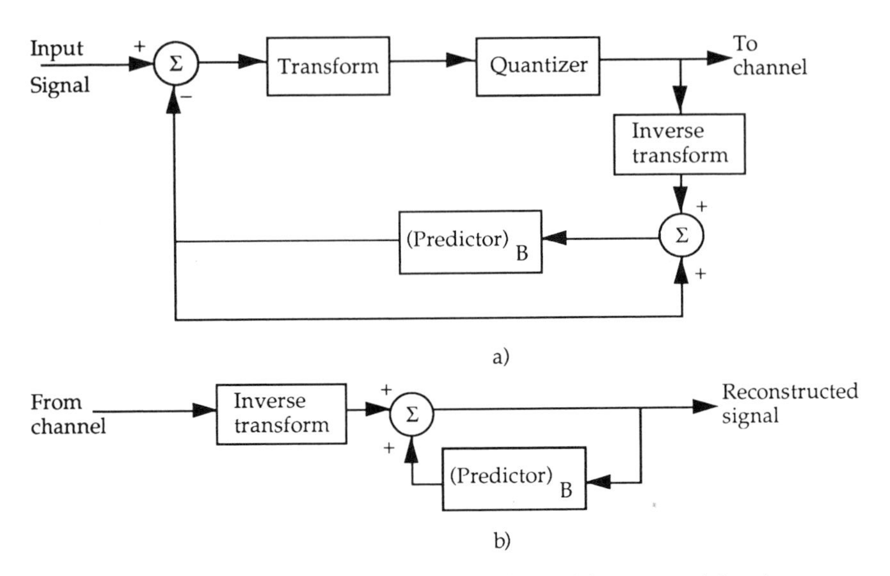

Fig. 7.43 Hybrid coding with prediction in the spatial or temporal domain.

the two quantizers are the same and

$$(\text{predictor})_B = (\text{transform})(\text{predictor})_A(\text{inverse transform}).$$

Even though Fig. 7.43 requires three transforms (one forward and two inverse), compared with only two transforms for the transform domain prediction (Fig. 7.42—one forward and one inverse), the former is invariably adopted in practice because

(1) in motion-compensated hybrid systems (see Section 7.10), motion is estimated in the spatial domain;

(2) the system allows staggered transform blocks for reducing the block structure;

(3) the system allows inclusion of nonlinear temporal filtering of the prediction error before transformation; and

(4) it is the best structure when some other space domain processing such as segmentation is carried out.

Traditionally the transform coefficients other than the dc component are assumed to follow the Gaussian distribution based on the central limit theorem [B-6]. The dc component, on the other hand, can be governed by the Rayleigh distribution, as it represents the average intensity of an image. Both [IC-1] and [IC-3] indicate that the ac coefficients, in general, deviate from the Gaussian assumption. Using test images and adaptive quantizers, Modestino, Farvardin, and Ogrinc [IC-1] have shown improved quantitative (SNR) and qualitative (subjective quality) performance compared with the Gaussian matched quantizers at rates of 1 BPP and above. Conducting the goodness of fit distribution tests on a large class of images, Reininger and Gibson [IC-3] have shown that the 2D-DCT ac coefficients, in general, can be more accurately modelled by Laplacian distribution compared with any other distribution. Similarly the dc component, in general, can be more accurately represented by Gaussian than by Rayleigh statistics. It is therefore safe to conclude that the dc and ac transform coefficients can be best represented by Gaussian and Laplacian statistics, respectively. On the other hand, the prediction errors in the DPCM case and the transform coefficients of the prediction errors in the hybrid (DPCM/transform) case are invariably modelled by the Laplacian distribution (Figs. 7.44 and 7.45). As selected transform coefficients can have all possible values within their bounds (see [LBR-5] on the bounds for the 2D-DCT coefficients), these coefficients need to be quantized to a finite number of levels

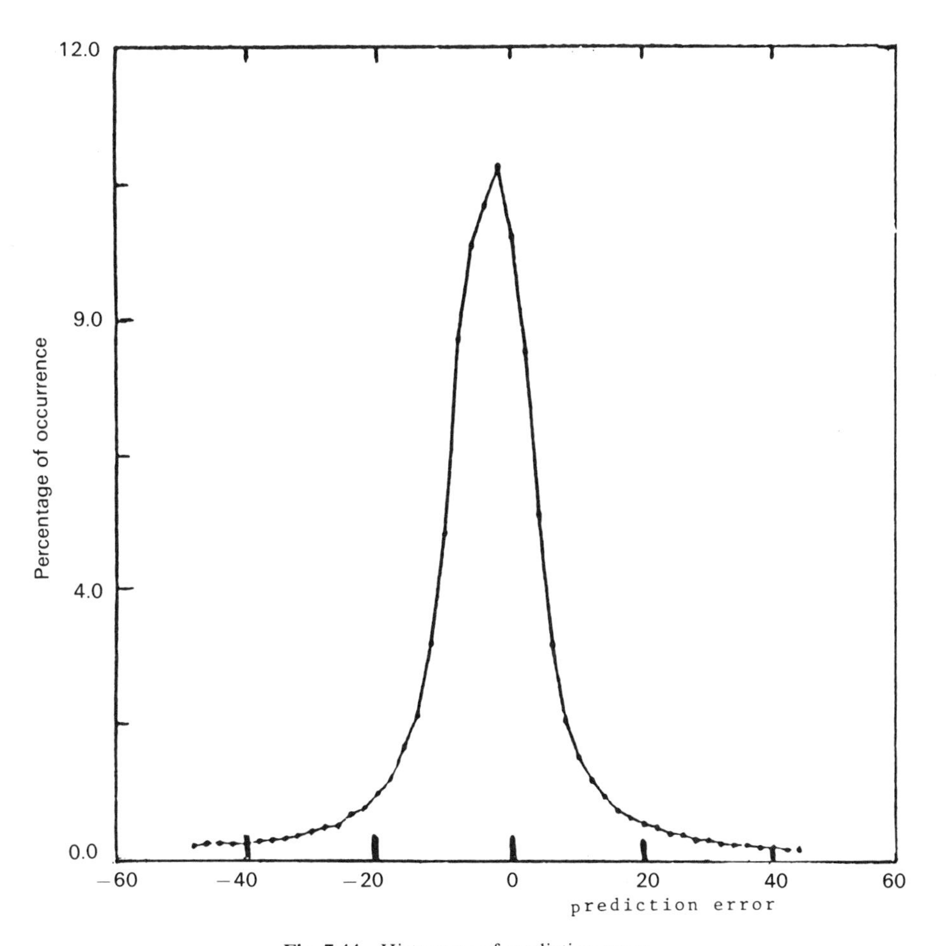

Fig. 7.44 Histogram of prediction error.

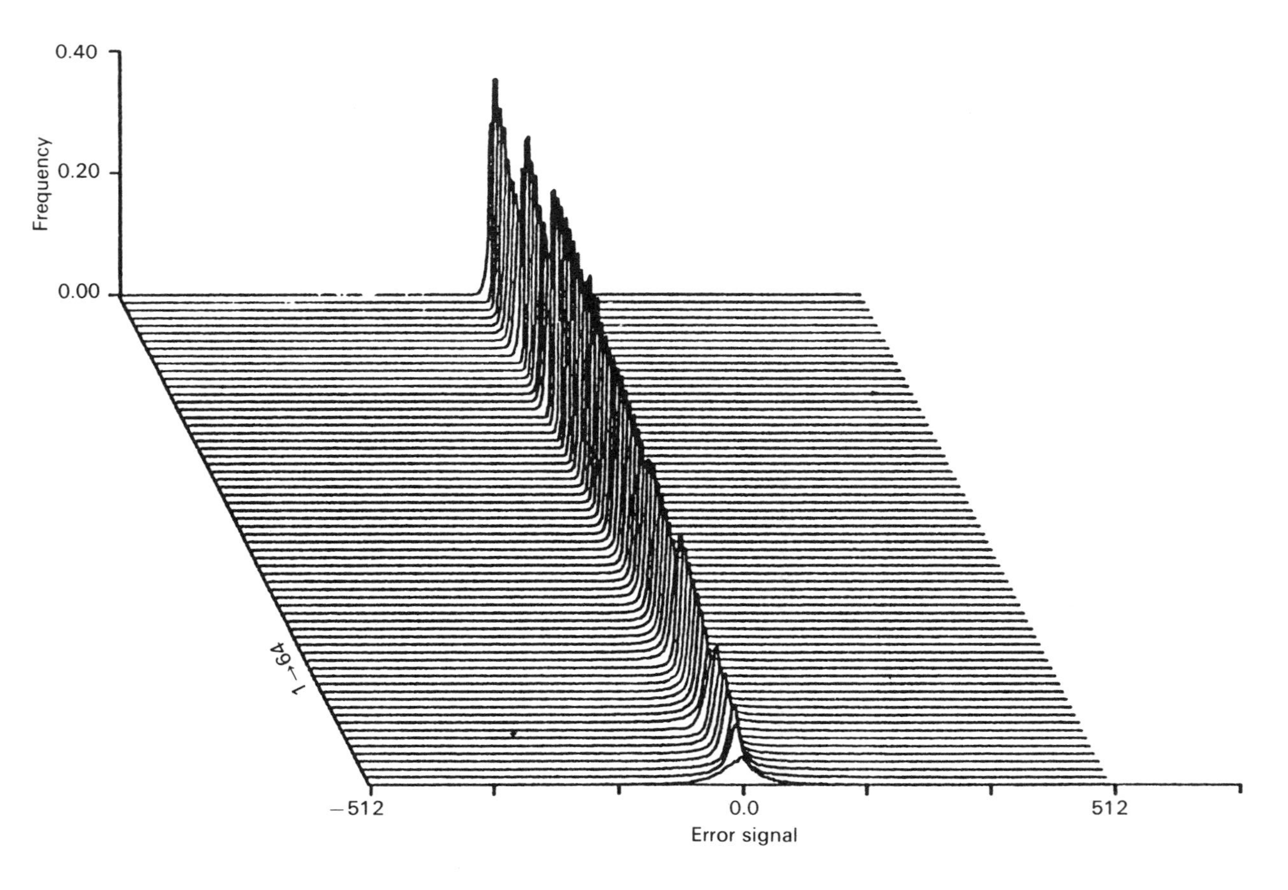

Fig. 7.45 Histograms of prediction errors (rearranged in a lexicographic order of 2D-DCT of 8×8 blocks).

before coding. A well-known quantizer design is based on the Lloyd–Max algorithm [Q-1, Q-3], which iteratively can develop the quantizer based on minimizing a specific distortion (quantization noise) for a particular distribution. Quantizers can be categorized as uniform or nonuniform, symmetric or nonsymmetric, and memoryless or with memory. Our discussion will be limited to memoryless symmetric quantizers. The former implies that quantization of the present sample is independent of the quantizer's previous history, whereas the latter denotes that the input (decision) levels and output (representative) levels are symmetric for both positive and negative ranges. For a symmetric quantizer, therefore, the positive (negative) levels completely describe the quantizer. A uniform quantizer (Fig. 7.46) can be described by the number of levels and step size (apart from midtreader or midriser).

Unlike the uniform quantizer, which has equal step sizes, a nonuniform quantizer is characterized by varying step sizes (Fig. 7.47). In general, for small input values the quantization is fine, becoming coarser as the |input| increases. An example of quantization distortion is the familiar mean square quantization error (MSQE) defined as

$$\text{MSQE} = \int (x - x_q)^2 p(x) dx = E[(x - x_q)^2], \tag{7.8.18}$$

where x and x_q are the input and its quantized output and $p(x)$ is the probability density function of x.

Other distortion measures similar to those described in Section 7.9 can be defined. The quantization tables available in the literature [Q-1 through Q-3] are based on zero mean and unit variance for standard distribution functions. Since the ac transform coefficients (these have zero means) are normalized (divided) by their standard deviations before quantization, the transform coefficient variance matrix has to be either transmitted to the receiver or estimated for denormalization of the quantized transform coefficients at the receiver. The dc coefficient, being an important parameter, is in general uniformly and finely quantized. The quantizer can also be matched to the actual statistics based on a large number of test sequences. An example of such a quantizer is shown in Table 7.8, in which the prediction errors are quantized prior to dual word-length (DWL) coding, which is a simple case of the variable word-length (VWL) coding.

7.8.3 *Interfield Hybrid Coding of Component Color TV Signals*

Following the 2D 8×8 DCT of luminance Y, and chrominance I and Q, interfield prediction of corresponding blocks in the 2D DCT domain is implemented [IC-7]. The transform coefficients of the previous field form the prediction of corresponding coefficients of the current field. Using 22-frame sequences of "Water Skier" and "Wheel of Fortune," the variances of the prediction errors of the DCT coefficients were utilized to obtain the bit

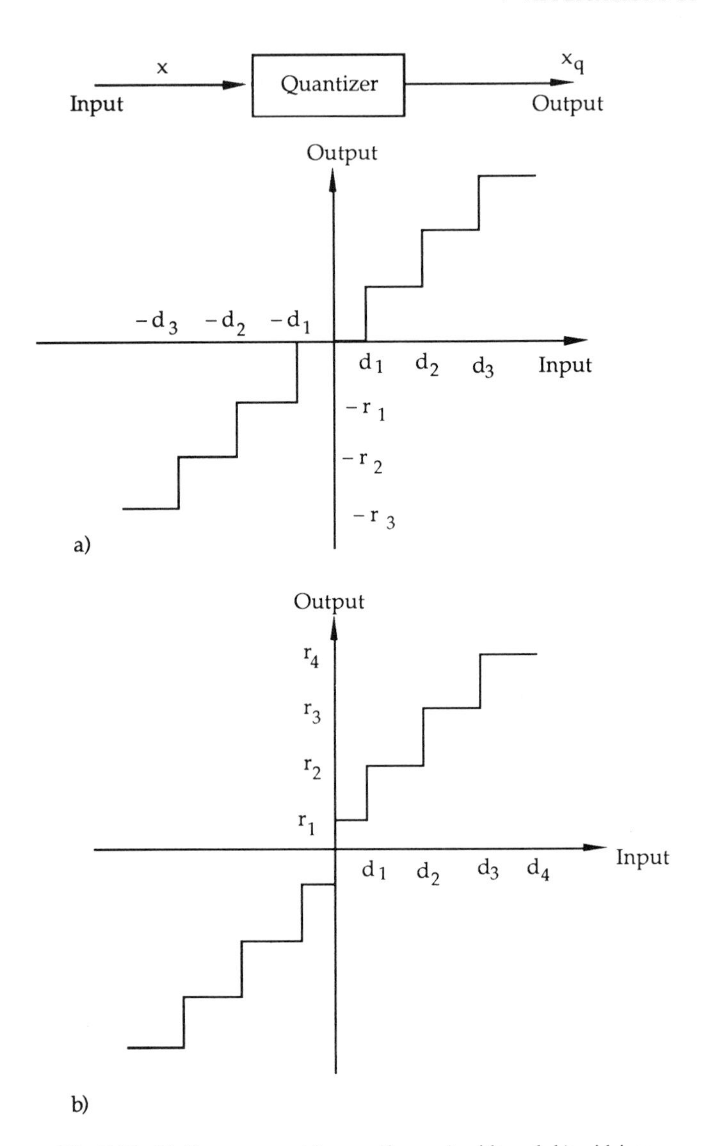

Fig. 7.46 Uniform symmetric quantizers: a) midtread; b) midriser.

allocation matrices (Fig. 7.48) for an overall average bit rate of 2 BPP. The bits in this figure indicate the number of levels that the corresponding prediction errors were quantized (nonuniform Max quantizer). To further improve the coding process, adaptive features were introduced into the hybrid scheme (Fig. 7.49). The luminance 8×8 block was divided into four sub-blocks with a minimum correlation between the sub-blocks based on an extensive statistical study of different groups (Fig. 7.50). The dc term is monitored separately to

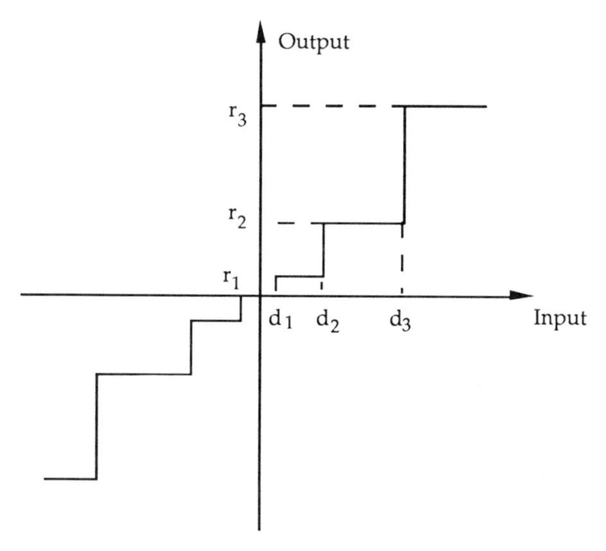

Fig. 7.47 A nonuniform symmetric quantizer. Similar to Fig. 7.46b, a midriser can be described for the nonuniform case.

determine the temporal activity (motion, scene change, etc.), and its cumulative probability distribution is obtained with the threshold for temporal activity set at 0.5 (Fig. 7.51). Similarly, the histograms of the ac energies of the sub-blocks were utilized to monitor the spatial activities (detail, edges, etc.) (Fig. 7.52). Bit allocation among the sub-blocks was adapted to these temporal and spatial activities. These techniques are described in detail in [IC-7]. The adaptive interfield hybrid coding has resulted in improved subjective sequences, especially during scene changes. This technique covers only the basic image coding scheme. Various other issues such as channel errors, multiplexers, network interfaces, pre and post processors, sync signals, etc, need to be considered in the overall system.

7.8.4 *Broadcast-Quality Video Codec at the (44.736 MBPS) DS-3 Level*

While the DCT-based codecs have been simulated, investigated, designed, and built for image coding at various bit rates starting from 56/64 KBPS (for progressive still-frame image coding, the bit rate at the first stage is as low as 0.08 BPP [PIT-18, PIT-20, PIT-24, PIT-26, PIT-27]), DCT also has been applied to broadcast-quality TV image coding at the DS-3 (44.736 MBPS) level [IC-53, IC-67, IC-75, IC-78]. Similarly, a motion-compensated adaptive intra/interframe 15 MBPS codec for CCIR recommendation 601 [G-21] 4:2:2 TV signals has been proposed [IC-61]. The algorithm is based on various modes, intraframe (2D DCT), interframe (DPCM/2D DCT), zig-zag scanning,

Table 7.8

Nonuniform quantizer matched to the prediction errors.

Quantizer		Coder Assignment	
Input (+ −)	Output (+ −)	(+)	(−)
A. Dual Mode (29 Levels)			
0	0	1111	1111
1	1	1010	0101
2	2	1110	0001
3	3	1100	0011
4 – 6	5	1000	0111
7 – 9	8	1001	0110
10 – 14	12	1011	0100
15 – 20	17	1101	0010
21 – 28	24	00001010	00000101
29 – 38	33	00001110	00000001
39 – 50	44	00001000	00000111
51 – 63	57	00001011	00000100
64 – 77	70	00001100	00000011
78 – 92	85	00001001	00000110
93 – 225	100	00001101	00000010
B. Forced Mode (15 Levels)			
0 – 1	0	1111	1111
2 – 5	3	1100	0011
6 – 11	8	1001	0110
12 – 19	15	1101	0010
20 – 28	24	1010	0101
29 – 38	33	1110	0001
39 – 50	44	1000	0111
51 – 225	57	1011	0100

DPCM quantizer levels and code assignment:
4-bit words: assigned to 15 most probable levels.
8-bit words: assigned to 14 least probable levels.
0000: long word prefix.
00001111: frame sync code.
000000000000: not assigned, used for stuffing.

(a)	(b)	(c)
5 4 3 2 1 0 1 1	4 2 1 0 0 0 0 0	3 1 0 0 0 0 0 0
4 3 3 2 1 0 1 1	3 1 0 0 0 0 0 0	2 1 0 0 0 0 0 0
4 3 2 1 1 0 1 1	2 1 0 0 0 0 0 0	2 0 0 0 0 0 0 0
4 3 2 1 1 0 1 1	2 1 0 0 0 0 0 0	1 0 0 0 0 0 0 0
4 2 2 1 1 0 0 1	2 0 0 0 0 0 0 0	1 0 0 0 0 0 0 0
4 2 2 1 0 0 0 0	1 0 0 0 0 0 0 0	1 0 0 0 0 0 0 0
3 2 2 1 0 0 0 0	0 0 0 0 0 0 0 0	0 0 0 0 0 0 0 0
3 2 2 1 0 0 1 1	0 0 0 0 0 0 0 0	0 0 0 0 0 0 0 0

Fig. 7.48 Bit allocation for the prediction errors for overall average bit rate of 2.0 bits/pel. (a) *Y*, (b) *I*, (c) *Q* [IC-7]. (© 1981 IEEE).

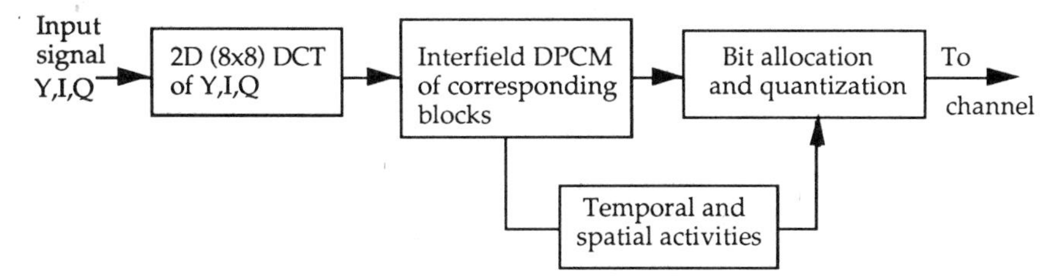

Fig. 7.49 Adaptive interfield hybrid (2D DCT/DPCM) component video coding scheme [IC-7]. (© 1981 IEEE).

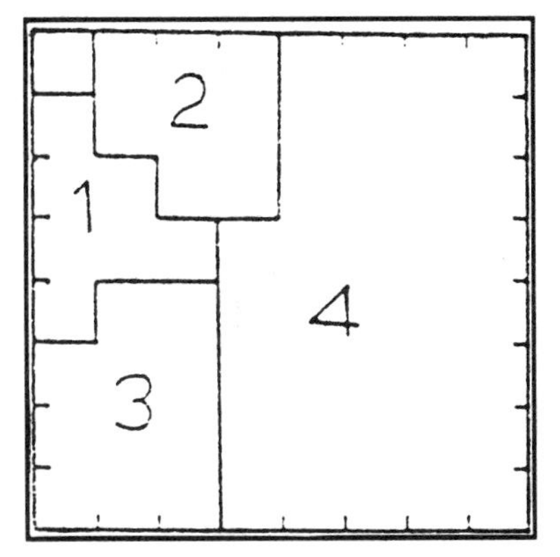

Fig. 7.50 Organization of sub-blocks in an 8×8 block of prediction errors in the 2D-DCT domain [IC-7]. (© 1981 IEEE).

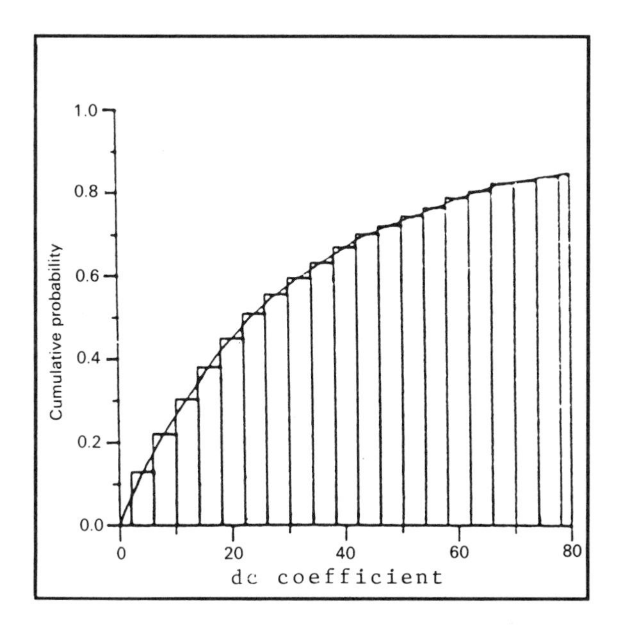

Fig. 7.51 Cumulative probability distribution of the prediction error of the dc coefficient [IC-7]. (© 1981 IEEE).

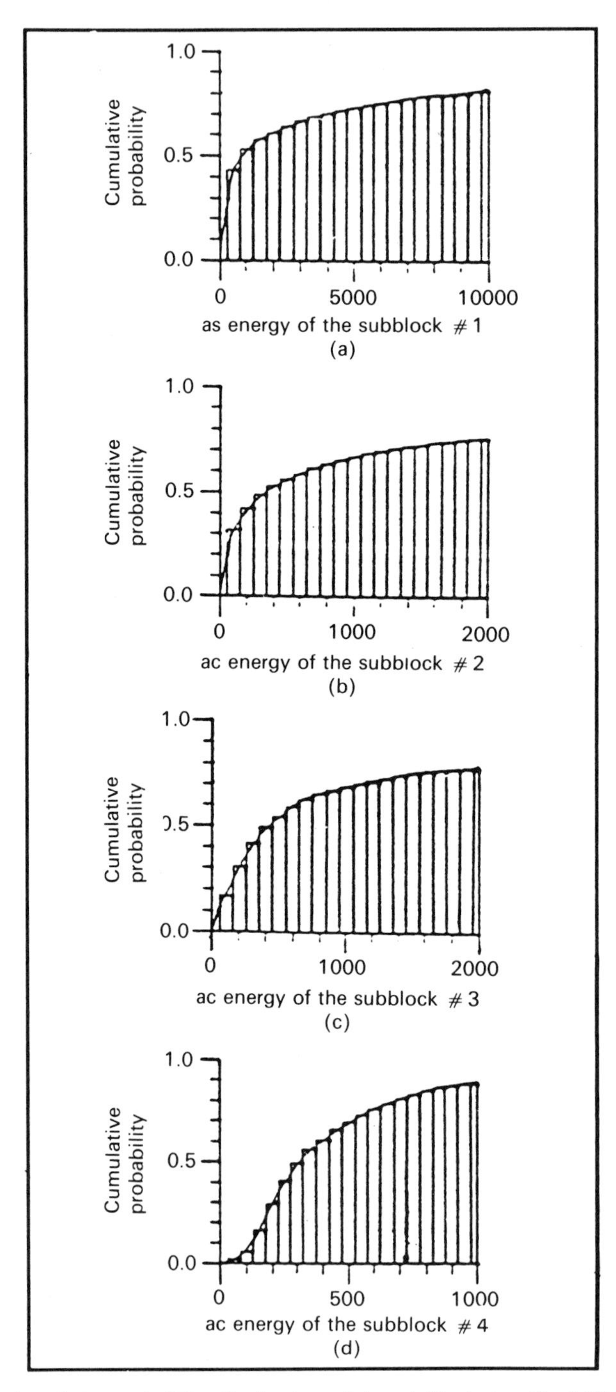

Fig. 7.52 Cumulative probability distributions of the sub-blocks (see Fig. 7.50) [IC-7]. (© 1981 IEEE).

VWL coding, and various other adaptive features. Motion estimation based on the three-step tree search [MC-28] leading to $\frac{1}{2}$-pel or $\frac{1}{2}$-line resolution of the motion vector is utilized in the hybrid mode. The DS-3 codec is based on the intrafield 2D 8×8 DCT of component signals followed by variable word-length coding of both the nonzero transform coefficients and the run length of zero coefficients scaled by buffer (fillness) status (Fig. 7.53). Other operations include error correction codes, multiplexing with audio, data, etc. network interface and preprocessing the input video. The corresponding block diagram for the decoder is shown in Fig. 7.54, where the functions are inverses of the coder operations. The coder can accept the TV signal in three formats, YUV, PAL, and NTSC, and transmit at two different rates, 34.368 Mbit/s (European digital hierarchy) [IC-73] and 44.736 Mbit/s (North American level, DS-3 rate). Other options include coding one or two TV signals together with one or two DS-1 data, all multiplexed into the DS-3 format. As indicated in Chapter 1, DCT, although significant, is only one of the many tools that are needed for achieving the bandwidth reduction. Details of these tools that comprise the algorithm (subject to change) are as follows:

7.8.4.1 VIDEO INTERFACE Depending on the particular video interface, the codec can accept the input TV signal in any of the three formats, YUV as per CCIR recommendation 601 [G-21], PAL, and NTSC (composite or component). Our discussion, henceforth, will be limited to the NTSC signal, although the basic compression techniques are valid for all the formats. The interface (Fig. 7.55) acts as a front-end preprocessor to the codec, i.e., low-pass filter, generation of various clock signals, A/D, component separation, etc. The clock signals include $8f_{SC}$, line clock ($f_h = 15.75$ KHz), and frame clock (30 frames/sec). The video output consists of pseudo components Y ($4f_{SC} = 14.3$ MHz) and C_B and C_R ($2f_{SC} = 7.15$ MHz) digitized to 10 BPP. The horizontal sync pulse is sampled at $(8/3)\,f_{SC}$, and this corresponds to 896 samples per horizontal sweep (868 for video + 28 for the sync pulse).

7.8.4.2 DCT 2D DCT is applied to 8×8 blocks of Y, C_R, and C_B. Two DCT processors operate in parallel, one on Y, and one on C_R and C_B alternately. 2D DCT is implemented using the 1D-DCT algorithm of Chen, Smith, and Fralick [FRA-1] with some modifications and the row-column approach (Fig. 5.1). The DCT output (11 bits) is scaled by $(\sqrt{2})^{-n}$, where n ranges from 1 to 12 controlled by the buffer (Fig. 7.56).

7.8.4.3 VWL CODING The transform coefficients of Y, C_R, and C_B are scanned based on the decreasing order of their energies computed from several NTSC video sequences (Fig. 7.57). These scanning orders are not necessarily optimum and are subject to change. Both the nonzero coefficients and the run lengths of zero coefficients are VWL coded on the basis of statistical computations performed on several test sequences.

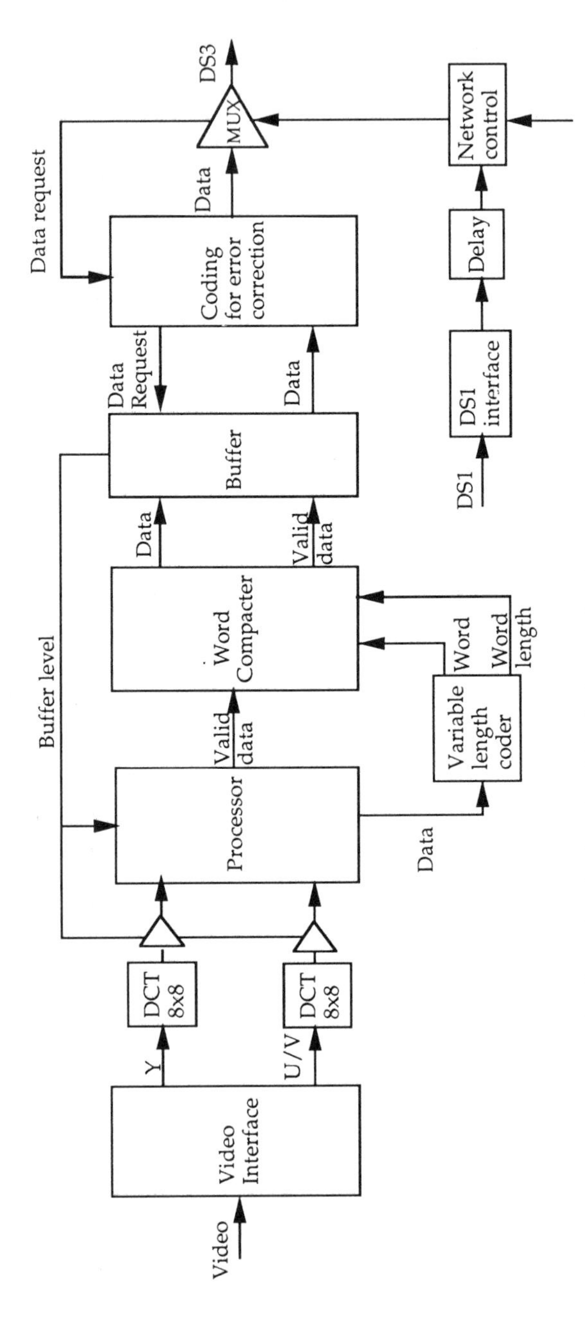

Fig. 7.53 General coder functional block diagram for the DS-3 NTSC video codec [IC-53].

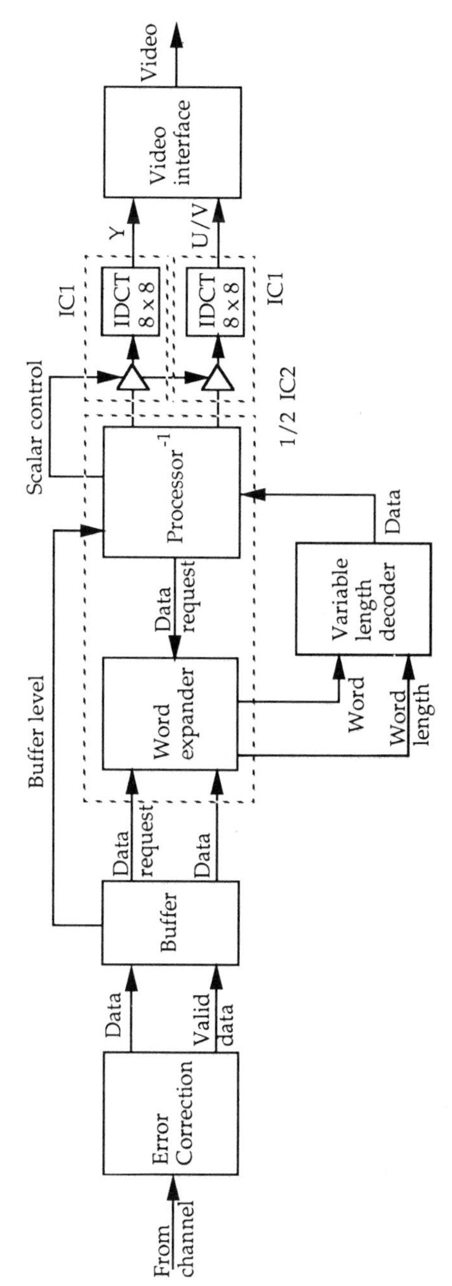

Fig. 7.54 Intrafield DCT video decoder corresponding to the codec shown in Fig. 7.53 [IC-53].

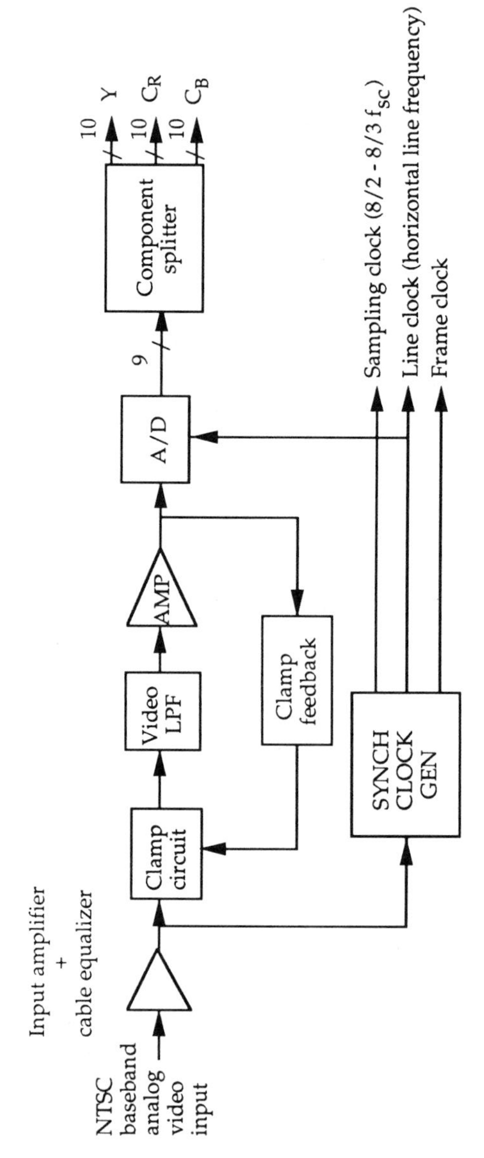

Fig. 7.55 Block diagram of DS-3 NTSC video codec front end (preprocessor) [IC-53].

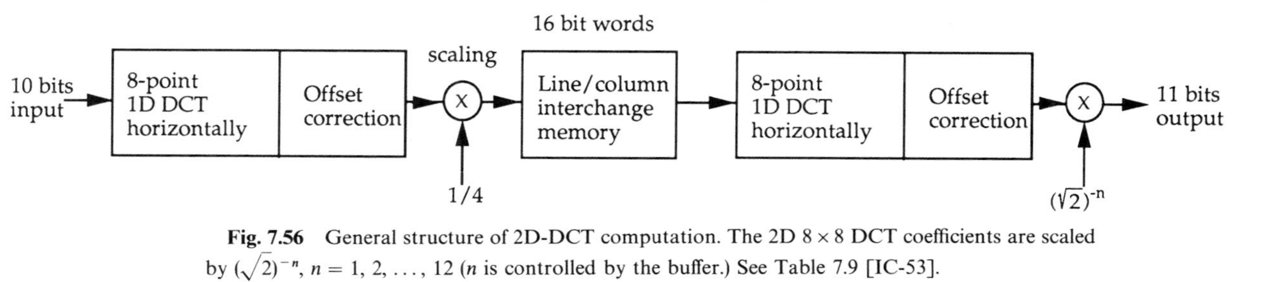

Fig. 7.56 General structure of 2D-DCT computation. The 2D 8×8 DCT coefficients are scaled by $(\sqrt{2})^{-n}$, $n = 1, 2, \ldots, 12$ (n is controlled by the buffer.) See Table 7.9 [IC-53].

0	2	4	8	22	16	25	20
1	7	11	18	33	30	36	45
3	10	15	21	37	31	48	54
5	14	19	26	40	42	53	59
6	17	24	29	51	44	56	60
9	23	28	38	50	47	57	61
12	27	34	41	52	49	58	62
13	32	35	39	43	46	55	63

Luminance

0	1	4	5	16	21	19	20
2	17	25	34	45	46	47	61
7	29	40	49	53	52	62	63
18	39	41	37	44	60	59	58
23	48	42	33	57	56	55	54
15	26	24	50	51	43	35	30
13	22	38	36	32	31	28	27
3	14	12	10	11	9	6	8

Chrominance

Fig. 7.57 Scanning paths of the 2D-DCT coefficients inside 8×8 blocks for luminance Y and chrominance (C_R and C_B) [IC-53, IC-67]. Reprinted with permission from *SMPTE Journal*, September 1989, "DCT-Based Television Codec for DS3 Digital Transmission," S. Cucchi and F. Molo.

7.8.4.4 LINE STRUCTURE OF VIDEO DATA The 896 samples per horizontal sweep are divided into 28 sets of 32 consecutive pels. The 32 samples of eight adjacent video lines are used to form two blocks of Y, one block of C_R, and one block of C_B, before applying the 2D 8×8 DCT. The 28 DCT transformed sets of data blocks are arranged as follows for transmission:

$$|Y|\mathrm{EOB}_0|C_R|\mathrm{EOB}_0|Y|\mathrm{EOB}_0|C_B|\mathrm{EOB}_1|\cdots|Y||\mathrm{EOB}_S|A|$$

27 sets

$$|C_R|\mathrm{EOB}_S|A||Y||\mathrm{EOB}_S|A|C_B|\mathrm{EOB}_S|A|$$

last set (28th),

where $|Y|C_R|C_B|$ are sequences of VWL coded data, scanned as shown in Fig. 7.57, $|\mathrm{EOB}_0|$ is the end of data in a block, $|\mathrm{EOB}_1|$ is the last 8×8 data block in a set of four blocks, $|\mathrm{EOB}_S|$ is the end of four blocks of 8×8 data at the end of a stripe of 8 lines, and $|A|$ indicates encoder scaling mode, transmission buffer status, and the number of the first line in a stripe of 8 lines.

7.8.5 *Buffer Control*

The variable-length codes coming into the buffer are processed and sent at a constant rate to the error correction coder. The buffer is of size 384 K bits and it is divided into 64 adjacent sectors S_i, $i = 0, 1, \ldots, 63$, each sector of size 64 Kbits. Buffer status and control are as follows:

S_0–S_5	underflow	Zero DCT coefficients are separately coded to increase the line bit rate.

S_5–S_{58} normal working condition

S_{59}–S_{61} overflow

S_{62}–S_{63} underflow

Only 32 coefficients in Y, C_R and C_B blocks are coded for transmission.

The buffer fillness sectors S_0–S_{63} are divided into subranges that control the DCT mode (Table 7.9). The DCT mode, in turn, controls the exponent n in the scale factor $(\sqrt{2})^{-n}$ (Table 7.10). The input to the multiplexer includes bits from encoded video, video synchronization, error correction, audio, data, interface to DS-3 framing, etc. Details on these and related aspects may be found in [IC-53, IC-67]. Only some basic highlights of the coder operations are presented here.

Table 7.9

Buffer fillness subranges [IC-53, IC-67].

Subrange	Encoder Mode	DCT mode (Y, C_R, C_B)
S5–S10	0	1
S11–S14	1	2
S15–S18	2	3
⋮	3	4
⋮	4	5
⋮	5	6
⋮	6	7
⋮	7	8
⋮	8	9
⋮	9	10
⋮	10	11
⋮	11	12
S55–S58	12	13
S0–S5, S62–S63	underflow	
S59–S61	overflow	

Reprinted with permission from *SMPTE Journal*, September 1989, "DCT-Based Television Codec for DS3 Digital Transmission," S. Cucchi and F. Molo.

Based on the intrafield coding described in Figs. 7.53 and 7.54, Telettra has designed and built video codecs for transmission and distribution at the DS-3 level (DTV-45) and at the 34.368 MBPS (DTV-34). Block diagrams of the coder and decoder for the DTV-34 are shown in Figs. 7.58 and 7.59, respectively. The original image and its reconstruction at 34.368 MBPS (processed by the intrafield DCT codec) are shown in Fig. 7.60 (color insert page I-1). This bit rate includes bits for audio, sync signals, framing, error correction, network interface, etc.

To achieve higher compression at the same quality, or improved quality at the same bit rate, the coding scheme described in Figs. 7.53 and 7.54 has been extended to that shown in Fig. 7.61 [IC-75, IC-76, IC-78]. This algorithm also

Table 7.10

Values of exponent n in $(\sqrt{2})^{-n}$ as function of the DCT mode and coefficient location inside an 8×8 block [IC-53, IC-67].

DCT mode	X_1	X_2	X_3
1	1	1	1
2	1	1	2
3	1	2	3
4	1	3	4
5	1	4	5
6	1	5	6
7	2	5	6
8	2	6	7
9	2	7	8
10	2	8	9
11	2	9	10
12	2	10	11
13	2	11	12

X_1 = dc coefficient.
X_2 = coefficient on and above the opposite principal diagonal (except X_1).
X_3 = coefficients other than X_1 and X_2.
Reprinted with permission from *SMPTE Journal*, September 1989, "DCT-Based Television Codec for DS3 Digital Transmission," S. Cucchi and F. Molo.

has been proposed by the CMTT/2 DCT Group [IC-76] for coding of TV signals (CCIR-601, Table 7.1) at both 30–34 MBPS and at the DS-3 level (44.736 MBPS). Some of the operations, such as DCT, scaling, quantization, buffer, and buffer control (also their inverse operations), are similar to those shown in Fig. 7.53. One of the new features is adaptive selection of intra/interfield and interframe modes. The selection is based on the AC energy (7.13.1) of each 8×8 luminance block. The intrafield mode is same as before (Fig. 7.58). For the interfield mode, DCT is applied to the differences between the pels of the current field and those interpolated from the previous field (Fig. 7.62).

For the interframe mode, prediction is based on the pels from the previous frame (second previous field) displaced by motion estimation, which is based on either semiglobal motion compensation (see Fig. 7.114) or 1 of 31 possible motion vectors, which include the nine semiglobal motion vectors. The motion vector range is ± 4 lines/frame with $\frac{1}{2}$-pel–line/frame resolution. Motion compensated interframe hybrid coding is described in Section 7.10. Details regarding VLC, scanning the DCT coefficients, FEC, etc. are described in [IC-75, IC-76, IC-78]. Various adaptive features introduced in this algorithm have contributed to coding of critical test (TV) sequences at 30 MBPS with high fidelity of the reconstructed images. Figure 7.64 (color insert page I-2) shows the

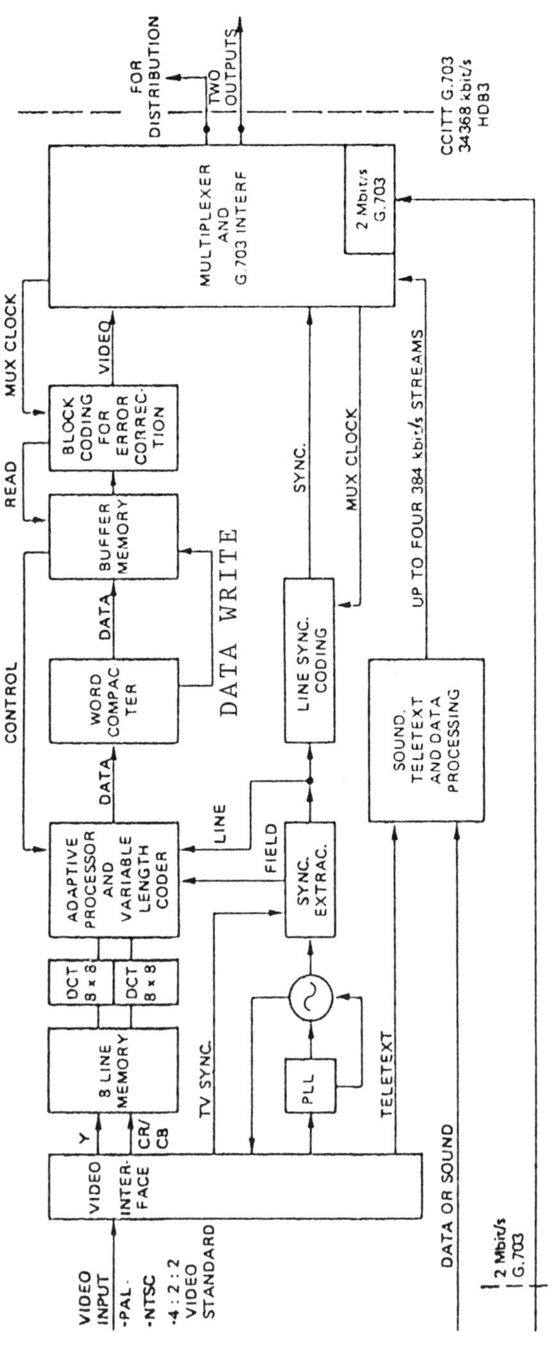

Fig. 7.58 Block diagram of intrafield video coder at 34.368 MBPS built by Telettra [IC-53].

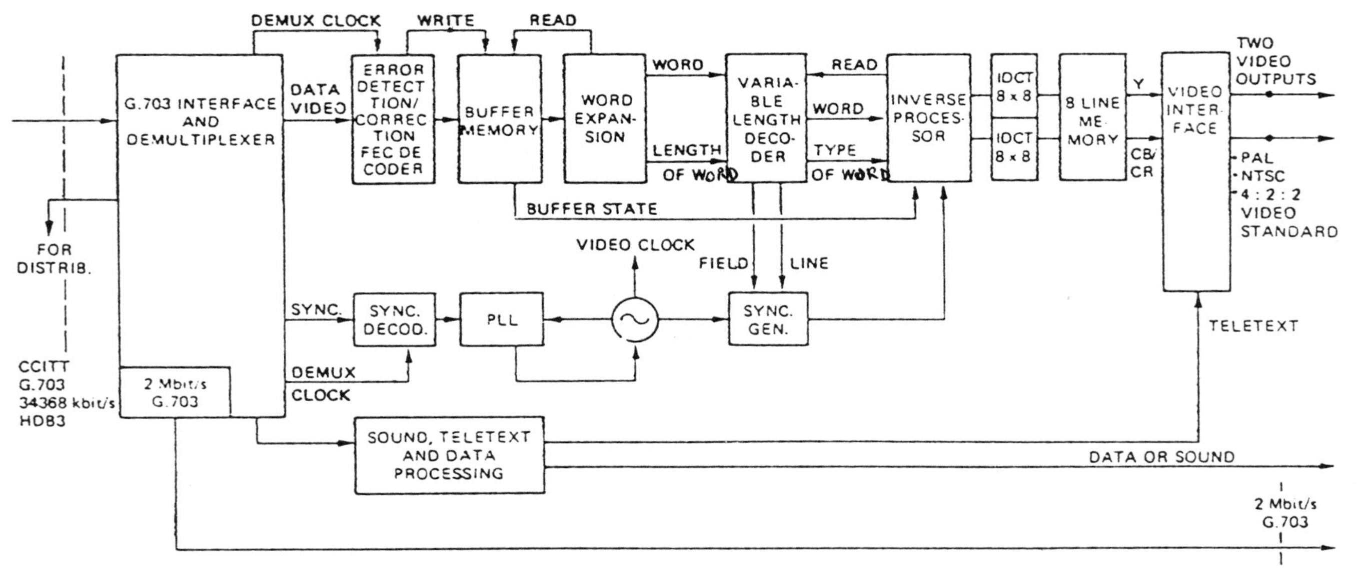

Fig. 7.59 Block diagram of intrafield video decoder at 34.368 MBPS built by Telettra.

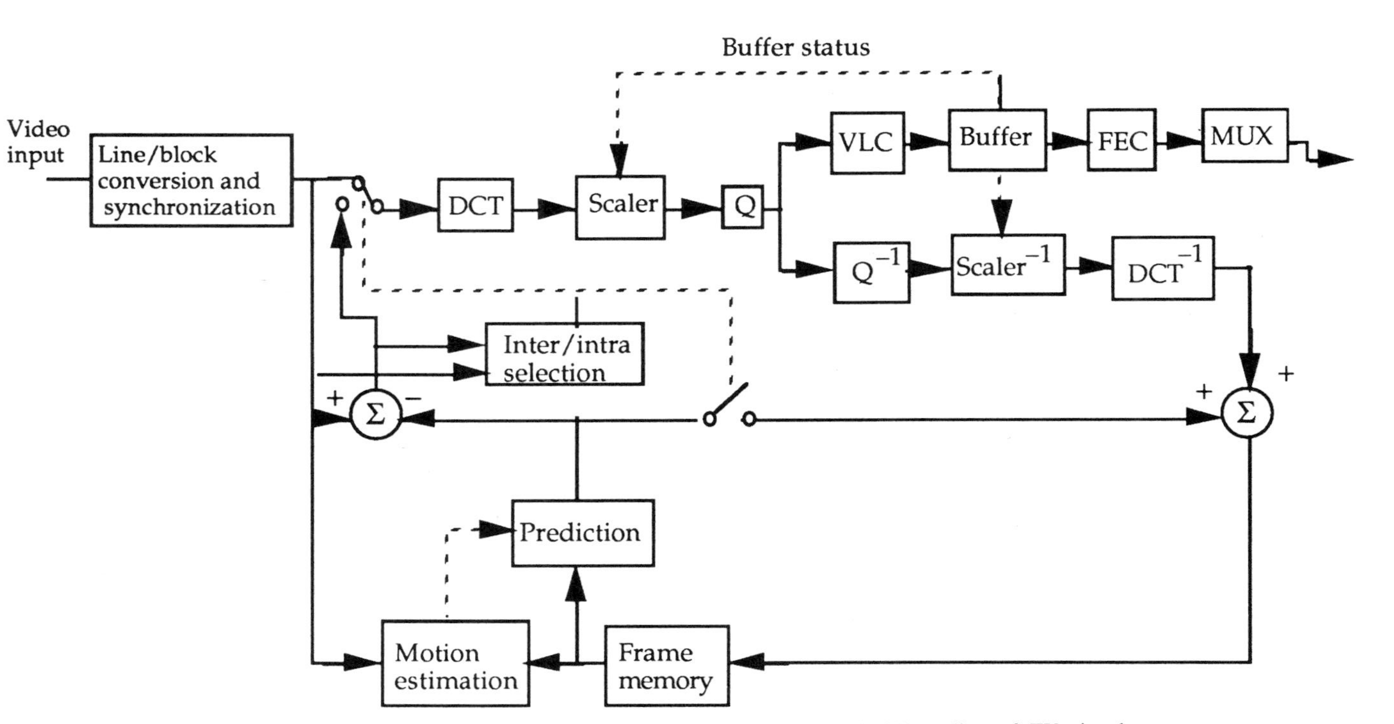

Fig. 7.61 Adaptive (intra/interfield, interframe) hybrid (DPCM/DCT) coding of TV signals [IC-75, IC-78]. Q = scanning and quantizing; VLC = variable-length coding, MUX = multiplexer.

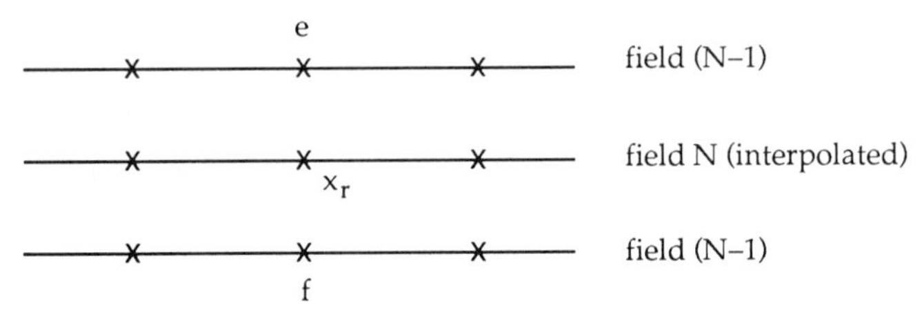

Fig. 7.62 Reference block for interfield mode (see Fig. 7.61). x_r, the nearest integer of $(e+f)/2$, is used for prediction [IC-75, IC-78].

0	2	6	12	20	28	36	44
1	5	11	19	27	35	43	51
3	7	13	21	29	37	45	52
4	10	18	26	34	42	50	57
8	14	22	30	38	46	53	58
9	17	25	33	41	49	56	61
15	23	31	39	47	54	59	62
16	24	32	40	48	55	60	63

Luminance

0	2	3	9	10	20	21	35
1	4	8	11	19	22	34	36
5	7	12	18	23	33	37	48
6	13	17	24	32	38	47	49
14	16	25	31	39	46	50	57
15	26	30	40	45	51	56	58
27	29	41	44	52	55	59	62
28	42	43	53	54	60	61	63

Chrominance

Fig. 7.63 Scanning paths for the 2D-DCT coefficients of prediction errors inside 8×8 blocks for luminance Y and chrominance C_R and C_B [IC-79].

processed images at 30 and 15 MBPS. It is proposed that this adaptive hybrid algorithm can be effectively applied to HDTV signals (Section 7.11), achieving contribution quality at 70 MBPS [IC-78]. The price for this superior performance is the codec complexity, i.e., adaptive selection of intra/intermode, additional IDCT at the coder, motion estimation, frame memory, additional side information, etc., as can be observed from a comparison of Figs. 7.53 and 7.61.

The scanning paths for luminance and chrominance components (2D 8×8 DCT coefficients of the prediction errors) of the hybrid coder (Fig. 7.61) are shown in Fig. 7.63. These are different from the corresponding paths for the intrafield coder (Fig. 7.57). Also the scanning path for C_R and C_B follows the zig-zag pattern similar to that shown in Fig. 7.129b.

The DCT-based NTSC video codec (Fig. 7.53) is presented to focus on the various facets that arise in the codec design. Several issues arise in evaluating a coder, such as noise in the encoding process, hardware complexity and cost, sensitivity to transmission errors, sensitivity to jitter of the video analog signal, sensitivity to microinterruption in the digital channel, compatibility with digital recording [IC-59], flexibility (composite or component signal, NTSC, PAL, CCIR rec. 601 [G-21], etc.), various appropriate bit rates, audio and data channel options, quality (subjective quality and fidelity), and modularity (extension to HDTV and other advanced TV systems [HDTV-6 through HDTV-9]).

This section has presented the basic principles of image coding with the objective of reducing the bandwidth for transmission or memory for storage. The performance of a coder can be evaluated from both subjective and objective criteria. Basic schemes that involve either DCT alone or the DCT/DPCM combination have been outlined. Various adaptive features can be integrated into these primary compression tools, leading to reduction of visual artifacts. While the adaptive features necessarily increase the coder complexity, compromise between the codec costs and picture fidelity has to be achieved. Choice of zonal filters, switched quantizers, sub-block classification, bit allocations, VWL coders, etc. offers an engineer a multitude of options in designing a codec appropriate for a specific application. With the rapid progress in the development and availability of both general purpose DSP chips and dedicated chips (for example, DCT chips), codec hardware can become simpler, and with an increasing market, the costs, hopefully, can be reduced. The following section discusses vector quantization (VQ), another powerful and increasingly popular data compression tool. This is combined with the DCT and, as with any other scheme, it is supplemented with various adaptive features, resulting in an effective and efficient coding scheme. For the low bit-rate codecs presented in Section 7.10, the compression concepts discussed in this section and the next are complemented with other tools, such as temporal and spatial subsampling, motion compensation, etc.

7.9 DCT/VQ

As indicated in the introduction (Section 7.1), DCT can be supplemented (or complemented) with other redundancy reduction methods such as DPCM, subband coding, VQ, etc. Also, additional features, such as motion compensation, HVS weighting, switched predictors and/or quantizers, and multiple scanning mechanisms (scanning and therefore prioritizing the transform coefficients), can be incorporated in the overall coding process. The concept of VQ data compression has become realistic and practical only recently, thanks to the powerful and fast codebook design, e.g., LBG algorithm [DV-30] and various modifications/improvements to the basic VQ coder (Fig. 7.65a). An excellent

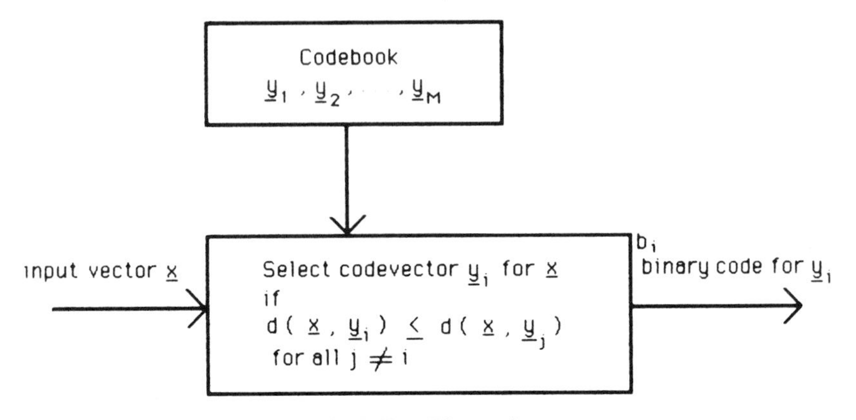

Fig. 7.65a VQ encoder.

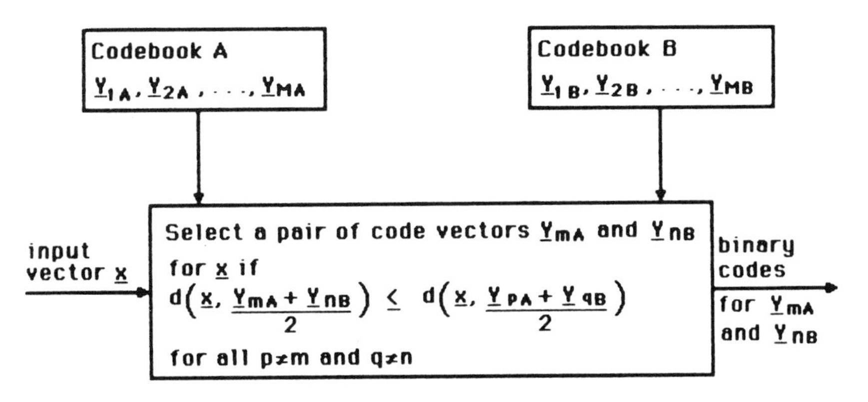

Fig. 7.65b Two-channel conjugate VQ encoder [DV-48].

review article on image coding using vector quantization has been published by Nasrabadi and King [DV-35].

The aim of this section is not so much to present a detailed presentation of VQ and many of its variations, but rather to show how VQ has been embedded with the DCT in image coding purely from a bit-rate reduction viewpoint. As with the other hybrid techniques, such as DCT/DPCM, DCT/sub-band, etc., the marriage of DCT with VQ is intended to mitigate the disadvantages of either of them, while at the same time strengthening their strong points. Not surprisingly, in the DCT/VQ consummation, there are many variations, adaptations, and options (this is also true of all other hybrid techniques).

Gray [DV-29] has published an excellent review article on vector quantization. The basic VQ process involves the following:

1. Developing a codebook based on a large training sequence, representative of the test sequences.
2. Formatting the input signal as an input vector.
3. Searching for the nearest codevector from the codebook, which is the most representative of the input vector. The selected codevector has the least distortion or the best match with the input vector. The distortion needs to be defined.
4. Sending a label (say a binary code) for the selected codevector over the communication channel.

Some of the variables in the basic VQ design are vector dimension (number of components of the input vector), codebook size (number of codevectors in the codebook), distortion criterion, search algorithm (to locate the best match for the input vector from the codebook), and how the codebook is developed. The codevector is alternatively called representative vector, pattern, member of the alphabet (code book), etc. Assuming that the input vector is of dimension K, i.e., $\mathbf{x} = \{x_1, x_2, \ldots, x_K\}$ and $\mathbf{y}_1, \mathbf{y}_2, \ldots, \mathbf{y}_M$ are members of the codebook of size M, the bit rate per vector is $\log_2 M$ or $(\log_2 M)/K$ bits/component. To achieve high performance i.e., the selected representative vector y_i has to closely resemble $\mathbf{x}$, the codebook size has to be large, and the input vector dimension has to be sufficiently large enough so that correlation with adjacent vectors is minimal. Increasing codebook size and/or input vector dimension implies large memory and exponentially increasing computations. For practical purposes, both M and K have to be finite and reasonable. For example, in image coding, $M = 64$, 128, 256, and 512, and $K = 4$, 9, 16 and 25 have been used. Also, many meaningful distortion measures (non-negative) can be ($\mathbf{x}$ is assumed to be real) used. Some of the distortion measures are listed below.

MSE

$$d_1(\mathbf{x}, \mathbf{y}_i) = \frac{1}{K} \sum_{m=1}^{K} (x_m - y_{im})^2. \tag{7.9.1}$$

MAE (mean absolute error)

$$d_2(\mathbf{x}, \mathbf{y}_i) = \frac{1}{K} \sum_{m=1}^{K} |x_m - y_{im}|. \tag{7.9.2}$$

l_n or holder norm

$$d_3(\mathbf{x}, \mathbf{y}_i) = \left[\sum_{m=1}^{K} |x_m - y_{im}|^n \right]^{1/n} = \|\mathbf{x} - \mathbf{y}_i\|_n. \tag{7.9.3}$$

nth law distortion

$$d_4(\mathbf{x}, \mathbf{y}_i) = \sum_{m=1}^{K} |x_m - y_{im}|^n = (\|\mathbf{x} - \mathbf{y}_i\|_n)^n. \tag{7.9.4}$$

Weighted squares distortion

$$d_5(\mathbf{x}, \mathbf{y}_i) = \sum_{m=1}^{K} w_m(x_m - y_{im})^2. \tag{7.9.5}$$

General quadratic distortion

$$d_6(\mathbf{x}, \mathbf{y}_i) = (\mathbf{x} - \mathbf{y}_i)[W](\mathbf{x} - \mathbf{y}_i)^T. \tag{7.9.6}$$

In these equations, $w_1, w_2, \ldots, w_k$ are the weights, and $[W]$ is a $(K \times K)$ positive definite symmetric matrix. This requirement guarantees that the distortion is non-negative and $\mathbf{y}_i = \{y_{i1}, y_{i2}, \ldots, y_{iK}\}$ is the ith member of the alphabet (codebook). By choosing $[W]$ appropriately, the auditory and perceptual properties of the speech and video, respectively, can be incorporated in the VQ coder. Needless to say, other weighted distortions such as weighted MAE or weighted nth law distortion can be defined. A variation of the basic VQ is the two-channel conjugate VQ [DV-48] in which two different codebooks are used (Fig. 7.65b). The two conjugate codebooks are designed to provide small distortion and to improve the robustness against channel errors. Also, for the MSE distortion (7.9.1), both the memory requirements and computational complexity are reduced compared with the basic VQ (Fig. 7.65a). Moriya and Suda [DV-48] have developed an iterative training algorithm for designing the conjugate codebooks for the two-channel VQ. Corresponding distortion measures for this encoder can be defined. For example, the MSE is

$$d_7\left(\mathbf{x}, \frac{\mathbf{Y}_{mA}}{\mathbf{Y}_{nB}}\right) = \frac{1}{K}\sum_{i=1}^{K}\left[X_i - \frac{(\mathbf{Y}_{mA} + \mathbf{Y}_{nB})_i}{2}\right]^2, \tag{7.9.7}$$

where $(\mathbf{Y}_{mA} + \mathbf{Y}_{nB})_i$ refers to the ith components of $\mathbf{Y}_{mA}$ and $\mathbf{Y}_{nB}$. As in (7.9.1), computation of (7.9.7) can be minimized as follows:

$$d_7\left(\mathbf{x}, \frac{\mathbf{Y}_{mA}}{\mathbf{Y}_{nB}}\right) = \frac{1}{K}\left[\sum_{i=1}^{K} X_i^2 - \sum_{i=1}^{K} X_i(\mathbf{Y}_{mA} + \mathbf{Y}_{nB})_i + \frac{1}{4}\sum_{i=1}^{K}[(\mathbf{Y}_{mA} + \mathbf{Y}_{nB})_i]^2\right]. \tag{7.9.8}$$

In (7.9.8), the last term can be precomputed and stored for every combination of m and n. Hence, only the second term needs to be computed. Note that the first term can be deleted from the distortion computation, as it is independent of the code vector pair selected.

An algorithm developed by Linde, Buzo, and Gray [DV-30], popularly called the LBG algorithm, for designing the codebook created the necessary incentive and impetus for applications of VQ in speech coding, image coding, multispectral imagery, color-print image coding, texture coding, etc. As is true of any concept, subsequent improvements and/or modifications [DV-32] to the LBG algorithm have been proposed. One dramatic improvement is that proposed by Equitz [DV-31], based on the nearest neighbor algorithm, also called the Equitz algorithm. Also, many variations of the basic VQ, such as product code (gain-shape) VQ, tree search VQ (TSVQ), multistage VQ (or cascaded VQ) (see Fig. 7.90) hierarchical VQ, (see Fig. 7.91) pyramid VQ (DV-14, DV-47], adaptive VQ[DV-6], conditional replenishment VQ, classified VQ [DV-11, DV-33], VQ-SQ, finite state VQ, differential (mean separated) VQ, predictive VQ, motion-compensated predictive VQ, two-channel conjugate VQ [DV-48], and interpolative VQ [BSI-10], have been proposed and investigated. By introducing adaptive features, the effectiveness of these variations can be enhanced. Incidentally, [BSI-10] has excellent summaries on VQ and VQ image coding. The objective of these variations, in general, is to reduce the memory and/or encoding complexity of the basic VQ or adaptive VQ, based on the input signal properties (local or global). Also, other compression concepts such as transform,

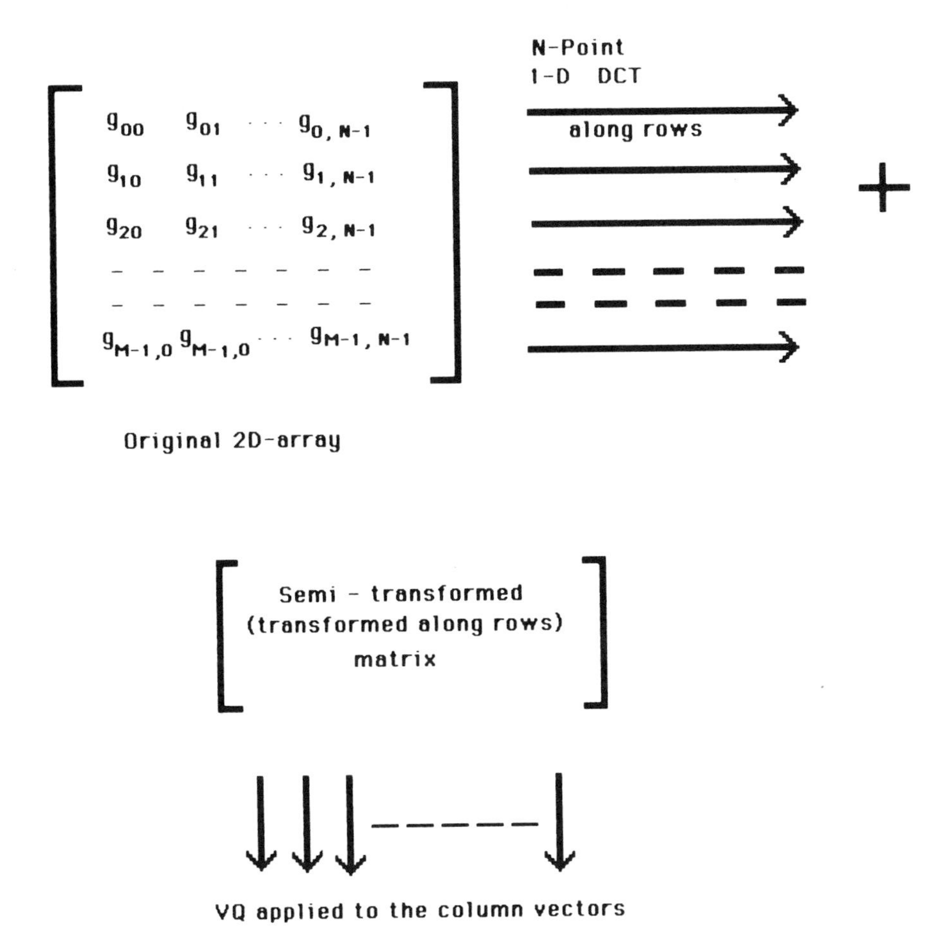

Fig. 7.66 DCT/VQ of a 2D array. DCT along rows followed by VQ along columns. (Operations on rows and columns can be interchanged.)

DPCM, and sub-band can be combined with the VQ to enhance the encoding process. The objective of this section, of course, is to summarize the DCT/VQ algorithms applied to the DSP. Similar to the DCT/DPCM, the DCT/VQ has many variations. DCT can be applied along the rows (columns) of a 2D block of an image followed by VQ along the columns (rows) of the semitransformed block (Fig. 7.66) [DV-2]. Another variation is 2D DCT of corresponding blocks of successive frames complemented by VQ along the temporal direction (Fig. 7.67).

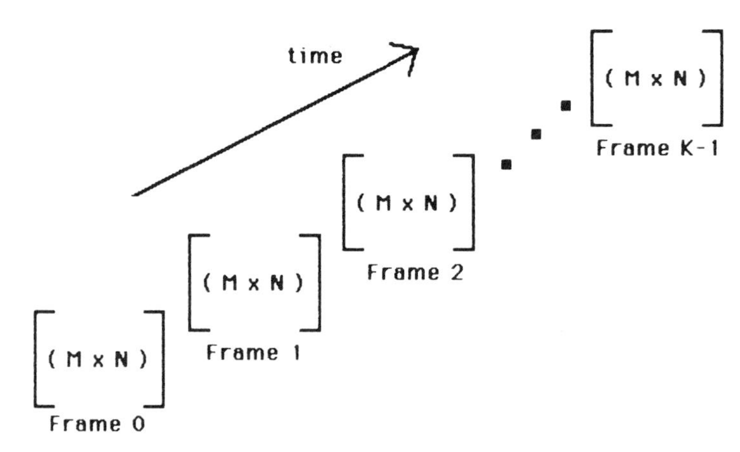

Fig. 7.67 DCT/VQ of 3D array. Obtain a 2D DCT of each frame as in Fig. 5.1. Follow this by VQ of MN vectors, each composed of the corresponding transform coefficients of the K frames.

By introducing geometrical zonal filtering (Fig. 7.29a), some of the high-frequency coefficients can be adaptively discarded. Instead of discarding these coefficients, they can be transmitted by a VQ scheme (Fig. 7.68), thus improving the image quality at the cost of a slightly higher bit rate [DV-35]. Also, the vector dimension K can be made adaptive based on the temporal correlation between the frames [DV-35]. Thus in an image sequence with little motion, K can be made quite large and vice versa. Several codebooks having different vector dimensions (K) and codebook sizes (M) can be designed with adaptivity based on motion, scene change, etc. (Fig. 7.69). Although the transform shown in this adaptive interframe VQ scheme is WHT, DCT or any other orthogonal transform can be used.

7.9.1 *Adaptive Transform Coding Using VQ/SQ*

Tu *et al.* [DV-18, DV-19] have applied an adaptive DCT/VQ coding scheme with scalar correction to still images (Fig. 7.70). Following the 2D DCT 8×8 of blocks of an image, the dc components of neighboring blocks are separately vector quantized to remove any correlation between these blocks. The ac

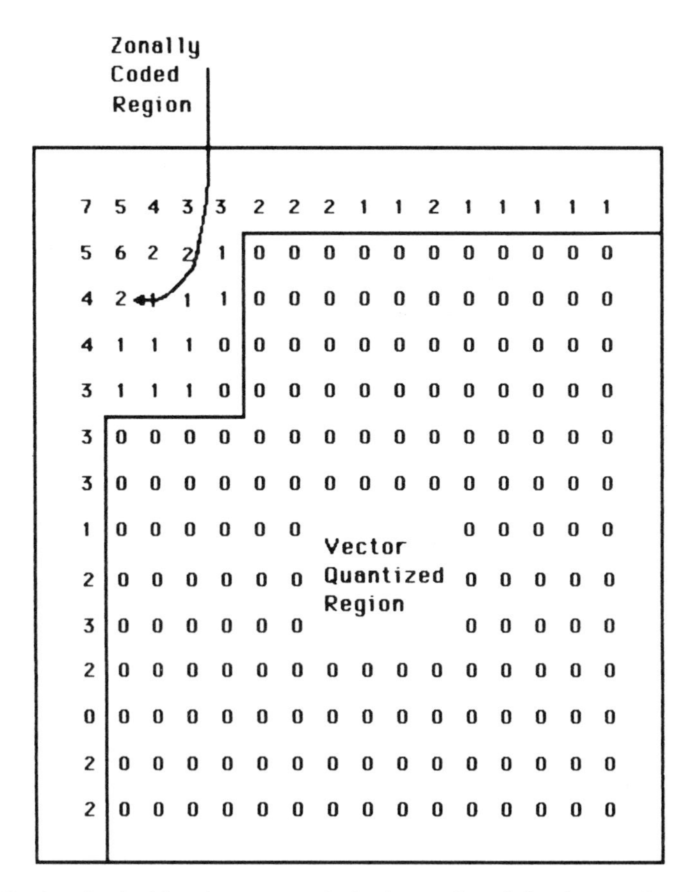

Fig. 7.68 An adaptive bit assignment matrix. In the zonally coded region, numbers indicate the number of bits allocated to the corresponding 2D-DCT coefficients. In the vector-quantized region, all the coefficients are vector quantized [DV-35]. (© 1988 IEEE).

coefficients of each block are divided into sub-blocks (Fig. 7.71). The vector composed of the coefficients in each sub-block is separately quantized based on the appropriate codebook. This operation is implemented only when the vector is above a predetermined perceptual threshold. Also, the vector representing the sub-block V is discarded. The control unit (CU) determines whether the error vectors representing the differences between the inputs and outputs of VQ need to be scalar quantized. The SQ also varies from fine to coarse depending on the frequencies of the transform coefficients. Although the adaptive VQ/scalar correction is complex and requires overhead, very good quality images are obtained at 0.4 BPP. Recently, motion estimation coupled with other changes has been incorporated in the DCT/SQ scheme for the coding of color image sequences [DV-36].

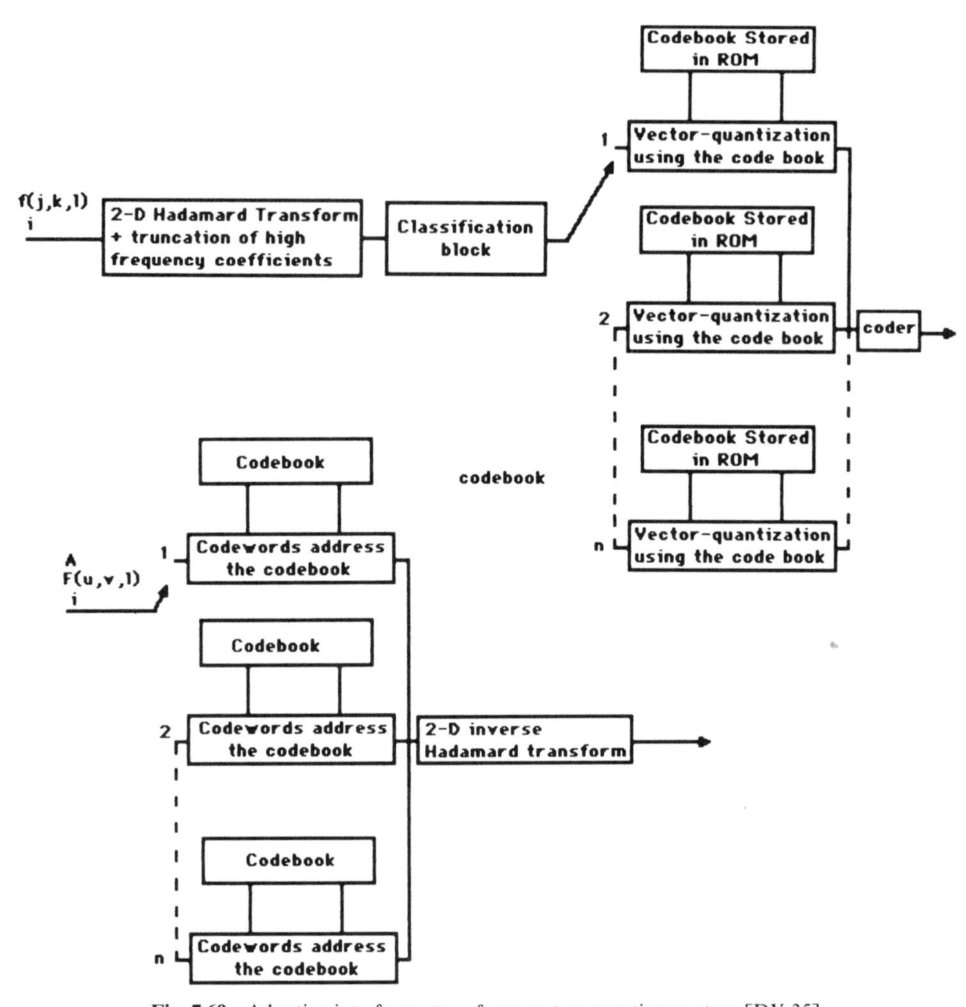

Fig. 7.69 Adaptive interframe transform vector quantizer system [DV-35].

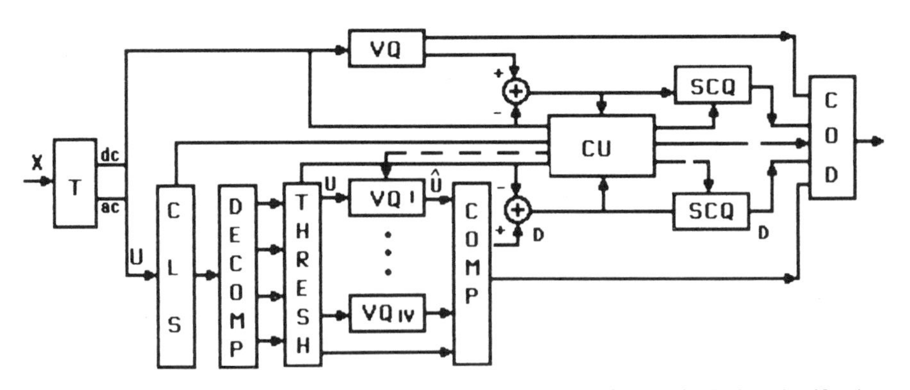

Fig. 7.70 Block diagram of the coding system. T = DCT transform unit: CLS = classification unit; DECOMP = decomposition unit; THRESH = threshold unit; CU = control unit; SCQ = scalar quantization [DV-18, DV-19].

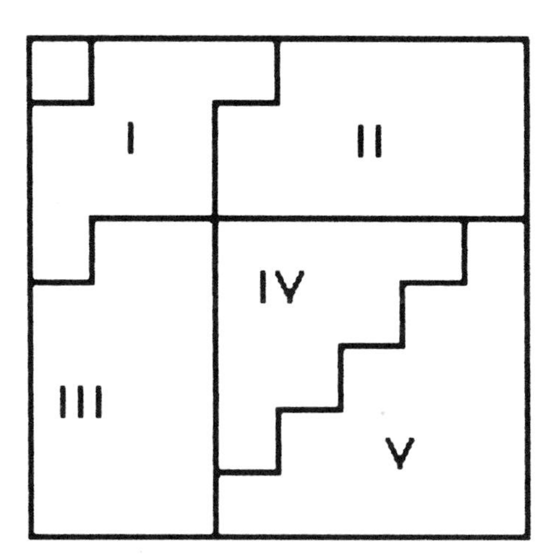

Fig. 7.71 Decomposition for the ac coefficients. The sub-blocks I, II, III, IV, and V are separably treated as vectors. Encoding sub-block I is often sufficient. Sub-blocks I and II represent the vertical structures. Sub-blocks I and III correspond to the horizontal structures [DV-18, DV-19].

7.9.2 *Classified VQ (CVQ)*

To preserve the perceptual features of an image, classified VQ (Fig. 7.72) was originally proposed by Ramamurthi and Gersho [DV-11]. In this scheme, each block is classified based on edges (orientation, location, and polarity), shade, midrange, and other categories (Fig. 7.73). Separate codebooks, called subcodebooks, for each class (the subcodebooks can be of different sizes) are generated based on the training vectors belonging to that class. For the midrange class, VQ is applied to the low-frequency coefficients of the 4×4 2D DCT of that class (Figs. 7.72 and 7.74).

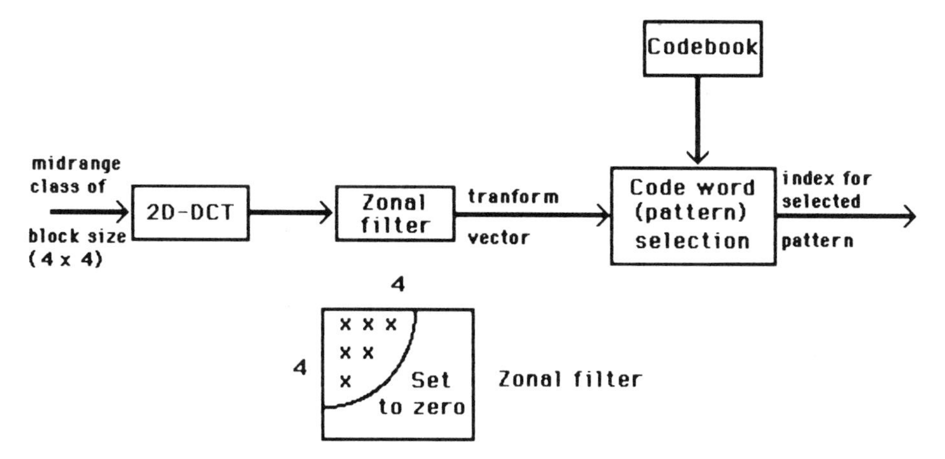

Fig. 7.72 2D-DCT CVQ image coding. The six low-frequency DCT coefficients of the 4×4 block constitute an input vector for the midrange class [DV-11]. (© 1986 IEEE).

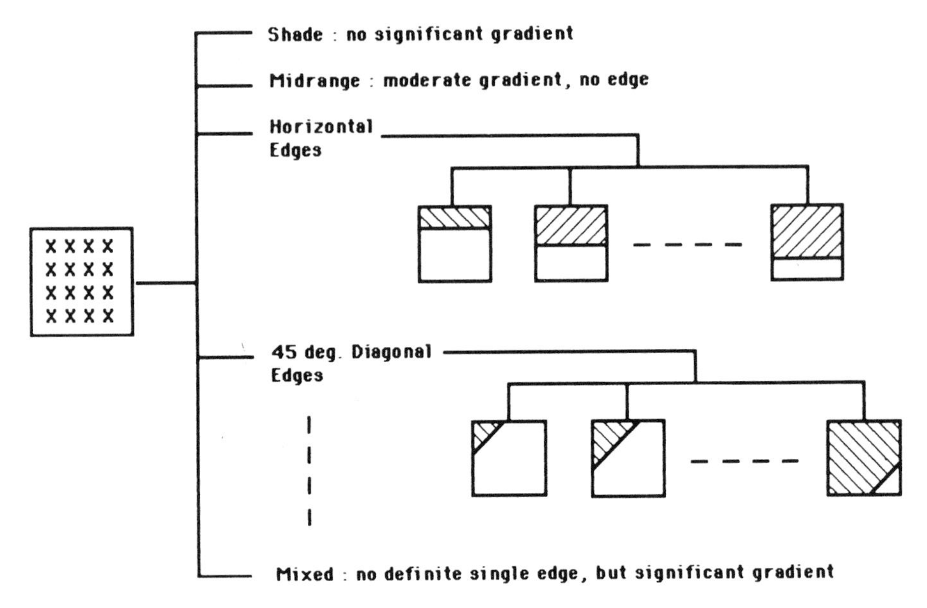

Fig. 7.73 Block classification [DV-11]. (© 1986 IEEE).

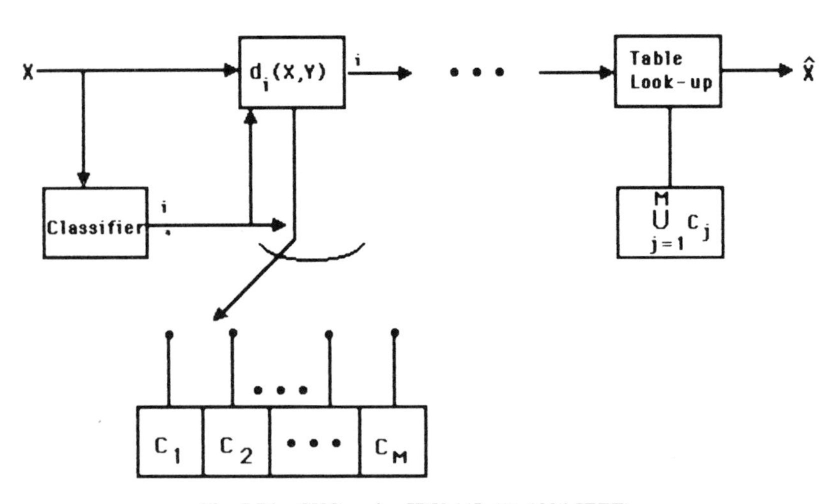

Fig. 7.74 CVQ coder [DV-11]. (© 1986 IEEE).

Initially the classifier selects the class of each block and the VQ process is based on the subcodebook of that class. Overhead bits identifying the block class, besides the index for the selected template, have to be transmitted. Overall, CVQ results in a better perceptual quality and considerably lower complexity compared with the basic VQ. A variation of CVQ (Fig. 7.75) is that the classification of a block is carried out in the transform domain based on the ac

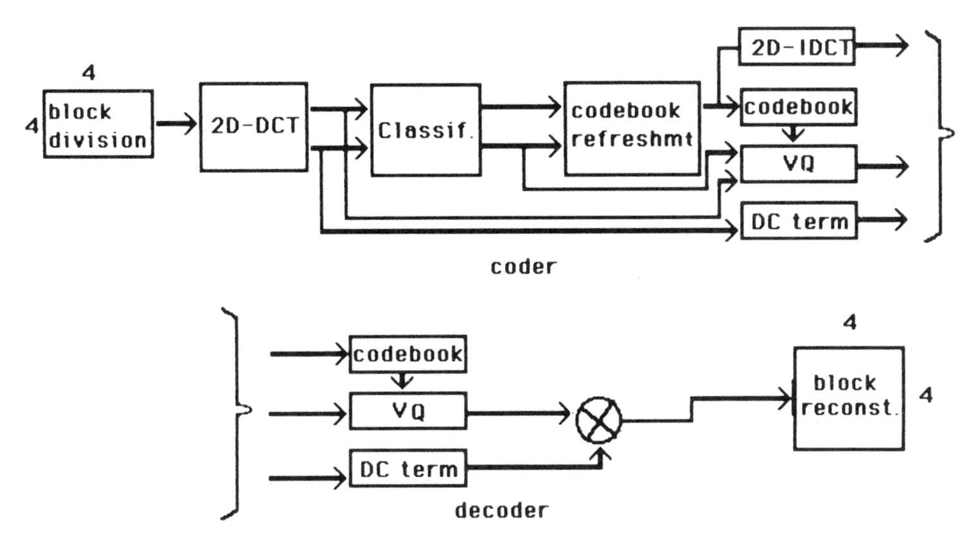

Fig. 7.75 Classified quantizer in a transformed domain applied to image sequence coding [DV-33].

energies of different regions (Fig. 7.76) [DV-33]. Adaptivity is built in by iteratively refreshing the subcodebooks, with the dc terms of all blocks coded separately (Fig. 7.75). It is shown that the 2D DCT adaptive CVQ of image sequences has resulted in increased visual quality of the processed images for a specified VQ complexity and bit rate.

C_0	C_1	C_2	C_3
C_4	C_5	C_6	C_7
C_8	C_9	C_{10}	C_{11}
C_{12}	C_{13}	C_{14}	C_{15}

2D DCT coefficients of a 4x4 pixel block

$$A_{tot} = \left(\sum_{i=0}^{15} C_i^2\right) - C_0^2$$

$$A_{diag} = C_5^2 + C_{10}^2 + C_{15}^2$$

$$A_{sup} = C_1^2 + C_2^2 + C_3^2 + C_6^2 + C_7^2 + C_{11}^2$$

$$A_{inf} = C_4^2 + C_8^2 + C_9^2 + C_{12}^2 + C_{13}^2 + C_{14}^2$$

$$A_{fin} = C_{11}^2 + C_{14}^2 + C_{15}^2$$

Fig. 7.76a Partial activities of 4×4 pixel blocks [DV-33].

Yim and Kim [DV-34] have developed a simple DCT-CVQ scheme in which the edge patterns of a spatial block (Fig. 7.77) are based on the two lowest frequency coefficients $X^{C(2)}(0,1)$ and $X^{C(2)}(1,0)$ (Fig. 7.78). They have shown that the relationship in magnitudes of these two coefficients reflects the distribution patterns of all the ac coefficients. Based on these two properties, the homogeneous and diagonal edge classes can be further classified (Fig. 7.79).

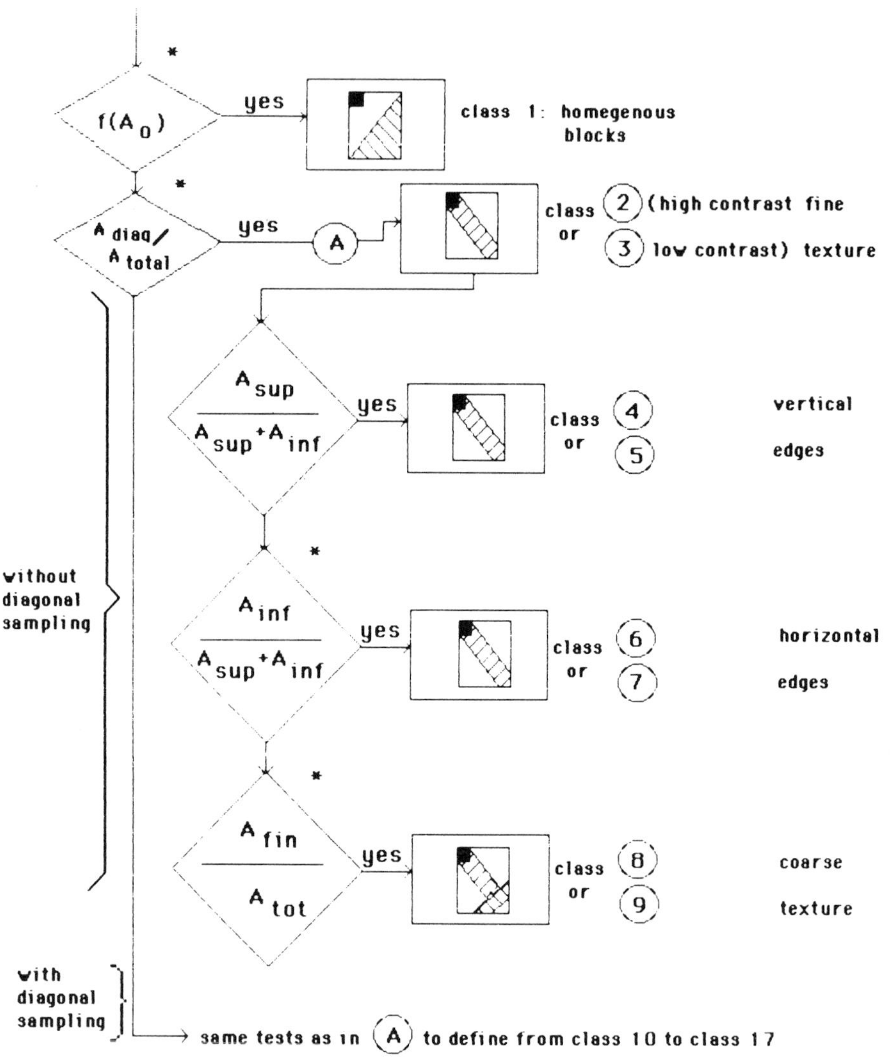

Fig. 7.76b Decision tree for the visual-based classifier. *Test functions are visual thresholdings [DV-33].

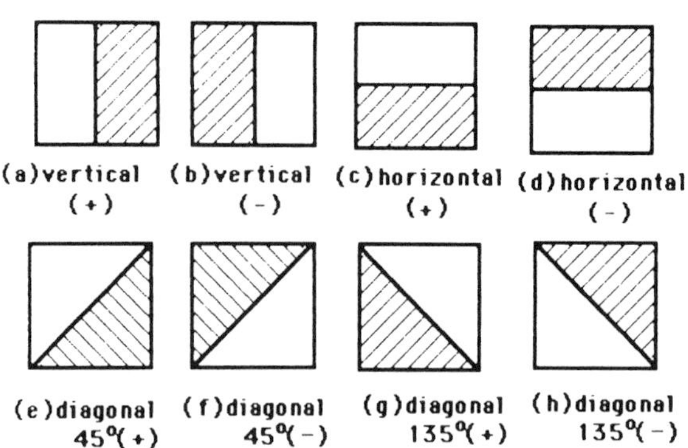

Fig. 7.77 Edge patterns [DV-34].

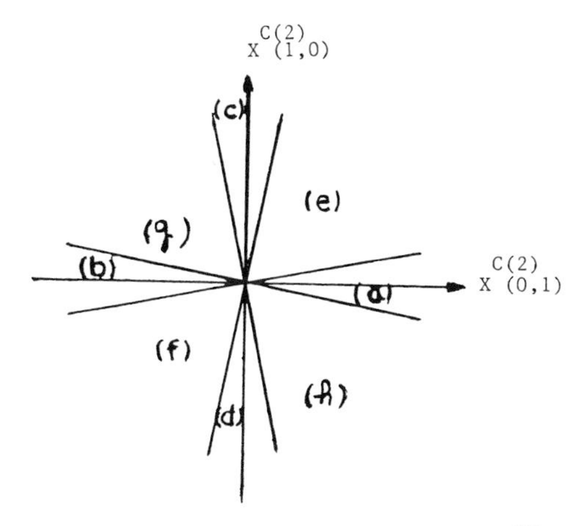

Fig. 7.78 Relation between edge patterns and DCT coefficients $X^{C(2)}_{(0,1)}$ and $X^{C(2)}_{(1,0)}$ [DV-34].

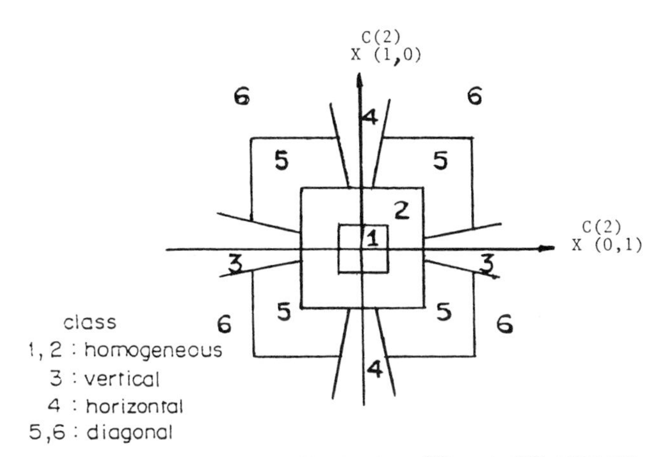

Fig. 7.79 Image block classification by $X^{C(2)}_{(0,1)}$ and $X^{C(2)}_{(1,0)}$ [DV-34].

7.9.3 *DCT/VQ Using Categorization with Adaptive Band Partition*

Similar to [DV-24], another ADCT/VQ scheme has been developed by Omachi, Takashima, and Okada [DV-17], based on the concept that locations of transform coefficients having large magnitudes change in the 2D-DCT domain according to the edge direction in the spatial domain. Highlights of this adaptive algorithm (Fig. 7.80) are as follows:

(1) As in [DV-24], divide an image into blocks, say, of size 8×8, and apply 2D DCT to each block.

(2) Categorize the blocks based on the locations of large magnitude coefficients. The category that has the largest absolute sum of coefficients in the

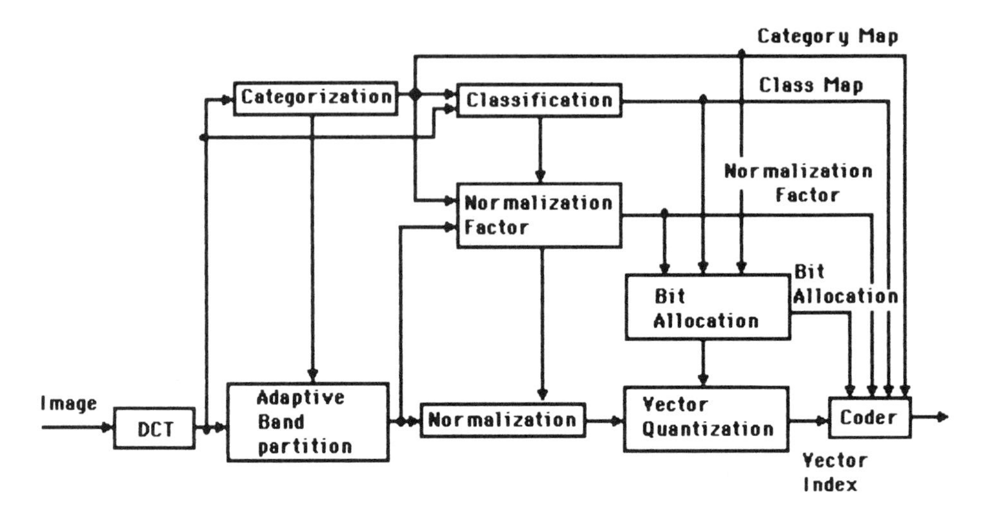

Fig. 7.80 Block diagram of DCT-VQ with adaptive band partition [DV-17].

hatched area (see Fig. 7.81 for patterns) is chosen for the block. Category 1 refers to non-edge image, whereas categories 2, 3, and 4 refer to horizontal, vertical, and diagonal edges, respectively.

(3) Band partition in each category is changed adaptively, as shown in Fig. 7.82.

(4) Blocks in each category are classified into four groups based on the ac energy. Normalization and bit allocation are developed for every band in each class.

(5) As in [DV-24], output indices of VQ and overhead bits are transmitted. Based on simulation of several still frames, it is claimed that this technique has a 0.5–2 dB SNR gain over the ordinary DCT/VQ process.

7.9.4 *2D-ADCT/VQ of Monochrome and Color Images*

By developing universal vector quantizers for the DCT coefficients, Aizawa, Harashima, and Miyakawa [DV-24] have simplified the VQ complexity and

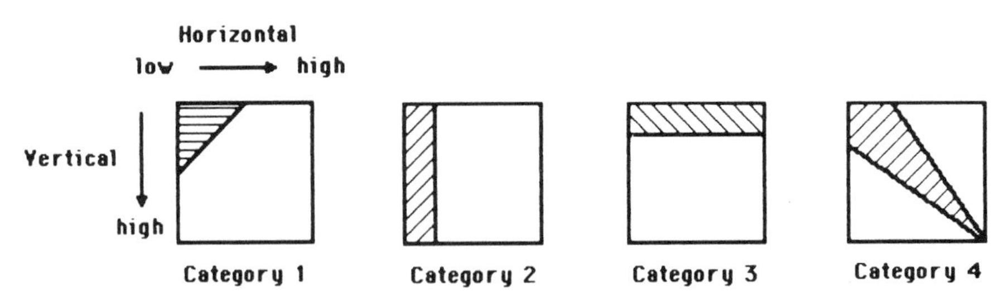

Fig. 7.81 Mask patterns for categorization [DV-17].

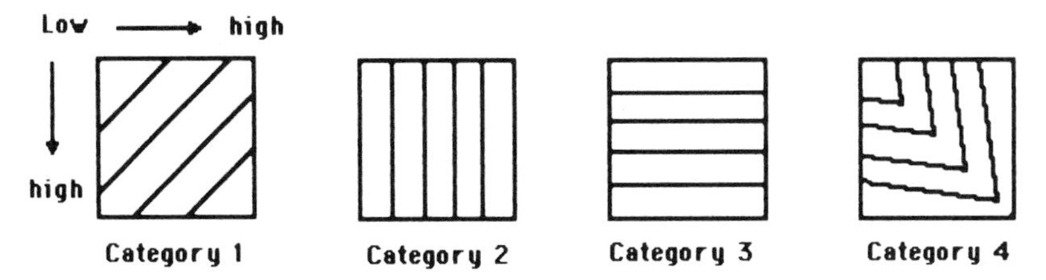

Fig. 7.82 Band partition in each category [DV-17].

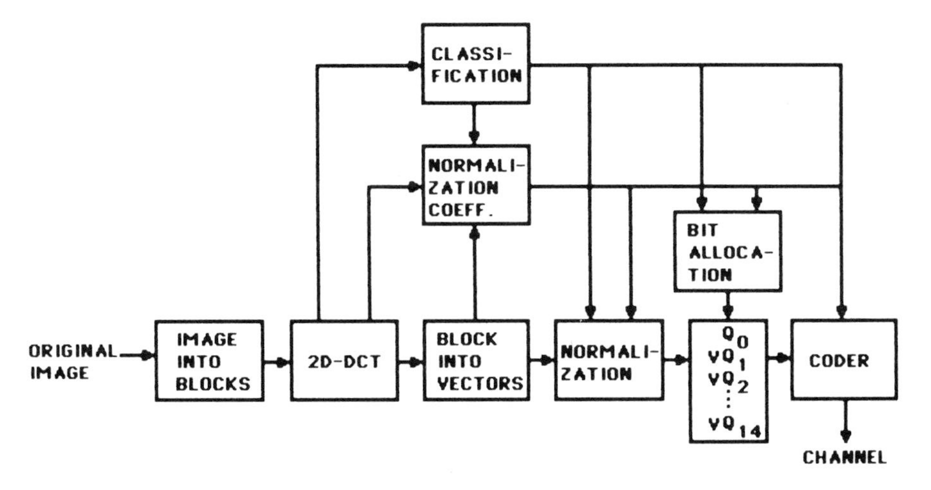

Fig. 7.83 A block diagram of adaptive DCT-VQ [DV-24]. (© 1986 IEEE)

improved the coding performance. Details of their algorithm (Fig. 7.83) are as follows:

(1) Divide an image into 8×8 blocks and apply the 2D DCT to each block.

(2) Classify the transformed blocks into four equally populated classes based on their ac energies (see Section 7.13).

(3) Partition the block into a dc term V_0 and 14 vectors ($\mathbf{V}_1, \mathbf{V}_2, \ldots, \mathbf{V}_{14}$), each composed of groups of transform coefficients (Fig. 7.84).

(4) The variance of each component of the 14 vectors in each class is computed. The geometric mean of the components of a vector is its normalization coefficient.

(5) Bit assignment (and therefore the codebook size) (Table 7.11) for each vector is similar to that of Chen and Smith [ACTC-3, ACTC-4].

(6) The dc term is uniformly quantized to 8 bits.

(7) Output indices of VQ and the overhead information (bit assignment, normalization coefficient, and the class) are transmitted.

(8) If the number of bits assigned to a vector exceeds 8 bits, then the

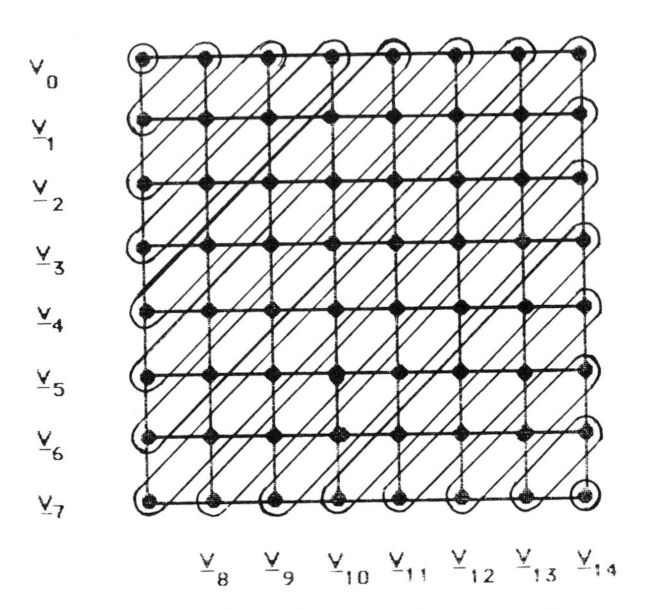

Fig. 7.84 Each 2D-DCT 8 × 8 block is partitioned into a dc term v_0 and 14 vectors $\mathbf{v}_1, \mathbf{v}_2, \ldots, \mathbf{v}_{14}$ [DV-24]. (© 1986 IEEE).

corresponding vector undergoes a two-stage VQ (first stage: 8 bits = 256 code-vectors; second stage: remaining bits) (Fig. 7.85) to reduce the overall VQ complexity. In the two-stage VQ, the error vector (difference between the input vector and the selected codeword) undergoes another VQ.

Table 7.11

An example of bit allocation for a monochrome image "GIRL." b_i refers to the number of bits assigned to vector $\mathbf{v}_i$ (see Fig. 7.84) [DV-24]. (© 1986 IEEE).

GIRL (rate = 0.4995 BPP)

	b_1	b_2	b_3	b_4	b_5	b_6	b_7	b_8	b_9	b_{10}	b_{11}	b_{12}	b_{13}	b_{14}
Class 1	2	0	0	0	0	0	0	0	0	0	0	0	0	0
Class 2	5	4	1	0	0	0	0	0	0	0	0	0	0	0
Class 3	7	7	5	3	2	0	0	0	0	0	0	0	0	0
Class 4	9	10	10	9	7	4	1	0	0	0	0	0	0	0

Adaptive DCT-VQ.

Since the transform coefficients are of zero mean and, in general, follow Laplacian distribution [IC-3], universal VQ based on multidimensional Laplacian distribution with zero mean and unit variance can be applied to the ADCT/VQ coding. The ADCT-VQ coding scheme has also been extended to

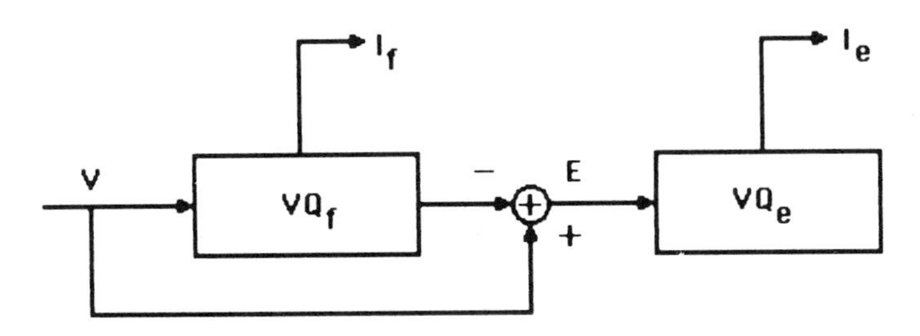

Fig. 7.85 Two-stage structure of VQ. VQ_f = first-stage vector quantizer; VQ_e = second-stage vector quantizer [DV-24]. (© 1986 IEEE).

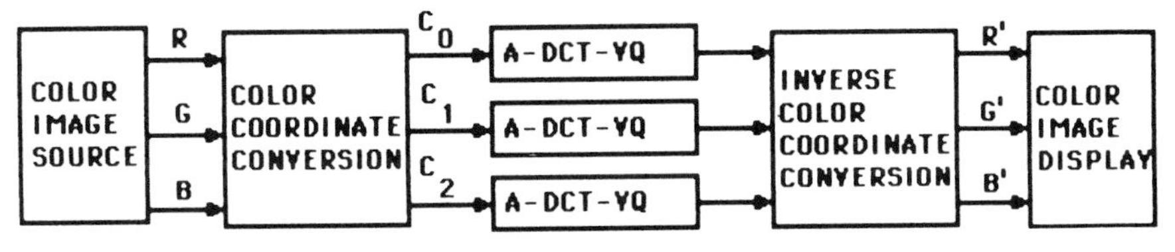

Fig. 7.86 Color image coding system. A-DCT-VQ = adaptive DCT-VQ [DV-24]. (© 1986 IEEE).

color images (Fig. 7.86). Before applying the ADCT/VQ, the *R*, *G*, and *B* components undergo a three-point DCT.

7.9.5 *2D-DCT/VQ of NTSC Color Images*

Abdelwahab and Kwatra [DV-10] have applied 2D DCT/VQ adaptively to NTSC TV composite color images sampled at $4f_{SC}$ (f_{SC} is the color subcarrier frequency). The coding involves 2D DCT of maximally correlated 4×4 blocks (pels with the same subcarrier phase), followed by their classification into one of eight classes (based on their ac energies). Whereas the dc term is uniformly quantized, the 15 ac coefficients of the block are vector quantized based on separate codebooks for each class (Fig. 7.87). The codes for the codevector index and the class are transmitted along with the code for the quantized dc term. Bit-rate reduction from 8 BPP to 0.75 BPP was achieved.

7.9.6 *Adaptive Threshold DCT-VQ Coding*

The activity parameters *A* (7.13.2) and *E* (7.13.5) were utilized in adaptively setting up perceptual thresholds for the 2D-DCT coefficients of 8×8 image blocks followed by sorting and division of the ac coefficients into groups of vectors $\mathbf{V}_1, \mathbf{V}_2, \ldots$ (Fig. 7.88). Also, an adaptive gain/shape VQ was applied to each group. It is shown that this adaptive DCT-VQ coding scheme has resulted in very low data rates with no visible impairments [DV-38].

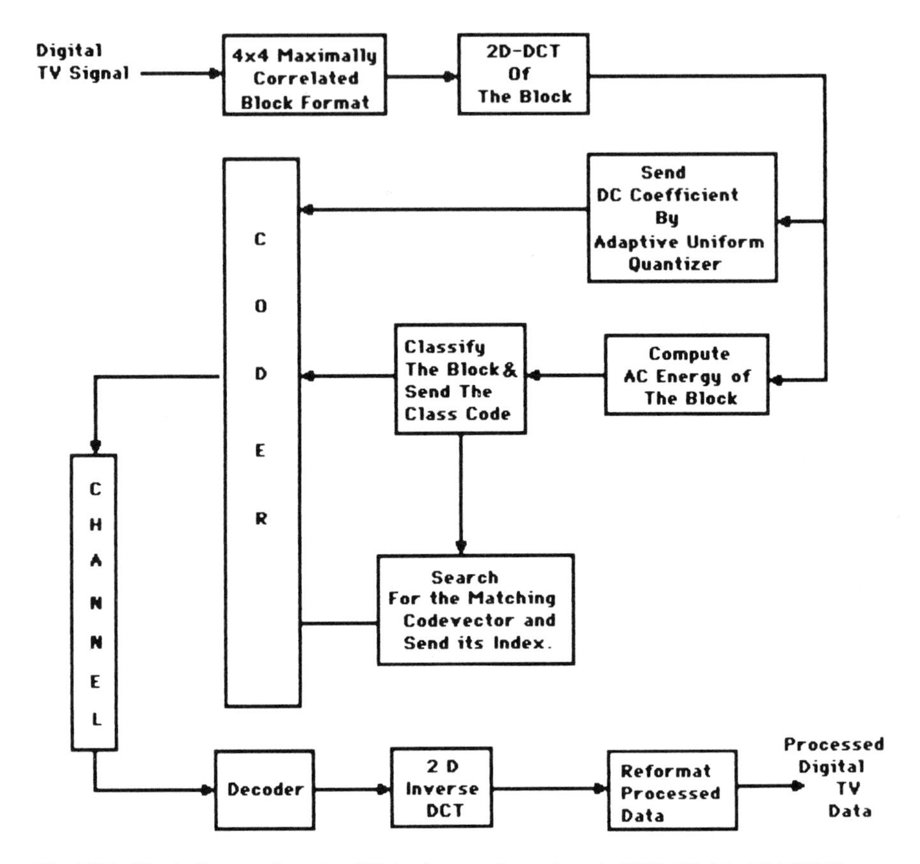

Fig. 7.87 Block diagram for using VQ in the transform domain [DV-10]. (© 1986 IEEE).

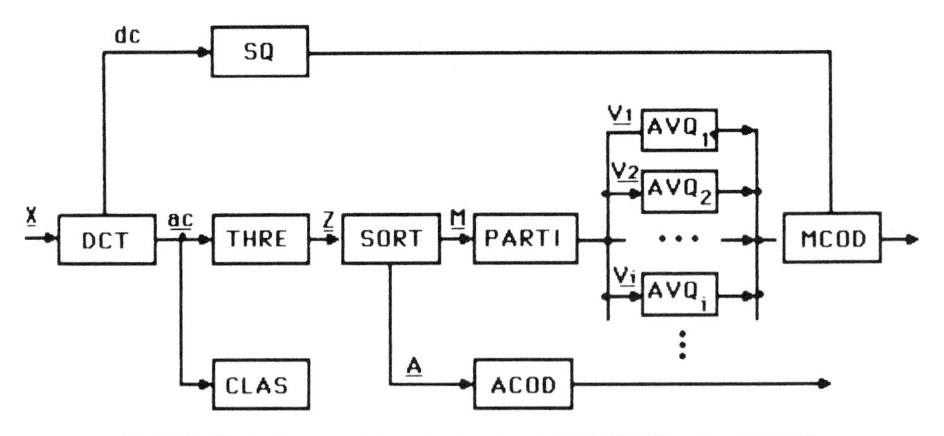

Fig. 7.88 Block diagram of the adaptive threshold DCT-VQ coding [DV-38].

7.9.7 *Classified Transform VQ Image Coding*

An adaptive hybrid scheme that incorporates multistage and hierarchical VQ and adaptive classification in the DCT domain appropriate for progressive image transmission and for visually pleasing pictures at low bit rates has been proposed by Ho and Gersho [DV-42] (Fig. 7.89). Other features, such as interblock interpolation to reduce the block structure, edge detectors to preserve the edges, and subvector classification of the DCT coefficients in each class to improve the overall coding efficiency, are incorporated in the coding process. This technique is called classified transform VQ (CTVQ). Both multistage VQ (MSVQ) and hierarchical VQ (HVQ) are briefly described.

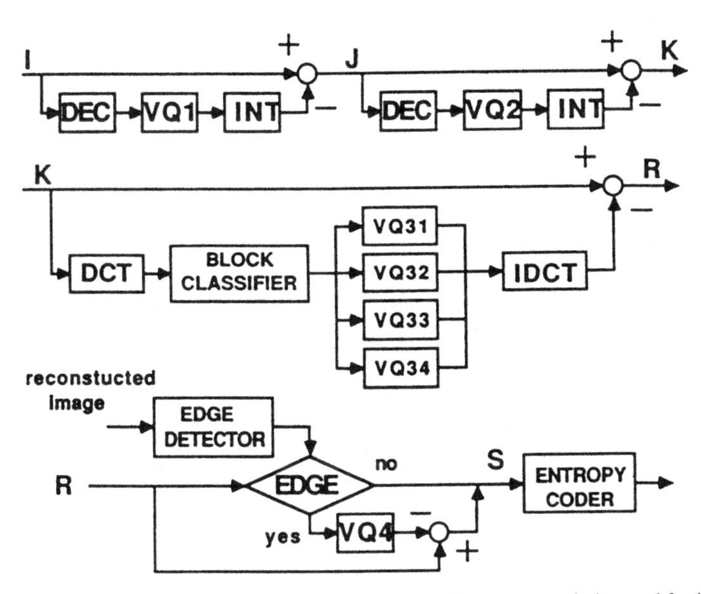

Fig. 7.89 Classified transform VQ coder for progressive image transmission and for low bit-rate image coding [DV-42]. (© 1989 IEEE).

The two-stage VQ (Fig. 7.85) is a simple version of the multistage VQ (Fig. 7.90), wherein the error vector between the input vector and its quantized output successively undergoes the vector quantization.

The binary indices of each stage are transmitted to the receiver. The signal can be reconstructed by summing up the output vectors of all the stages. This scheme is also appropriate for the progressive image transmission, as summation of the output vectors of successive stages represents a gradual build-up of image quality. The codebook size in each stage can be considerably less compared with the codebook size of single-stage VQ, and the codebook in each stage is designed based on the training sequence of the input vector to that stage.

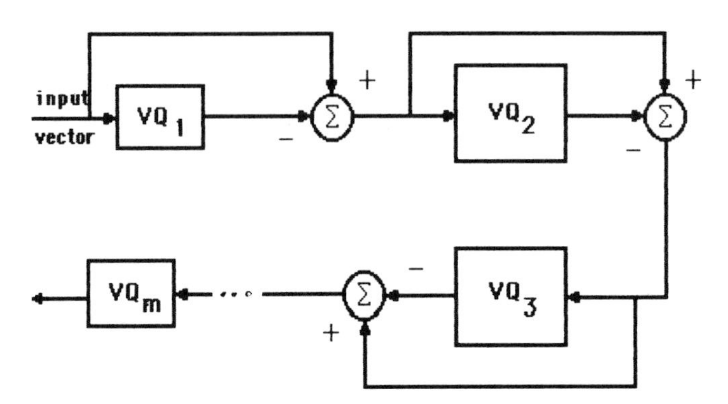

Fig. 7.90 Multistage (m-stage) VQ encoder.

In hierarchical VQ (HVQ), originally proposed by Gersho and Shoham [DV-44], a large dimensional input vector, called a supervector, is split into lower dimensional subvectors. In the next level each of these subvectors, in turn, is split into minivectors (vector dimension decreases at each level). This process is repeated until the lowest level subvectors, called cells, are obtained. At each level, a feature vector composed of the features of all the subvectors at that level is vector quantized. The quantized feature vectors not only represent the side information, but also control the bit allocation of lower level feature vectors and finally the cells. For example, in a three-level HVQ (Fig. 7.91a) a K-dimensional supervector $\mathbf{X} = \{x_1, x_2, \ldots, x_K\}$ is split into M subvectors each of dimension K/M, i.e., $\mathbf{X} = \{\mathbf{X}_1, \mathbf{X}_2, \ldots, \mathbf{X}_M\}$. A feature vector $\mathbf{S} = \{s_1, s_2, \ldots, s_M\}$, representing one feature per subvector, is quantized to $\hat{\mathbf{S}}$. The feature of a subvector can be mean, Euclidean norm, or any other meaningful measure. In the next level, each subvector is split into L cells, i.e., $\mathbf{X}_i = \{\mathbf{X}_{i1}, \mathbf{X}_{i2}, \ldots, \mathbf{X}_{iL}\}$, $i = 1, 2, \ldots, M$. Similar to the first level, a feature vector $\mathbf{P}_{ij} = \{P_{ij1}, P_{ij2}, \ldots, P_{ijL}\}$, representing features from the L cells of $\mathbf{X}_i$, is extracted. At this stage the M feature vectors, each of dimension L, are quantized. The cells are quantized by variable-size codebooks controlled by these feature vectors. The quantized outputs of the cells constitute the main information, whereas the corresponding outputs of the feature vectors represent the side information. The feature vectors control the bit allocation (codebook sizes) and normalization of the subvectors. This control mechanism preserves the supervector characteristics, and the signal is reconstructed from the side and main information at the receiver. In [DV-44] three-stage VQ was applied to speech segments (range of K from 256 to 768, $M = 16$, and $L = 4$).

In another example of hierarchical VQ (HVQ), an input vector of large dimension pk is divided into p subvectors, each of dimension k, and a feature value (say, the mean of the subvector) is extracted from each subvector to form a

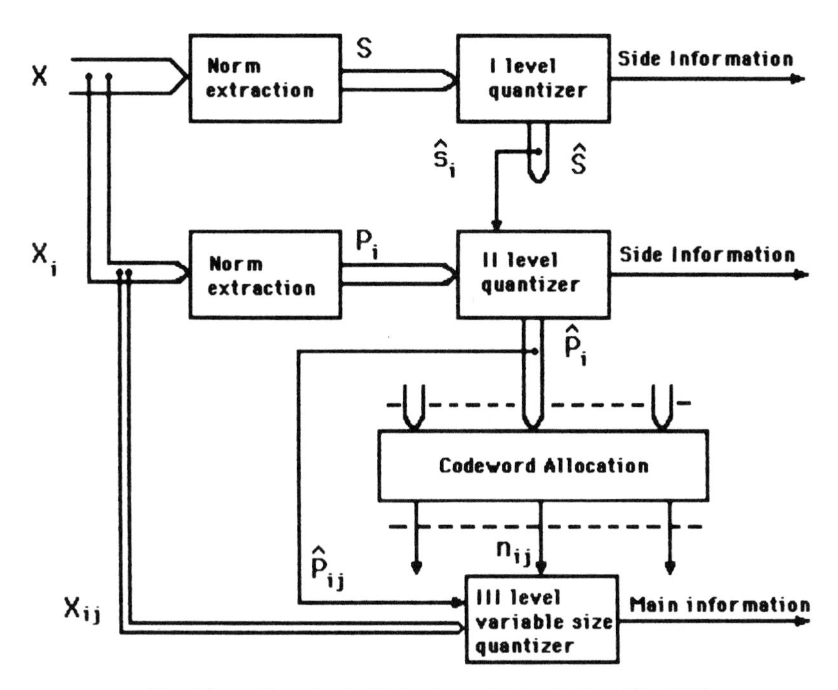

Fig. 7.91a Three-level HVQ scheme [DV-44]. (© 1984 IEEE).

feature vector of dimension p that undergoes a p-dimensional VQ. The output of this feature vector is used to remove the redundancy partially from the subvectors. Each of these resulting residual subvectors undergoes a k-dimensional VQ (Fig. 7.91b).

In the CTVQ, an image is divided into 16×16 nonoverlapping blocks and the averages of the four 8×8 subblocks constitute the feature vector. The four-dimensional feature vector is vector quantized and the 16×16 block is reconstructed using the linear interblock interpolation technique over the quantized feature vector space (see Fig. 7.152 for interpolative VQ). As in the multistage VQ, the difference between the original and interpolated images is fed into the second stage, in which the sub-blocks are of size 4×4. The feature vector dimension is now 16, and the same process as in the first stage is repeated. DCT is applied to 8×8 blocks of the error image of the second stage. The 8×8 transform blocks are divided into four classes based on the ac energies (7.13.1) of groups of coefficients corresponding to the directional features of the block in the spatial domain. The coefficient grouping and the classification scheme is illustrated in Fig. 7.92, where TH and L denote threshold and low activity class, respectively. In each class, the DCT coefficients are partitioned into subvectors such that the coefficients with nearly equal variances belong to the same subvector (Fig. 7.93). Bit allocation among the subvectors is based on the total

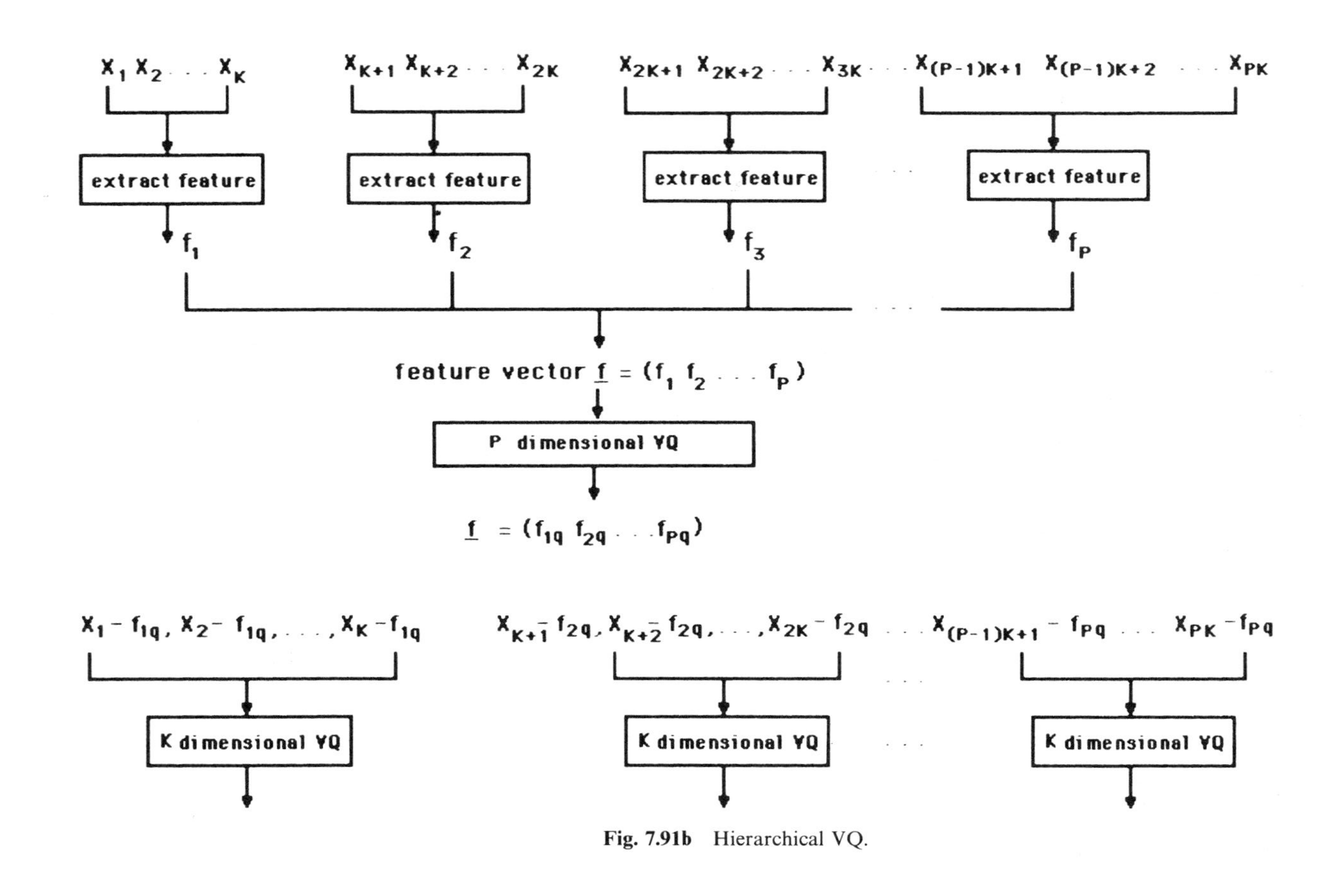

Fig. 7.91b Hierarchical VQ.

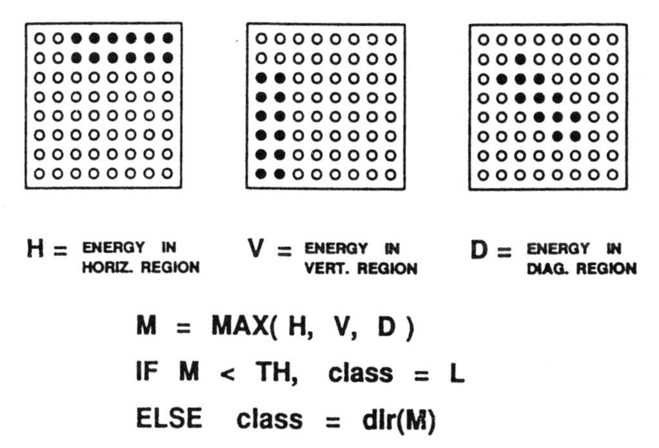

Fig. 7.92 Classification of 8×8 DCT blocks based on ac energies of DCT coefficients (solid circles) [DV-42]. (See Figs. 7.81 and 7.161.) (© 1989 IEEE).

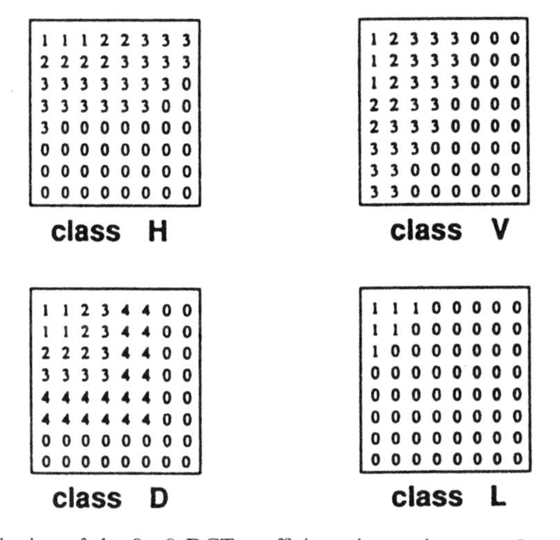

Fig. 7.93 Partitioning of the 8×8 DCT coefficients into subvectors. In each class (Fig. 7.92) coefficients with the same digit belong to a subvector. Coefficients indicated by 0 are discarded [DV-42]. (© 1989 IEEE).

variance of each subvector. Hence, the subvector dimension and the codebook size vary based on the variance distribution in each class.

An edge detector is applied to the image reconstructed from the first three stages to detect blocks with significant edges. The selected blocks go through a fourth-stage VQ. The final operation of the coder involves Huffman (VWL) coding of the error image with the objective of reconstructing a lossless image. Because of the multiple stages, this technique is suitable for progressive image transmission. Multistage VQ, interblock interpolation, adaptive perceptual classification of the 8×8 DCT sub-blocks and subvector groups, and bit allocations and edge detection are designed to yield good-quality images at low bit rates, with the final stage (entropy coding) leading to a lossless image. Details of the decimation factors and codebook sizes for the first two stages are shown in Table 7.12. The overall bit rates and SNR for the two test images "Lena" and "boat" (Fig. 7.94) are described in Table 7.13. The reconstructed images "Lena" and "boat" at stages two and four are shown in Figs. 7.95 and 7.96, respectively. Various innovative features built into the CTVQ have resulted in images with structural information at about 0.055 BPP and good quality at rates between 0.3 and 0.4 BPP. Needless to say, there are a number of options regarding the number of stages, block and sub-block sizes, DCT block classification, sub-vector groups, codebook sizes, and other adaptive features.

Table 7.12

Decimation factors, feature vector dimensions, and codebook sizes (Fig. 7.89) [DV-42]. (© 1989 IEEE)

Stage	Decimation factor	Codebook size	
		Dimension	No. of codevectors
1	8×8	4	512 (9 bits)
2	4×4	16	256 (8 bits)

Table 7.13

Overall bit rates and SNR for the test images (Figs. 7.89 and 7.94) [DV-42]. (© 1989 IEEE).

Stage	"Lena"		"Boat"	
	Bit rate (bpp)	SNR (dB)	Bit rate (bpp)	SNR (dB)
1	0.029	22.86	0.029	21.60
2	0.055	24.70	0.057	23.05
3	0.279	30.61	0.348	29.86
4	0.302	31.05	0.377	30.40
5	5.067	errorless	5.262	errorless

7.9.8 *MC Hybrid Coder Using VQ with Scene Adaptive Codebook*

Holzlwimmer, Brandt, and Tengler [MC-36] have developed a 64 KBPS video codec based on MC DPCM/transform augmented by adaptive VQ of selected transform coefficients (Fig. 7.97). The adaptive coding unit is shown in Fig. 7.98. Algorithmic details are as follows:

(1) Image sequence is temporally subsampled, coded, and transmitted at 8.33 frames/sec. The full frame rate is obtained at the receiver by motion adaptive interpolation.

Fig. 7.94 Original images (a) Lena and (b) boat [DV-42].

Fig. 7.95 Reconstructed images "Lena" based on CTVQ (Fig. 7.89). a) second stage, 0.055 BPP; b) fourth stage, 0.302 BPP [DV-42]. (© 1989 IEEE).

Fig. 7.96 Reconstructed images "boat" based on CTVQ (Fig. 7.89). a) second stage, 0.057 BPP; b) fourth stage, 0.377 BPP. (© 1989 IEEE).

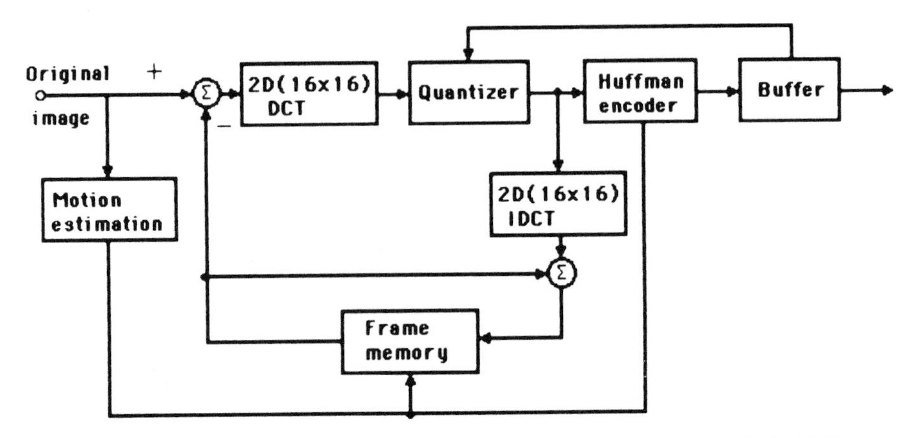

Fig. 7.97a General block diagram of the DCT hybrid coder [MC-36]. (© 1987 IEEE).

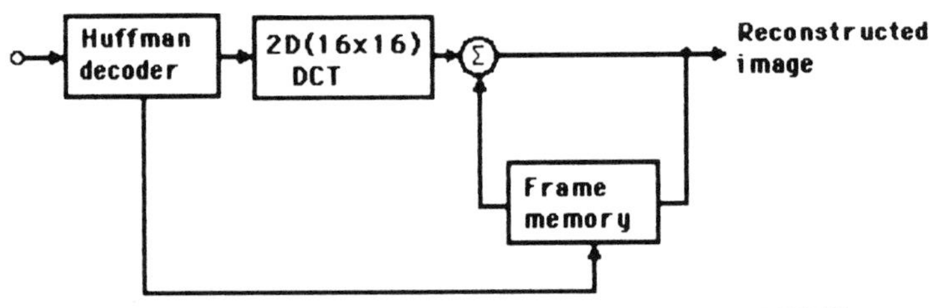

Fig. 7.97b Block diagram of the decoder [MC-36]. (© 1987 IEEE).

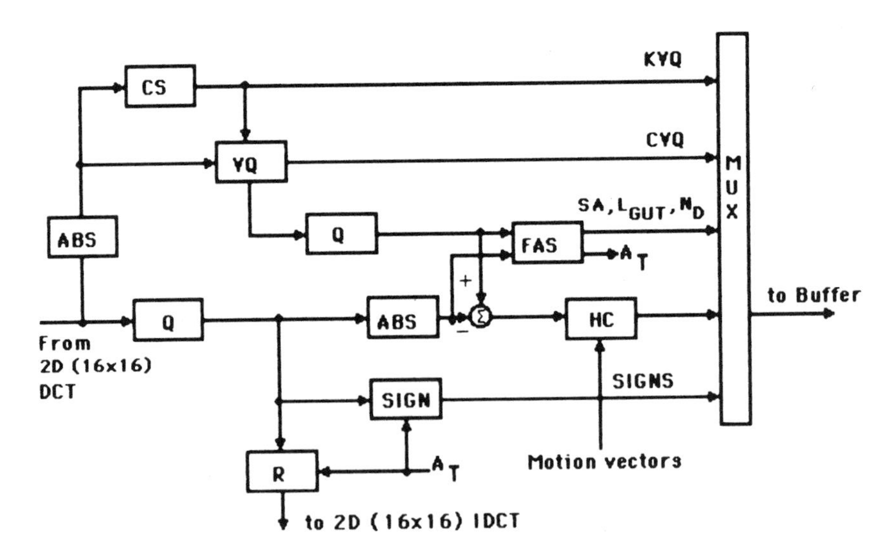

Fig. 7.98 The adaptive block coding unit [MC-36]. CS = codebook selector; FAS = final area selection; Q = quantization; R = selection of DCT coefficients and inverse scaling; HC = Huffman coder; CVQ = classified VQ; ABS = absolute value. (© 1987 IEEE).

(2) 2D 16×16 DCT is applied to MC interframe prediction errors. Hierarchical motion estimation based on BMA is described in Fig. 7.117a.

(3) Only the changed blocks indicated by the DCT coefficients' energy above a threshold controlled by the buffer are coded.

(4) Block classification and adaptive VQ coding is described in Figs. 7.162–7.164.

(5) The components of the error vector (difference between the input and output of VQ) are coded provided its energy is above a threshold. Only a few of these components based on subarea energies are actually coded.

(6) Overhead representing block classification, item (5), motion vectors, and other relevant information is transmitted.

Because of the various adaptive features, test sequences processed through this codec have been reconstructed at good videophone quality.

7.9.9 *ATC-VQ Image Coding*

Another ATC-VQ scheme proposed for image coding (Fig. 7.99) involves adaptive classification of the 8×8 blocks in the 2D-DCT domain into m disjoint sub-blocks SBK_i, $i = 1, 2, \ldots, m$, based on the activity and directional feature (angle orientation of the local image structure) of the block followed by VQ of the sub-block coefficients [DV-46].

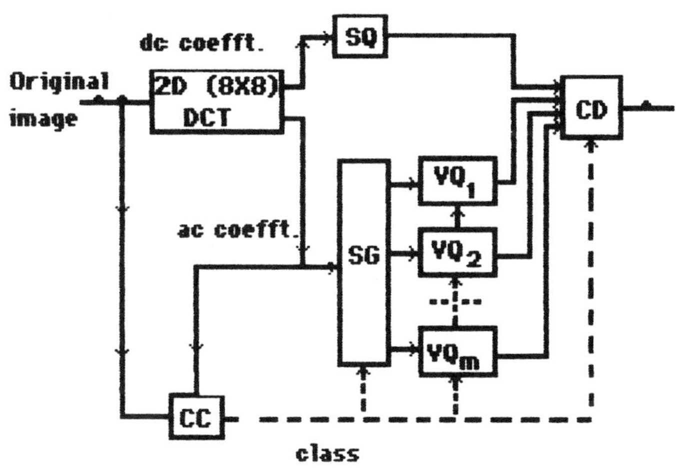

Fig. 7.99 Block diagram of the ATC-VQ system. CC = classification unit; CD = coding unit; SG = sub-blocks generator; SQ = scalar quantizer; VQ_i = VQ for ith sub-block [DV-46].

The directional classification of the 8×8 block is based on mean absolute differences between adjacent pels in the horizontal and vertical directions Δ_h and Δ_v, respectively, and corresponding differences along the principal (major) diagonal and secondary (minor) diagonal Δ_{pd} and Δ_{sd}, respectively. The features

$$\Delta_M = (\bar{\Delta}_h, \bar{\Delta}_v)$$

and

$$\mathbf{\Delta}_{\mathrm{D}} = (\bar{\Delta}_{\mathrm{pd}}, \bar{\Delta}_{\mathrm{sd}})$$

control the classification areas of the block (Fig. 7.100). The perceptibility of luminance change in the sub-block is determined from μ, defined as

$$\begin{aligned} \mu &= 0 \quad \text{if } |y_{ij}| \leqslant w_{ij} \qquad \forall\, i, j \text{ except } i = j = 0, \\ &= 1 \quad \text{otherwise,} \end{aligned}$$

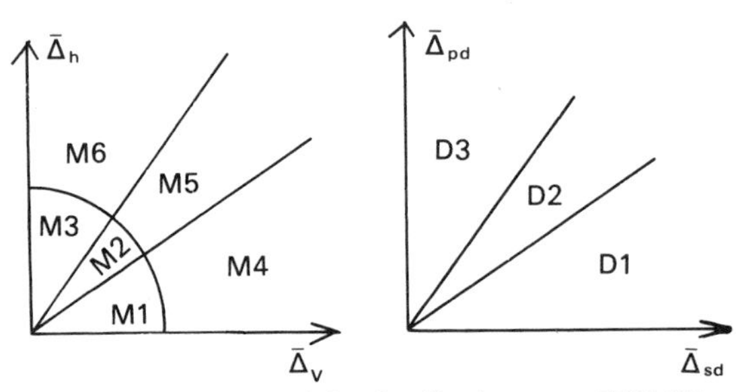

Fig. 7.100 Division of the classification areas [DV-46].

where w_{ij} is the perception threshold of the (i, j)th 2D-DCT coefficient. When $\mu = 0$ only the dc term is transmitted. Using Fig. 7.100 and μ, the 8×8 blocks are classified into 13 categories representing activity and directional structure (Table 7.14). For each of the 13 classifications, the 8×8 block is divided into m disjoint sub-blocks such that the DCT coefficients within a sub-block are highly

Table 7.14

Classification areas and corresponding structures [DV-46].

Classification area	Activity and direction of structure (in degree)
$\mathrm{K1} = \mathrm{M6} \times \mathrm{D2}$	high activity; 0
$\mathrm{K2} = \mathrm{M6} \times \mathrm{D3}$	high activity; 20
$\mathrm{K3} = \mathrm{M5} \times \mathrm{D3}$	high activity; 45
$\mathrm{K4} = \mathrm{M4} \times \mathrm{D3}$	high activity; 70
$\mathrm{K5} = \mathrm{M4} \times \mathrm{D2}$	high activity; 90
$\mathrm{K6} = \mathrm{M4} \times \mathrm{D1}$	high activity; 110
$\mathrm{K7} = \mathrm{M5} \times \mathrm{D1}$	high activity; 135
$\mathrm{K8} = \mathrm{M6} \times \mathrm{D1}$	high activity; 160
$\mathrm{K9} = \mathrm{M5} \times \mathrm{D2}$	high activity; mixed
$\mathrm{K10} = \mathrm{M3} \times (\mu = 1)$	medium activity; 0
$\mathrm{K11} = \mathrm{M2} \times (\mu = 1)$	medium activity; mixed
$\mathrm{K12} = \mathrm{M1} \times (\mu = 1)$	medium activity; 90
$\mathrm{K13} = (\mu = 0)$	low activity

correlated, whereas the coefficients representing different sub-blocks are weakly correlated. An example of this sub-block division is shown in Fig. 7.101 for the first three classes. The coefficients indicated by dark squares in each sub-block are vector quantized. The codebook sizes for the sub-blocks reflect the spectral variances and correlations. For all classes, the dc coefficient is separately (uniform) quantized. The performance of the ATC-VQ is compared with ATC and VQ [DV-16].

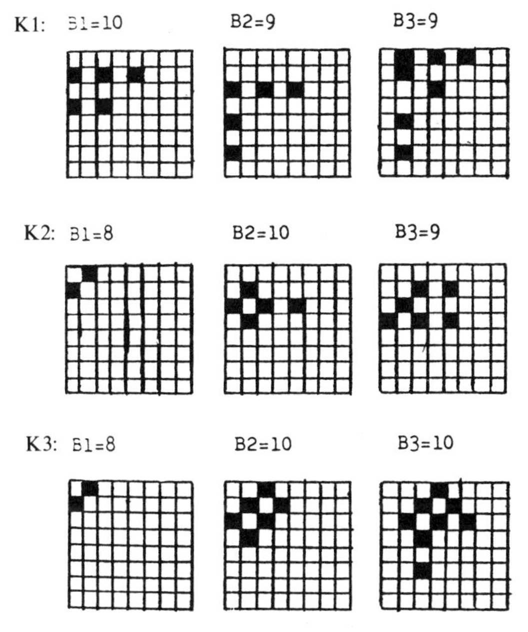

Fig. 7.101 Results of decomposition and bit assignment [DV-46]. Bi = number of bits for SBKi.

7.9.10 *DCT-Adaptive CVQ*

Whereas Ramamurthi and Gersho [DV-11] have developed an edge-oriented classifier in the spatial domain, Kim and Lee [DV-41] have extended the classification to the 2D-DCT domain (Fig. 7.102). Similar to the technique of Y_{im} and Kim [DV-34] (see Figs. 7.78 and 7.79), the coefficients $X^{C(2)}(0,1)$ and $X^{C(2)}(1,0)$ indicate the edge patterns of the 8×8 block (Fig. 7.103). Using the test images with artificially generated edges and orientations, the reflected feature vectors (Fig. 7.104) and eight classes (Fig. 7.105) consisting of shade (class 0), weak diagonal edges (classes 1–4), horizontal edges (class 5), vertical edges (class 6), and strong diagonal edges (class 7) are developed. Using $X^{C(2)}(0,1)$ and $X^{C(2)}(1,0)$ as the features (these are shown as V1 and H1, respectively, in Figs. 7.103 and 7.104), the classifier identifies (based on the nearest neighbor search) the 8×8 block as one of the eight classes.

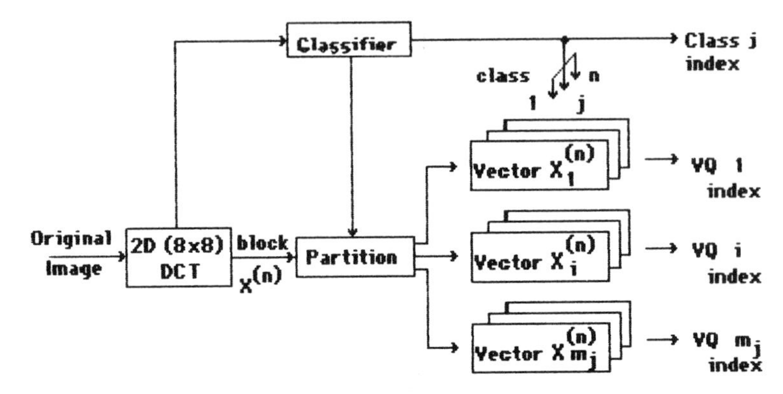

Fig. 7.102 Block diagram of DCT-CVQ [DV-41]. (© 1989 IEEE).

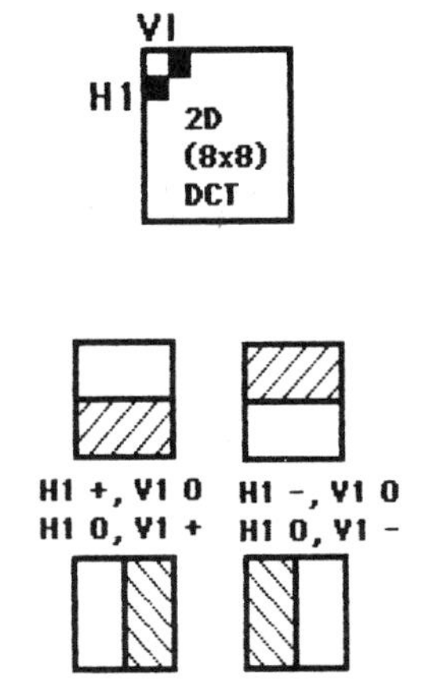

Fig. 7.103 V1, H1 feature vector [DV-41] +, −, and 0 denote positive, negative, and zero, respectively. (© 1989 IEEE).

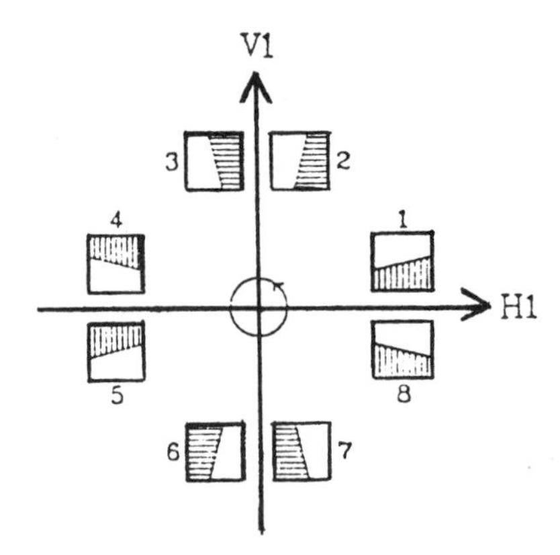

Fig. 7.104 Reflected feature vectors [DV-41]. (© 1989 IEEE).

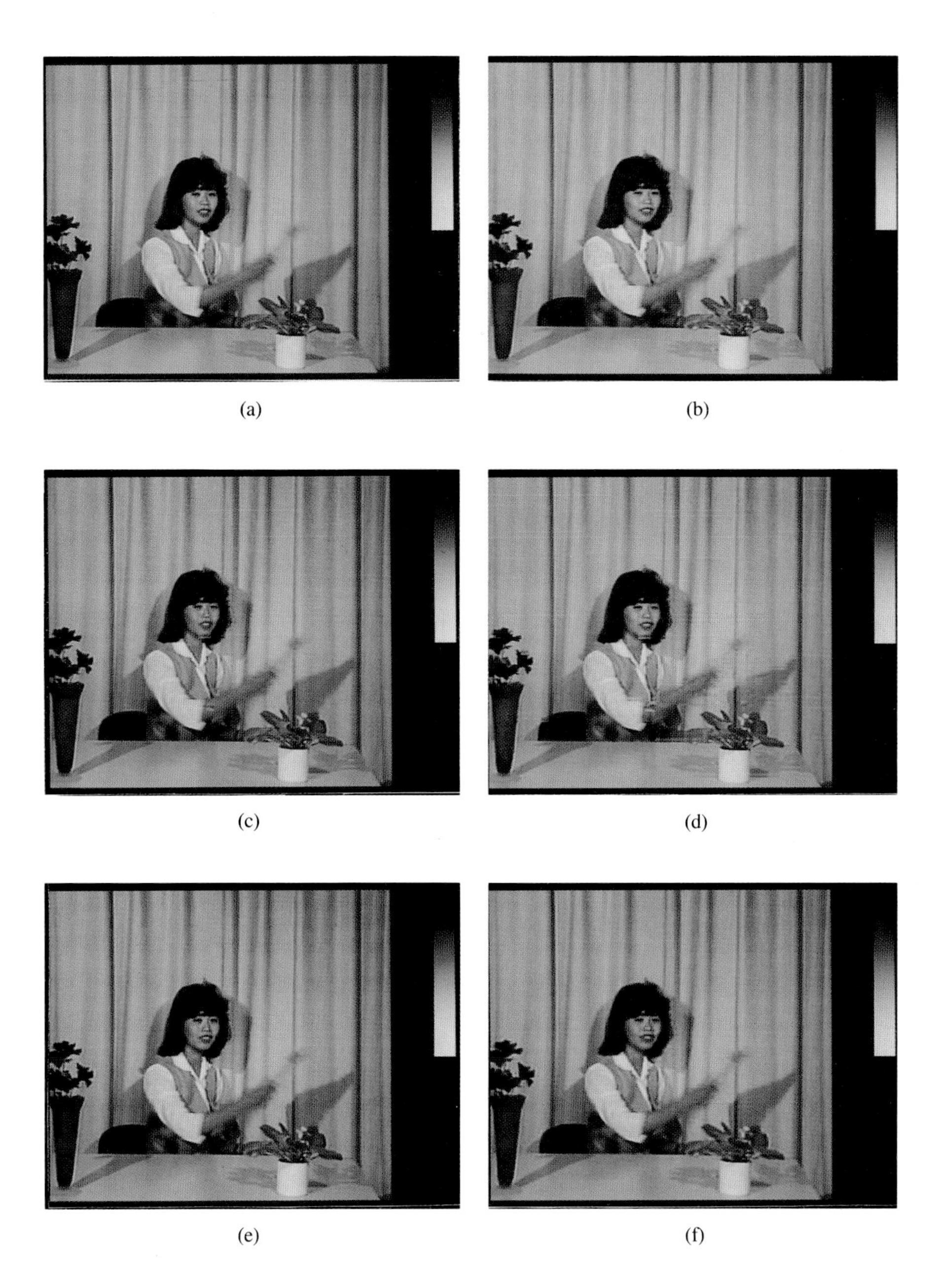

(a) (b) (c) (d) (e) (f)

Fig. 7.198 The effects of random and layered packet losses in LBR video coding [PV-1, PV-2]. (b) through (f) are reconstructed images at 0.7 BPP (5.1 MBPS). (a) original image, 512×480 pel, 8-bit PCM, 30 noninterlaced frames/sec. (b) No packet loss; SNR 41 dB. (c) 1% random packet loss; SNR 37 dB. (d) 10% random packet loss; SNR 35 dB. (e) layered loss 15%; SNR 39 dB. (f) layered loss 20%; SNR 36 dB.

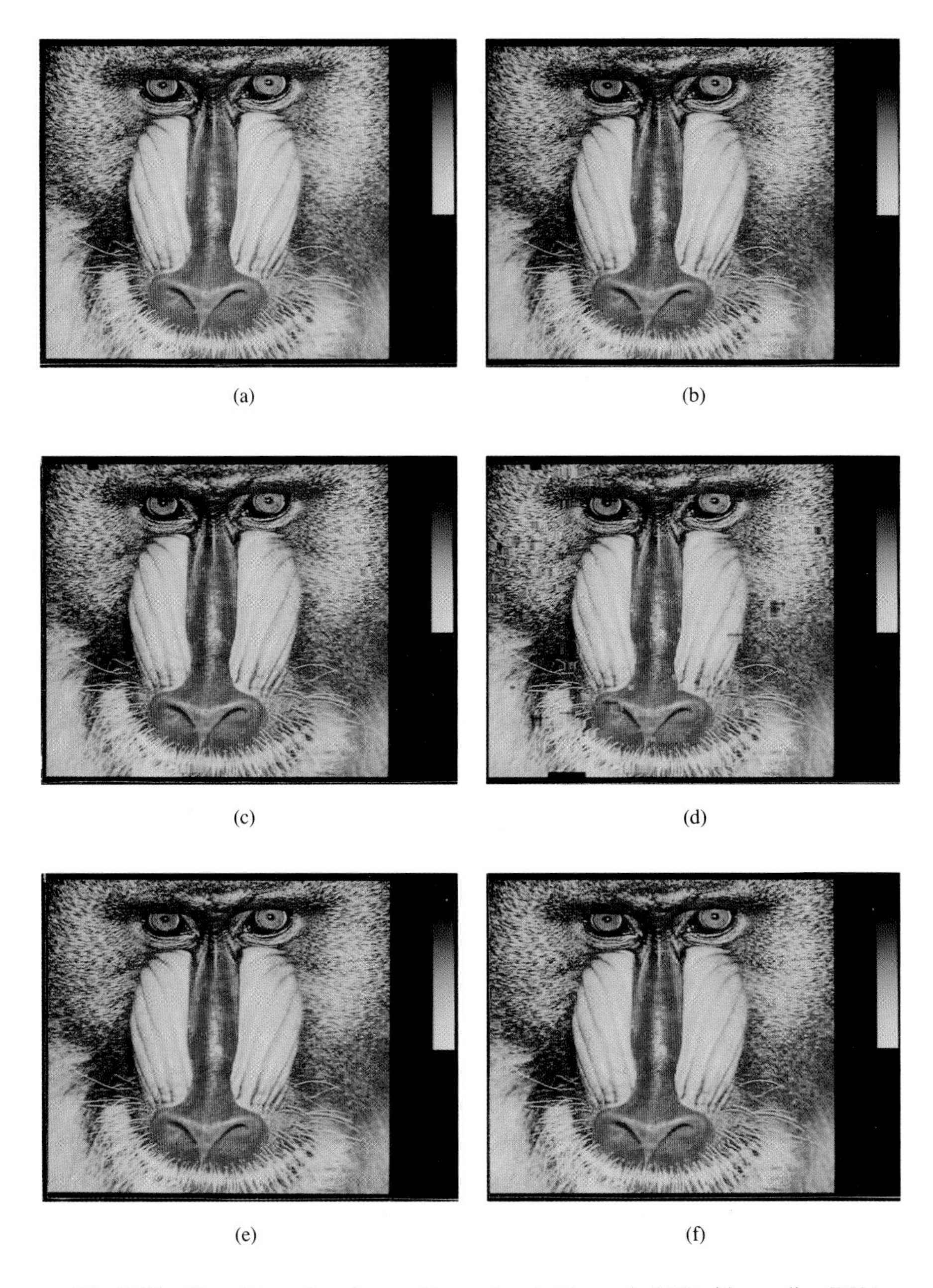

Fig. 7.199 The effects of random and layered packet losses in LBR video coding [PV-1, PV-2]. (b) through (f) are reconstructed images at 2.4 BPP (17.8 MBPS). (a) original image, 512×480 pel, 8-bit PCM, 30 frames/sec. (b) No packet loss; SNR 29 dB. (c) 1% random packet loss; SNR 28 dB. (d) 10% random packet loss; SNR 24 dB. (e) layered loss 15% SNR 27 dB. (f) layered loss 23%; SNR 24 dB.

(a)

(b)

Fig. 7.60 Original image (a) and its reconstruction (b) by the intrafield DCT codec described in Fig. 7.58.

(a)

(b)

Fig. 7.64 Images processed by the adaptive interfield hybrid (DPCM/DCT) codec described in Fig. 7.61. (a) 30 MBPS; (b) 15 MBPS.

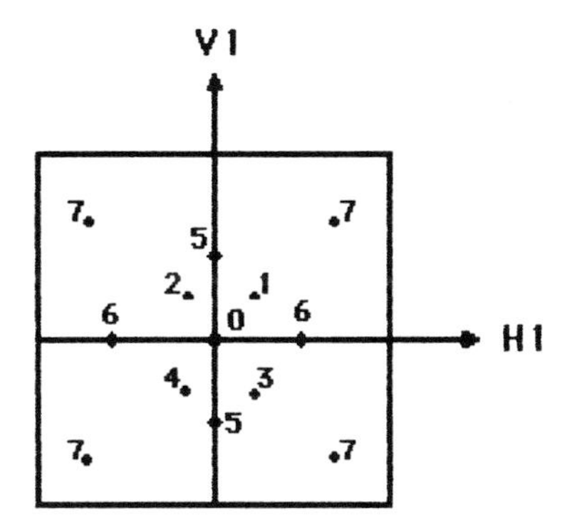

Fig. 7.105 Scatter diagram in V1, H1 feature domain proposed classifier [DV-41]. (© 1989 IEEE).

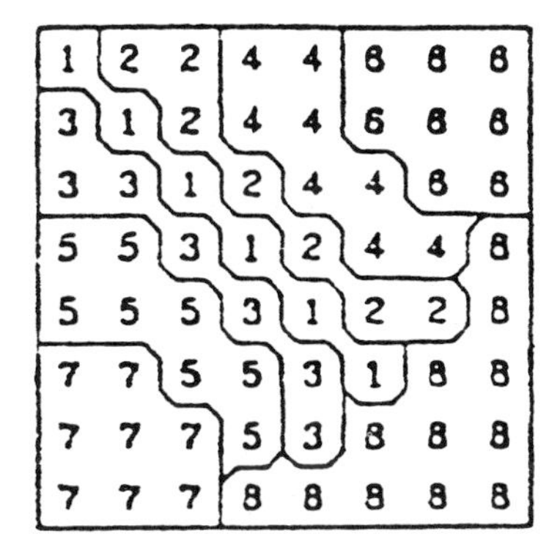

Fig. 7.106 Partitioning the 2D 8×8 DCT into sub-blocks. Coefficients with the same digit belong to a vector [DV-41]. (© 1989 IEEE).

The 2-D 8×8 DCT coefficients are divided into eight sub-blocks (Fig. 7.106) based on small geometric mean vector variances. The coefficients in each sub-block represent a vector, and separate codebooks for each sub-block and for each class are developed. Details regarding the sub-block classification, codebook generation, and performance of the DCT-adaptive CVQ scheme are described in [DV-41].

This section has presented an introduction to the DCT/VQ of images. The reader may recall the DCT/VQ techniques applied to speech coding in Section 7.9. The objective of the hybrid scheme is to arrive at a compromise from the viewpoint of a) reducing the complexity of VQ, b) approaching its rate distortion bound performance, and c) reducing the suboptimality of transform (DCT) coding compared with the VQ. The hybrid method has been applied to a wide spectrum of images (teleconferencing, videophone, monochrome, color, broadcast TV). Teleconferencing video codecs with either DCT or VQ alone as the main compression tool have been built and marketed. These developments indicate that practical application of DCT/VQ in video codecs is almost a certainty. This section also has presented a summary of some of the adaptive features applied to the DCT/VQ. This summary is intended to provide a flavor for the reader regarding the limitless adaptations that can be envisioned. No doubt any adaptive feature automatically increases the coder complexity, even though perceptually improved images can be reconstructed. One, therefore, has to strike a balance between the increased complexity and the improved subjective quality. This presentation is by no means exhaustive, and many other efficient DCT/VQ adaptive algorithms, such as the adaptive transform VQ (ATVQ) [DV-20] and transform/adaptive TSVQ [DV-21, DV-26, DV-37], have been left out purely due to space constraints.

7.10 Low Bit-Rate Coding

The two previous sections have dealt with image coding based on DCT and DCT/DPCM (Section 7.8) and DCT/VQ (Section 7.9). The object of this section is to apply these tools, along with other aspects, to low bit-rate (LBR) video coding. This classification is somewhat arbitrary. One can probably assign teleconferencing and videophone as applications of LBR coding. In any case, the activity in these two fields appears to be intense, as evidenced from the standards being formulated by the international organization [LBR-23, LBR-35, LBR-37]. Also, various industries, research labs, and government agencies are designing and developing codecs (videoconferencing and/or videophone) that are based on these standards. In North America, western Europe, and Japan, some of these codecs (although different algorithms, different bit rates, and different resolutions) have been in the market since the early 1980s and even as early as the late 1970s. Their growth has been inhibited not only by the apparent incompatibility between the codecs manufactured by different companies, but also by the high costs consistent with a small market. A combination of factors, such as the introduction of ISDN and consequent availability of public digital networks, formulation of codec standards (algorithm, resolution, etc.), integration of the algorithm or its major portion into VLSI, decreasing costs of logic and memory (also increasing speed and increasing memory size), proliferation of desktop multifunction/multiservice terminals, and introduction of interactive multimedia communication services (teletext, videotext, photographic-quality color facsimile, etc.) can provide the incentive and impetus for the videophone to reach the consumer market. Similarly, the videoconferencing codec can be aimed at (affordable to) even medium and small industries and businesses. These include entertainment, recreation, and other service sectors. Because of video/audio interactive communication, the videophone will be a powerful and popular tool in the electronic consumer market. Thus LBR coding is destined to play a significant role in industrial and consumer electronics. As in the previous sections, an algorithmic approach to LBR coding will be presented with the objective that the reader can be exposed to DCT-based tools.

7.10.1 *Universal Block Diagram*

A universal block diagram for LBR video codecs is shown in Fig. 7.107. Some of the functions, such as preprocessor, quantizer, coder, buffer, etc., are in principle the same as in any video coding scheme. In applications such as videoconferencing and videophone, characterized by the absence of fast motion, special effects, etc., reduced resolution (both spatial and temporal) can be

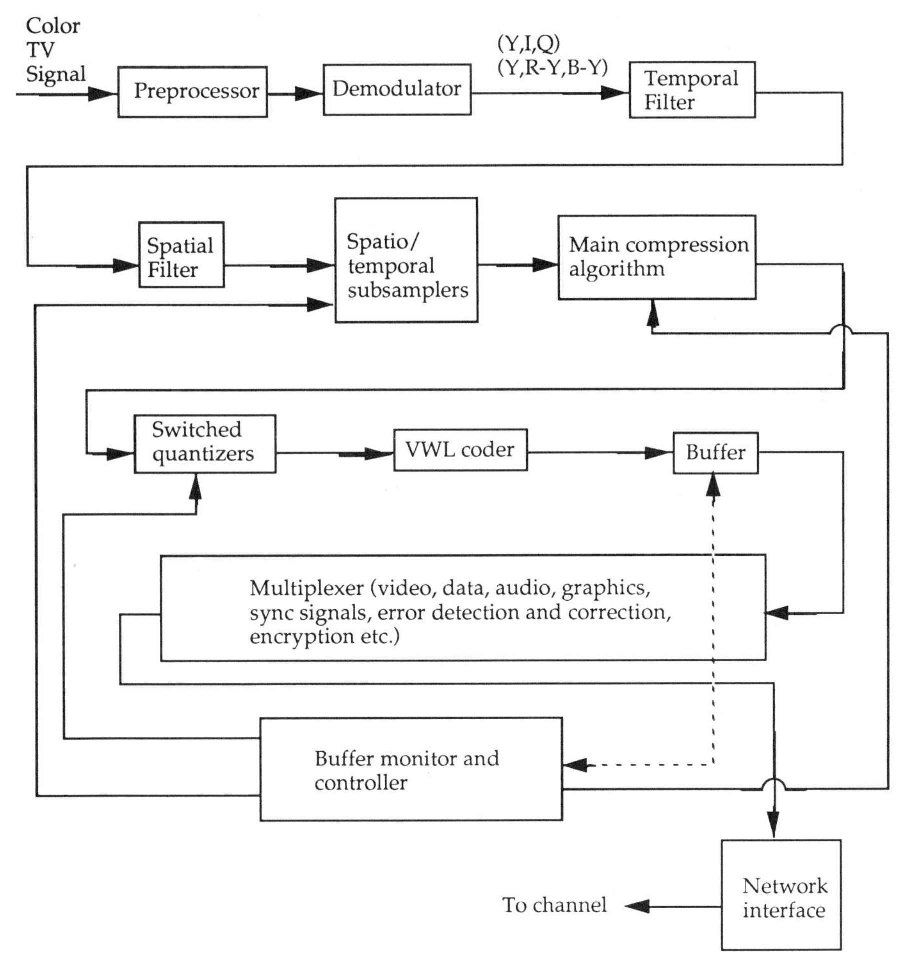

Fig. 7.107 Universal block diagram for low bit-rate TV codecs. Compression algorithm can utilize prediction (DPCM), transform, sub-band, VQ (with or without motion compensation), and their combinations. In each of these techniques there are a number of variables.

combined with motion compensation to achieve additional bit-rate reduction. Also, the subjective quality of the reconstructed images need not be as stringent as in broadcast-quality TV. Of course, the processed video needs to be visually pleasing from the end-user (human) perspective. With this scenario in mind, codec designers have adopted human visual system (HVS) weighting of the transform coefficients, designed quantizers with subjective thresholds, adaptively scanned the transform coefficients, and switched to intraframe or interframe modes. With increasing understanding of the HVS, additional gains in codec performance can be achieved.

7.10.2 *Motion Compensation* (*MC*)

In LBR coding, interframe/interfield prediction can be improved by motion estimation. If the motion of a pel or block of pels can be reasonably estimated between successive fields/frames, then interframe/interfield prediction accuracy improves. Also, for temporally subsampled images, missing frames/fields can be better recovered at the receiver by interpolation along the motion trajectory [MC-1, LBR-13]. Both pel recursive algorithms (PRA) and block matching algorithms (BMA) have been developed. The former relates to motion estimation of individual pels, whereas the latter deals with the motion of a block (group) of pels. PRA is computationally intensive and has been seldom used in practice, even though it is very effective. BMA, on the other hand, in spite of the inherent assumption that all the pels in the block have the same motion, has been utilized in practice. VLSI implementations of PRA are described in [MC-38, MC-44]. Also VLSI design/architecture leading to ASICs for implementing BMA, including full-search and fractional pel/line accuracy, is progressing rapidly [MC-30, through MC-33, MC-37, MC-39 through MC-43]. A list of motion-estimation VLSI chip manufacturers is presented in Appendix B.3. Codecs have been built using MC based on BMA. Also, the videoconferencing codec standards being developed include BMA-MC as an optional feature [LBR-23, LBR-35, BR-37, LBR-38, BR-40, LBR-41)].

In BMA, the block of pels of size $(M \times N)$ is compared with a corresponding block within a search area of size $(M + 2p \times N + 2p)$ in the previous frame (Fig. 7.108), and the best match is found based on a cost function such as minimum MSE, minimum MAE, or maximum cross correlation, defined as follows:

Cross-correlation function (*CCF*)

$$M_1(i, j) = \frac{\sum_{m=1}^{M} \sum_{n=1}^{N} U(m, n) U_R(m + i, n + j)}{\left[\sum_{m=1}^{M} \sum_{n=1}^{N} U^2(m, n)\right]^{1/2} \left[\sum_{m=1}^{M} \sum_{n=1}^{N} U_R^2(m + i, n + j)\right]^{1/2}}$$

$$-p \leqslant i, j \leqslant p. \tag{7.10.1}$$

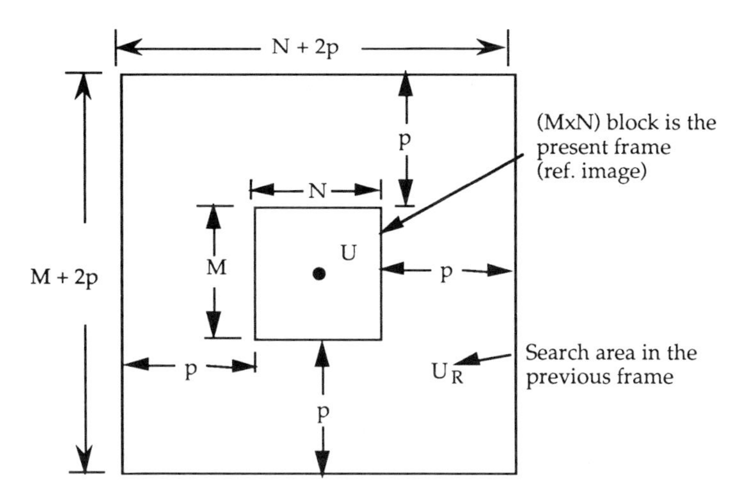

Fig. 7.108 Geometry for manipulation in $M \times N$ reference image U with $(M + 2p) \times (N + 2p)$ image U_R (search area) in the previous frame.

Normalized MSE (NMSE)

$$M_2(i, j) = \frac{\sum_{m=1}^{M} \sum_{n=1}^{N} [U(m, n) - U_R(m + i, n + j)]^2}{\sum_{m=1}^{M} \sum_{n=1}^{N} U^2(m, n)} \qquad -p \leqslant i, j \leqslant p \tag{7.10.2}$$

Mean absolute error (MAE)

$$M_3(i, j) = \frac{1}{MN} \sum_{m=1}^{M} \sum_{n=1}^{N} |U(m, n) - U_R(m + i, n + j)| \qquad -p \leqslant i, j \leqslant p. \tag{7.10.3}$$

Some other cost functions utilized in BMA are defined as follows:

$$M_4(i, j) = M_3(i, j) + \theta \|\mathbf{v}\|_{L_1} \qquad -p \leqslant i, j \leqslant p, \tag{7.10.4}$$

where $\|\mathbf{v}\|_{L_1}$ is the L_1 norm and θ is some control parameter. The second component in (7.10.4) tends to bias towards small-motion vectors, as it is directly proportional to the magnitude of the motion vector. For $\theta = 0.2$, it was shown [MC-36] that this component effectively suppressed spurious nonzero motion vectors (due to noise) in the stationary background:

$$M_5(i, j) = \sum_{m=1}^{M} \sum_{n=1}^{N} [f(T_{0'}|U(m, n) - U_R(m + i, n + j)|] \qquad -p \leqslant i, j \leqslant p, \tag{7.10.5}$$

where

$$f(T_0, a) = \begin{matrix} 1 & T_0 < a \\ 0 & T_0 \geqslant a \end{matrix}$$

and T_0 is a preset threshold. $U(m, n)$ is the intensity of pel in row m and line n in the current frame/field and $U_R(m + i, n + j)$ is the intensity of pel in the reference image (previous frame/field) displaced from $U(m, n)$ by i rows and j lines (columns).

In practice both MSE and MAE have been used as the criteria for BMA. In this process the motion estimation is limited to $\pm p$ pels or lines/frame and brute force search requires $(2p + 1)^2$ computations of the cost function, considering every pel shift (horizontally) and every line shift (vertically). Even though the full-search BMA is computationally very intensive, ASICs for motion estimation with fractional pel/line precision have been developed [MC-30 through MC-33, MC-37, MC-39 through MC-43]. A flexible VLSI architecture for motion estimation based on full-search BMA and MAE (7.10.3) as the distortion function has been proposed [MC-33]. The same chip can be designed for different block sizes (8×8, 16×16, $32 \times 32, \ldots$). Also four chips can be cascaded to extend the tracking range from -8 to $+7$ pels-lines/frame to -16 to $+15$ pels-lines/frame. Cascading is also useful for processing higher sampling rate (up to 25 MHz) video sources [MC-42]. Various efficient algorithms that minimize this complexity, but at the same time track the motion, have been developed. For example, the logarithmic search procedure proposed by Jain and Jain [MC-1] tracks the direction of minimum distortion, resulting in considerable computational simplicity (Fig. 7.109). A modification of this is suggested in [MC-6] (Fig. 7.110). Another technique called a three-step (3-2-1) search (Fig. 7.111) requires only 25 computations compared with the 225 required for the brute force when $p = 7$ [MC-11]. Another BMA that has been implemented after extensive simulations is based on a fixed search pattern with 32 possible motion vectors (Fig. 7.112) for an 8×8 block [MC-8]. If a nonzero motion vector is estimated, then the MC interframe predictor uses five displaced pels from the previous frame for filtering the high-frequency noise and for improving the prediction. Similar loop filters have been proposed and utilized in the videoconferencing codecs [MC-19].

Another efficient BMA, based on optimization, called the conjugate direction search (CDS), involves determining the minimum distortion locations along the conjugate directions successively (Fig. 7.113) [MC-27]. A simplified version of this, called one at a time search (OTS), involves searching for the location of minimum distortion once along the horizontal direction and once along the vertical direction. Extensive simulations using various test sequences, and also MSE and MAE, as the cost functions have shown that the OTS using MAE as the distortion function requires on the average of seven searches ($p = 8$) for an 8×8 block. Thus OTS-MAE is an effective BMA and is simple to implement in hardware. Ghanbari [MC-29] has proposed a BMA based on a logarithmic step

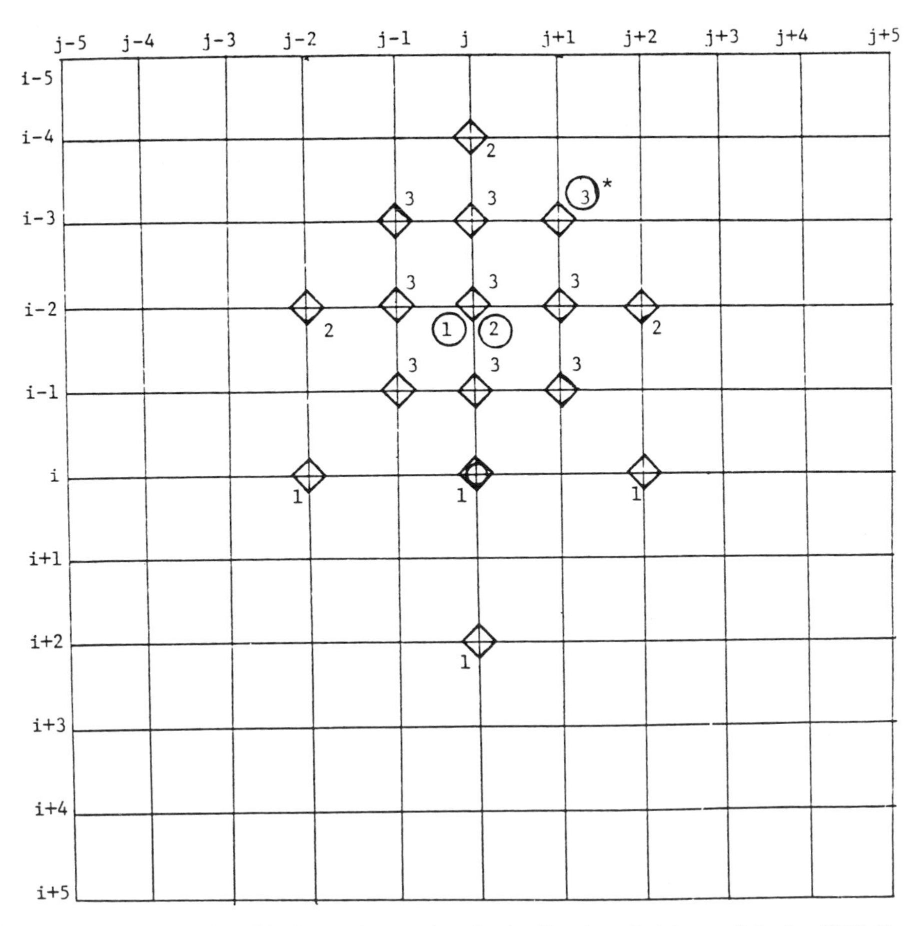

Fig. 7.109 A 2D logarithmic search procedure for the direction of minimum distortion [MC-1]. The figure above shows the concept of the 2D logarithmic search to find a pixel in another frame that is registered with respect to the pixel (i, j) of a given frame, such that the mean square error over a block defined around (i, j) is minimized. The search is done step by step with ◇ indicating the direction searched at a step number marked. The numbers circled show the optimum directions for that search step and the * shows the final optimum direction $(i-3, j+1)$ in the above example. This procedure requires only searching 13–21 directions for the above grid, as opposed to 121 total possibilities. (© 1981 IEEE).

search with only four locations tested in each step. This algorithm requires only $(5 + 4 \log_2 p)$ computations of the cost function. It has a performance based on interframe prediction errors comparable with the three-step search [MC-11] and the CDS [MC-27] with less computational complexity. In general, BMAs have been utilized for estimating the motion of 8×8 blocks. An extension of BMA is object matching. Here motion vectors are estimated for the entire moving object(s) [LBR-13]. Assumption of uniform motion for all pels in a

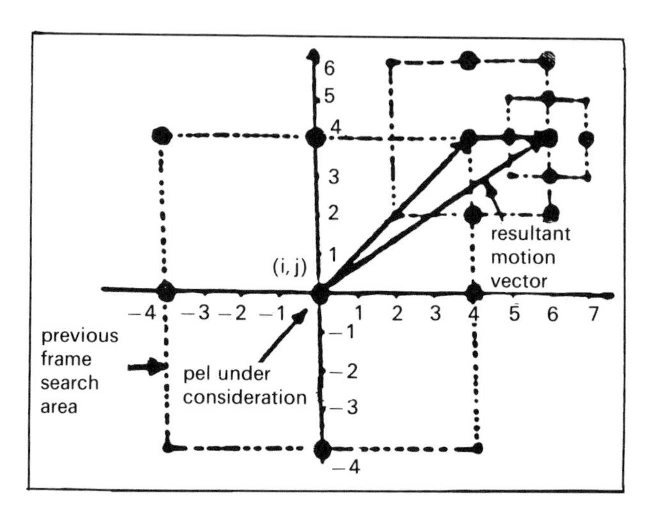

Fig. 7.110 Modified motion estimation algorithm [MC-6]. ● denotes pel position searched; • denotes optimum direction for a particular step. (© 1985 IEEE).

block implies that smaller blocks (say, 4×4) should be used to keep this assumption valid. However, the overhead (information bits for motion vector) increases for smaller blocks. Thus the choice of block size for BMA is a compromise between these two conflicting requirements. An exception to this is global motion estimation performed on each frame for applying 3D DCT to

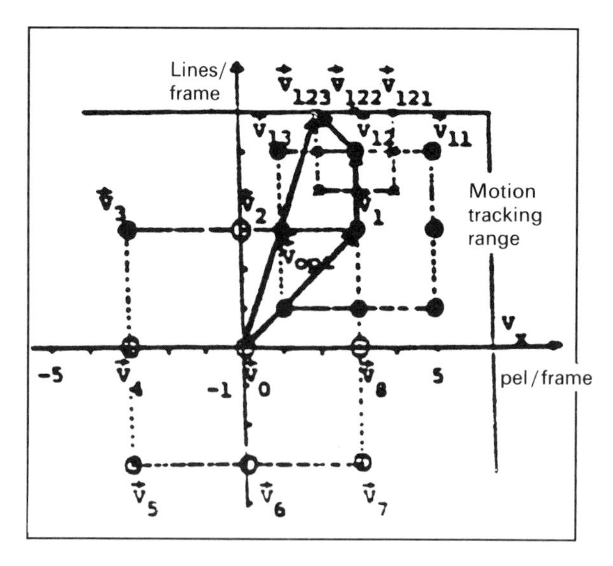

Fig. 7.111 The three-step search procedure [MC-28]. The arrows denote the optimum direction of the minimum cost function. (© 1981 IEEE).

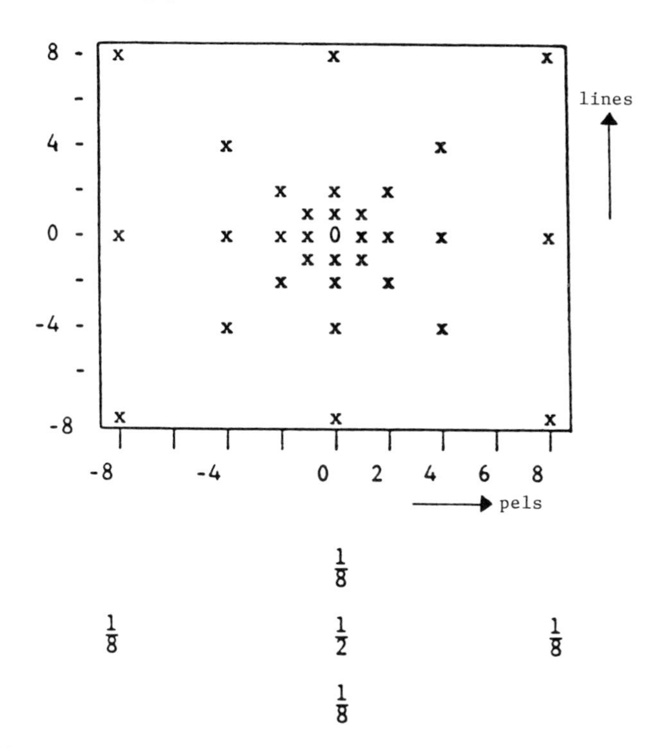

Fig. 7.112 BMA for 8×8 blocks. Motion vector range is up to ± 8 pel-lines/frame. Only 32 possible motion vectors ($\times$) are considered [MC-8].

motion-oriented blocks generated from noninterlaced HDTV frames [HDTV-10]. To reduce the computational complexity in estimating the motion vector, the distortion function (7.10.3) is computed based on an orthogonal subsampling pattern of pels (16:1 horizontally and 8:1 vertically). The range of motion vector is ± 8 pels/frame and ± 4 lines/frame. Another exception is the semiglobal motion estimation recommended by the CMTT/2 DCT group [IC-75, IC-76] for the video codecs designed for 30–34 MBPS and 45 MBPS. In this system a field is divided into nine zones and motion vectors are estimated for each zone (Fig. 7.114). The range of the motion vectors is ± 7 pels (horizontal) and ± 4 lines (vertical) with $\frac{1}{2}$-pel and $\frac{1}{2}$-line accuracy. Another option for these codecs is selection of 1 of 31 specified motion vectors (see Fig. 7.112 for another example) for each quadblock (two adjacent Y blocks, one C_R block, and one C_B block each of size 8×8), which include the nine motion vectors of the nine zones in the semiglobal motion estimation mode. To reduce the overhead for transmitting the motion vectors, techniques such as 2D variable-length coding (VLC) of motion vectors and predictive coding of motion vectors have been proposed [LBR-23, LBR-35].

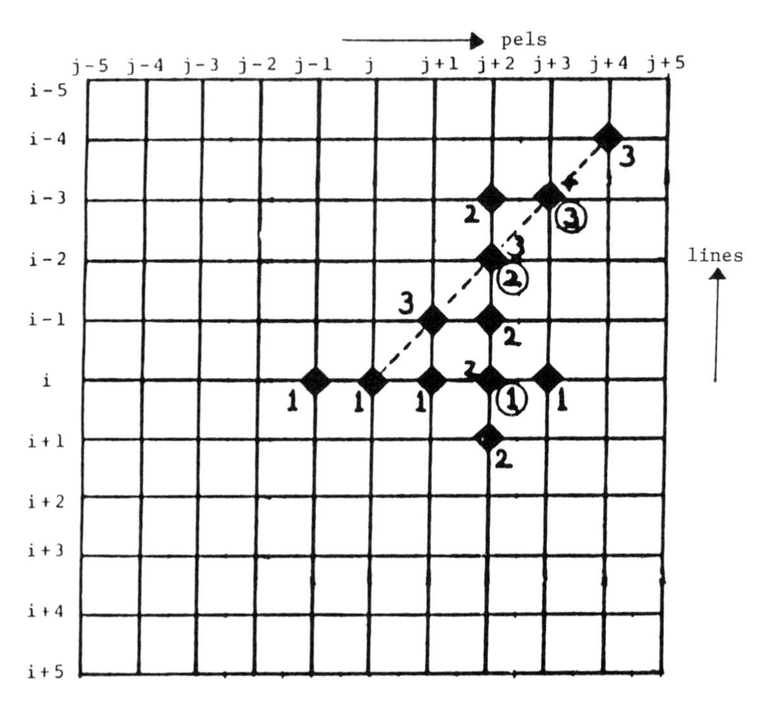

Fig. 7.113 A conjugate direction search for the direction of minimum distortion [MC-27]. The illustration shows the conjugate direction search to find a pel position in another frame corresponding to position (i, j) of the current frame such that the mean of absolute error is minimized. The ◆ indicates directions searched in any given step. Circled numbers are optimum directions for that search step and the + shows the final optimum direction. The one-at-a-time search concludes at ②. Shown here is a case where the final optimum direction differs in the two methods. (© 1984 IEEE).

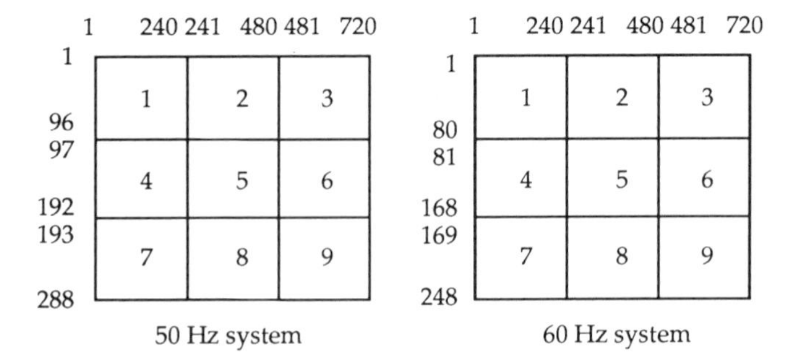

Fig. 7.114 Semiglobal motion estimation for nine zones in a field [IC-76].

7.10.3 *Orthogonal Search*

A BMA, called orthogonal search, having good properties in terms of convergence, fewer search points, fewer search steps, and noise immunity has been developed by Puri, Hang, and Schilling [MC-34]. Details are as follows:

$$\text{Search region (SR)}(M + 2p) \times (N + 2p)$$

(see Fig. 7.108).

Initial step size $\ell = \lceil p/2 \rceil$, where $\ell = \lceil x \rceil$ indicates the smallest integer greater than x.

(a) Step 1: Three search points are placed horizontally in the center of SR. The distance between every two neighboring points equals the step size, as shown in Fig. 7.115. The minimum point will be selected as the center for the next step. Step number $i \leftarrow (i + 1)$.

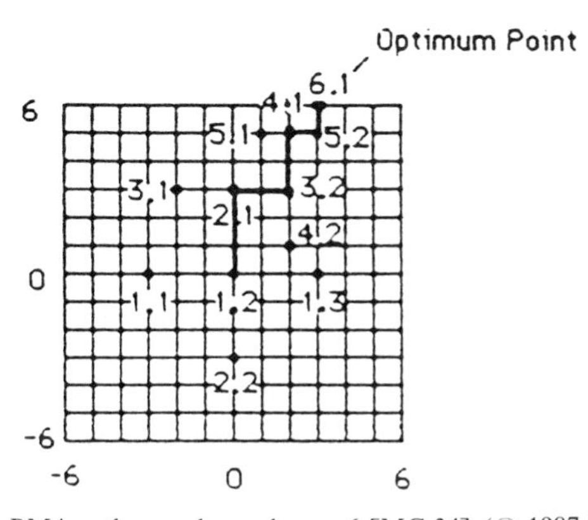

Fig. 7.115 BMA orthogonal search, $p = 6$ [MC-34]. (© 1987 IEEE).

(b) Vertical step: Two more search points are placed vertically around the minimum from the previous step. The distance between every two neighboring points equals the step size, as shown by the example in Fig. 7.115. The minimum point will be the center at the next step.

(c) Stopping rule: The remaining region of uncertainty (RUC) now has an area of $4(\ell - 1) \times (\ell - 1)$. If $\ell = 1$, stop. Otherwise, $\ell \leftarrow \lceil \ell/2 \rceil$. $i \leftarrow (i + 1)$ and continue.

(d) Horizontal step: Two or more search points are placed horizontally around the minimum from the vertical step. The minimum point will be the center for the next step. $i \leftarrow (i + 1)$, and go back to step b.

The total number of search points for the orthogonal search is $4\lceil \log_2 p \rceil + 1$. A comparison of the number of search points and the number of search steps for various BMA is shown in Table 7.15.

Table 7.15

Worst case performance of fast search algorithms [MC-34] (© 1987 IEEE)

	Max. Disp. $p = 3$		Max. Disp. $p = 6$	
Algorithm	Search Points	Search Steps	Search Steps	Search Steps
Exhaustive search	49	1	169	1
2D logarithmic [MC-1]	15	3	21	7
3- step [MC-28]	17	2	25	3
One-at-a-time [MC-27]	9	6	15	12
Modified 2D logarithmic [MC-6]	13	4	19	6
Orthogonal search [MC-34]	9	4	13	6

A modification of the orthogonal search [MC-34] is choosing the motion estimation of the identically located block in the previous frame as the center of the search region (step 1) for the block in the current frame. The orthogonal search is then applied to this shifted search region. Some other changes, such as a smaller search area, are also introduced in this process. Another modification [MC-35] is applying an exhaustive search in a small search region around the block shifted by the motion estimate of the corresponding block in the previous frame. It is shown that this simplified search performs almost as well as the exhaustive search [MC-35]. It may be observed that additional adaptive features can be incorporated into the MC hybrid (DPCM/transform) coding to improve the overall efficiency. For example, in the simplified version of the orthogonal search (BMA), Puri, Hang, and Schilling [MC-35] have introduced the following adaptive concepts:

7.10.3.1 NONMOVING (STATIONARY) BLOCKS The 8×8 image block is tested to see if it is a stationary or a moving block based on

$$M_5(0, 0) = \sum_{m=1}^{M} \sum_{n=1}^{N} [f(T_0, |U(m, n) - U_R(m, n)|].$$

If $M_5(0, 0) \geqslant N_0$, a preset threshold, then the block is moving, otherwise it is stationary. For the latter case, only the overhead information is transmitted. For the 8×8 block, $T_0 = 3$ and $N_0 = 10$ worked well (8-bit PCM).

7.10.3.2 COMPENSABLE MOVING BLOCKS All moving blocks go through the motion estimation process. The MC frame difference is evaluated to see if the moving block can be replaced at the receiver by the displaced block of the previous frame. For the compensable moving block, only the motion vectors need to be coded.

7.10.3.3 UNCOMPENSABLE MOVING BLOCKS All the remaining blocks are grouped into this category. In this case the MC frame difference goes through the 2D 8×8 DCT, quantization, and coding. There are several options in scanning, quantization, and coding of the 2D-DCT coefficients. Codes for the transform coefficients and motion vector need to be transmitted. These operations are summarized in Fig. 7.116.

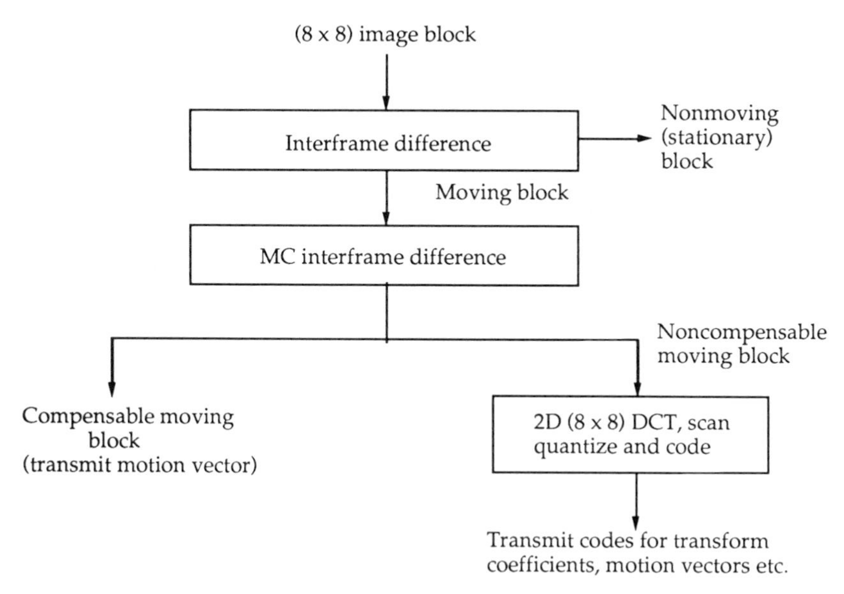

Fig. 7.116 Frame difference block types in adaptive MC DPCM/transform coding [MC-35]. (© 1987 IEEE).

7.10.3.4 HIERARCHICAL MOTION ESTIMATION For low bit-rate coding, Holzlwimmer, Brandt, and Tengler [MC-36] have proposed a hierarchical motion estimation (Fig. 7.117a) similar to the three-step search [MC-28]. The original image sequence is successively low-pass filtered (2D separable Gaussian FIR filter) and subsampled 2:1 both horizontally and vertically at each stage. For example, after the second stage, BMA using (7.10.4) as the distortion function is applied to 4×4 blocks. The starting point for the BMA at the first stage applied to 8×8 blocks is the motion vector (appropriately scaled) detected during the second stage. Repetition of this process one more time yields the motion vector for 16×16 blocks of the original image sequence. Also, by predicting the motion vector $\mathbf{v}$ based on the motion vectors of the neighborhood blocks (Fig. 7.117b), only the prediction error of the motion vector, $\mathbf{v} - \hat{\mathbf{v}}$ is encoded (VWL code) and transmitted.

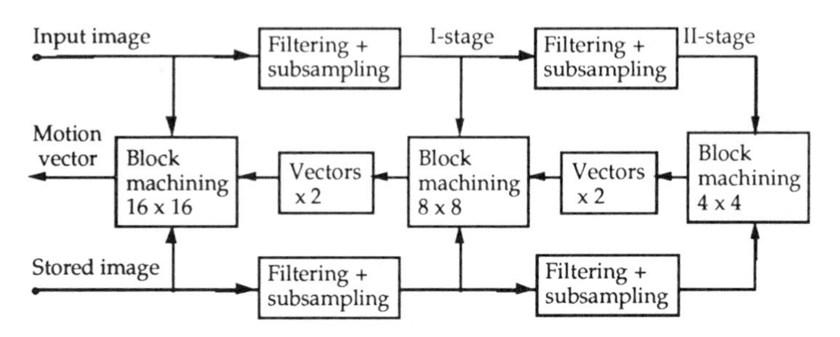

Fig. 7.117a Hierarchical block matching algorithm for motion estimation [MC-36]. (© 1987 IEEE).

$\underline{v}_B$	$\underline{v}_C$	$\underline{v}_D$
$\underline{v}_A$	$\underline{v}$	

Fig. 7.117b Neighborhood blocks used for predicting the motion vector **v** [MC-36]. (© 1987 IEEE).

A variable motion vector search [LBR-33] that has been implemented in a prototype 64 KBPS video codec is shown in Fig. 7.118. This scheme is similar to that of the three-step search [MC-28], except that the number of searches/steps is dependent on the magnitude of the estimated motion vector.

Another variation of the three-step search [MC-28] that reduces the computational complexity has been proposed by Kummerfeldt, May, and Wolf [MC-14]. The number of steps for block motion estimation is reduced by a coarse-fine search strategy (Fig. 7.119). This pattern indicates the growing

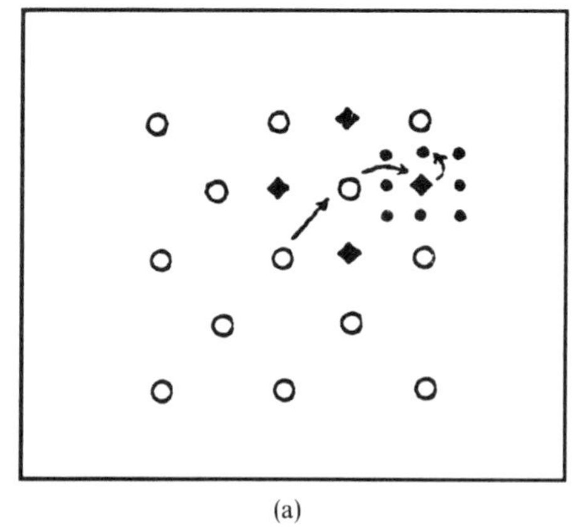

(a)

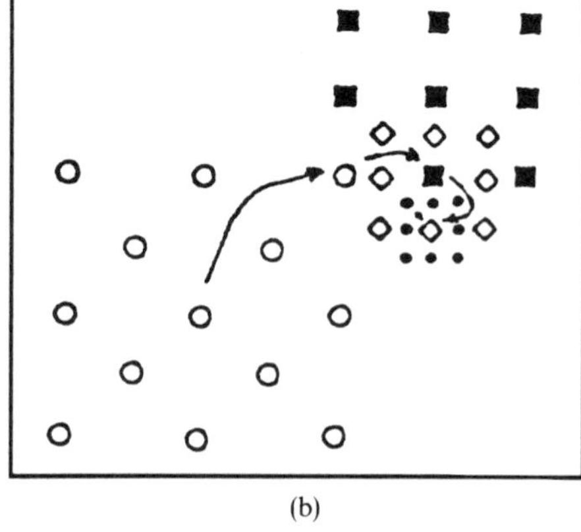

(b)

Fig. 7.118 Proposed motion vector search (a) short motion vector (○→■→●). (b) Long motion vector (○→■→●) [LBR-33].

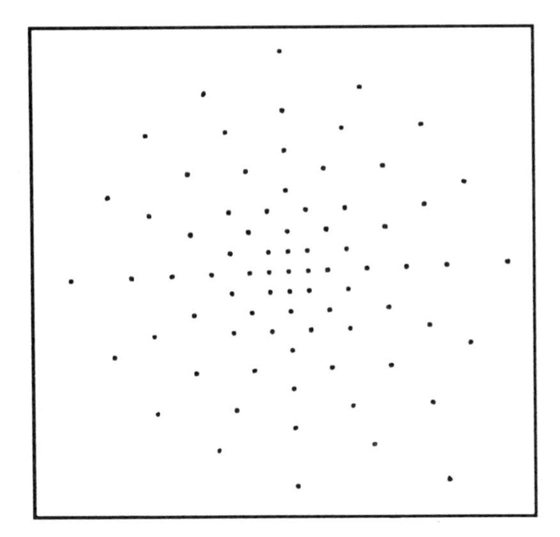

Fig. 7.119 Fine coarse search pattern for block motion estimation [MC-14].

coarseness outwardly from the center. A fine search in the 3×3 neighborhood of the motion vector estimated in the first stage is performed to obtain the desired motion vector. The distortion function (MAE or MSE) for both stages is based on only a subset of the 16×16 block (Fig. 7.120). These pels, randomly distributed in the block, represent about 12% of all the pels in the block.

```
   1 2 3 4 5 6 7 8 9 0 1 2 3 4 5 6
1  . . O . . . . . . . O . . . . .
2  . . . . . . O . . . . . . . O .
3  O . . . . . . . O . . . . . . .
4  . . . . O . . . . . . . O . . .
5  . O . . . . . . . O . . . . . .
6  . . . . . O . . . . . . . O . .
7  . . . O . . . . . . . O . . . .
8  . . . . . . . O . . . . . . . O
9  . . O . . . . . . . O . . . . .
0  . . . . . . O . . . . . . . O .
1  O . . . . . . . O . . . . . . .
2  . . . . O . . . . . . . O . . .
3  . O . . . . . . . O . . . . . .
4  . . . . . O . . . . . . . O . .
5  . . . O . . . . . . . O . . . .
6  . . . . . . . O . . . . . . . .
```

Fig. 7.120 Subset of pels selected in a 16×16 block for computing the distortion function, MAE or MSE [MC-14].

Computation of the distortion function is considerably simplified by this process.

So far, BMA discussion applied to MC has been limited to integer pel and integer line displacements. Motion estimation accuracy can be improved by including fractional pel and fractional line displacements in the BMA, at the cost of increased complexity. MC methods have been applied to both predictive (interframe DPCM) [MC-27, MC-28] and hybrid (DPCM/transform) coding [MC-1, MC-2, MC-3, MC-5, MC-6, MC-8, MC-11, MC-13, MC-17, MC-19, MC-21, MC-26, LBR-11, LBR-13, LBR-22, LBR-23, LBR-28, LBR-35, LBR-37, IC-75, IC-76]. In component video coding, motion estimation of the luminance block is, in general, applied (appropriately scaled) to the chrominance components. Also, modifications such as motion estimation of a larger block followed by the transform (2D DCT) of its sub-blocks have been proposed and simulated [LBR-22, LBR-23, LBR-28, LBR-35, LBR-37, IC-75, IC-76, LBR-48]. In addition, experimental (prototype) codecs have been built that incorporate the "flexible hardware" [LBR-23, LBR-35] for implementing the standard algorithms being recommended by the CCITT/SGXV Working Party XV/1 Specialists Group on Coding for Visual Telephony, leading to teleconferencing and videophone codecs.

7.10.4 *MC Interframe Hybrid Coding*

As stated earlier, in interframe hybrid (transform/DPCM) coding motion compensation can yield additional data compression. As described in Figs. 7.42 and 7.43, the MC interframe prediction can be in the transform domain (Fig. 7.121a) or in the spatial domain (Fig. 7.121b) [MC-1]. In both cases block motion estimation is implemented in the spatial domain. The former (Fig. 7.121a) requires three 2D DCTs (two forward and one inverse), as opposed to two 2D DCTs (one forward and one inverse) for the latter (Fig. 7.121b). Also, for the decoder the former requires two 2D IDCTs, whereas the latter requires only one 2D IDCT. This is also illustrated in the hybrid coder (Fig. 7.122a) and decoder (Fig. 7.122b) based on the C-matrix transform (CMT) [IC-11]. Because of this simplicity, in MC interframe hybrid coding, temporal prediction in the spatial domain has been invariably utilized in all practical systems. The basic algorithm being recommended by the CCITT SGXV Working Party XV/1 Specialists Group on Coding for Visual Telephony [LBR-23, LBR-35, LBR-37, LBR-40, LBR-44] also follows this pattern. A skeleton description of this algorithm can illustrate its data compression power.

The specialists group has been charged with producing draft recommendations on the low bit-rate coding, leading to the design of the codecs for videoconferencing and videophones so that these are compatible worldwide, even though the originating video can be NTSC, PAL, or SECAM. The coder bit rate, including audio and optional data channels, is $p \times 64$ KBPS ($p = 1, 2, \ldots, 30$). The codecs are not restricted to all these values of p. Desirable values of

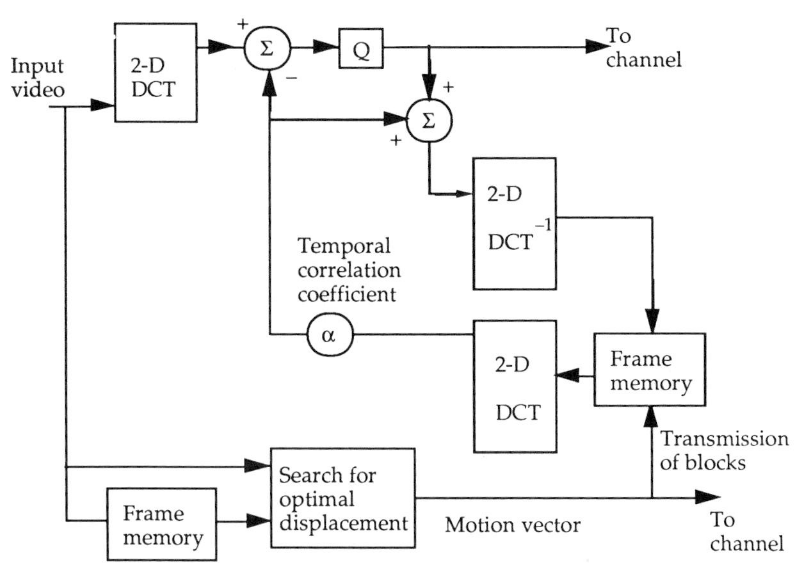

Fig. 7.121a MC interframe DCT/DPCM coding with prediction in the transform domain [MC-1]. (© 1981 IEEE).

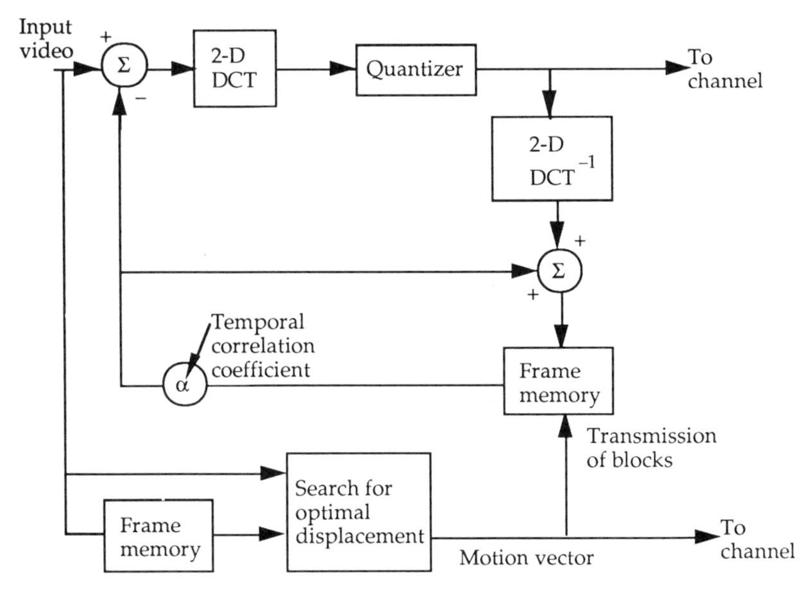

Fig. 7.121b MC interframe DPCM/DCT coding with prediction in the temporal domain [MC-1]. (© 1981 IEEE).

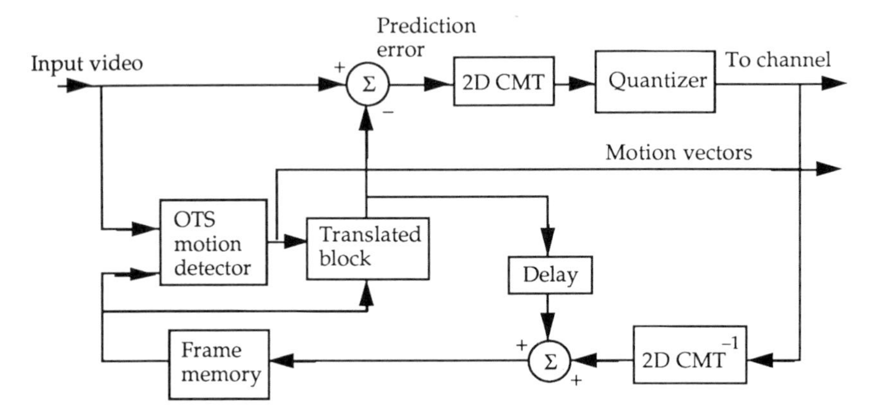

Fig. 7.122a Block diagram of proposed motion-compensated hybrid CMT coder.

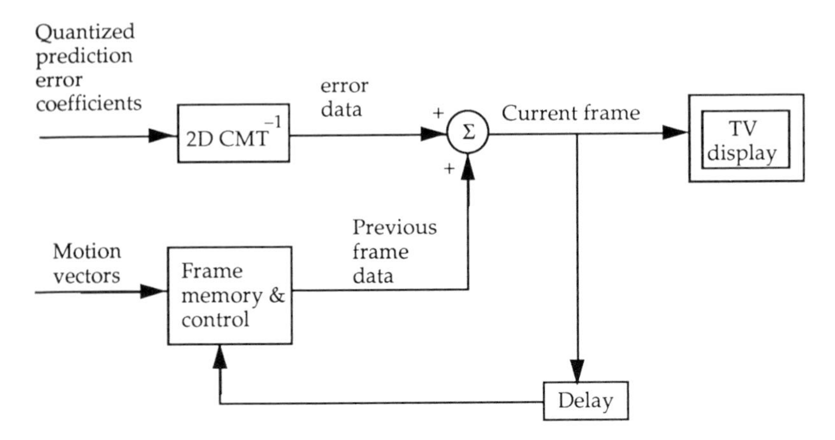

Fig. 7.122b Block diagram of receiver for proposed motion-compensated hybrid CMT coder.

p, however, are 1, 2, 6, 24, and 30. Even though the basic algorithm [LBR-23] (see Figs. 7.123a and 7.123b for coder and decoder, respectively) has been agreed upon, it is speculated that many detailed operations are subject to change, and invariably some of the internal functions will be updated/refined. This algorithm has been arrived at after extensive evaluation of various techniques proposed by leading experts (mainly core members of the study group).

Initially, as there are essentially two TV resolutions (625 line-50 Hz and 525 line-60 Hz), a common source intermediate format (CSIF) has been agreed upon as the input to the coder (Table 7.16). The number of pels/line is compatible with sampling the active portions of Y, C_B, and C_R signals from 525- or 625-line sources at 6.75 MHz and 3.75 MHz, respectively. These frequencies are one half of those specified in CCIR rec. 601 [G-21]. The $\frac{1}{4}$ CSIF has been tentatively

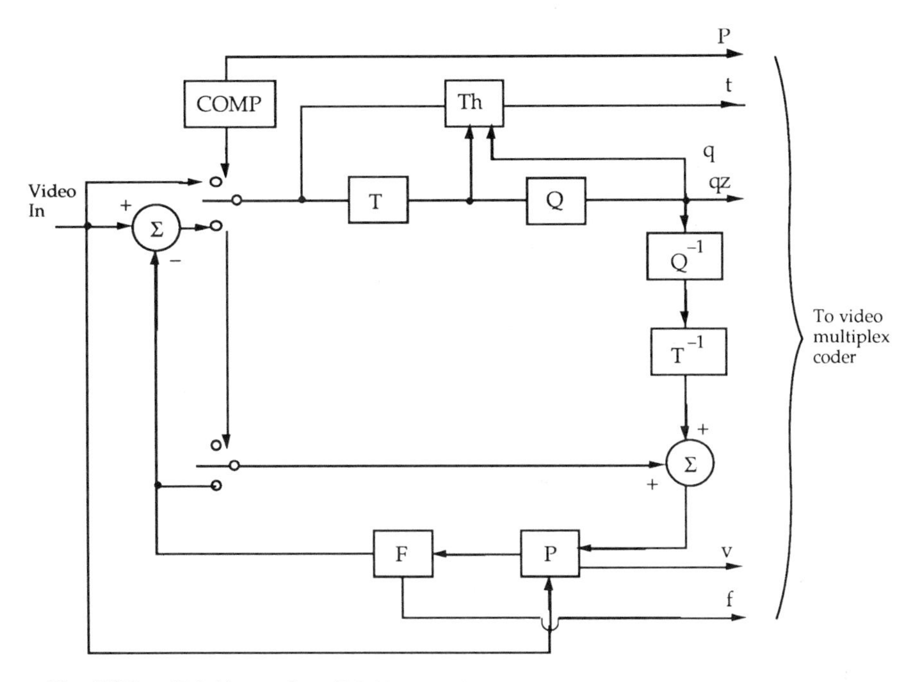

Fig. 7.123a Hybrid transform/DPCM encoder [LBR-37, LBR-40]. The transform and its inverse refer to the 2D 8×8 DCT. COMP = comparator for intra/inter; Th = threshold; T = transform; Q = quantizer; P = picture memory with motion-compensated variable delay; F = loop filter; p = flag for intra/inter; t = flag for transmitted or not; q = quantizing index for transform coefficients; qz = quantizer indication; v = motion vector; f = switching on/off of the loop filter.

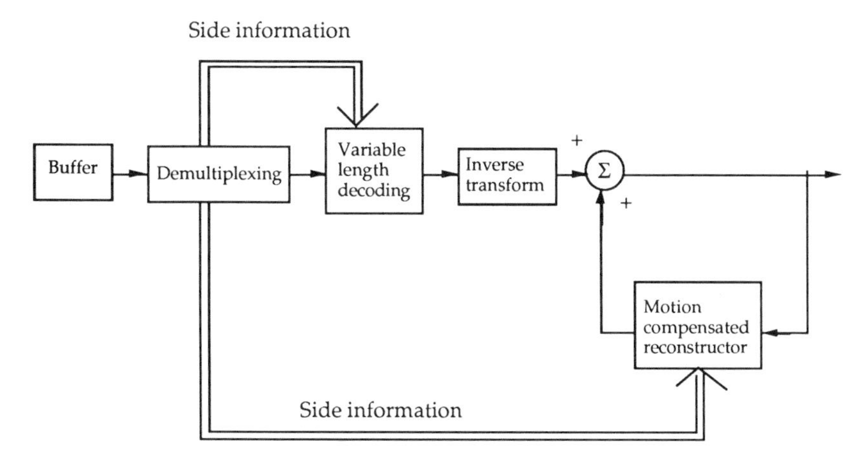

Fig. 7.123b Hybrid transform /DPCM decoder [LBR-23].

Table 7.16

Source format (full CSIF and $\frac{1}{4}$ CSIF) [LBR-23].

	Full CSIF	($\frac{1}{4}$) CSIF
Number of active lines		
Luminance (Y)	288	144
Chrominance (C_B, C_R)	144	72
Number of active pixels per line		
Luminance (Y)	360	180
Chrominance (C_B, C_R)	180	90

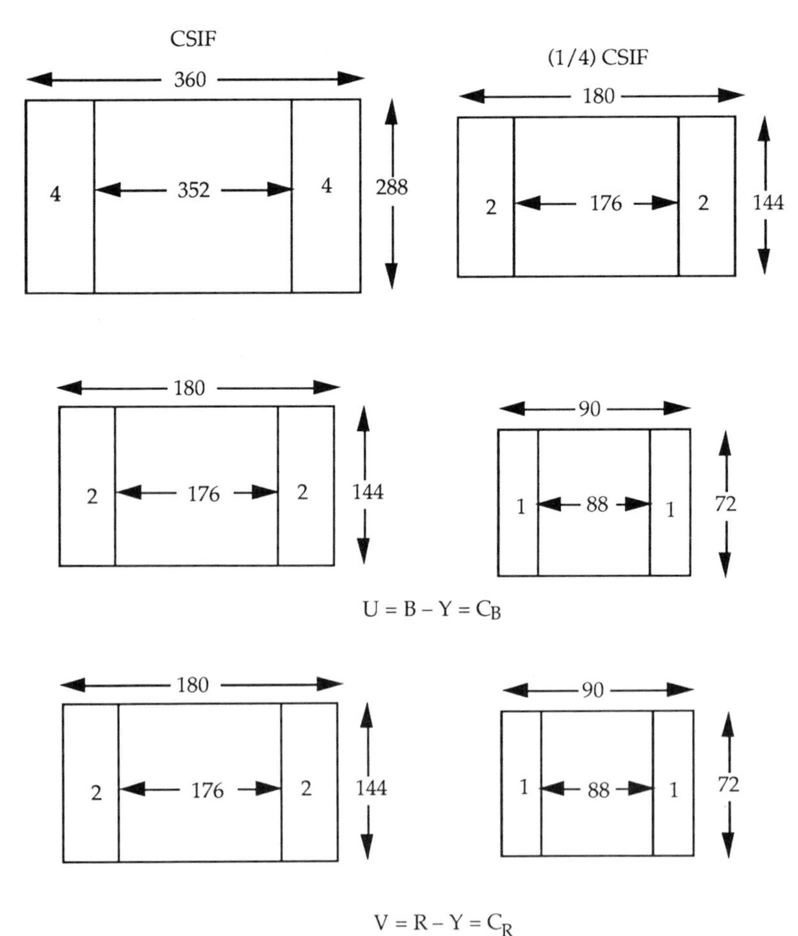

Fig. 7.124 Definition of significant pel area for the CSIF [LBR-23].

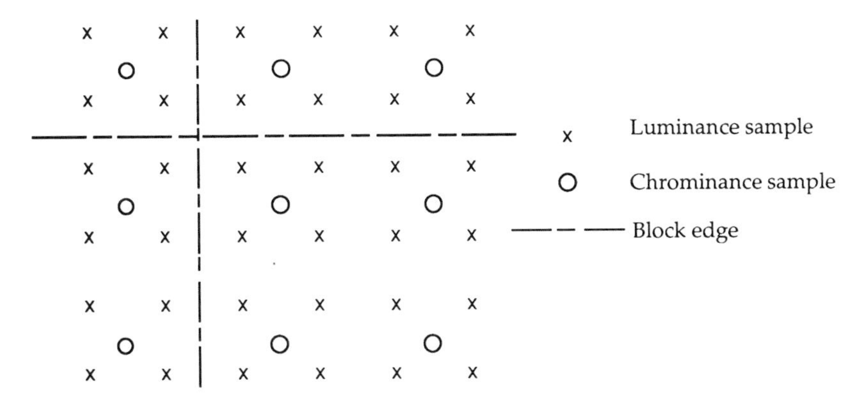

Fig. 7.125 Positioning of luminance and chrominance samples (CSIF) [LBR-23].

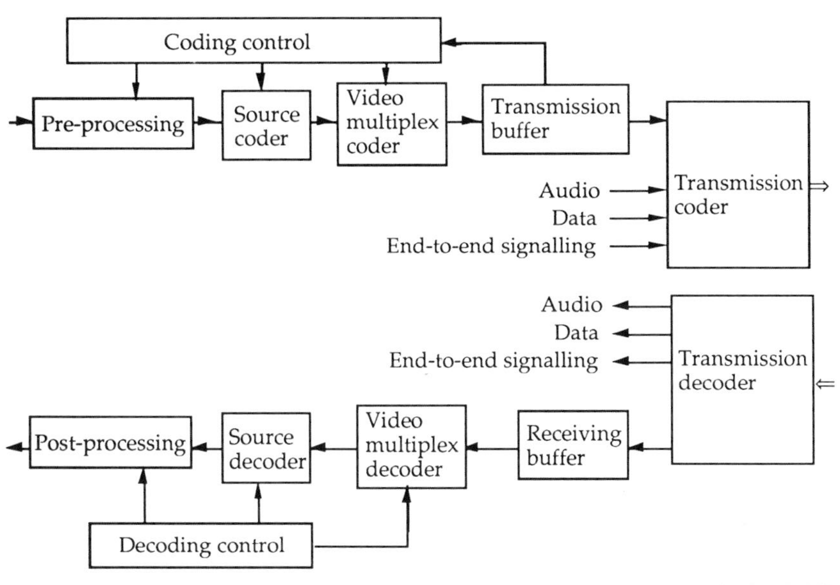

Fig. 7.126 Outline of block diagram of the videoconferencing/videophone codec [LBR-23].

agreed as the standard for LBR (say, 2×64 KBPS) codecs designed for videophone applications. The number of coded pels per line is reduced, because 360 divided by 16 does not yield an integer value. The obtained format is called significant pel area (SPA) (Fig. 7.124). The temporal resolution of the input video is 29.97 pictures (noninterlaced)/sec. Note that the chrominance components C_B and C_R are subsampled 2:1 both horizontally and vertically. All three components are orthogonally sampled, and they share the same block boundaries (Fig. 7.125). An overall block diagram of the codec is shown in Fig. 7.126. Highlights of the block functions in the coder (Fig. 7.123a) are as follows:

7.10.4.1 Video Source Coding Algorithm The source coding algorithm (Fig. 7.123a) includes intraframe (intra mode) or interframe difference with or without motion compensation (inter mode). In both modes the data is divided into eight lines by eight pel blocks and classified as transmitted or not transmitted. Those classified as transmitted undergo a 2D 8×8 DCT. The transform coefficients are quantized and VWL coded. The criteria for intra/inter mode selection and for transmitting an 8×8 block may be varied dynamically as part of data rate control strategy. In the inter mode, the prediction error may be augmented by motion compensation and a spatial (loop) filter. For panning and zooming sequences, the macroblocks are motion compensated but not spatially filtered. The VWL codes for the different modes are shown in Table 7.17.

Table 7.17

VWL codes for different modes (NF: not filtered).

Mode	MC	Filtered	Coded	Intra	Q	VLC	Length
1. No MC coded			X			1	1
2. MC coded	X	X	X			01	2
3. MC not-coded	X	X				001	3
4. Intra				X		0001	4
5. No MC coded Q			X		X	0 0001	5
6. MC coded Q	X	X	X		X	00 0001	6
7. Intra + Q				X	X	000 0001	7
8. MC coded NF	X		X			0000 0001	8
9. MC not coded NF	X					0 0000 0001	9
10. MC coded Q NF	X		X		X	00 0000 0001	10

7.10.4.2 Motion Estimation The motion estimation is based on a 16×16 Y block and is applied to four 8×8 Y blocks and to one C_B block and one C_R block, each of size 8×8, all of these constituting a macroblock (Fig. 7.127).

Motion compensation, however, is optional in the encoder. The decoder will accept one motion vector per macroblock. Both horizontal and vertical components of these motion vectors have integer values not exceeding ± 15. The

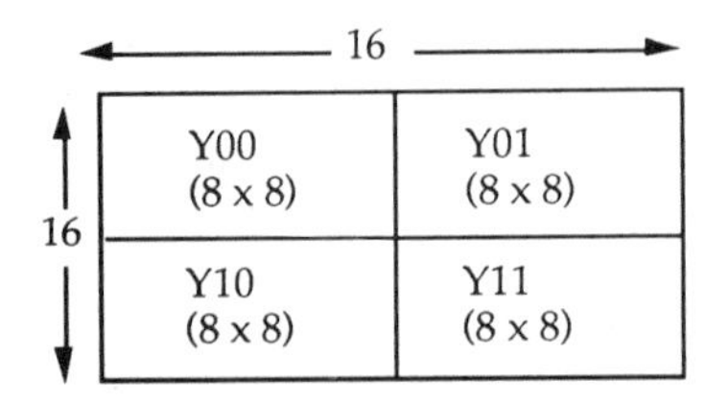

Fig. 7.127 Macroblock and its structure [LBR-23].

motion vector is used for all four luminance blocks in the macroblock (Fig. 7.127). The motion vector for both color difference blocks is derived by halving the component values of the macroblock motion vector and truncating towards zero. A positive value of the horizontal or vertical component of the motion vector signifies that the prediction is formed from pels in the previous picture that are spatially to the right or below the pels being predicted. Motion vectors are restricted such that all pels referenced by them are within the coded picture area. Motion vector data (MVD) is included for all MC macroblocks. MVD is obtained by subtracting the motion vector of the preceding MB from that of the current MB. MVD consists of a VWL code for the horizontal component followed by a VWL code for the vertical component (Table 7.18).

Table 7.18

VLC table for motion vector data (MVD) [LBR-40].

MVD			Code		
−16	&	16	0000	0011	001
−15	&	17	0000	0011	011
−14	&	18	0000	0011	101
−13	&	19	0000	0011	111
−12	&	20	0000	0100	001
−11	&	21	0000	0100	011
−10	&	22	0000	0100	11
−9	&	23	0000	0101	01
−8	&	24	0000	0101	11
−7	&	25	0000	0111	
−6	&	26	0000	1001	
−5	&	27	0000	1011	
−4	&	28	0000	111	
−3	&	29	0001	1	
−2	&	30	0011		
−1			011		
0			1		
1			010		
2	&	−30	0010		
3	&	−29	0001	0	
4	&	−28	0000	110	
5	&	−27	0000	1010	
6	&	−26	0000	1000	
7	&	−25	0000	0110	
8	&	−24	0000	0101	10
9	&	−23	0000	0101	00
10	&	−22	0000	0100	10
11	&	−21	0000	0100	010
12	&	−20	0000	0100	000
13	&	−19	0000	0011	110
14	&	−18	0000	0011	100
15	&	−17	0000	0011	010

7.10.4.3 QUANTIZER A nearly uniform quantizer (Fig. 7.128) is applied to Y, C_R, and C_B. When the threshold T (also called dead zone) is equal to the step size $g/2$, this becomes a uniform quantizer. In this case the threshold T is equal to the step size g. Control of the quantizer step size is left to the hardware design.

The selected quantizer is used for all the coefficients, except the dc coefficient in the intra coded blocks. For the first image intra mode is used. In this mode, 2D 8×8 DCT is applied to all the three components. The quantized ac coefficients are VWL coded, whereas the dc component is uniformly quantized to 8 bits (constant word length).

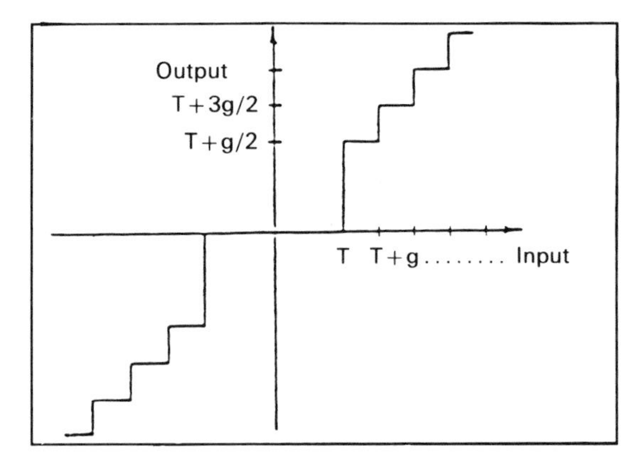

Fig. 7.128 Characteristic of the quantizer. g = stepsize; T = threshold. In RM5, T = g [LBR-23].

7.10.4.4 CODING THE DCT COEFFICIENTS The 2D-DCT coefficients are scanned in a zig-zag (Fig. 7.129b) fashion and are coded using a 2D-VWL code. This means that "events" are coded, where an "event" is defined as a combination of a magnitude (nonzero quantization index) and a run (number of zero indexes preceding the current nonzero index). Nonzero coefficients defining the end of the run length are considered as a composite rather than as a separate statistical event. The run length and the magnitude of composite events define the entries of the 2D-run-length table, which contains the code words for the composite events. Events are coded with Huffman's algorithm. However, events with low probabilities are coded using fixed length codes. These codes consist of the following three parts:

(1) Escape (6 bits) for indicating the use of fixed length codes
(2) Run (6 bits)
(3) Level (8 bits; here, it is assumed that the number of quantization indexes is less than 128)

$$
\begin{array}{cccccccc}
3 & 0 & 0 & 0 & 0 & 0 & 0 & 0 \\
2 & 0 & 0 & 0 & 0 & 0 & 0 & 0 \\
0 & 0 & 0 & 0 & 0 & 0 & 0 & 0 \\
0 & 0 & 0 & 0 & 0 & 0 & 0 & 0 \\
1 & 0 & 0 & 0 & 0 & 0 & 0 & 0 \\
0 & 0 & 0 & 0 & 0 & 0 & 0 & 0 \\
0 & 0 & 0 & 0 & 0 & 0 & 0 & 0 \\
0 & 0 & 0 & 0 & 0 & 0 & 0 & 0
\end{array}
$$

Fig. 7.129a Example of 2D VLC [LBR-23]. The 2D 8×8 DCT coefficients are zig-zag scanned (Fig. 7.129b).

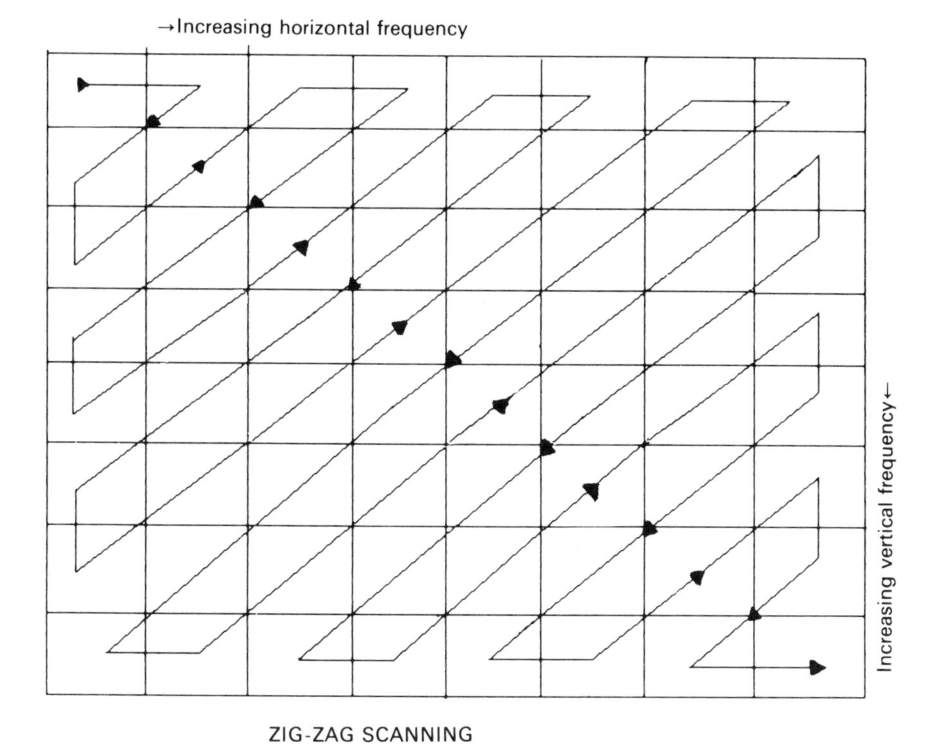

Fig. 7.129b Zig-zag scanning of 2D 8×8 DCT coefficients.

NOTE 1: Note that clipping must be introduced for the quantized coefficients $-128\,g < F < 128\,g$. The maximum range for the nonzero coefficients is now $+128\,g$ and $-128\,g$. The length of the EOB word is two bits.

NOTE 2: $0 < \text{run} < 64$ (for block size 8×8). After the last nonzero coefficient, an end-of-block (EOB) marker is sent indicating that all other coefficients are zero. An example of the two-dimensional VLC is given in Fig. 7.129a and the table is annexed.

EVENT = (RUN, LEVEL)

Example: (0, 3)(1, 2)(7, 1) EOB

This means

(0, 3) the DC component that has the value +3 (1, 2) is the next nonzero component according to the zig-zag scanning (Fig. 7.129b) and the number of zeros is 1. The next component is 1 preceded by 7 zeros, resulting in (7, 1). EOB is an end-of-block marker that indicates there are no more nonzero components. The VWL code for the nonzero transform coefficients and "zero run" is described in Table 7.19.

7.10.4.5 BUFFER Buffer size is set at $p \times 6400$ bits, and the average number of bits per macroblock are as follows for the $p \times 64$ KBPS codec. This description relates to reference model 8.

Frame rate	Bits/macro block
(Hz)	
30	$p \times 5$
15	$p \times 10$
10	$p \times 15$
7.5	$p \times 20$

The number of bits shown above are subtracted from the buffer content after each macroblock. If the buffer overflows during the coding of a MB, the next MB is forced to be fixed. For the recommendation, buffer size and design are under study.

7.10.4.6 LOOP FILTER A LPF inside the DPCM loop is activated when the motion vector is nonzero and is applied to all six blocks in a macroblock. For panning and zooming scenes, however, macroblocks are motion compensated but not filtered. The loop filter can be applied even if the motion vector is zero. An encoder without MC can switch on the loop filter by signalling a MB as motion compensated with a zero-motion vector. The filter reduces both the high-frequency artifacts caused by MC and the quantization noise in the feedback loop. The 2D filter is a separable filter. The 1D filters (horizontal and vertical) are nonrecursive with coefficients $\frac{1}{4}$, $\frac{1}{2}$, and $\frac{1}{4}$. At block edges these coefficients change to 0, 1, and 0.

Table 7.19

VWL code for nonzero transform coefficients and "zero run" [LBR-40]. The most commonly occurring combination of zero run and the following value are encoded with variable length codes as listed in the table below. End of block (EOB) is in this set. Because the coded block pattern (CBP) indicates those blocks with no coefficient data, EOB cannot occur as the first coefficient. Hence, EOB can be removed from the VLC table for the first coefficient. The last bit "s" denotes the sign of the level, "0" for $+ve$ and "1" for $-ve$

RUN	|LEVEL|	CODE
EOB		10
0	1	1s If first coefficient
0	1	11s Not first coefficient
0	2	0100 s
0	3	0010 1s
0	4	0000 110s
0	5	0010 0110 s
0	6	0010 0001 s
0	7	0000 0010 10s
0	8	0000 0001 1101 s
0	9	0000 0001 1000 s
0	10	0000 0001 0011 s
0	11	0000 0001 0000 s
0	12	0000 0000 1101 0s
0	13	0000 0000 1100 1s
0	14	0000 0000 1100 0s
0	15	0000 0000 1011 1s
1	1	011s
1	2	0001 10s
1	3	0010 0101 s
1	4	0000 0011 00s
1	5	0000 0001 1011 s
1	6	0000 0000 1011 0s
1	7	0000 0000 1010 1s
2	1	0101 s
2	2	0000 100s
2	3	0000 0010 11s
2	4	0000 0001 0100 s
2	5	0000 0000 1010 0s
3	1	0011 1s
3	2	0010 0100 s
3	3	0000 0001 1100 s
3	4	0000 0000 1001 1s
4	1	0011 0s
4	2	0000 0011 11s
4	3	0000 0001 0010 s
5	1	0001 11s
5	2	0000 0010 01s
5	3	0000 0000 1001 0s
6	1	0001 01s
6	2	0000 0001 1110 s

Table 7.19 continued)

7	1	0001 00s
7	2	0000 0001 0101 s
8	1	0000 111s
8	2	0000 0001 0001 s
9	1	0000 101s
9	2	0000 0000 1000 1s
10	1	0010 0111 s
10	2	0000 0000 1000 0s
11	1	0010 0011 s
12	1	0010 0010 s
13	1	0010 0000 s
14	1	0000 0011 10s
15	1	0000 0011 01s
16	1	0000 0010 00s
17	1	0000 0001 1111 s
18	1	0000 0001 1010 s
19	1	0000 0001 1001 s
20	1	0000 0001 0111 s
21	1	0000 0001 0110 s
22	1	0000 0000 1111 1s
23	1	0000 0000 1111 0s
24	1	0000 0000 1110 1s
25	1	0000 0000 1110 0s
26	1	0000 0000 1101 1s
ESCAPE		0000 01

The remaining combinations of (RUN, LEVEL) are encoded with a 20-bit word consisting of 6 bits ESCAPE, 6 bits RUN, and 8 bits LEVEL.

RUN is a 6-bit fixed length code

0	0000	00
1	0000	01
2	0000	10
●	●	
●	●	
63	1111	11

LEVEL is an 8-bit fixed length code

−127	1000	0001	
●	●		
●	●		
−2	1111	1110	
−1	1111	1111	
0	0000	0000	Forbidden
1	0000	0001	
2	0000	0010	
●	●		
127	0111	1111	

7.10.4.7 FORWARD ERROR CORRECTION Of the three forward error correction (FEC) codes suggested, (511,493) BCH code is chosen. The other two codes are (511, 484) BCH code and (1024, 492) Reed–Solomon code. All these codes correct both random and burst errors.

7.10.4.8 DCT/IDCT MISMATCH The IDCT implementation at both the coder and decoder has to be exact if there is to be no difference between the reconstructed images at both ends in the absence of channel errors. However, because of the finite word-length effects (also different algorithms may be used for the IDCT at the coder and decoder), there can be mismatch between the two reconstructed images. The mismatch, therefore, is caused by different accuracies and/or different algorithms for the IDCT at the coder and the decoder. The forward DCT at the coder does not contribute to the mismatch. It has been shown that this mismatch progressively increases, leading to visible image degradations [LBR-34]. To counteract the mismatch error accumulation at the decoder, after extensive simulations, specifications for implementing the IDCT have been formulated. These specifications not only can eliminate the incompatibility problem between the coder and decoder, but also can reduce the cost of the DCT/IDCT chip. Also, intraframe DCT coding (cyclic refresh) can be periodically implemented so that the visible artifacts can be avoided. Too low a refresh rate would let the mismatch error dominate. On the other hand, too high a refresh rate requires too many bits for intraframe coding. A refresh period between 45 and 135 frames at 15 frames/sec is a compromise between these two effects.

The minimum accuracy for forward and inverse DCTs that does not degrade the image quality is shown in Fig. 7.130. The two numbers in each bracket represent the bit size above and below the binary point. When mismatch exists, the reconstructed images at the decoder degrade rapidly, no matter what the IDCT accuracy is at the decoder. Cyclic refresh at a rate faster than the rate of image quality degradation is needed to counteract this artifact.

7.10.4.9 MODELING OF MISMATCH ERROR The mismatch error between the coder and decoder can be modelled (Fig. 7.131) by introducing the error after each operation, i.e., DCT, quantizer, IDCT. The decoded signals in the coder and the decoder are:

$$\begin{aligned}\hat{X}(i) &= [X(i) - \hat{X}(i-1)] + E1(i) + Q(i) + E2(i) + \hat{X}(i-1) \\ &= X(i) \qquad\qquad\qquad + E1(i) + Q(i) + E2(i)\end{aligned} \tag{7.10.6}$$

$$\tilde{X}(i) = [X(i) - \hat{X}(i-1)] + E1(i) + Q(i) + E2(i) + \tilde{X}(i-1) + E3(i), \tag{7.10.7}$$

where $X(i)$ = original signal in the ith frame, $\hat{X}(i)$ = local decoded signal in the ith frame (coder), $\tilde{X}(i)$ = decoded signal in the ith frame (decoder), $E1(i)$ = DCT calculation error in the ith frame, $E2(i)$ = IDCT calculation error in the coder ith

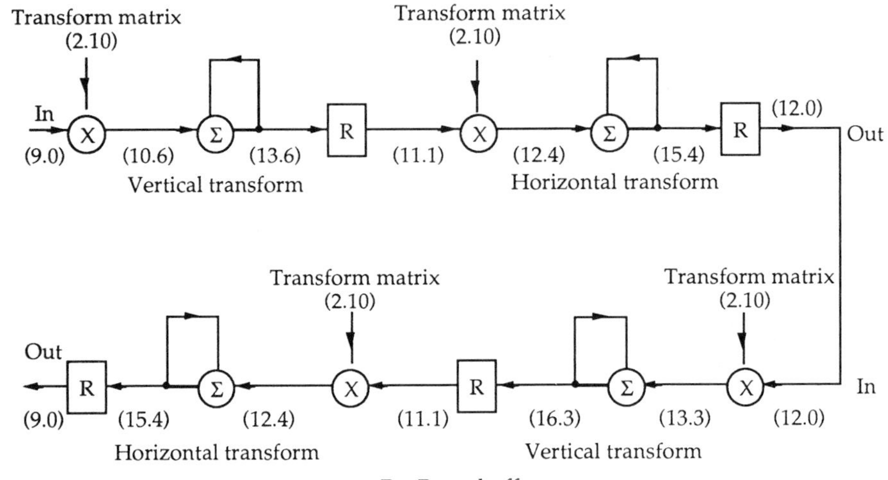

Fig. 7.130 Minimum accuracy for forward and inverse DCT [LBR-39]. The two numbers in each bracket represent the bit size above and below the binary point. These numbers do not represent the CCITT specification.

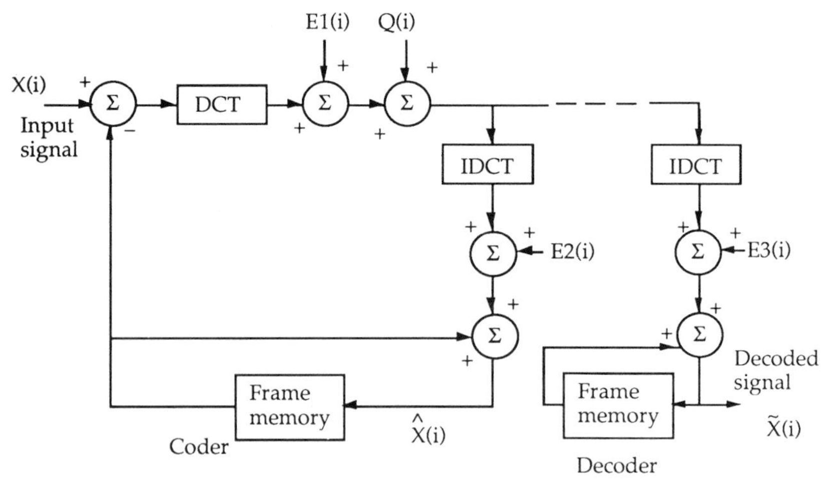

Fig. 7.131 Model for mismatch error [LBR-34].

frame, $E3(i)$ = IDCT calculation error in the decoder ith frame, and $Q(i)$ = quantization error in the ith frame.

The mismatch error between the reconstructed images at the coder and decoder is

$$\tilde{X}(i) - \hat{X}(i) = \tilde{X}(i-1) - \hat{X}(i-1) + E3(i) - E2(i)$$

$$= \sum_{i} [E3(i) - E2(i)]. \qquad (7.10.8)$$

The error between the input signal and the decoded signal is

$$\tilde{X}(i) - X(i) = E1(i) + Q(i) + E2(i) + \sum_{i} [E3(i) - E2(i)]. \qquad (7.10.9)$$

The accumulation of mismatch error in the reconstructed image at the decoder is evident from (7.10.8). To eliminate this accumulation, the following IDCT specification [LBR-40] has been set by the CCITT/SGXV specialists group. Also IEEE circuits and the systems standards committee is working on sponsoring this specification as an IEEE standard.

7.10.5 *IDCT Specification*

(1) Generate random integer pixel data values in the range $-L$ to $+H$ according to the attached random number generator (C version). Arrange into 8×8 blocks by allocating each set of consecutive eight numbers in a row. Data sets of 10,000 blocks each should be generated for ($L = 256$, $H = 255$), ($L = H = 5$), and ($L = H = 300$).

(2) For each 8×8 block, perform a separable, orthonormal, matrix multiply, forward discrete cosine transform (FDCT) using at least 64-bit floating point accuracy.

(3) For each block, round the 64 resulting transformed coefficients to the nearest integer values. Then clip them to the range -2048 to $+2047$. This is the 12-bit input data to the inverse transform.

(4) For each 8×8 block of 12-bit data produced by step 3, perform a separable, orthonormal, matrix multiply, inverse discrete cosine transform (IDCT) using at least 64-bit floating point accuracy. Round the resulting pixels to the nearest integer and clip to the range -256 to $+255$. These blocks of 8×8 pixels are the "reference" IDCT output data.

(5) For each 8×8 block of 12-bit data produced by step 3, use the proposed IDCT chip or an exact-bit simulation thereof to perform an inverse discrete cosine transform. Clip the output to the range -256 to $+255$. These blocks of 8×8 pixels are the "test" IDCT output data.

(6) For each of the 64 IDCT output pixels and for each of the 10,000 block data sets generated above, measure the peak, mean, and mean square error between the "reference" and "test" data.

(7) For any pixel, the peak error should not exceed 1 in magnitude. For any pixel, the mean square error should not exceed 0.06. Overall, the mean square error should not exceed 0.02. For any pixel the mean error should not exceed 0.015 in magnitude. Overall, the mean error should not exceed 0.0015 in magnitude.

(8) All zeros in must produce all zeros out.

(9) Rerun the measurements using exactly the same data values of step 1, but change the sign on each pixel.

```
                /*L and H must be long, ie, 32 bits*/
long rand(L,H)
long  L.H;
{
    static long randx = 1;  /*long is 32 bits*/
    static double  z = (double)0 × 7fffffff;

    long   i,j;
    double x;              /*double is 64 bits*/

    randx = (randx * 1103515245) + 12345;
    i = randx & 0 × 7ffffffe;              /*keep 30 bits*/
    x = ((double)i)/z;                     /*range 0 to 0.99999...*/
    x* = (L + H + 1);                      /*range 0 to <L + H + 1*/
    j = x ;                                /*truncate to integer*/
    return(j − L);                /*range −L to H*/
}
```

Standards regarding the detailed operation of the codec (Figs. 7.123a and 7.123b) are still under evolutionary and evaluation stages, although the basic algorithm, interframe hybrid (DPCM/2D DCT), has been identified. Transmission methods for control signals for several items such as freeze-picture request, fast-update request (to encode the next picture in intra mode), picture freeze release, and video data buffering are under study. CSIF $\Leftrightarrow$ ($\frac{1}{4}$) CSIF conversion is left to each hardware design. Details regarding arrangement of macroblocks in a GOB (group of blocks), macroblock address, coded block pattern (CBP), corresponding VWL coder, and various other operations are described in [LBR-40]. The draft recommendation H.261 "Codec for audio-visual services at p × 64 Kbit/s" is to be forwarded to the CCITT SGXV Working Party XV/1 by early 1990 for approval.

Similar to the basic adaptive MC interframe hybrid (DPCM/DCT) coding being standardized [LBR-23 through LBR-26, LBR-35, LBR-37, LBR-40], another algorithm [LBR-22, LBR-28] with a number of adaptive features has been investigated for videophone application at the transmission rates of 48, 64, and 128 KBPS. This algorithm is designed such that motion video and audio can be transmitted over one or two B channels (each B channel = 64 KBPS) of ISDN, as follows:

Rate 1: one B channel for both audio and video (e.g., 48 KBPS for video and 16 KBPS for audio)
Rate 2: one B channel each for video and audio (the audio B channel can have both audio and data)
Rate 3: Video occupies more than one B channel (e.g., 112 KBPS for video and 16 KBPS for audio)

The basic algorithm is shown in Fig. 7.132, where the block classification and BMA motion estimation (full search) is based on the $m \times m$ block and the 2D

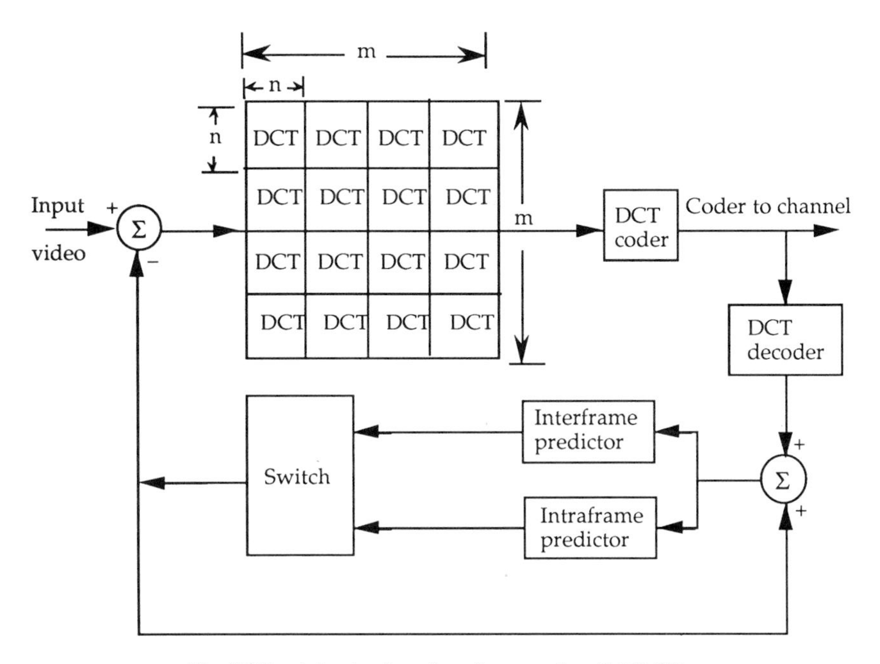

Fig. 7.132 Adaptive intra/interframe coding [LBR-28].

DCT is applied to $n \times n$ sub-blocks. The block classification is: each $m \times m$ block is classified as static interframe (interframe prediction without MC), dynamic interframe (MC interframe prediction), or intraframe block based on absolute average block differences. If the block is classified as intraframe, sub-block prediction based on the dc values of the previously decoded subblocks in both horizontal and vertical directions is carried out.

2D DCT OF ($n \times n$) SUB-BLOCKS 2D DCT of the predicted sub-blocks (static interframe or dynamic interframe or intraframe) is followed by adaptive quantization and adaptive scanning.

7.10.6 *Quantization*

A nearly uniform quantizer (see Fig. 7.128), with the parameters dead zone T and step size g adaptively varied, is applied. In general, the low-frequency coefficients are finely quantized and vice versa. The two-dimensional distribution of various quantizers for the transform coefficients is performed by a masking file. The function of the quantized masking file is to assign a set of quantizers to each coefficient. For example, $Q_{h\ell}$ is the quantizer that is assigned to the coefficient in the hth column and the ℓth row in the sub-block of $n \times n$ pels. These quantizers differ in the value of their dead zone. The dead zone is same as the threshold T (Fig. 7.128). Different quantization masking files are used in the coding process, and their selection is controlled by the buffer fullness.

7.10.7 *Scanning and Coding*

The nonzero transform coefficients of the predicted sub-blocks are VWL coded along with their corresponding run-length coded positional information. Also, the coding efficiency is improved by exploiting the inter-sub-block correlation. In this mechanism, the sub-blocks are scanned in three ways: (a) zig-zag, b) vertical, and c) horizontal (Fig. 7.133). Along each scanning, the first (dc) coefficients of all sub-blocks are grouped, followed by groups of second coefficients of all sub-blocks, and so on. The scanning mechanism that requires the least number of bits is chosen for transmission. As usual, overhead information indicates the scanning chosen.

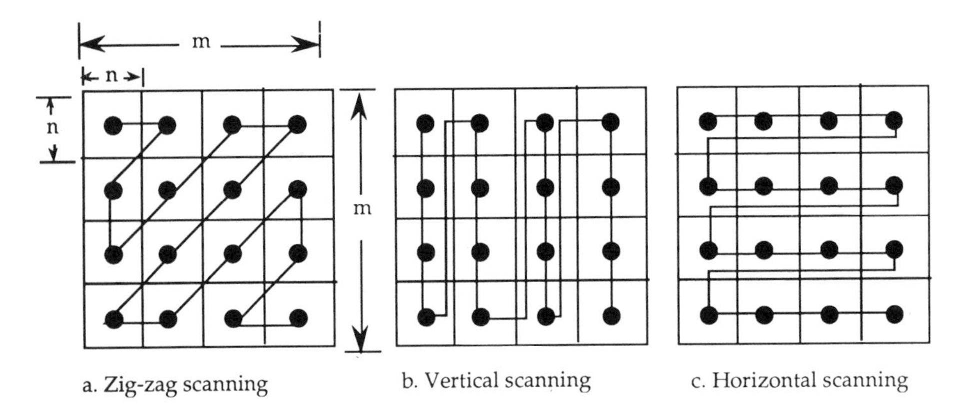

Fig. 7.133 Inter sub-block scanning [LBR-28].

Of the two main block sizes, 16×16 and 32×32, and the sub-block sizes 2×2, 4×4, and 8×8 considered, the block size of 16×16 with a subblock size of 4×4 was chosen based on SNR and subjective quality. Video sequences with 256×240 spatial resolution at 15 frames/sec were reconstructed at rates of 48 KBPS, 64 KBPS, and 128 KBPS.

As described at the end of Section 7.8, the techniques (DCT alone or DCT/DPCM) adopted for broadcast-quality codecs are, in principle, valid for LBR codecs. As LBR codecs require large compression, invariably both spatial and temporal resolutions are reduced. Also, in a teleconferencing-videophone scenario, MC can be utilized, taking advantage of the absence of rapid motion and less stringent visual fidelity requirements. While the LBR codecs are designed for real-time motion, additional features, such as high-resolution document-graphics capabilities in a non-real-time format, are invariably introduced to meet the multiservice and multifunction aspects of these codecs. Various other options, such as speaker insert in a document mode, conference background, zoom in of the speaker, audio and/or video mute, etc., further enhance the codec utility from a marketable strategy. Also, the codecs are equipped with preprocessors (to accept any of the standard formats—NTSC, PAL/SECAM, CCIR-601) and corresponding postprocessors for eventual display. Design and development of LBR codecs depend on the technology, applications, marketability, and various complementing services, e.g., ISDN. One particularly significant factor is the development of international standards (Fig. 7.186), which will ensure the compatibility among the codecs. As described in Sections 7.8 and 7.9, any number of adaptive features can be introduced in the LBR codecs, the overriding factor being the cost-quality-capability compromise.

7.11 HDTV Image Coding

7.11.1 *Introduction*

Various national and international organizations are developing standards for advanced TV systems [HDTV-6, HDTV-7, HDTV-13 through HDTV-15, HDTV-17, HDTV-19, HDTV-20] that are characterized by higher spatial (horizontal and vertical) resolution, greater aspect ratio, stereophonic digital audio, decreased cross couplings between luminance and chrominance components, decreased chroma-to-chroma crosstalk, and subjectively improved high quality (both video and audio). These advanced TV systems include both improved NTSC (enhanced picture fidelity while retaining the 525 line/59.94 fields per second and 4:3 aspect ratio) and high-definition television (HDTV), which has anywhere from 1050 to 1200 horizontal scanning lines/frame and a wider aspect ratio (16:9 or 5:3, frame width to height). These systems are also described as advanced compatible TV (ACTV) (for reception also by current NTSC/PAL/SECAM receivers), improved-definition TV (IDTV), multiplex analog component (MAC), enhanced-definition TV (EDTV), extended-quality TV (EQTV), and HDTV. For compatibility with the current broadcast systems (NTSC/PAL/SECAM), processing can be either at the receiver alone or at both the source and the receiver, or processing is applicable to channels using MAC color coding [HDTV-17]. While the worldwide standards for advanced TV (ATV) studio production are yet to be agreed upon (Table 7.20), consumer

Table 7.20.

Proposed HDTV format [HDTV-14]. (© 1989 IEEE).

	NBC proposal	Japan Broadcasting Corp (NHK)	European (PAL-SECAM)
No. of scanning lines/frame	1050	1125	1152
No. of fields/frame	59.94	60	50
Interlaced scanning	2:1 Also 1:1 progressive scanning	2:1	2:1 progressive scanning to be added
Aspect ratio	16:9	16:9	16:9

electronics in terms of cameras, HDTV CRT (cathode ray tubes), projection TV, VCR (video cassette recorder), VDP (video disc player), optical video disc, video juke box, CD ROM, etc are being developed. The activity in HDTV can be gauged from the special issues of the IEEE Transactions and the international workshops devoted to HDTV (see references on HDTV). The recent ruling (Sept. 1, 1988) of the U.S. Federal Communications Commission (FCC) [HDTV-14] has stipulated that ATV signals be compatible with NTSC or simulcast ATV and NTSC signals on separate channels (applicable only to terrestrial broadcast services such as UHF and VHF, not applicable to other media such as cable, satellite, fiber, VCR, video disc, etc.) and share parts of the spectrum currently allocated to broadcast TV. The Advanced Television Systems Committee (ATSC) composed of IEEE, EIA, NAB, NCTA, and SMPTE is to coordinate the development of voluntary national technical standards for ATV and to make recommendations to the U.S. Department of State for its use in developing U.S. positions on various standards issues formulated by international organizations [HDTV-6]. An excellent discussion

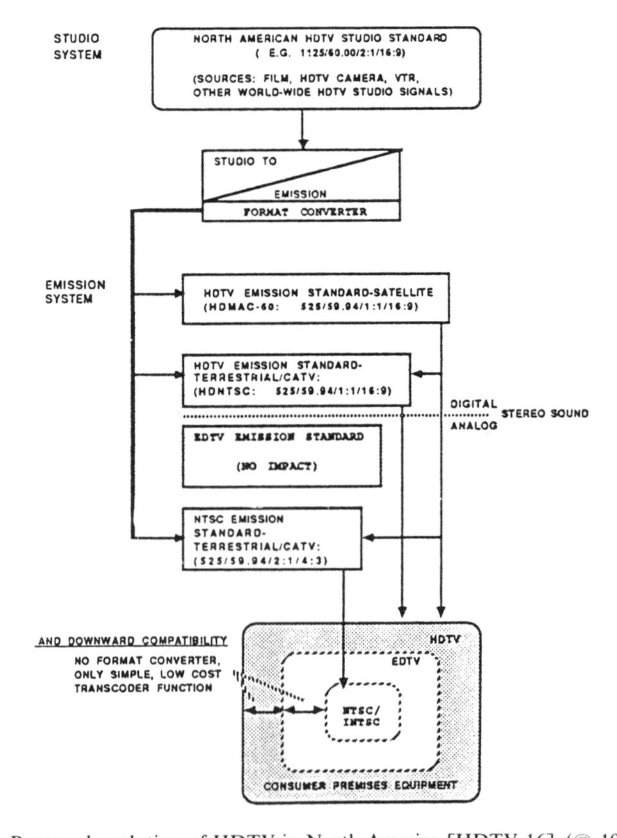

Fig. 7.134 Proposed evolution of HDTV in North America [HDTV-16]. (© 1988 IEEE).

on the proposals for HDTV transmission in the United States and compatibility (with NTSC) issues is presented by Hopkins [HDTV-15] for HDTV transmission in the United States. It seems likely that the 1125/60/2:1/16:9 HDTV system will become the 60-Hz studio standard. Toth, Tsinberg, and Rhodes [HDTV-16] suggest a hierarchical evolution of HDTV compatible with present NTSC receivers and VCRs (Fig. 7.134). In this system, HDMAC-60 is a feeder signal for use between the program origination and program distribution points and direct broadcast satellite (DBS) (Fig. 7.135). The HDNTSC is a delivery signal for terrestrial broadcast (Fig. 7.136), CATV distribution (Fig. 7.137), and recording media (VCR, laser video disc, etc.). A broadband optical fiber access network is shown in Fig. 7.138.

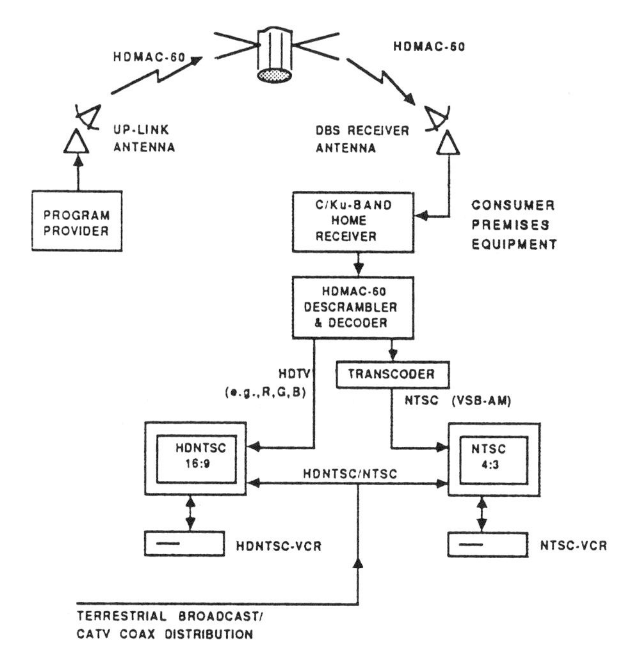

Fig. 7.135 DBS distribution of HDTV [HDTV-16]. (© 1988 IEEE).

As the horizontal and vertical resolutions of the HDTV are about twice those of the NTSC signal, and, as the former has a larger aspect ratio (16:9 or 5:3 compared with 4:3 for the NTSC), transmission and storage of digital HDTV may require at least five times the bit rate of the NTSC digital signal. Digital processing of HDTV in a component form may require as much as 1.2 Gbit/sec (GBPS) (Fig. 7.139).

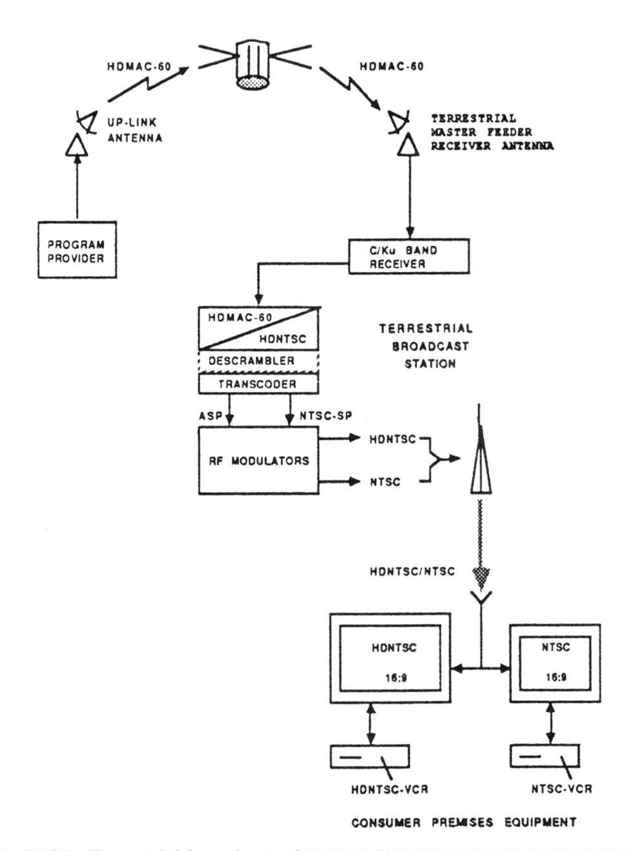

Fig. 7.136 Terrestrial broadcast of HDTV [HDTV-16]. (© 1988 IEEE).

For transmission of HDTV at the H4 level of ISDN (broadband ISDN, 140 MBPS), a compression ratio of 8 : 1 is required. Whereas special issues of the *IEEE Transactions* (also other journals such as the SMPTE) and workshops on HDTV deal with various aspects of HDTV, the discussion henceforth will be limited to DCT-based compression of ATV signals. The compression techniques may include provisions for processing the HDTV in a multichannel format, so that the present 525/60 (NTSC) and 625/50 (PAL-SECAM) systems and the HDTV are simultaneously available to the consumer. A standard receiver can display the former, whereas the HDTV receiver can display the super-resolution signal. Such a coding technique is called compatible coding. The digital transmission of these signals is at different bit rates [for example, the NTSC video at 44.736 MBPS (H22), the PAL-SECAM signal at 32.768 MBPS (H21), and the HDTV at 135 MBPS (H4)].

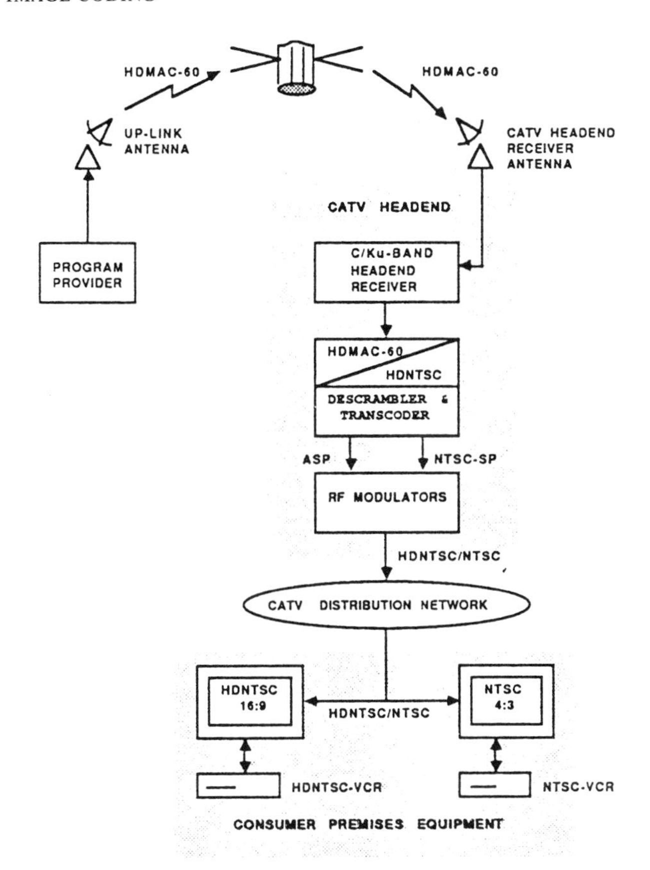

Fig. 7.137 CATV distribution of HDTV [HDTV-16]. (© 1988 IEEE).

7.11.2 *DCT-based HDTV Image Compression*

Although there are any number of techniques [HDTV-3, HDTV-6, HDTV-7, HDTV-12, HDTV-13] for bit-rate reduction of HDTV signals, we will limit our discussion to DCT-based systems. In HDTV image coding, DCT is supplemented with other compression tools such as DPCM, VQ, sub-band, motion compensation, etc., so that the high fidelity and resolution of the HDTV signal are preserved in spite of the large compression required. By multichannel coding, the HDTV signal is processed such that both NTSC (or PAL/SECAM) and HDTV receivers can display the TV signals at the appropriate resolution [HDTV-4, HDTV-8, HDTV-12]. This additional feature can enhance the utility of the compression scheme. For example, the technique (Fig. 7.140) proposed by Le Gall, Gaggioni, and Chen [HDTV-12] involves dividing the HDTV spectrum (both luminance and chrominance components) into four equal bands

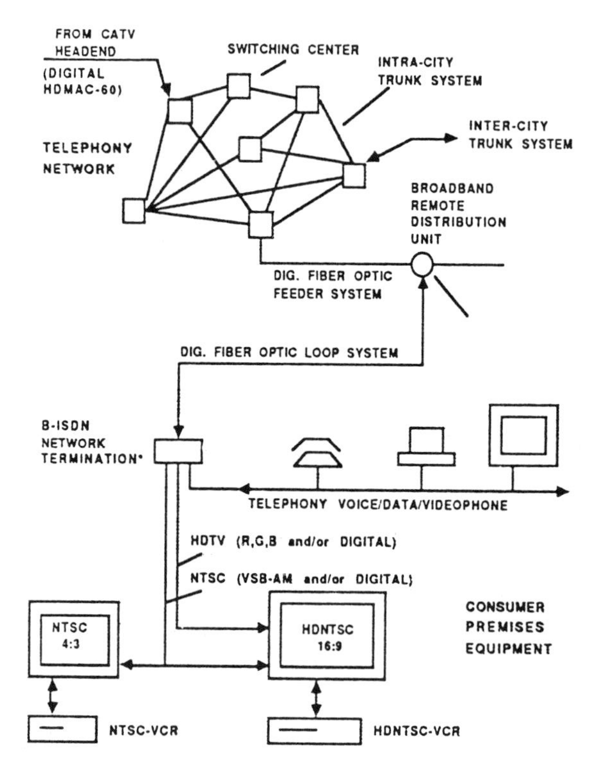

Fig. 7.138 Telephony network distribution of HDTV [HDTV-16]. (© 1988 IEEE).

in the 2D-frequency domain (Fig. 7.141) by sub-band coding. This is followed by 2D DCT of luminance (4×4 and 8×4 blocks) and color (8×8 blocks) components of the baseband signal, uniform quantization of the transform coefficients, and Huffman coding of the quantized coefficients and run lengths of zero coefficients (variation of the scene adaptive coder [LBR-3, LBR-5]). All the other bands are quantized directly, followed by run lengths of zero values and VWL coding of nonzero values (Fig. 7.142). The output bit rate is kept constant by the buffer, which controls the DCT quantizer step size and the dead zones (Fig. 7.128) for the higher bands.

Decoding the baseband signal (65 MBPS) and higher band signals (55 MBPS) facilitates the reception of a standard TV signal and an HDTV signal at 65 MBPS and 120 MBPS, respectively. Besides the compatible coding, the intrafield sub-band/DCT technique has minimized the high-speed circuitry (all the bands are processed after 2:1 horizontal and vertical decimation).

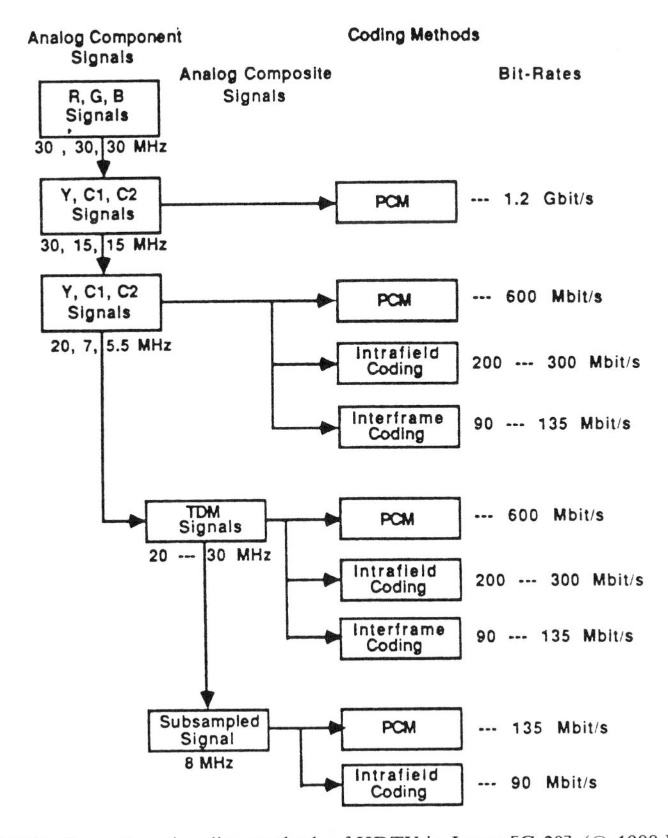

Fig. 7.139 Formats and coding methods of HDTV in Japan [G-20]. (© 1988 IEEE).

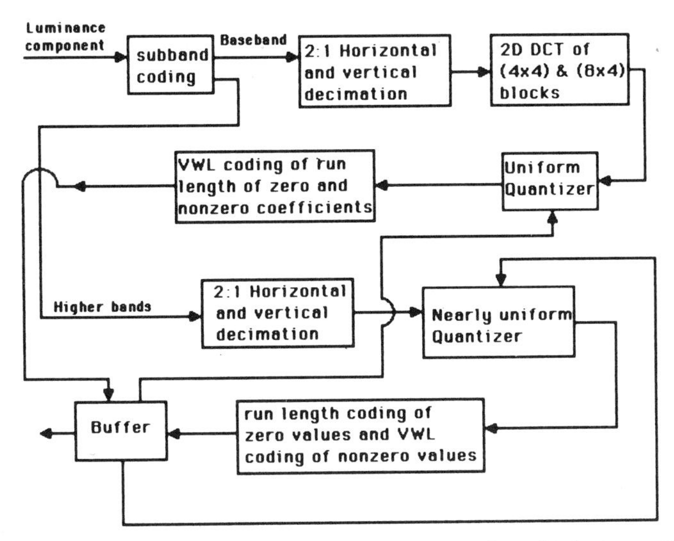

Fig. 7.140 Sub-band/DCT coding of HDTV signal. Only coding of Y is shown. Coding of C_R and C_B is similar to that of Y. Decoding of baseband signal is used for standard NTSC receiver [HDTV-12].

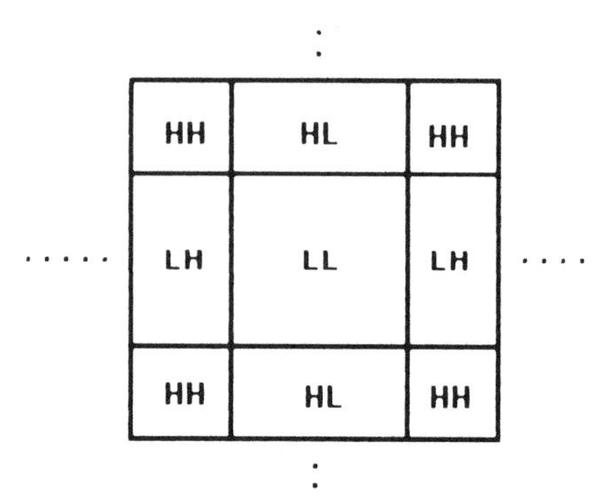

Fig. 7.141 Four-band split with separable filter bank. LL = baseband; LH = horizontal high and vertical low; HL = horizontal low and vertical high; HH = horizontal high and vertical high [HDTV-12].

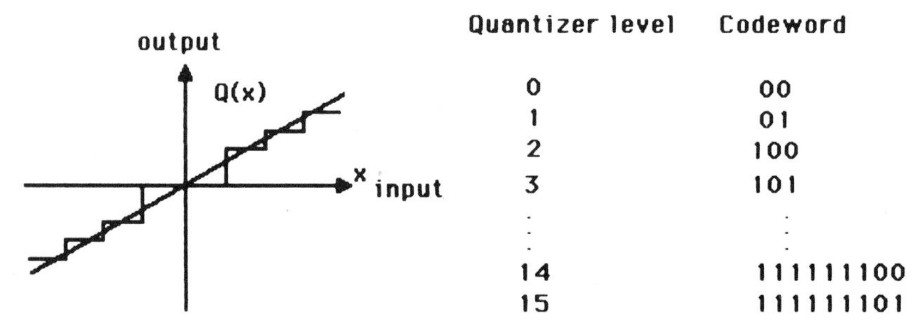

Fig. 7.142 Quantizer and variable length code for the higher bands [HDTV-12].

7.11.3 *Compatible HDTV Coding for Broadband ISDN*

Tzou *et al.* [HDTV-8] have developed a compatible HDTV coding technique (Fig. 7.143) in which part of the HDTV signal is extracted to provide component video with spatial resolution comparable with the CCIR 601 recommendation [G-21] at the DS3 level (44.736 MBPS). This signal is refined during a second pass and is combined with the remainder of the HDTV signal for transmission at the H4 level (135 MBPS) of the broadband ISDN. The simulation is based on two test sequences containing pan and/or zoom (Table 7.21).

The compatible coding consisting of a number of adaptive features, such as switched quantizers with different step sizes and dead zones (Fig. 7.128), multiple Huffman codes for run lengths of zeros and nonzero coefficients in low-frequency and high-frequency zones, various scanning patterns (Fig. 7.144), and multipass coding of low-frequency coefficients, has resulted in good quality images at both rates, i.e., EQTV at 44.736 MBPS and HDTV at 135 MBPS. Original and reconstructed images based on this algorithm are shown in Figs. 7.145–7.147.

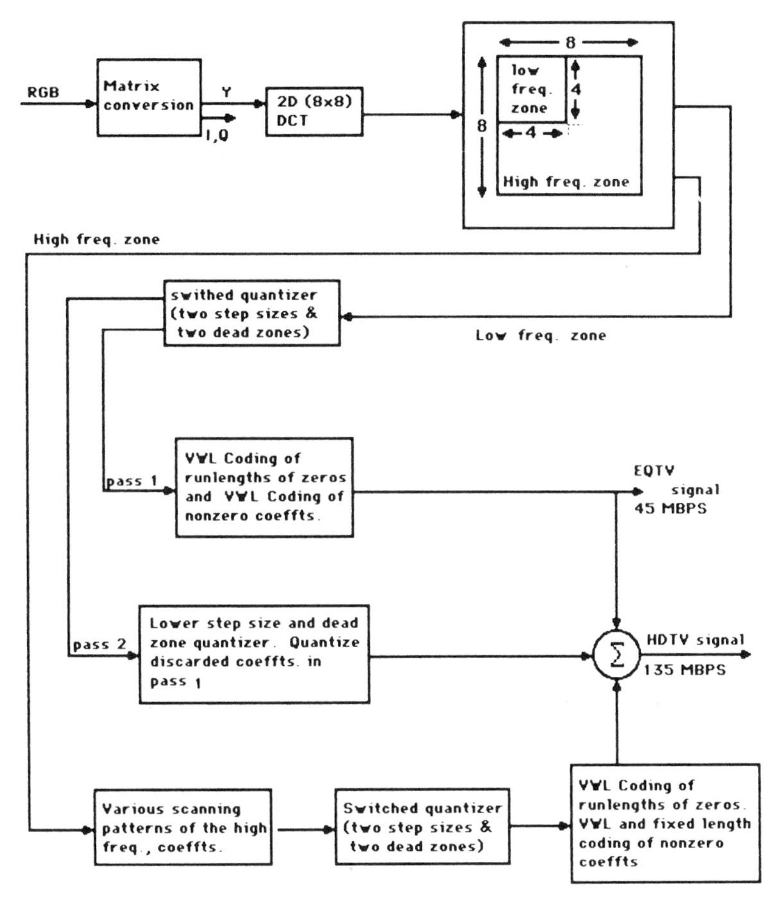

Fig. 7.143 Compatible HDTV coding for broadband ISDN. I and Q signals are subsampled 2:1 horizontally and are coded similar to the Y signal. After decoding, I and Q signals are interpolated 1:2 horizontally. The bit rates for EQTV and HDTV signals include Y, I, and Q [HDTV-8]. (© 1988 IEEE).

Table 7.21

HDTV test sequences used for simulation (see Fig. 7.143) [HDTV-8].

Sequence source	Scanning	Pels/line	Lines/frame	Aspect ratio	Frames /sec	Coding
MIT	progressive (noninterlaced)	1600	960	5:3	30	intraframe
HHI	2:1 interlaced	1440	1152 (576 lines/field)	4:3	25	intrafield

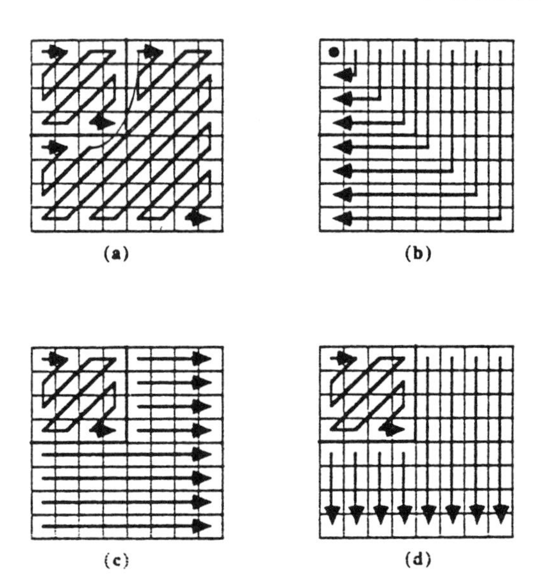

Fig. 7.144 Various scanning patterns studied in compatible HDTV coding [HDTV-8]. (© 1988 IEEE).

An adaptive scheme [HDTV-1] composed of 2D-DCT blocks classified based on their ac energies weighted by the dc coefficients and pyramid VQ of ac coefficients appears to be promising in HDTV image coding at the H4 rate. Another technique proposed by Barbero, Cucchi, and Stroppiana [HDTV-9] involving coding of the 4:2:2 signal (CCIR-601, [G-21]) at 20 MBPS along the lines suggested in [IC-53, IC-75] (see Fig. 7.53), with additional features such as intrafield for moving areas, interfield for moderately moving areas, and interframe for stationary areas, has demonstrated the feasibility of HDTV coding in the 70–140 MBPS range.

Various data compression techniques implemented for still-frame, broadcast TV, videophone, and videoconferencing images are also applicable to HDTV images. These techniques, however, have to be appropriately modified and strengthened by other tools, considering the high resolution and high quality of the HDTV. Also, another consideration is processing the HDTV signal such that the reception can be at various levels (NTSC/PAL/SECAM, EQTV, HDTV, etc.). With the developments in submicron technologies, various signal processing operations inherent in HDTV compression can be feasible with ASICs. The development and formulation of international standards [HDTV-19] for both contribution and distribution of HDTV will speed up the hardware aspects of these codecs.

Fig. 7.145 Simulation results for an HHI picture: original (top) and compressed-decompressed (bottom) [HDTV-8]. (© 1988 IEEE).

Fig. 7.146 Simulation results for an MIT picture: original (top) and compressed-decompressed (bottom) [HDTV-8]. (© 1988 IEEE).

(a)

(b)

Fig. 7.147 Simulation results of derived EQTV (top) and compressed-decompressed EQTV (bottom): (a) HHI picture and (b) MIT picture [HDTV-8]. (© 1988 IEEE).

7.12 Block Structure/Distortion in Transform Image Coding

Transform image coding, in general, is implemented on small blocks, such as 8×8 or 16×16, not only from a complexity point of view (both processing and storage requirements increase with block size), but also from adaptivity considerations, as coding of individual blocks can be tailored to local activity or detail. At low bit rates, however, as block coding assumes no correlation across the block boundaries, visual artifacts such as block boundaries tend to appear. Several techniques have been developed to minimize or eliminate this blocky structure. These are outlined below.

7.12.1 *Block Overlapping*

By overlapping adjacent blocks (Figs. 7.148 and 7.149), the block effect can be reduced. As neighborhood pels are common to adjacent blocks, both the processing and the bit rate increase. For pels common to adjacent blocks, after processing their averages represent the reconstructed values. In speech coding, overlapped trapezoidal windows have been adopted (Fig. 7.11).

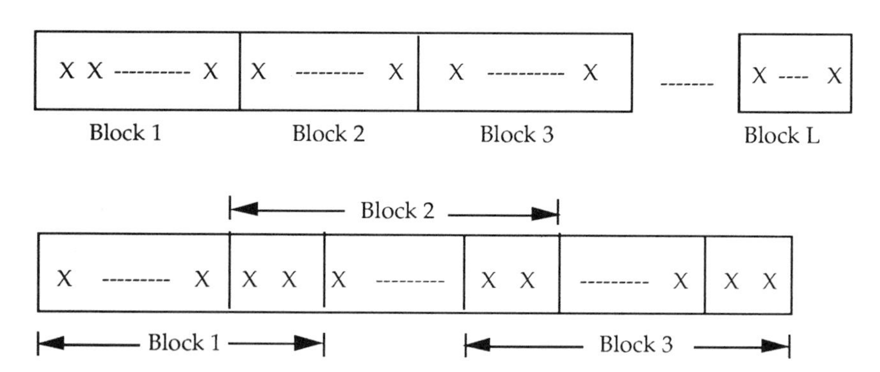

Fig. 7.148 One-dimensional block overlapping: Two neighborhood pels are common to adjacent blocks.

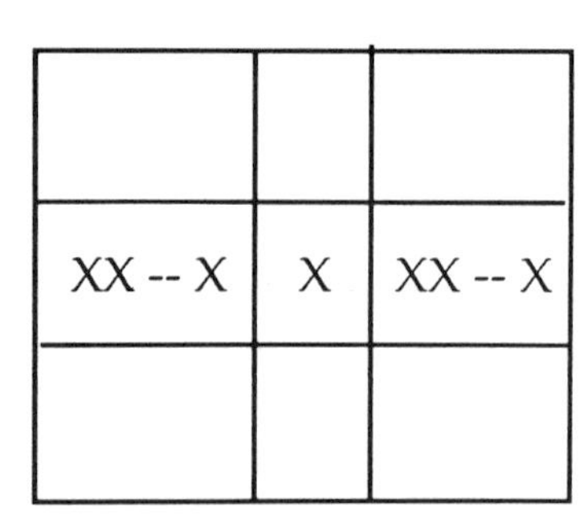

Fig. 7.149 Two-dimensional block overlapping. Pels near the block boundaries are common to adjacent blocks.

7.12.2 *Low-Pass Filtering at the Block Edges*

By low-pass filtering the neighborhood pels after reconstruction, the block discontinuity can be reduced. As this smoothing is implemented at the receiver, no additional bit rate is required. An example of such a filter (symmetrical Gaussian filter) is shown in Fig. 7.150 [BSI-1].

0.0751	0.1239	0.0751
X	X	X
0.1239	0.2042	0.1239
X	X	X
0.0751	0.1239	0.0751
X	X	X

Fig. 7.150 3×3 symmetrical Gaussian filter. The center pel is filtered by the neighborhood pels weighted by the filter matrix [BSI-1]. (© 1983 IEEE).

7.12.3 *Recursive Block Coding*

In recursive block coding (RBC), developed by Jain and Farrelle [BSI-2], the original data is divided into overlapping blocks, with one pel overlapping for practical images. In this scheme the original data is decomposed into two mutually uncorrelated processes: a) boundary response, i.e., prediction of pels within the block based on overlapped pels and b) the residual process or prediction error, which is transform coded. Farrelle and Jain [BSI-4] have shown that the RBC reduces or even suppresses the block structure and simply blurs the edges, resulting in a sharper image compared with the DCT coding.

7.12.4 *Discrete Legendre Transform*

Miyahara and Kotani [BSI-7] have described the general requirements for a discrete transform that can suppress the block structure and have shown, based on simulation of test images, that the discrete Legendre transform (DLT) is superior to the DCT, purely from a block structure viewpoint. The DLT, however, has neither a fast algorithm nor a recursive structure, [DW-5].

7.12.5 *Lapped Orthogonal Transform*

The lapped orthogonal transform (LOT) [BSI-5, BSI-6, SC-29, T-7] combines the DCTs of adjacent blocks in such a way that the resulting transform has basis functions that overlap adjacent blocks. The basis functions decay smoothly towards zero at their boundaries, so that the LOT is virtually free from block boundary mismatch errors, as shown in [BSI-5] for image coding and in [SC-29] for speech coding.

The flowgraph of the fast LOT is shown in Fig. 7.151 [SC-29], where $x(n)$ is the input signal and $X_k(m)$ is the mth transform coefficient of the kth block.

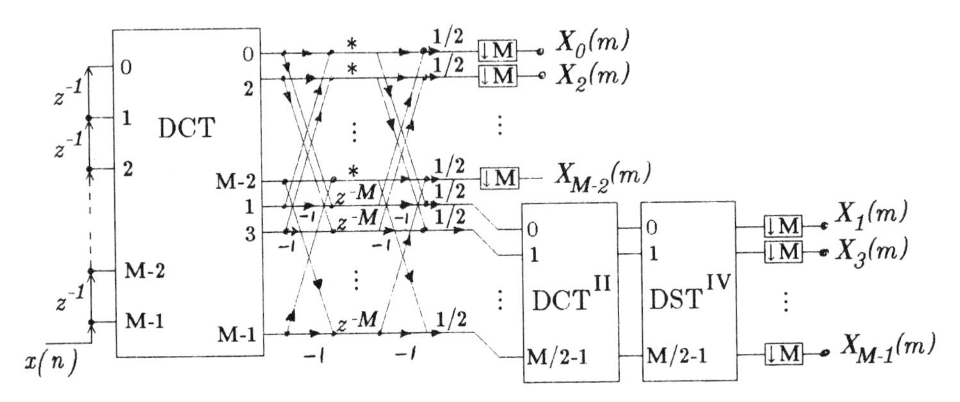

Fig. 7.151 Flowgraph of the last direct LOT (analysis filter bank). The flowgraph of the inverse LOT is just the transpose, but the z^{-M} delay units should be moved to the branches marked by asterisks. The symbol —[↓M]— implies $M:1$ decimation [SC-29]. (© 1989 IEEE).

Besides the computation of the DCT of the current block, the LOT requires a few extra butterflies and the computation of inverse LOT and a DST-IV [FRA-8]. Therefore, the LOT requires roughly twice the number of operations needed for the DCT (for small block sizes, the alternative LOT implementations described in [BSI-5, BSI-6] are more efficient). Since the number of transform coefficients is identical to the DCT length, no extra bit rate is required.

There are several versions of LOT. A specific version optimized in terms of energy compaction has been applied to image coding [BSI-13]. In intraframe coding at bit rates between 0.1 and 0.35 BPP, this transform led to improved subjective quality compared with the DCT. In interframe head-and-shoulders motion video, this improvement was maintained at 28 and 56 KBPS. At 56 KBPS with motion compensation, there was apparently no gain in comparison to the DCT.

7.12.6 *Adaptive Block Distortion Equalization*

Whenever there is a large distortion gradient between adjacent blocks that shows up as a visible block structure, an adaptive technique that equalizes the distortion between the blocks, resulting in a reduced blocky effect, has been developed [HTC-5 through HTC-7, HTC-10]. The price for the improved image is the additional complexity, involving computation of distortion gradient, buffer monitoring, and distortion equalization.

7.12.7 *Interpolative VQ*

Hang and Haskell [BSI-10] have developed interpolative VQ (IVQ) and have shown that the IVQ alleviates the visible block structure of the coded images. In IVQ, a representative value of each block of an image is sent by PCM, and the remaining pels of the block are estimated by interpolation. VQ is applied to the

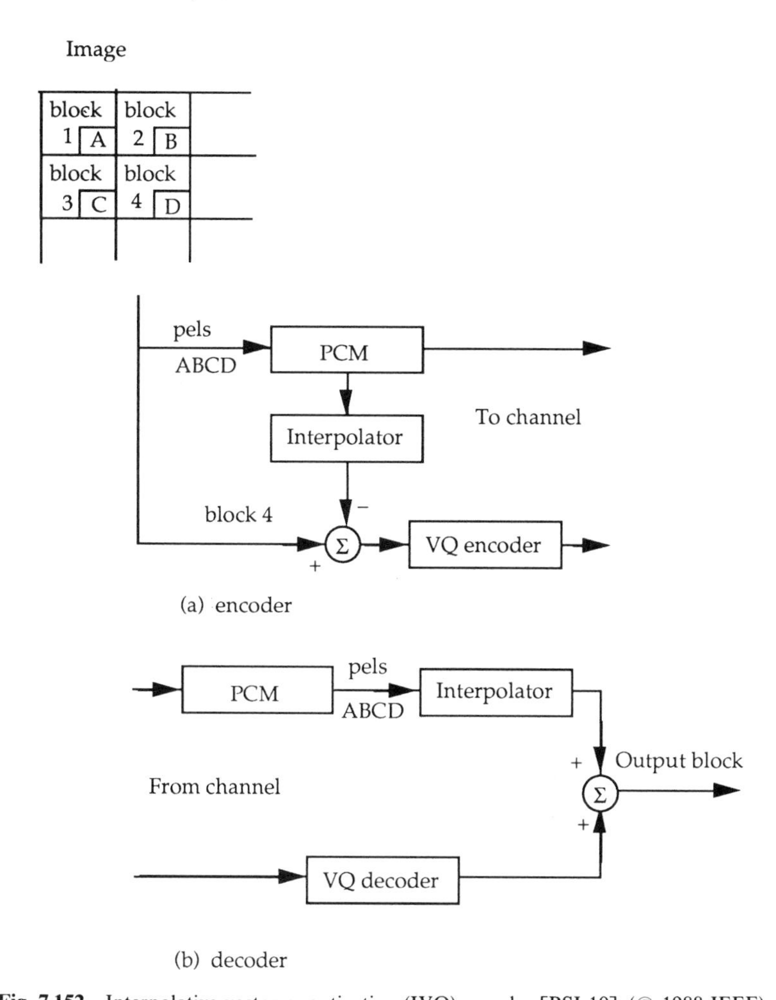

Fig. 7.152 Interpolative vector quantization (IVQ) encoder [BSI-10]. (© 1988 IEEE).

interpolation error, i.e., the difference between the original and interpolated pels (Fig. 7.152). The objective of IVQ is not only data compression, but also to produce perceptually pleasing pictures.

7.12.8 *Anisotropic Adaptive Filter*

To adapt to the different frequency characteristics of the horizontal and vertical boundaries in block transform coding, Tzou [BSI-11] has developed an adaptive circular-asymmetric 2D filter that eliminates the block distortion at bit rates above $\frac{1}{4}$ BPP for DCT coding. For coding at lower bit rates, however, block overlapping is required to reduce this distortion.

DCT Coding with Average Separation

2D DCT of average, separated image blocks, as developed by Nemoto and Omachi [BSI-12], not only reduces the block distortion, but also makes this method suitable for hierarchical coding. Block diagram representations of this technique, together with the generation of the interpolation image, are shown in Figs. 7.153a and 7.153b, respectively.

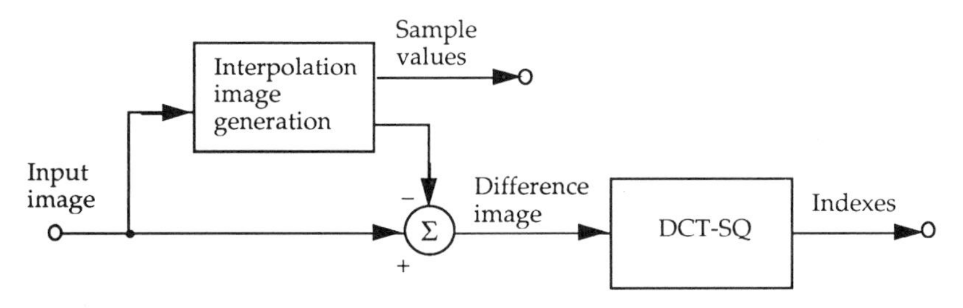

Fig. 7.153a DCT coding with average separation [BSI-12].

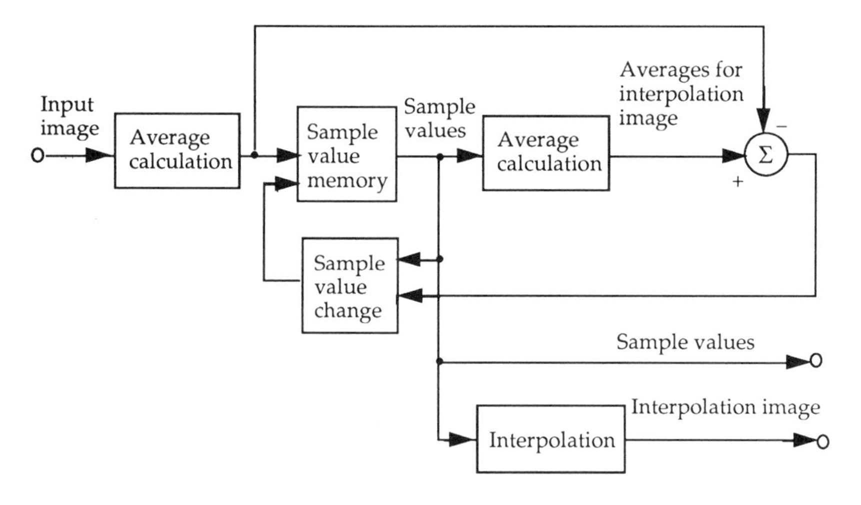

Fig. 7.153b Interpolation image generation [BSI-12].

Another technique that removes the block boundary mismatch errors is the constrained transform (DCT) coding developed by Watson, Haralick, and Zuniga [BSI-8]. The block boundary mismatch is, however, eliminated at the expense of slightly defocusing the image within each block. Nasrabadi and King [BSI-9] have applied 2D overlapping of blocks adaptively to 2D-DCT image coding. If a block is highly correlated with its four nearest neighbors, then overlapping is applied. This correlation is based on conditional entropies of the block being coded with its four nearest neighbors. Except for the RBC and DLT, all other techniques involve DCT as the basic compression tool. The latter techniques require either additional bit rates and/or additional processing hardware. The improvement is the smoothness and continuity across the block boundaries, even at low bit rates.

To eliminate the blocking artifact, an adaptive hybrid technique for coding of videophone/videoconference sequences at low bit rates (56 and 112 KBPS) has been developed [LBR-48]. Block diagrams of the coder and decoder are shown in Fig. 7.154. Highlights of this method are:

(1) Decomposition of the image into low and high frequency bands using a low pass filter.

(2) Interframe motion estimation of 16 * 16 blocks of the combined low and high frequency bands based on the block matching algorithms (Fig. 7.108).

(3) 2D 8 * 8 DCT of luminance and chrominance interframe MC frame differences representing the low frequency band.

(4) Uniform quantization, zig-zag scanning and 2-dimensional (run length of zero coefficients and amplitudes of nonzero coefficients) variable length coding.

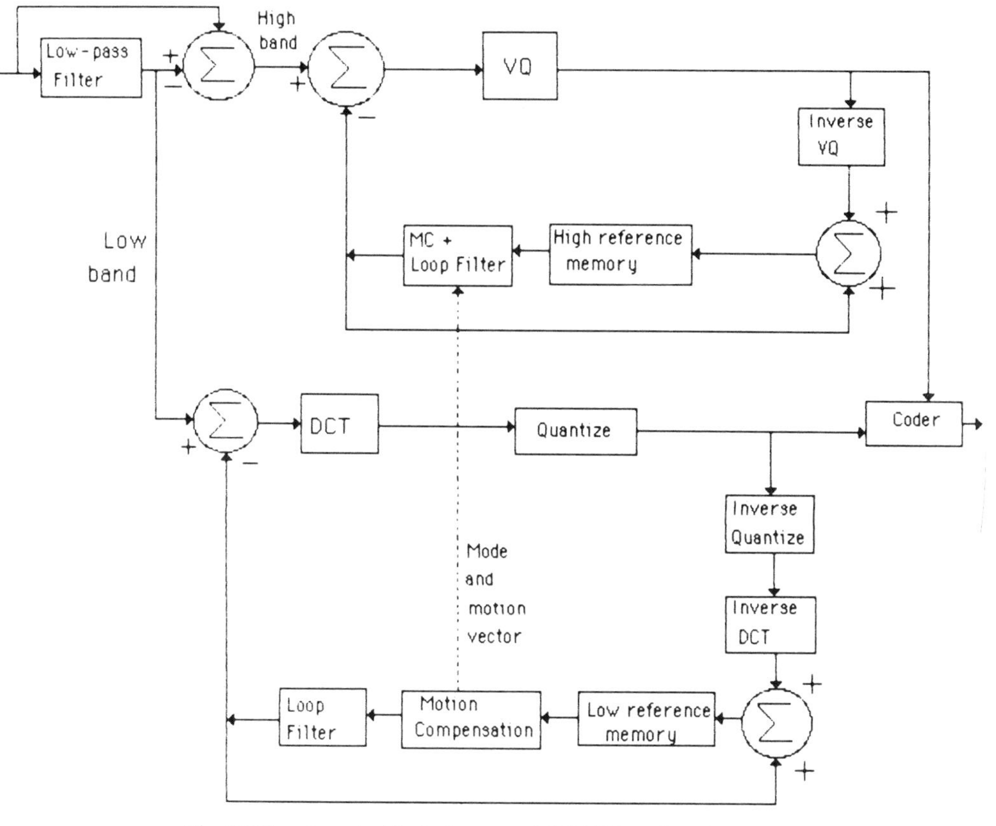

Fig. 7.154a Encoder block diagram of DCT/VQ coding, [LBR-48].

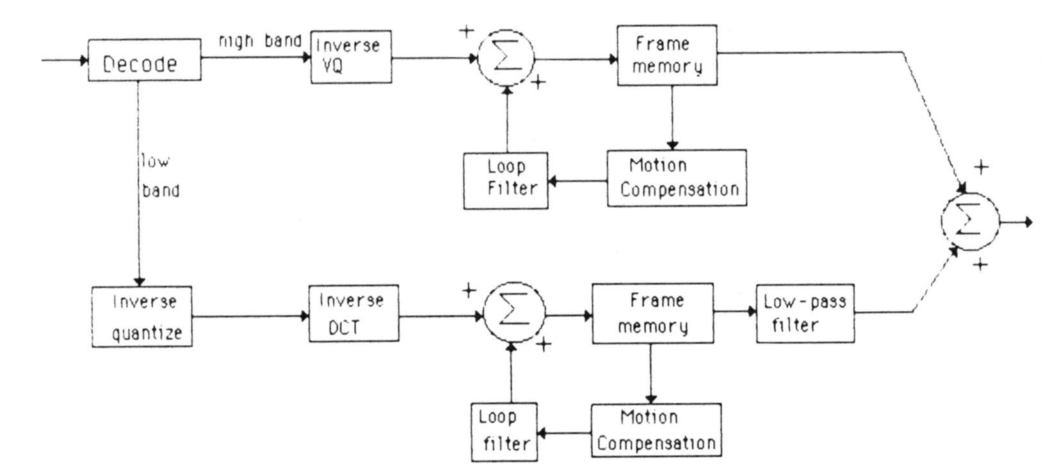

Fig. 7.154b Decoder block diagram of DCT/VQ coding, [LBR-48].

(5) VQ of $4*4$ blocks formed from the MC interframe differences.
(6) Loop filters in the feedback loops of both the frequency bands.
(7) Adaptive classification of each macroblock consisting of on $16*16$ luminance block and two $8*8$ chrominance blocks as fixed, MC coded, MC but not coded and no MC but coded. This is similar to various modes described in Table 7.17.

Based on simulation of test sequences Maeng and Hein [LBR-48] have shown that this adaptive DCT/VQ scheme results in subjectively better picture quality (edge sharpness, smooth texture, and almost no blocking artifact) compared to the RM-5 model [LBR-23].

7.13 Activity Classification in Adaptive Transform Coding

In adaptive transform coding of images, the activity of a block can be described in several ways. The general concept is to allocate more bits to a high activity class and vice versa. Assuming that an $N \times N$ image is divided into blocks of size $L \times L$, the 2D-DCT coefficients of each block are $X^{C(2)}(u, v)$, u, $v = 0, 1, \ldots, L-1$. Based on the test image or a class of images, the blocks can be divided into several classes for adaptive coding. Although the classification concepts described below are based on the 2D DCT, these schemes can, in general, be extended to the 3D DCT. Some of these classes are as follows:

7.13.1 *AC Energy*

Whereas the dc coefficient represents the average brightness of a block, the ac energy (defined below) is an indication of activity or detail of a block.

$$\text{ac energy of a block} = \sum_{u=0}^{L-1} \sum_{v=0}^{L-1} [X^{C(2)}(u, v)]^2 - [X^{C(2)}(0, 0)]^2. \quad (7.13.1)$$

The ac energies of all the blocks in an image or class of images can be computed and the blocks can be divided into several equally populated classes. For example, a 1024×1024 monochrome image is divided into 16×16 blocks and, based on 2D DCT of these blocks and their ac energy distribution (Fig. 7.155), four equally populated classes have been developed [ACTC-4]. The variances of the transform coefficients in each class can be computed, and, using this variance distribution, the bit allocation matrix (see Problem 7.10) for each class can be developed. For example, the classification map and the bit allocation matrices based on four classes are shown in Figs. 7.156 and 7.157, respectively [ACTC-4]. Not only are the higher activity classes allocated more bits, but also the dc and low-frequency coefficients in each class, in general, are allocated more bits compared with the high-frequency coefficients, with some of the later even being discarded (zero bits).

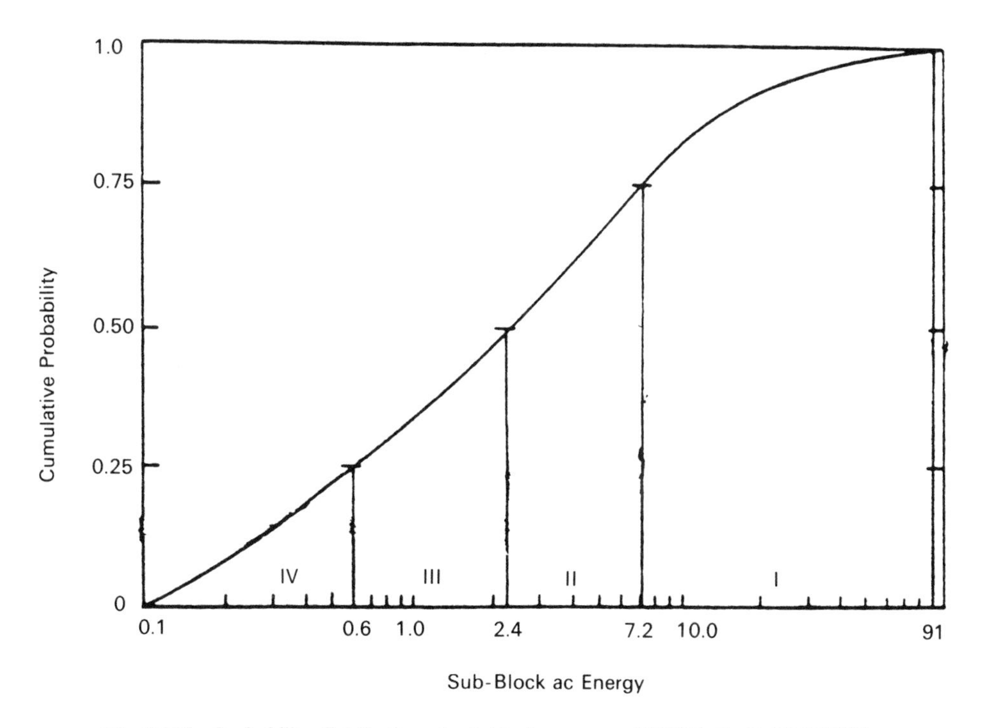

Fig. 7.155 Probability distribution of sub-block ac energy [ACTC-4]. © 1977 IEEE).

The number of bits allocated to each class can be based on the total ac energy of each class. Let

$$E_k = \sum_{j=1}^{M_k} E_{kj}$$

1	2	2	3	4	4	4	3	2	2	2	3	4	2	1	1	1	2	4	4
3	2	3	4	4	4	4	3	1	3	2	3	2	1	2	2	2	4	4	4
4	2	4	4	4	4	4	3	3	3	3	3	2	1	1	2	3	4	1	1
4	1	4	4	3	3	4	3	3	3	3	2	2	1	2	3	3	1	3	3
4	2	3	3	3	3	4	3	3	3	1	2	1	2	2	1	3	4	3	3
4	2	2	2	4	4	4	3	4	1	2	1	1	1	3	3	4	3	3	3
4	3	3	3	4	4	2	3	2	2	2	1	2	1	2	3	3	3	3	3
4	4	2	4	3	4	3	2	1	2	1	3	2	1	3	4	3	3	3	3
4	4	2	3	4	3	4	1	2	1	2	3	1	1	1	1	3	3	3	3
4	4	2	3	4	3	1	2	1	2	3	1	3	2	2	1	3	2	3	2
4	4	3	2	4	1	2	1	2	3	1	3	1	3	1	3	3	3	1	2
3	4	4	2	1	2	1	2	2	1	3	2	3	3	3	1	1	2	2	2
4	4	4	1	1	2	1	2	2	3	2	3	1	3	2	1	2	1	3	3
4	4	4	2	1	2	2	2	1	2	1	2	3	1	1	2	1	1	2	3
4	4	3	2	3	2	3	1	2	1	1	2	2	2	2	1	1	1	1	4
2	4	2	3	2	2	1	2	3	2	1	2	2	2	1	1	1	1	1	1
3	1	3	2	1	1	2	3	2	3	3	1	3	1	1	1	2	2	1	1
1	3	2	2	1	2	2	1	1	3	2	2	1	1	2	1	2	2	1	3
2	1	2	4	1	4	3	1	3	2	1	1	1	2	3	1	1	1	1	2
2	2	4	4	4	3	3	3	4	1	2	2	1	2	2	1	1	1	1	2

Fig. 7.156 Classification of the central 20 × 20 transform sub-blocks for the monochrome image. 1 specifies the highest activity sub-blocks; 4 specifies the lowest activity sub-blocks [ACTC-4]. (© 1977 IEEE).

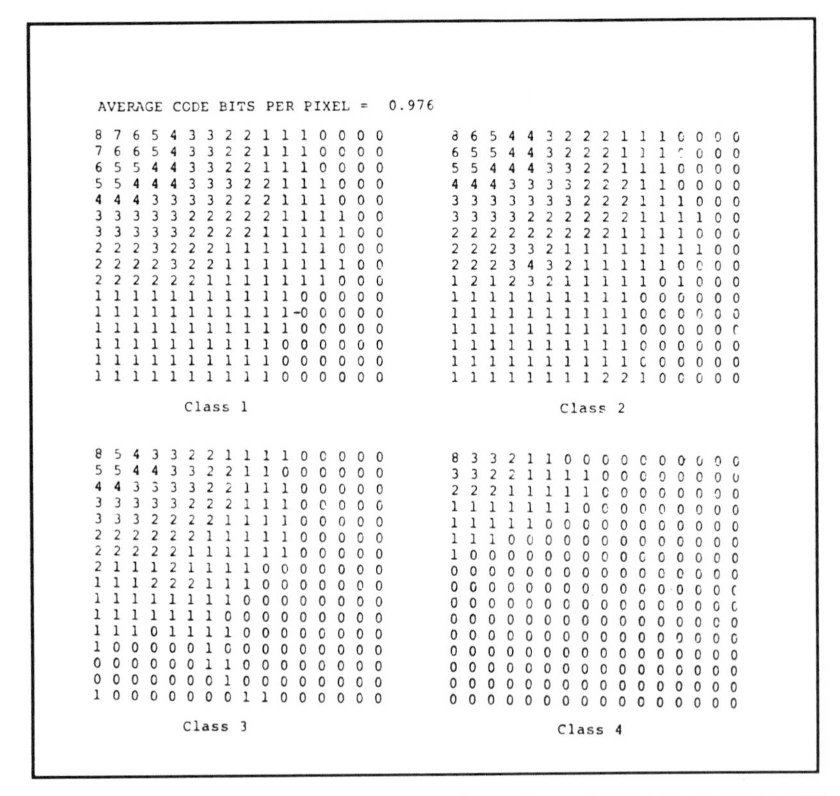

AVERAGE CODE BITS PER PIXEL = 0.976

8	7	6	5	4	3	3	2	2	1	1	1	0	0	0	0
7	6	6	5	4	3	3	2	2	1	1	1	0	0	0	0
6	5	5	4	4	3	3	2	2	1	1	1	0	0	0	0
5	5	4	4	4	3	3	3	2	2	1	1	1	0	0	0
4	4	4	3	3	3	3	2	2	2	1	1	1	0	0	0
3	3	3	3	3	2	2	2	2	2	1	1	1	1	0	0
3	3	3	3	3	2	2	2	2	1	1	1	1	1	0	0
2	2	2	3	2	2	2	1	1	1	1	1	1	0	0	0
2	2	2	2	3	2	2	1	1	1	1	1	1	1	0	0
2	2	2	2	2	2	1	1	1	1	1	1	1	0	0	0
1	1	1	1	1	1	1	1	1	1	1	0	0	0	0	0
1	1	1	1	1	1	1	1	1	1	1	-0	0	0	0	0
1	1	1	1	1	1	1	1	1	1	1	0	0	0	0	0
1	1	1	1	1	1	1	1	1	1	0	0	0	0	0	0
1	1	1	1	1	1	1	1	1	1	0	0	0	0	0	0
1	1	1	1	1	1	1	1	1	1	0	0	0	0	0	0

Class 1

8	6	5	4	4	3	2	2	2	1	1	1	0	0	0	0
6	5	5	4	4	3	2	2	2	1	1	1	0	0	0	0
5	5	4	4	4	3	3	2	2	1	1	1	0	0	0	0
4	4	4	3	3	3	3	2	2	2	1	1	0	0	0	0
3	3	3	3	3	3	3	2	2	2	1	1	1	0	0	0
3	3	3	3	2	2	2	2	2	2	1	1	1	1	0	0
2	2	2	2	2	2	2	2	2	1	1	1	1	0	0	0
2	2	2	3	3	2	1	1	1	1	1	1	1	1	0	0
2	2	2	3	4	3	2	1	1	1	1	1	0	0	0	0
1	2	1	2	3	2	1	1	1	1	1	0	1	0	0	0
1	1	1	1	1	1	1	1	1	1	0	0	0	0	0	0
1	1	1	1	1	1	1	1	1	1	0	0	0	0	0	0
1	1	1	1	1	1	1	1	1	1	0	0	0	0	0	0
1	1	1	1	1	1	1	1	1	1	0	0	0	0	0	0
1	1	1	1	1	1	1	1	1	1	0	0	0	0	0	0
1	1	1	1	1	1	1	1	2	2	1	0	0	0	0	0

Class 2

8	5	4	3	3	2	2	1	1	1	1	0	0	0	0	0
5	5	4	4	3	3	2	2	1	1	0	0	0	0	0	0
4	4	3	3	3	3	2	2	1	1	1	0	0	0	0	0
3	3	3	3	3	2	2	2	1	1	1	0	0	0	0	0
3	3	3	2	2	2	2	1	1	1	1	0	0	0	0	0
2	2	2	2	2	2	1	1	1	1	1	0	0	0	0	0
2	2	2	2	2	1	1	1	1	1	1	0	0	0	0	0
2	1	1	1	2	1	1	1	1	0	0	0	0	0	0	0
1	1	1	2	2	2	1	1	1	0	0	0	0	0	0	0
1	1	1	1	1	1	1	1	0	0	0	0	0	0	0	0
1	1	1	1	1	1	1	0	0	0	0	0	0	0	0	0
1	1	1	0	1	1	1	1	0	0	0	0	0	0	0	0
1	0	0	0	0	0	1	0	0	0	0	0	0	0	0	0
0	0	0	0	0	0	1	1	0	0	0	0	0	0	0	0
0	0	0	0	0	0	0	1	0	0	0	0	0	0	0	0
1	0	0	0	0	0	0	0	1	1	0	0	0	0	0	0

Class 3

8	3	3	2	1	1	0	0	0	0	0	0	0	0	0	0
3	3	2	2	1	1	1	1	0	0	0	0	0	0	0	0
2	2	2	1	1	1	1	1	0	0	0	0	0	0	0	0
1	1	1	1	1	1	1	0	0	0	0	0	0	0	0	0
1	1	1	1	1	0	0	0	0	0	0	0	0	0	0	0
1	1	1	0	0	0	0	0	0	0	0	0	0	0	0	0
1	0	0	0	0	0	0	0	0	0	0	0	0	0	0	0
0	0	0	0	0	0	0	0	0	0	0	0	0	0	0	0
0	0	0	0	0	0	0	0	0	0	0	0	0	0	0	0
0	0	0	0	0	0	0	0	0	0	0	0	0	0	0	0
0	0	0	0	0	0	0	0	0	0	0	0	0	0	0	0
0	0	0	0	0	0	0	0	0	0	0	0	0	0	0	0
0	0	0	0	0	0	0	0	0	0	0	0	0	0	0	0
0	0	0	0	0	0	0	0	0	0	0	0	0	0	0	0
0	0	0	0	0	0	0	0	0	0	0	0	0	0	0	0
0	0	0	0	0	0	0	0	0	0	0	0	0	0	0	0

Class 4

Fig. 7.157 Bit allocation matrices for the monochrome image [ACTC-4]. (© 1977 IEEE).

be the total energy of class k. E_{kj} is the ac energy of block j in class k, and M_k is the number of blocks in this class. The total ac energy of an image or set of images is then

$$E_{\text{total}} = \sum_{k=1}^{K} E_k,$$

where K is the total number of classes. If B_{total} is the total number of bits allocated to an image, then the number of bits allocated to any class can be based on the total energy of that class. Ferreyra, Picco, and Sobrero [IC-74] have investigated this bit allocation based on the following three models;

$$\log(E_k + 1),\ (E_k)^{1/3}, \qquad \text{and } \frac{[\log(E_k) + (E_k)^{1/2}]}{2},$$

and have concluded that the middle one gave the best results from the image compression-quality viewpoint.

The classification technique described in [ACTC-4] was also utilized in adaptive interfield hybrid (DPCM-DCT) coding of component color TV signals (Fig. 7.49) [IC-7]. In this scheme, the 2D DCT of 8×8 blocks representing Y, I, and Q is followed by interfield prediction in the transform domain and adaptive classification of the sub-blocks based on the ac energies of these sub-blocks (Fig. 7.50). The number of sub-blocks and their groupings (no optimality is claimed) are developed after extensive investigation of the correlations among the DCT coefficients using a large number of test sequences as the statistical basis. As in the adaptive coding scheme of Chen and Smith [ACT-3, ACTC-4], bit allocation among the interfield prediction errors of the transform coefficients reflects their variances and energy distributions. Activity classification is based on Y only.

Whereas the ac energy classification scheme originally proposed by Chen and Smith [ACTC-4] is based on equally populated classes (Fig. 7.155), other criteria, such as classes with equal energies, classes with a half-maximum block energy ratio (all blocks in a class have at least half the energy of the block with maximum energy in that class), and classes with a fixed but not equal number of blocks, have also been used in image coding [HDTV-1]. In this simulation both 8 and 16 classes were used in the HDTV image coding. Even though overhead identifying the classification of the blocks and the complexity (storage of bit allocation matrices for all the classes, classification, etc.) increases with the number of classes, the degree of adaptation based on block activity, detail, etc. is improved.

7.13.2 *Magnitude Sum of the AC Coefficients*

This parameter can be expressed as

$$A = \sum_{u=0}^{L-1} \sum_{v=0}^{L-1} |X^{C(2)}(u, v)| - |X^{C(2)}(0, 0)|. \tag{7.13.2}$$

Compared with the ac energy, the magnitude sum of the ac coefficients of a block is computationally much simpler. In four-color, printed-image coding [PIC-2], the parameter A, together with $|X^{C(2)}(0,0)|$ was utilized in classifying the blocks. Large values of A indicate blocks with strong edges, whereas the magnitude of the dc coefficient can indicate dark blocks with little ac energy. The activity index defined by (7.13.2) was originally proposed by Gimlett [ACTC-5]. He has also suggested weighted activity measures such as

$$\text{weighted ac energy of a block} = \sum_{u=0}^{L-1} \sum_{v=0}^{L-1} w_1(u, v)[X^{C(2)}(u, v)]^2 - [X^{C(2)}(0, 0)]^2, \tag{7.13.3}$$

$$\left.\begin{array}{l}\text{weighted magnitude sum} \\ \text{of the ac coefficient}\end{array}\right] = \sum_{u=0}^{L-1} \sum_{v=0}^{L-1} w_2(u, v)|X^{C(2)}(u, v)| - X^{C(2)}(0, 0), \tag{7.13.4}$$

where $w_1(u, v)$ and $w_2(u, v)$ are the weights for application in image coding.

The parameter A (7.13.2) and $X^{C(2)}(0,0)$ formed the basis for classifying the 2D DCT of 8×8 blocks of the four-color space—yellow, magenta, cyan, and black—in printed-image coding [PIC-1, PIC-2]. This classification determined the number of bits allocated, the threshold (transform coefficients above this threshold were coded), and also the number of coefficients set to zero in a block. The same parameters were also utilized in classifying the 2D DCT of 8×8 blocks formed from the KLT of the four-color components [PIC-2].

7.13.3 *Spectral Entropy—AC Energy*

Mester and Franke [ACTC-2] have proposed block activity based on a spectral entropy-ac energy combination [in this case (7.13.2) is used as the ac energy]. The spectral entropy is to indicate the degree of energy compaction (e.g., a few of the transform coefficients having the most energy of the block). The spectral entropy is defined as

$$E = -\sum_{u=0}^{L-1} \sum_{v=0}^{L-1} \frac{|X^{C(2)}(u, v)|}{A} \log_2 \left[\frac{|X^{C(2)}(u, v)|}{A}\right], \tag{7.13.5}$$

where A is defined in (7.13.2).

Low entropy implies that the block energy is compacted into a few transform coefficients. Based on the ranges of A and E, a suggested classification scheme (100 classes) is proposed in Table 7.22. Using monochrome images and 2D DCT of 8×8 blocks, for each class $[A_i, E_j]$ the maximum threshold that yields no visible degradation of the blocks belonging to that class has been determined. The |ac coefficients| below this threshold are discarded. For practical application, the number of classes need to be reduced to, e.g., 4, 8, or 16. The parameters A and E were also utilized in an adaptive DCT-VQ coding of monochrome images [DV-38].

Table 7.22

Classification intervals using logarithmic quantization of the features A (activity) and E (spectral entropy) [ACTC-2].

Activity class	Interval bounds for activity A	Entropy class	Interval bounds for spectral entropy E
A_1	[100.0*, 140.5)	E_1	[3.993*, 3.739)
A_2	[140.5, 197.4)	E_2	[3.739, 3.502)
A_3	[197.4, 277.4)	E_3	[3.502, 3.280)
A_4	[277.4, 389.8)	E_4	[3.280, 3.071)
A_5	[389.8, 547.7)	E_5	[3.071, 2.876)
A_6	[547.67, 769.6)	E_6	[2.876, 2.693)
A_7	[769.6, 1081.4)	E_7	[2.693, 2.522)
A_8	[1081.4, 1519.5)	E_8	[2.522, 2.362)
A_9	[1519.5, 2135.1)	E_9	[2.362, 2.212)
A_{10}	[2135.1, 3000.0*]	E_{10}	[2.212, 2.072*]

*The actual feature values are limited to the value range 100.0–3000.0 for the activity and 3.993–2.072 for the spectral entropy, respectively. Almost all transform blocks of natural images have feature values in this range.

7.13.4 *MACE, Direction, and Fineness*

To reduce the differences between blocks belonging to the same class, Wu and Burge [ACTC-1] proposed block classification based on its features defined as follows:

7.13.4.1 Middle Frequency AC Energy (MACE)

$$\mathrm{MACE}(p, q) = \sum_{u=0}^{p-1} \sum_{v=0}^{p-1} [X^{C(2)}(u, v)]^2 - \sum_{u=0}^{q-1} \sum_{v=0}^{q-1} [X^{C(2)}(u, v)]^2. \quad (7.13.6)$$

MACE represents the ac energy within a spatial frequency window with MACE $(L-1, 0)$ as the ac energy of the block.

7.13.4.2 Directional Features (DIR)

$$\mathrm{DIR} = \frac{\sum_{u=0}^{12} \sum_{v=0}^{12} \tan^{-1}\left(\frac{u}{v}\right) |X^{C(2)}(u, v)|}{\sum_{u=0}^{12} \sum_{v=0}^{12} |X^{C(2)}(u, v)|}. \quad (7.13.7)$$

Note that (7.13.7) is defined for $u = v = 16$ and can be modified for other block sizes. DIR exhibits strong directional features.

7.13.4.3 Fineness of a Subimage (FIN)

$$\mathrm{FIN} = \frac{\displaystyle\int_{0}^{\pi/2}\int_{\rho_1}^{\rho_2} X^{C(2)}(\rho, \theta)\, d\rho d\theta \Big/ \int_{0}^{\pi/2}\int_{\rho_1}^{\rho_2} d\rho d\theta}{\displaystyle\int_{0}^{\pi/2}\int_{\rho_3}^{\rho_4} X^{C(2)}(\rho, \theta)\, \mathrm{d}\rho\, \mathrm{d}\theta \Big/ \int_{0}^{\pi/2}\int_{\rho_3}^{\rho_4} d\rho\, d\theta}. \tag{7.13.8}$$

8	6	5	4	3	3	2	2	1	1	0	0	0	0	0	0
6	5	4	3	3	2	2	1	1	0	0	0	0	0	0	0
5	4	4	3	3	2	2	1	1	0	0	0	0	0	0	0
4	3	3	2	2	2	1	1	0	0	0	0	0	0	0	0
3	3	3	2	2	2	1	1	0	0	0	0	0	0	0	0
3	3	3	2	2	1	1	0	0	0	0	0	0	0	0	0
3	3	2	2	2	1	1	0	0	0	0	0	0	0	0	0
3	2	2	2	1	1	0	0	0	0	0	0	0	0	0	0
2	2	2	1	1	1	0	0	0	0	0	0	0	0	0	0
2	2	1	1	0	0	0	0	0	0	0	0	0	0	0	0
1	1	1	0	0	0	0	0	0	0	0	0	0	0	0	0
0	0	0	0	0	0	0	0	0	0	0	0	0	0	0	0
0	0	0	0	0	0	0	0	0	0	0	0	0	0	0	0
0	0	0	0	0	0	0	0	0	0	0	0	0	0	0	0
0	0	0	0	0	0	0	0	0	0	0	0	0	0	0	0
0	0	0	0	0	0	0	0	0	0	0	0	0	0	0	0

(a) 0.719 bit/pixel

8	4	3	3	3	2	2	2	2	1	1	1	0	0	0	0
4	3	3	2	2	2	2	2	2	1	1	1	0	0	0	0
3	3	3	2	2	2	2	2	2	1	1	1	0	0	0	0
3	3	3	2	2	2	2	2	1	1	1	0	0	0	0	0
3	3	2	2	2	2	2	2	1	1	1	0	0	0	0	0
2	2	2	2	2	2	2	2	1	1	1	0	0	0	0	0
2	2	2	2	2	3	3	2	1	1	0	0	0	0	0	0
2	2	2	2	2	2	2	1	1	0	0	0	0	0	0	0
2	2	2	2	1	1	1	0	0	0	0	0	0	0	0	0
1	1	1	1	1	1	0	0	0	0	0	0	0	0	0	0
1	1	1	1	0	0	0	0	0	0	0	0	0	0	0	0
0	0	0	0	0	0	0	0	0	0	0	0	0	0	0	0
0	0	0	0	0	0	0	0	0	0	0	0	0	0	0	0
0	0	0	0	0	0	0	0	0	0	0	0	0	0	0	0
0	0	0	0	0	0	0	0	0	0	0	0	0	0	0	0
0	0	0	0	0	0	0	0	0	0	0	0	0	0	0	0

(b) 0.773 bit/pixel

8	6	5	5	4	4	4	4	4	4	3	3	3	3	2	2
4	4	4	3	3	3	2	2	2	1	1	1	0	0	0	0
2	2	2	1	1	1	0	0	0	0	0	0	0	0	0	0
0	0	0	0	0	0	0	0	0	0	0	0	0	0	0	0
0	0	0	0	0	0	0	0	0	0	0	0	0	0	0	0
0	0	0	0	0	0	0	0	0	0	0	0	0	0	0	0
0	0	0	0	0	0	0	0	0	0	0	0	0	0	0	0
0	0	0	0	0	0	0	0	0	0	0	0	0	0	0	0
0	0	0	0	0	0	0	0	0	0	0	0	0	0	0	0
0	0	0	0	0	0	0	0	0	0	0	0	0	0	0	0
0	0	0	0	0	0	0	0	0	0	0	0	0	0	0	0
0	0	0	0	0	0	0	0	0	0	0	0	0	0	0	0
0	0	0	0	0	0	0	0	0	0	0	0	0	0	0	0
0	0	0	0	0	0	0	0	0	0	0	0	0	0	0	0
0	0	0	0	0	0	0	0	0	0	0	0	0	0	0	0
0	0	0	0	0	0	0	0	0	0	0	0	0	0	0	0

(c) 0.398 bit/pixel

8	7	6	5	4	4	4	3	3	3	3	2	2	1	1	0
7	6	5	4	4	4	3	3	3	2	2	1	1	1	0	0
6	5	4	4	4	3	3	2	2	2	1	1	1	0	0	0
5	4	4	4	3	3	3	2	2	1	1	1	1	0	0	0
4	4	4	3	3	3	3	2	2	1	1	1	0	0	0	0
4	4	4	3	3	3	3	2	2	1	1	0	0	0	0	0
4	3	3	3	3	3	2	2	1	1	1	0	0	0	0	0
3	3	3	2	2	2	2	2	1	1	0	0	0	0	0	0
3	3	3	2	2	2	2	1	1	0	0	0	0	0	0	0
3	3	3	2	2	2	1	1	0	0	0	0	0	0	0	0
2	2	2	2	1	1	0	0	0	0	0	0	0	0	0	0
2	2	1	1	1	0	0	0	0	0	0	0	0	0	0	0
1	1	1	1	0	0	0	0	0	0	0	0	0	0	0	0
0	0	0	0	0	0	0	0	0	0	0	0	0	0	0	0
0	0	0	0	0	0	0	0	0	0	0	0	0	0	0	0
0	0	0	0	0	0	0	0	0	0	0	0	0	0	0	0

(d) 1.334 bit/pixel

Fig. 7.158 Four types of bit assignment matrices. The matrices refer to (a) and (b) for fine image detail, (c) for horizontal direction, and (d) for higher middle frequency ac energy [ACTC-1].

In (7.13.8), (ρ, θ) are the polar coordinates of the block in the transform domain. Using these three features, Wu and Burge [ACTC-1] have developed 10 classes and four bit assignment matrices (Fig. 7.158). Because of the extensive computations involved in obtaining the features, and also the classifications, adaptive image coding based on these features may be applicable to still-frame images where pattern recognition and feature classification in a compressed domain (reduced bit-rate transmission and storage) is desirable.

7.13.5 *Coefficient Power Distribution*

As videoconferencing and videophone, in general, are characterized by low resolution, absence of violent motion, abrupt scene changes, and special effects, they can be transmitted by narrow band channels in real time. One technique to achieve a large compression for transmitting these signals, say at 384 KBPS, is to apply adaptive transform coding to motion-compensated interframe blocks (see Section 7.10). Mukawa and Okubo [MC-12] proposed that the transform

class 1	class 2	class 3	class 4
3 4 4 2 1 1 0 0	3 4 4 3 3 2 1 0	3 0 4 2 2 1 1 0	3 4 0 3 2 3 2 1
4 4 4 3 2 1 1 0	0 4 4 3 2 1 1 0	4 4 3 3 2 1 1 0	2 2 2 3 3 2 1 1
1 3 2 2 1 1 0 0	4 3 2 2 1 1 1 0	4 4 2 2 1 1 0 0	4 2 2 3 3 1 1 1
1 2 2 1 1 1 0 0	2 3 2 2 1 1 0 0	3 3 2 1 1 1 0 0	2 3 2 2 2 1 1 1
1 2 1 1 1 0 0 0	2 2 1 1 1 0 0 0	3 2 1 1 1 1 0 0	2 2 2 1 1 1 1 1
1 1 1 1 0 0 0 0	1 1 1 1 1 0 0 0	2 1 1 1 0 0 0 0	2 1 1 1 1 1 0 0
0 0 0 0 0 0 0 0	1 1 0 0 0 0 0 0	1 1 1 1 0 0 0 0	1 1 1 1 1 0 0 0
0 0 0 0 0 0 0 0	0 0 0 0 0 0 0 0	0 0 0 0 0 0 0 0	1 1 0 1 0 0 0 0
class 5	**class 6**	**class 7**	**class 8**
3 4 1 4 3 1 1 1	3 1 4 4 2 1 1 1	3 4 4 4 4 4 3 1	3 2 4 2 1 1 0 0
4 4 1 4 2 1 1 0	4 4 2 4 1 1 1 1	2 2 1 1 1 1 1 0	0 2 4 2 1 1 1 0
3 2 1 1 1 1 1 0	2 2 2 2 2 1 1 0	3 2 2 2 1 1 1 0	1 1 1 2 1 1 0 0
1 1 1 1 1 0 0 0	1 1 1 1 1 1 0 0	2 1 1 1 1 0 0 0	1 1 1 1 1 1 0 0
1 1 1 1 0 0 0 0	1 0 1 1 0 0 0 0	1 1 1 1 0 0 0 0	0 0 1 0 0 0 0 0
0 0 0 0 0 0 0 0	0 0 0 0 0 0 0 0	1 1 0 0 0 0 0 0	0 0 0 0 0 0 0 0
0 0 0 0 0 0 0 0	0 0 0 0 0 0 0 0	0 0 0 0 0 0 0 0	0 0 0 0 0 0 0 0
0 0 0 0 0 0 0 0	0 0 0 0 0 0 0 0	0 0 0 0 0 0 0 0	0 0 0 0 0 0 0 0
class 9	**class 10**	**class 11**	**class 12**
3 4 3 1 1 0 0 0	3 4 2 1 1 0 0 0	3 2 3 2 1 1 0 0	3 0 1 1 0 0 0 0
4 4 2 1 1 0 0 0	1 4 2 1 0 0 0 0	4 2 2 1 1 1 0 0	1 2 1 1 0 0 0 0
1 1 1 1 1 0 0 0	4 2 2 1 1 0 0 0	4 1 2 1 1 0 0 0	4 4 1 1 0 0 0 0
4 4 1 1 1 0 0 0	4 3 2 1 1 0 0 0	4 1 2 1 1 0 0 0	3 2 1 1 0 0 0 0
3 2 1 1 0 0 0 0	2 1 2 1 0 0 0 0	4 1 1 1 0 0 0 0	1 1 1 1 0 0 0 0
1 1 1 0 0 0 0 0	2 2 1 1 0 0 0 0	4 1 1 0 0 0 0 0	1 1 1 0 0 0 0 0
1 1 1 0 0 0 0 0	1 1 1 0 0 0 0 0	3 1 1 0 0 0 0 0	0 1 0 0 0 0 0 0
1 0 0 0 0 0 0 0	1 1 0 0 0 0 0 0	1 1 0 0 0 0 0 0	0 0 0 0 0 0 0 0
class 13	**class 14**	**class 15**	**class 16**
3 2 2 1 1 1 0 0	3 1 1 0 0 0 0 0	3 4 0 4 2 1 1 0	3 0 0 0 0 0 0 0
4 4 2 2 1 1 0 0	4 1 1 0 0 0 0 0	1 1 1 1 1 1 0 0	0 0 0 1 0 0 0 0
1 1 1 0 0 0 0 0	0 1 0 0 0 0 0 0	1 1 0 1 0 0 0 0	0 0 1 1 1 0 0 0
0 0 0 0 0 0 0 0	4 1 0 0 0 0 0 0	0 0 0 0 0 0 0 0	1 1 1 1 0 0 0 0
0 0 0 0 0 0 0 0	2 1 0 0 0 0 0 0	0 0 0 0 0 0 0 0	1 1 1 1 0 0 0 0
0 0 0 0 0 0 0 0	1 0 0 0 0 0 0 0	0 0 0 0 0 0 0 0	0 1 1 0 0 0 0 0
0 0 0 0 0 0 0 0	1 0 0 0 0 0 0 0	0 0 0 0 0 0 0 0	0 0 0 0 0 0 0 0
0 0 0 0 0 0 0 0	0 0 0 0 0 0 0 0	0 0 0 0 0 0 0 0	0 0 0 0 0 0 0 0

Fig. 7.159 Examples of VWLC set selection matrices (16 example classes in 65 classes). 1–4: VWLC set index; 0: not transmitted [MC-12]. (© 1987 IEEE).

blocks can be classified based on positions and powers of significant coefficients in the block (*significant coefficient* means a coefficient with power large enough to be transmitted). The ac energy as defined in (7.13.1) does not consider the positional characteristics of significant coefficients. In [MC-12] the blocks are classified according to their coefficient power distributions (CPD). Each significant coefficient is uniformly quantized and VWLC is assigned to the quantization index. Adaptivity is provided by changing the VWLC set fitting the probability density function of each quantized coefficient (Fig. 7.159).

7.13.6 *Adaptive Segmentation of 2D DCT Blocks into Regions*

Another scheme proposed for activity classification is based on segmentation of 2D DCT of image blocks into regions such that the statistics for each region are almost stationary [ACTC-6]. The regions are segmented such that the magnitudes of the transform coefficients in each region are within a threshold set for that region. Both the number of regions and the thresholds can be varied. An example of block segmentation is shown in Fig. 7.160.

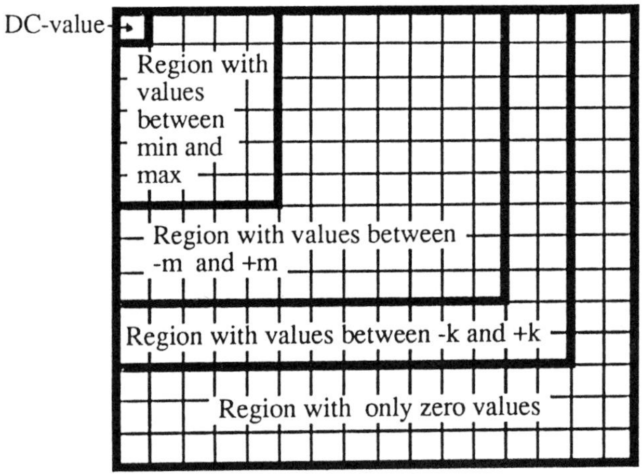

DC-value
Region with values between min and max
Region with values between -m and +m
Region with values between -k and +k
Region with only zero values

Fig. 7.160 Adaptive segmentation of 2D 16 × 16 DCT of image blocks into regions [ACTC-6].

7.13.7 *Activity Classification According to the Dominant Structure*

An activity classification based on the dominant structure has been incorporated in MC DPCM/DCT coding for developing teleconferencing codecs at rates of 64–320 KBPS [LBR-13]. In this scheme the total ac energy in each region (Fig. 7.161) is used to classify the block (horizontal, vertical, and diagonal structures). This classification followed by further subclasses in each of these three regions, along with object matching, has effectively responded to translation, panning, and zooming of video, resulting in improved subjective quality

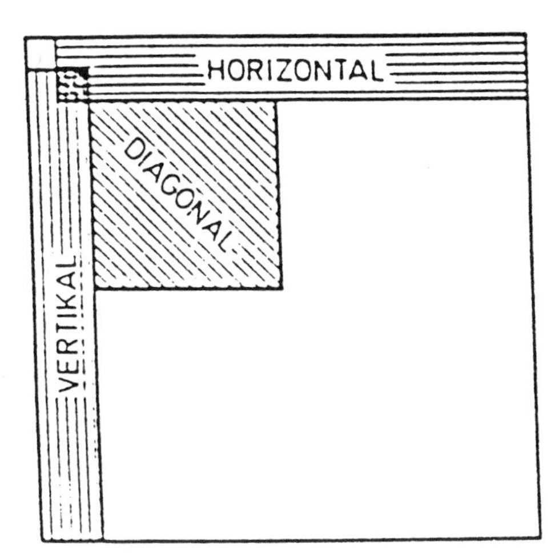

Fig. 7.161 Regions of classification according to the dominant structure in the 2D 16 × 16 DCT blocks [LBR-13].

of the processed images. This structural information can be observed from the 2D 8 × 8 DCT basis images shown in Fig. 7.25. A similar classification (Fig. 7.92) proposed in [DV-42] has been utilized in classified transform VQ coding of images (Fig. 7.89). Another application of this structural classification is in HDTV image coding [HDTV-1].

7.13.8 *Activity Classification Based on Maximum Magnitude of AC Coefficient*

In the adaptive interframe (3D DCT or 2D DCT/DPCM) and intraframe (2D DCT) HDTV coding [HDTV-10], the 3D blocks are classified based on their ac coefficients. Block activity is determined by

$$\max|X^{C(2)}(u, v, \ell)| \qquad (u, v, \ell) \neq (0, 0, 0) \tag{7.13.9}$$

in a block. Based on this criterion the blocks are divided into four classes. Both the weighting function and the step size of the uniform quantizer for the ac coefficients are determined by the block classification. For the 2D DCT blocks, a criteria similar to (7.13.9) can be used. By multilevel thresholding the number of classes can be chosen as, for example, 4 or 8.

7.13.9 *Activity Classification Based on Horizontal, Vertical, and Total AC Energies*

In the 2D 16 × 16 DCT of MC interframe prediction errors applied to a 64 KBPS video codec [MC-36], the vector composed of the absolute values of the 135 ac coefficients (Fig. 7.16.2) on and above the minor diagonal (the rest are discarded) undergoes a VQ. (The signs, i.e., positive or negative, of the

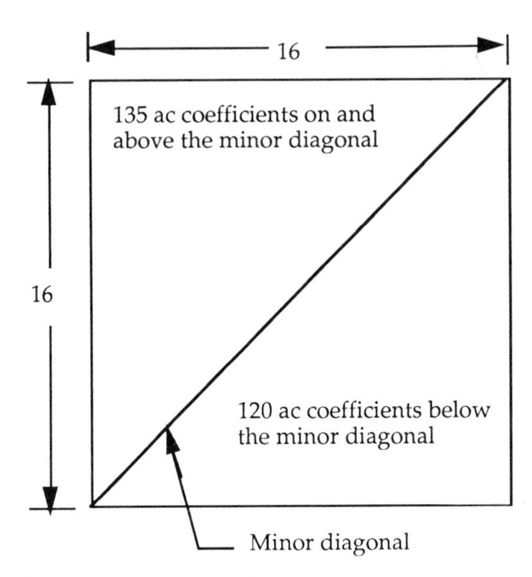

Fig. 7.162 Division of ac (2D DCT) coefficients above and below the minor diagonal.

coefficients are coded separately.) Adaptive coding is introduced by classifying the 16×16 block as belonging to one of eight classes based on horizontal energy E_H, vertical energy E_V, and total ac energy (Fig. 7.163). These parameters represent the horizontal structure, vertical structure, and block detail, respectively (Fig. 7.164). Based on this classification, separate codebooks, each of size 128, are developed.

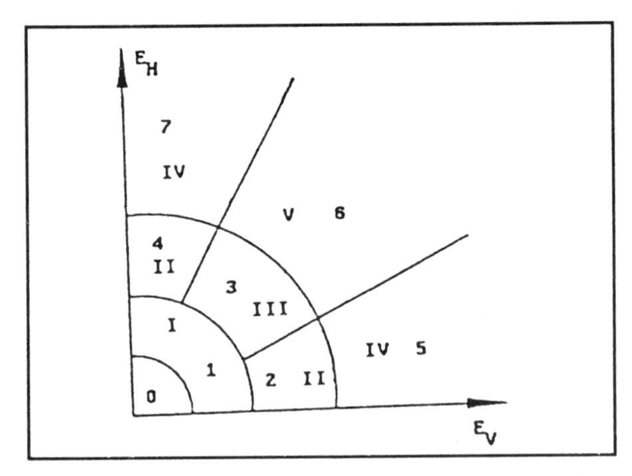

Fig. 7.163 Classification for codebook selection. I . . . V = codebook number; 0 . . . 7 = VQ class number [MC-36]. (© 1987 IEEE).

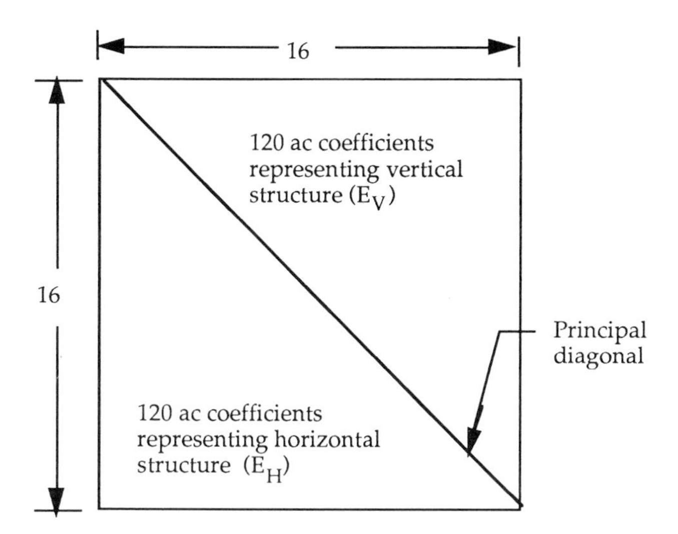

Fig. 7.164 Classification scheme representing horizontal structure, vertical structure, and block detail [MC-36]. (© 1987 IEEE).

7.14 HVS in Transform Coding

Human visual sensitivity (HVS) has been incorporated in transform coding of images by several researchers. Mannos and Sakrison's work [HTC-1] may be the first major breakthrough in image coding incorporating the HVS. Using the assumption that the HVS is isotropic, they modelled the HVS as a nonlinear point transformation followed by the modulation transfer function (MTF) of the form

$$H(f) = a(b + cf)\exp(-cf)^{d}, \tag{7.14.1}$$

where f is the radial frequency in cycles/degree of visual angle subtended and a, b, c, and d are constants. By varying these constants, the shape and the peak frequency of the MTF are altered. After carrying out extensive experiments, they have arrived at the following HVS model:

$$H(f) = 2.6(0.192 + 0.114f)\exp[-(0.114f)^{1.1}]. \tag{7.14.2}$$

This MTF has a peak at $f = 8$ cycles/degree.

Incorporation of HVS in the transform domain implied multiplying the transform (DFT) coefficients by $H(f)$ at the corresponding frequencies. Earlier research in transform coding incorporating the HVS applied the DFT to the entire frame [HTC-1]. This was followed by applying the DCT to small blocks [IC-10, HTC-2, HTC-3, PIT-10, PIT-2]. Most researchers (except [PIT-2, PIT-10]) used Mannos and Sakrison's transfer function (7.14.2) as the weighting function. The nonlinear function is ignored in [HTC-2, HTC-3, PIT-2].

Recently new MTFs have been proposed for using with the DCT. Nill [HTC-4] has shown that in order to use the HVS-DCT in image coding, the original image has to be even extended. This change, however, causes the loss of physical significance, as the viewer is not observing this altered image. To overcome this problem, Nill [HTC-4] proposed that the MTF be multiplied by the weighting function $A(f)$ defined as follows:

$$|A(f)| = \left(\frac{1}{4} + \frac{1}{\pi^2}\left[\log_e\left(\frac{2\pi f}{\alpha} + \sqrt{\frac{4\pi^2 f^2}{\alpha^2} + 1}\right)\right]^2\right)^{1/2}, \qquad (7.14.3)$$

where α has a typical value of 11.636/degree. He has also proposed the following MTF as a good working representation of the HVS:

$$H(f) = (0.2 + 0.45f)\exp(-0.18f). \qquad (7.14.4)$$

This function has a peak value at a spatial frequency around 5 cycles per degree.

Ngan *et al.* [HTC-5] used Nill's multiplicative function $A(f)$ with their MTF. It has been found in this work that the peak frequency of the transfer function of

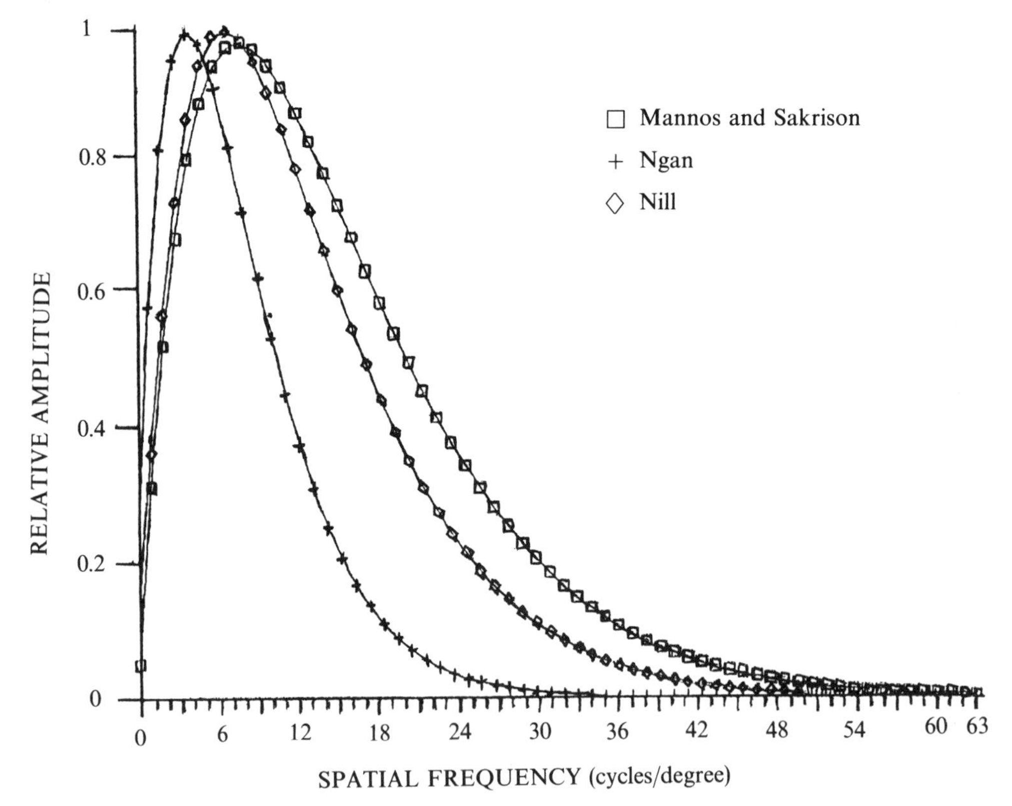

Fig. 7.165 Comparison of the MTFs [PIT-12]. (© 1989 IEEE).

3 cycles per degree gives the best results. The transfer function takes the form of

$$H(f) = (0.31 + 0.69f)\exp(-0.29f). \tag{7.14.5}$$

After multiplying $H(f)$ by $A(f)$, the resulting function has the peak frequency around 4 cycles per degree. Using a zig-zag scanning sequence (Fig. 7.174) for the DCT coefficients, they achieved an acceptable reconstructed image at the bit rates of 0.2–0.3 BPP depending on the images. The MTFs of Mannos and Sakrison [HTC-1], Nill [HTC-4], and Ngan [HTC-5] are shown in Fig. 7.165. Note that the latter two functions are obtained after multiplying (7.14.4) and (7.14.5) by $A(f)$. Based on extensive experiments, Chitraprasert and Rao [PIT-16] developed the following MTF (this has a peak at 3.75 cycles/degree), which yields good results for a variety of test images.

$$H(f) = 2.46(0.1 + 0.25f)\exp(-0.25f). \tag{7.14.6}$$

Using the convolution-multiplication property of the DCT-II [F-7] for a sampling density of 64 pels/degree (this corresponds to a viewing distance at which 64 pels are subtended in 1 degree of vision), the HVS weighting function based on (7.14.6) for an 8×8 matrix is developed (Fig. 7.166). The 8×8 2D-DCT coefficients are multiplied by the corresponding elements of this matrix, resulting in their HVS weighting. For another development of the HVS weighting matrix, the reader is referred to [HTC-3].

0.4942	1.0000	0.7023	0.3814	0.1856	0.0849	0.0374	0.0160
1.0000	0.4549	0.3085	0.1706	0.0845	0.0392	0.0174	0.0075
0.7023	0.3085	0.2139	0.1244	0.0645	0.0311	0.0142	0.0063
0.3814	0.1706	0.1244	0.0771	0.0425	0.0215	0.0103	0.0047
0.1856	0.0845	0.0645	0.0425	0.0246	0.0133	0.0067	0.0032
0.0849	0.0392	0.0311	0.0215	0.0133	0.0075	0.0040	0.0020
0.0374	0.0174	0.0142	0.0103	0.0067	0.0040	0.0022	0.0011
0.0160	0.0075	0.0063	0.0047	0.0032	0.0020	0.0011	0.0006

Fig. 7.166 The HVS weighting function, $H_F(k, m)$ based on a sampling density of 64 pels/degree [PIT-16]. (© 1988 IEEE).

7.15 Data Compression

It may be recalled from Section 6.2 that the transform coefficients with large variances can be retained as significant features for purposes of pattern recognition or classification. Also, either by discarding the transform coefficients with small variances and/or quantizing the coefficients based on their variances (Problem 7.10), a signal can be almost fully recovered. Data compression is an important application of transform coding when retrieval of a signal from a large database is required. Significant transform coefficients are stored in a central database, and a specific signal can be nearly reconstructed by inverse transform of these coefficients. Compression ratios on the order of 10:1 can be obtained depending on the required fidelity of the retrieved signal.

It was shown in Section 3.2 that the mse between the original signal and the reconstructed signal based on a truncated set of the transform coefficients (for the KLT) is equal to the sum of the eigenvalues of the auto-covariance matrix corresponding to the discarded coefficients [see (3.2.13)]. Hence it is logical to retain those transform coefficients (for the KLT) having large variances so that the mse is minimized when a specified number of transform coefficients are set to zero. This concept is adapted for all the other discrete transforms, even though the auto-covariance matrix is not diagonalized (see Section 6.4). As pointed out earlier, because of the statistical dependency of the KLT (apart from the implementation problems), orthogonal transforms other than KLT, although suboptimal, have been used in many aspects of digital signal processing, including data compression. As signal energy tends to be compacted into a few transform coefficients (see Chapter 8 in [B-12]), storage and/or transmission of large databases is much more efficient (reduced storage and/or reduced bandwidth) in the transform domain than in the original data domain. Needless to say, some information is invariably lost. However, its impact can be minimized by adaptive processing. This involves perceptual (such as auditory or visual) considerations. An application of the DCT relating to data compression is in the storage and retrieval of a biomedical signal such as electrocardiogram (ECG) [DC-1] (Fig. 7.167) and vectorcardiogram (VCG) data [DC-2] (Fig.

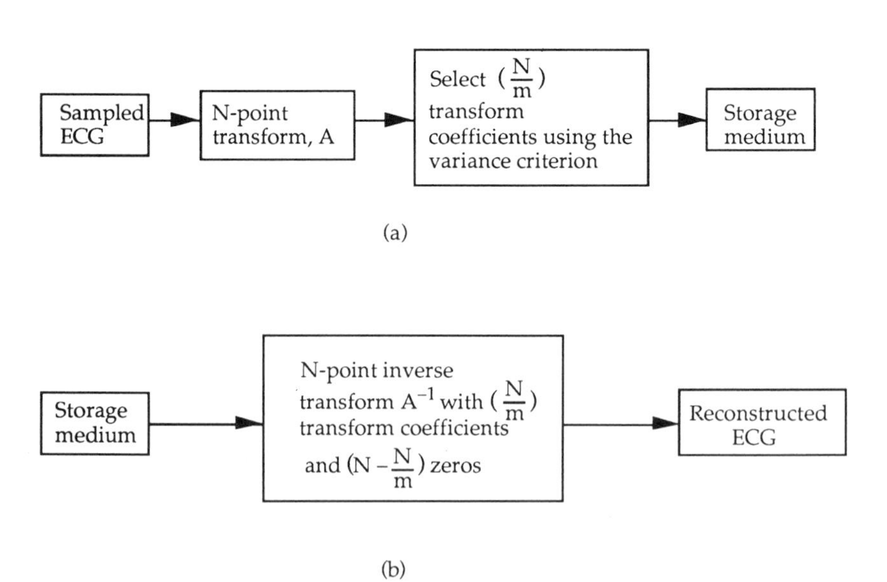

Fig. 7.167 Pertaining to an m:1 data compression. (a) storage; (b) retrieval [DC-1]. (© 1975 IEEE).

7.168). Using a large number of canine ECGs, Ahmed, Milne, and Harris [DC-2] have shown that the reconstructed ECGs based on 3:1 compression retain the significant features for proper diagnosis by a radiologist. Similar investigation based on human VCG data has shown that compression ratios in the range of 3:1 to 5:1 can be obtained by using the 2D DCT (Fig. 7.168). For the 2D DCT, either discarding the low variance coefficients or setting them to their means made no discernable difference in the fidelity of recovered VCGs for compression ratios of up to 6:1. A VCG can indicate problem areas such as chronic obstructive pulmonary disease (COPD) and myocardial infarction (MI).

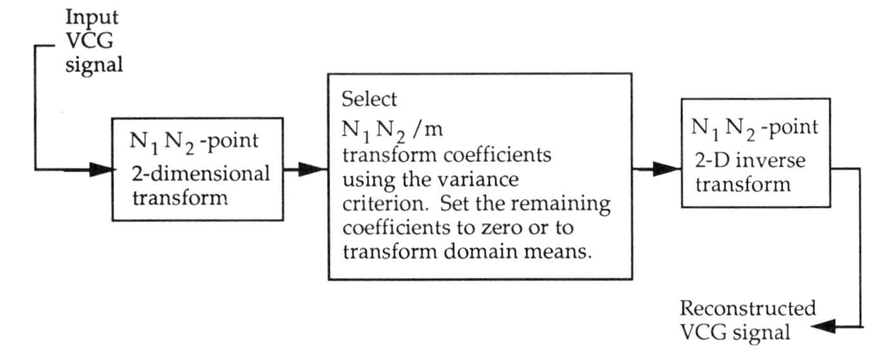

Fig. 7.168 Model for $m:1$ compression of VCG data [DC-2]. (© 1985 IEEE).

7.16 Classification

Whereas signal reconstruction from reduced but relevant information is the objective of transform data compression, the concept of classification is to reduce the dimensionality of feature or decision space such that a signal can be classified with some degree of certainty based on the significant features of the signal. The decision space and classification can be implemented in the transform domain (more specifically, in the DCT domain). Such an investigation has been carried out by Dyer, Ahmed, and Hummels [C-1] for classifying human vectorcardiograms (VCGs) based on transform coefficients with large variances (variance criterion) (Fig. 7.169). Based on a large traning set, the $N-M$ transform coefficients (these have large variances) of each class are stored in the feature space. Corresponding transform coefficients of a test signal are compared with the coefficients of each class, and the test signal is identified with the class with which it has the minimum distance. An example of distance measure is the mse (7.9.1) ($L2$ norm) between the test signal and a member of the class in the feature space based on the corresponding $N-M$ transform

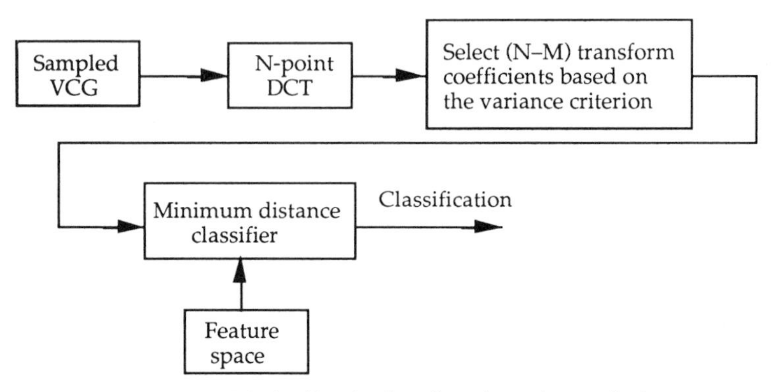

Fig. 7.169 VCG classification based on the variance criterion.

coefficients, i.e.,

$$\sum_{k=0}^{N-M-1} [X_{\mathrm{t}}^{\mathrm{C(2)}}(k) - X_{\mathrm{cl}(i)}^{\mathrm{C(2)}}(k)]^2. \tag{7.16.1}$$

In (7.16.1) the subscripts t and cl(i) refer to test signal and class i in the feature space, respectively.

7.17 Surface Texture Analysis

Mulvaney, Newland, and Gill [STA-1] have compared various discrete transforms such as the DFT, DCT, $(\mathrm{WHT})_{\mathrm{W}}$, HT, and BIFORE transforms [B-1, B-2] based on their ability to characterize surface texture data. $(\mathrm{WHT})_{\mathrm{W}}$ and BIFORE stand for Walsh- or sequency ordered Walsh–Hadamard transform (WHT) and binary Fourier representation. BIFORE is obtained by rearranging the basis functions of $(\mathrm{WHT})_{\mathrm{W}}$. It is also called the Hadamard or natural ordered WHT, i.e., $(\mathrm{WHT})_{\mathrm{h}}$. BIFORE has a compacted power spectrum. The comparison of the discrete transforms is based on two criteria: a) subjective: inspection of the spectral estimates of all the machined specimens; b) analytical: rate of convergence of the spectral estimates, which is important in assessing the amount of data required in order to characterize the surface texture band being investigated. They have concluded that both the DCT and DFT can be utilized in surface texture analysis because of their rapid rates of convergence and also their ability to characterize the data and, in particular, the machining peaks.

7.18 Topographic Classification

Watson, Laffey, and Haralick [PCL-1] have used generalized splines and DCT for describing the topographical primal sketch of digital image intensity surfaces. The major topographic categories are peak, pit, ridge, ravine, saddle, flat, and hillside, with inflection point, slope, convex hill, concave hill, and saddle hill as hillside subcategories. This topographic classification is useful in computer vision systems. The authors conclude that the performance of the generalized splines and DCT is similar and suggest that further research with regard to better basis functions, use of fitting error, solving the ridge continuum problem, and grouping of the topographic structures is needed.

7.19 Photovideotex

In Section 7.21 progressive transmission of images over low data-rate channels is discussed. The objectives in this case are early recognition of an image for either rejection or subsequent build-up to almost its original fidelity, and also reduced storage (memory) requirements in the central database. In the videotex and teletext configuration, both the memory and transmission time are reduced by processing the images and/or text, based on data compression techniques [P-1, P-2]. In this interactive two-way information retrieval system, a host of data, such as newspaper pictures, still frames, office documents, text, etc., can access text and pictorial information from central databases via narrow bandwidth digital links such as the telephone network or the ISDN B channel (64 KBPS).

One possible scenario [P-2] is for the subscriber to log on to the videotex computer and request page(s) of information via the telephone network. Passwords, identity codes, etc. will be built into the system to preclude unauthorized use. The videotex computer not only has a large database stored in the compressed form, but also processes the subscriber's request, and retrieves and transmits the desired information along with the adaptor identity code. Whereas the subscriber to the videotex computer link is through the ordinary telephone lines, the return communication medium can be a television channel (fiber, microwave, satellite) so that the videotex pages are transmitted and displayed almost immediately at the subscriber's monitor (Fig. 7.170). The adaptor's function includes decoding, storage, and high transfer rate for display on the monitor. The basic system can be supplemented with various peripherals (VTR, printers, floppy disk drives, etc.) to further enhance the capabilities of this service. Bit-rate reduction is obtained by 2D DCT of image blocks, followed by their ac energy (7.13.1) related classification, variance-dependent bit allocation, quantization, and coding (Fig. 7.171). This technique was originally proposed by Chen [LBR-3] and Chen and Pratt [LBR-5]. For proper decoding, overhead information regarding the bit allocation matrices, classification maps, and transform coefficient variances (overhead bits for the variances can be reduced by dynamically estimating these variances [IC-20]) has to be transmitted. Simulation results presented by Ngan, Hui, and Lim [P-2] indicate that, depending on the images, compression ratios from 4:1 through 13:1 can be achieved in the videotex arena.

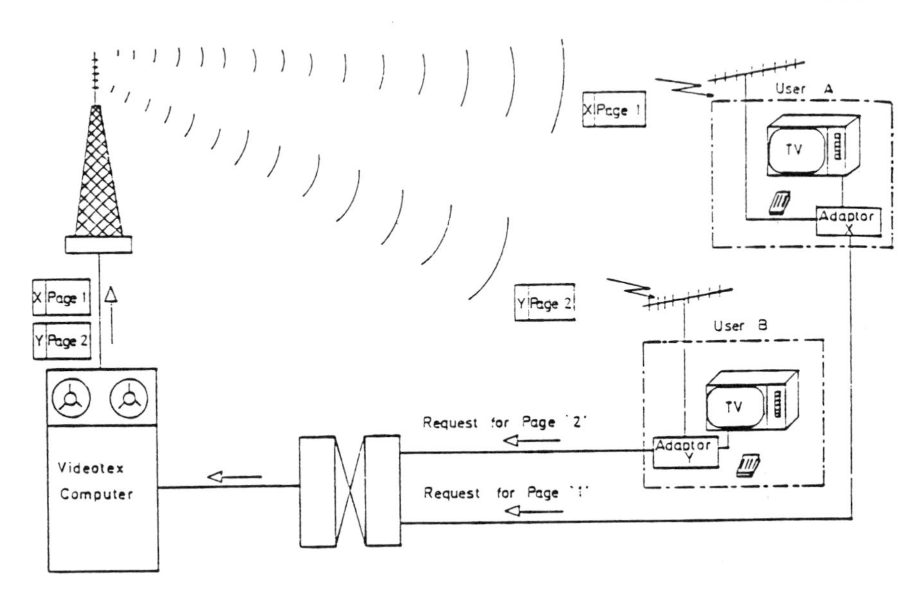

Fig. 7.170 Typical configuration of the videotex system [P-2]. (© 1985 IEEE).

In the ISDN environment, the two-way communication (transmit and receive) is over the integrated network, and the concept of compression plays a vital role in reducing the memory for storage and the time for transmission of data (audio, video, text, etc.). In the integrated visual communication system, DCT has been proposed, investigated, and recommended as the primary compression device for teleconferencing and videophone ($p \times 64$ KBPS, $p = 1, 2, \ldots, 30$) [LBR-23, LBR-35], and for still-image communication services [PIT-13, PIT-14, PIT-17, PIT-18, PIT-20, PIT-21, PIT-22, PIT-24 through PIT-27, PIT-30], and is one of the possible candidates for broadcast-quality TV at the DS-3 level (44.736 MBPS) [IC-53, IC-67]. Also, the CMTT/2-DCT group (CMTT is a joint CCIR/CCITT study group) has presented detailed specifications [IC-75, IC-76, IC-78] for digital transmission of TV signals at 30–34 MBPS and 45 MBPS with the 2D 8×8 DCT of component signals (Y, C_R, and C_B) as the main compression tool. Several adaptive features such as intrafield, interfield, and MC interframe coding; switched quantizers; semiglobal or set of motion vectors; different scanning patterns for the luminance and chrominance components; multiple VWL codes; etc. are introduced to meet the distribution and contribution applications of the 4:2:2 signal CCIR recommendation 601 [G-21]. CMTT stands for Committee for Mixed Telephone and Television. Several companies, such as Picturetel, Video Telecom. Corp, and Compression Labs, have developed and marketed DCT-based teleconferencing codecs.

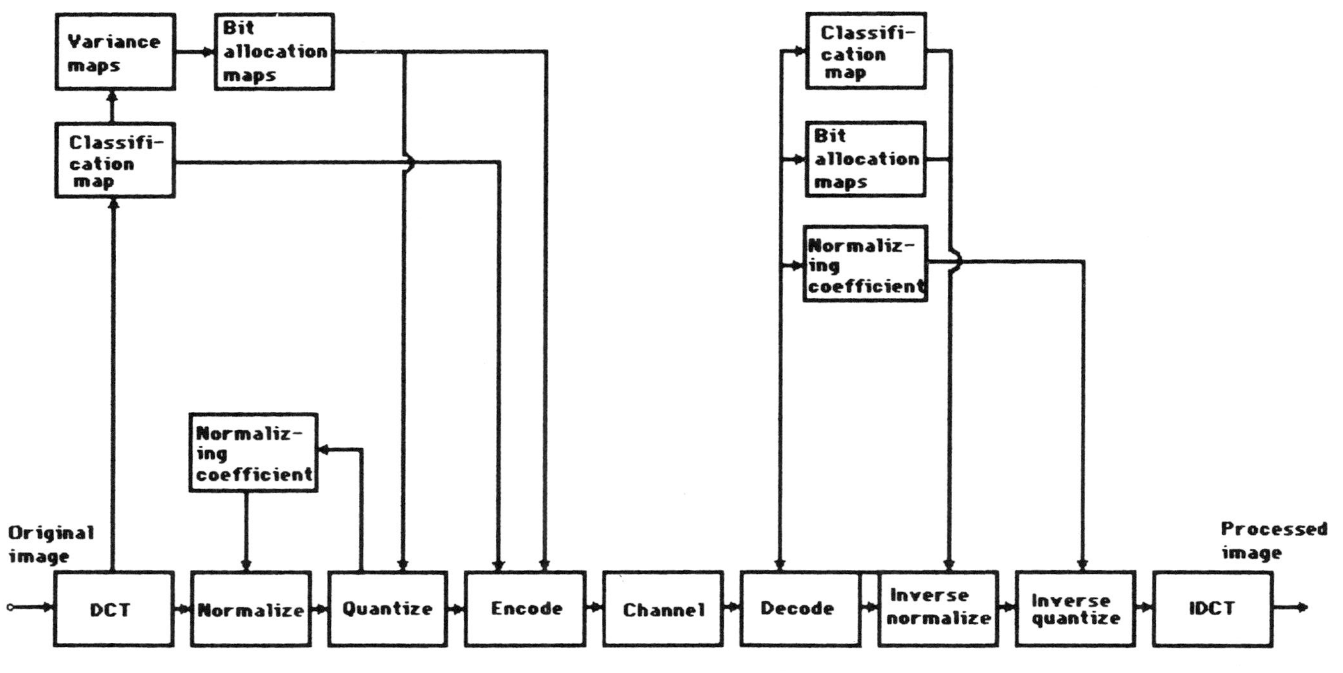

Fig. 7.171. Block diagram of adaptive transform coding system [P-2]. (© 1985 IEEE).

7.20 Pattern Recognition

In Section 7.16 classification of signals based on reduced dimensionality in the transform domain has been discussed. A pattern recognition experiment in which the features are transform coefficients rearranged as Mandala groups has been developed by Hsu et al. [PR-1]. The mandala transform can be defined based on any discrete transform. The Mandala/DCT is described in Fig. 7.172. As in image coding, the original image is divided into blocks of size $m \times n$ and each block is mapped into the transform domain. The coefficients of these blocks are rearranged as Mandala blocks (Fig. 7.172c) such that the top left block is composed of dc values of the transform coefficients, whereas the bottom right block constitutes the highest frequency content of the transform coefficients. Each of the remaining Mandala blocks has coefficients of a specific horizontal frequency and a specific vertical frequency, gathered from all the transform blocks. These blocks progress horizontally to the right and vertically to the bottom based on increasing horizontal and vertical spatial frequencies, respectively. Depending on the discrete transform used, this particular mapping is called Mandala/Fourier, Mandala/Walsh, Mandala/cosine, etc. transforms. The relationship between the Mandala blocks (Fig. 7.172c) and transforms blocks (Fig. 7.172b) is simply a sorting of the coefficients from one form to the other, whereas each transform block is a mapping of the corresponding spatial block into the transform domain. Common frequencies are grouped as Mandala blocks.

Using various features drawn from the Mandala blocks, Hsu *et al.* [PR-1] have simulated a target identification/pattern recognition experiment (based on the Mandala/DCT) that successfully discriminates between natural and manmade objects. This experiment includes ranking of the features, reducing the feature space, and implementation of different classifiers (statistical decision surfaces). Extending to a wide range of images, using different block sizes, developing and ordering the features, and applying various distance measures can constitute further research areas.

7.21 Progressive Image Transmission

Progressive transmission of images involves an approximate reconstruction of an image whose fidelity is built up gradually until the viewer decides either to abort the transmission sequence or to allow further reconstruction [PIT-11, PIT-31]. The applications are in transmitting images over low bandwidth channels such as telephone lines. If an image is sent in its original form in a raster scan fashion over narrowband channels, transmission of the entire image can take several minutes. Several schemes in which the image quality is built up hierarchically have been developed. Tzou [PIT-31] has presented an excellent review and comparison of techniques applied to progressive image transmission.

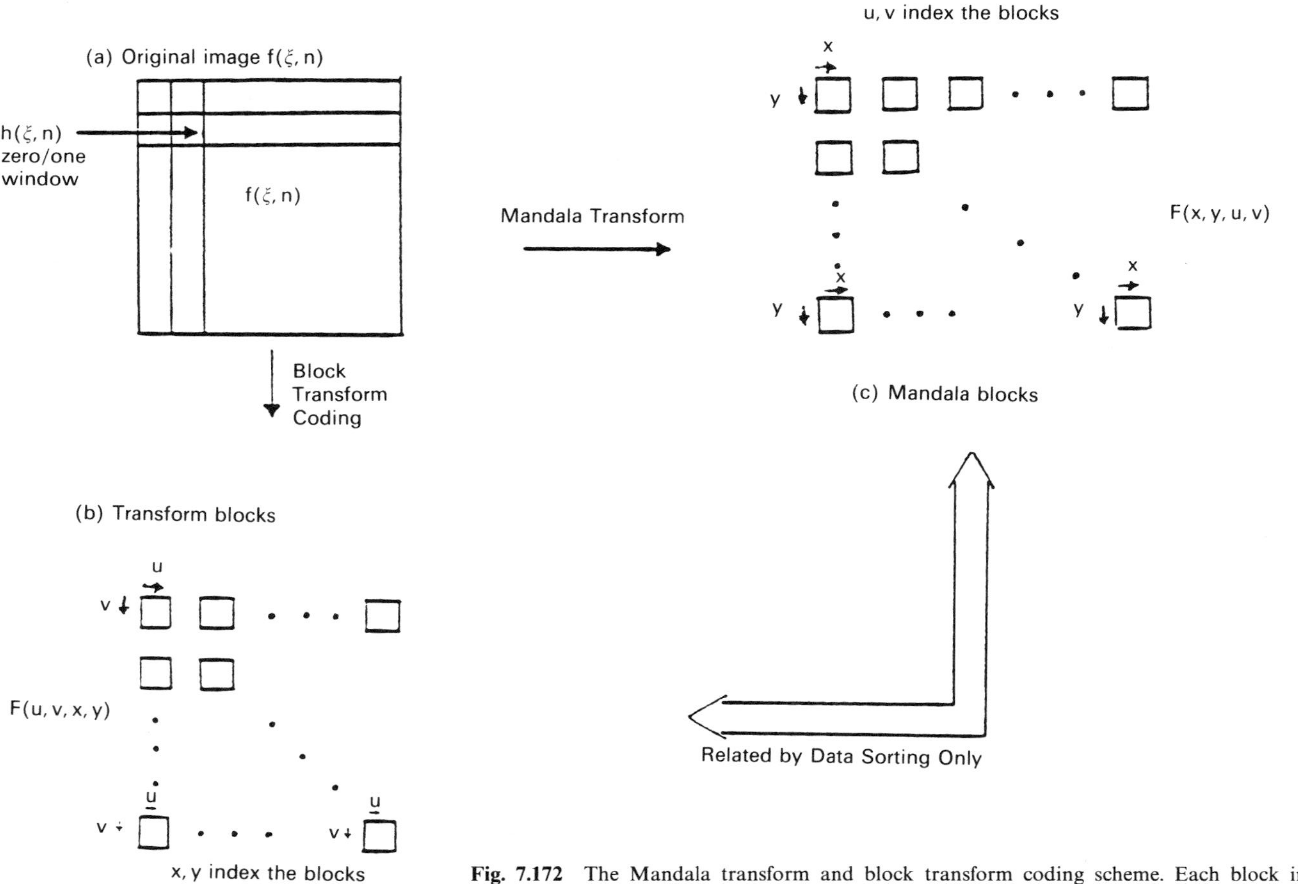

Fig. 7.172 The Mandala transform and block transform coding scheme. Each block in (b) represents the 2D transformation of the corresponding block in the spatial domain (a). Each Mandala block (c) is composed of the same coefficient from all the transform blocks (b) [PR-1]. (© 1983 IEEE).

The objective in all these schemes is to develop the significant features of an image at an early stage so that a viewer can interactively respond. Other considerations, such as the complexity of the technique and robustness to channel noise, play significant roles in the merits of a scheme.

Applications of progressive image transmission (PIT) include teleconferencing, telebrowsing, medical diagnostic imaging, videotex, security services, electronic shopping in mail order companies, and access to a large database [PIT-2, PIT-5, PIT-9, PIT-11]. Telebrowsing or videobrowsing is the system in which the recipient wishes to browse through remotely stored images and quickly abort transmission of the unwanted ones as soon as they are recognized. Progressive transmission can play an important role in picture archiving and communication systems (PACS) whose functions are transmitting, storing, processing, and displaying large data such as radiological images [PIT-9].

Progressive transmission of images can be classified into two categories as a) spatial or pel domain and b) transform or spectral domain. The latter has the advantage of information packing, and the image build-up can be achieved adaptively based on the significance of the transform coefficients. Furthermore, selection of coefficients can be altered by HVS weighting in the spectral domain.

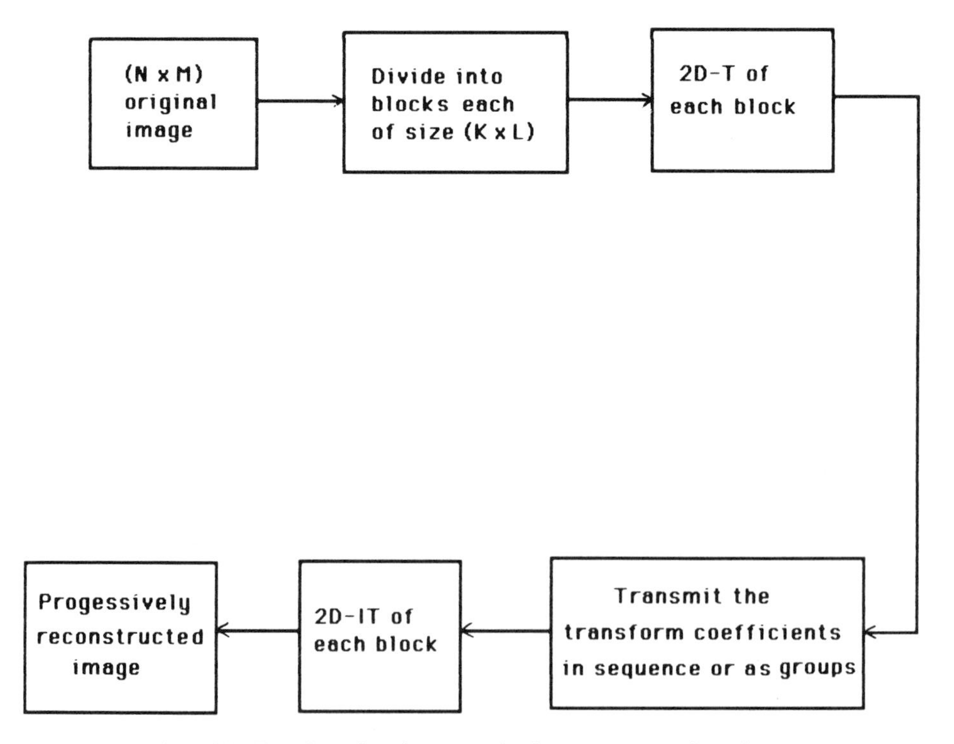

Fig. 7.173 Transform domain progressive image reconstruction scheme.

By either weighting or rearranging the transmission sequence of transform coefficients by human visual sensitivity, the image build-up can be perceptually pleasing, even at earlier stages [PIT-12, PIT-16].

Several schemes for the transform domain progressive transmission of images have been proposed. The discussion in this section will necessarily be limited to the PIT based on the DCT. In general an $N \times M$ image is divided into blocks, each of size $K \times L$. After mapping the blocks into the transform domain, the quantized transform coefficients are transmitted in some sequence (either adaptively, such as controlled by their variances, or in a fixed manner) to the receiver. The receiver reconstructs the image by inverse mapping groups of transform coefficients and successively adding to the image build-up based on the previous stages (Fig. 7.173). The details and fidelity of the image are successively refined with almost a lossless image at the final stage. There are, of course, a number of modifications to the original scheme, all of them with the objective of reducing the bit rate and improving the image quality during the intermediate stages.

Ngan [PIT-1] has proposed five schemes for the transmission sequence of transform coefficients based on: a) the mse the particular transform coefficient contributes to the reconstructed image, b) zig-zag scanning (Fig. 7.174), c) vertical scanning (left to right), d) horizontal scanning (top to bottom), and e) mix of vertical and horizontal scanning (Fig. 7.175). For the first scheme, the hierarchy map of each image has to be computed and transmitted as overhead to the receiver. Ngan [PIT-1] has shown that for normal images (smooth features), zig-zag scanning (Fig. 7.174) is quite effective, whereas for images with edge details, scheme (e) (Fig. 7.175) is superior. By slicing the full bit-assignment map into layers of incremental bit-assignment maps and sending the information corresponding to a slice of the full bit-assignment map at each stage (Fig. 7.176), Tzou and Elnahas [PIT-4, PIT-6] have shown that their approach is superior to zig-zag scanning during almost the entire course of progression.

Chen and Smith [ACTC-3, ACTC-4] have developed an adaptive coding of images based on equally populated classes (each block is classified according to its ac energy). A modified version of the scheme (Figs. 7.155–7.157) has been applied to the progressive transmission of radiological images by Elnahas *et al.* [PIT-8, PIT-9]. Subsets of transform coefficients are transmitted in a specific order (Fig. 7.177) for progressively reconstructing the images. The bit-assignment matrices and the transmission sequence for each image or for groups of images need to be stored at the receiver for orderly reconstruction. They have applied this technique to interactive browsing of computer tomography (CT) and magnetic resonance imaging (MRI) by a radiologist. The objective here is to abort the unwanted images at an early stage and to identify and build the details of the desired image such that a clinical diagnosis can be made.

By incorporating human visual system (HVS) weighting in the transmission sequence (hierarchy) of the transform (DCT-II) coefficients [PIT-12, PIT-16] (Fig. 7.178), it is shown that visually more pleasing images can be obtained, even

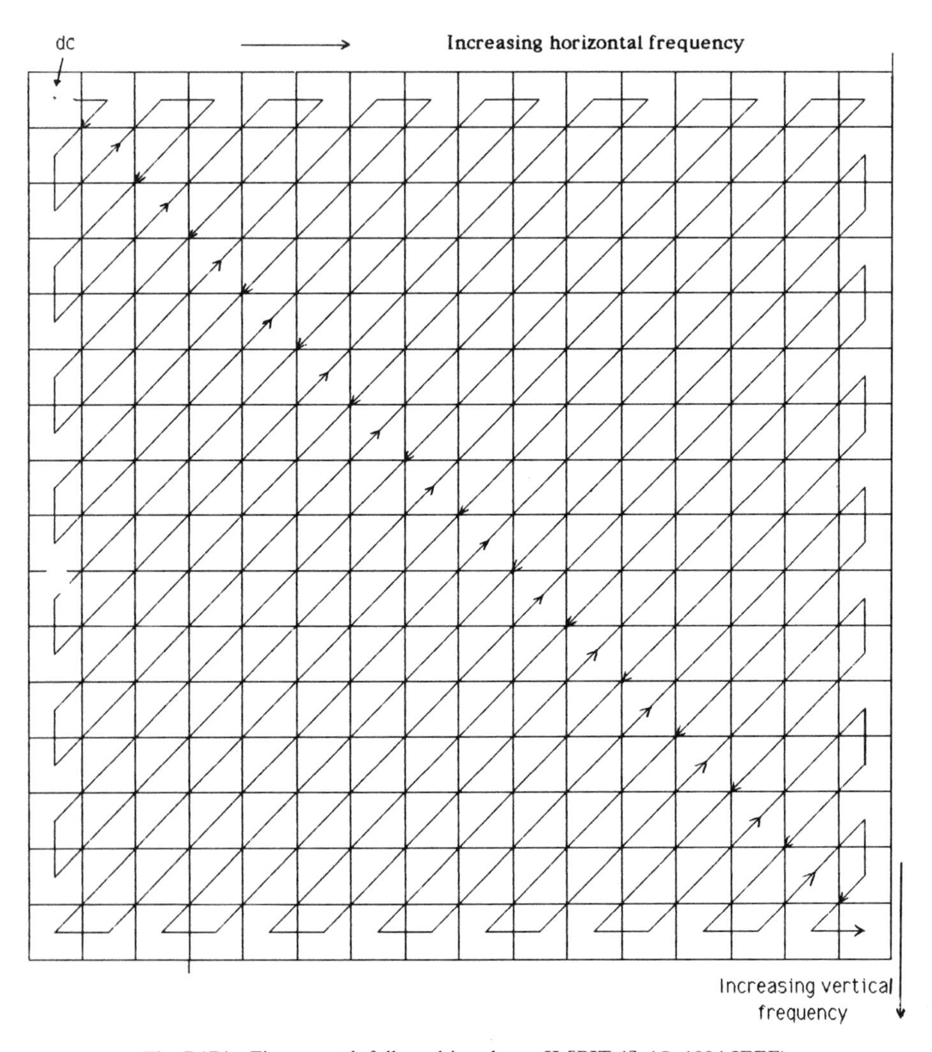

Fig. 7.174 Zig-zag path followed in scheme II [PIT-1]. (© 1984 IEEE).

at the early stages of reconstruction. After the 2D DCT of 8×8 blocks of an image, the blocks are classified into four equally populated classes based on their ac energies [ACTC-3, ACTC-4]. The dc coefficients of all blocks are transmitted first for the initial image build-up. Subsequent build-up, however, is dependent on the transmission sequence of the DCT coefficients controlled by their HVS weighted variances (Fig. 7.179). For comparison, the transmission sequence for the unweighted case is shown in Fig. 7.180. It can be seen that more ac coefficients of the lower activity classes of the HVS weighted method are sent and also they are sent earlier than those of the unweighted method. Figures

1	2	4	6	8	10	12	14
3	16	17	19	21	23	25	38
5	18	27	28	30	34	40	45
7	20	29	32	36	42	47	51
9	22	31	35	43	49	53	56
11	24	33	41	48	54	58	60
13	26	39	46	52	57	61	63
15	37	44	50	55	59	62	64

Fig. 7.175 Transmission sequence in scheme V [PIT-1]. (© 1984 IEEE).

7.181–7.184 illustrate the original image "Baboon" and the reconstructed images at various stages based on the unweighted and HVS weighted methods. It can be observed that the HVS scheme retains the sharp edges in the high-activity regions and smooths the noise in the low-activity areas. The underlying philosophy here is that by adapting the progressive transmission to the human eye, the visual perception of the image build-up can be substantially improved.

A novel technique for lossless coding applied to progressive image transmission has been developed by Wang and Goldberg [PIT-7, PIT-11]. By recoding the residual errors caused by the quantization of the transform coefficients on a stage-by-stage basis (Fig. 7.185), it is theoretically possible to recover exactly the original image. By using an entropy coder on the final residual error image, in practice, lossless recovery can be obtained with a small

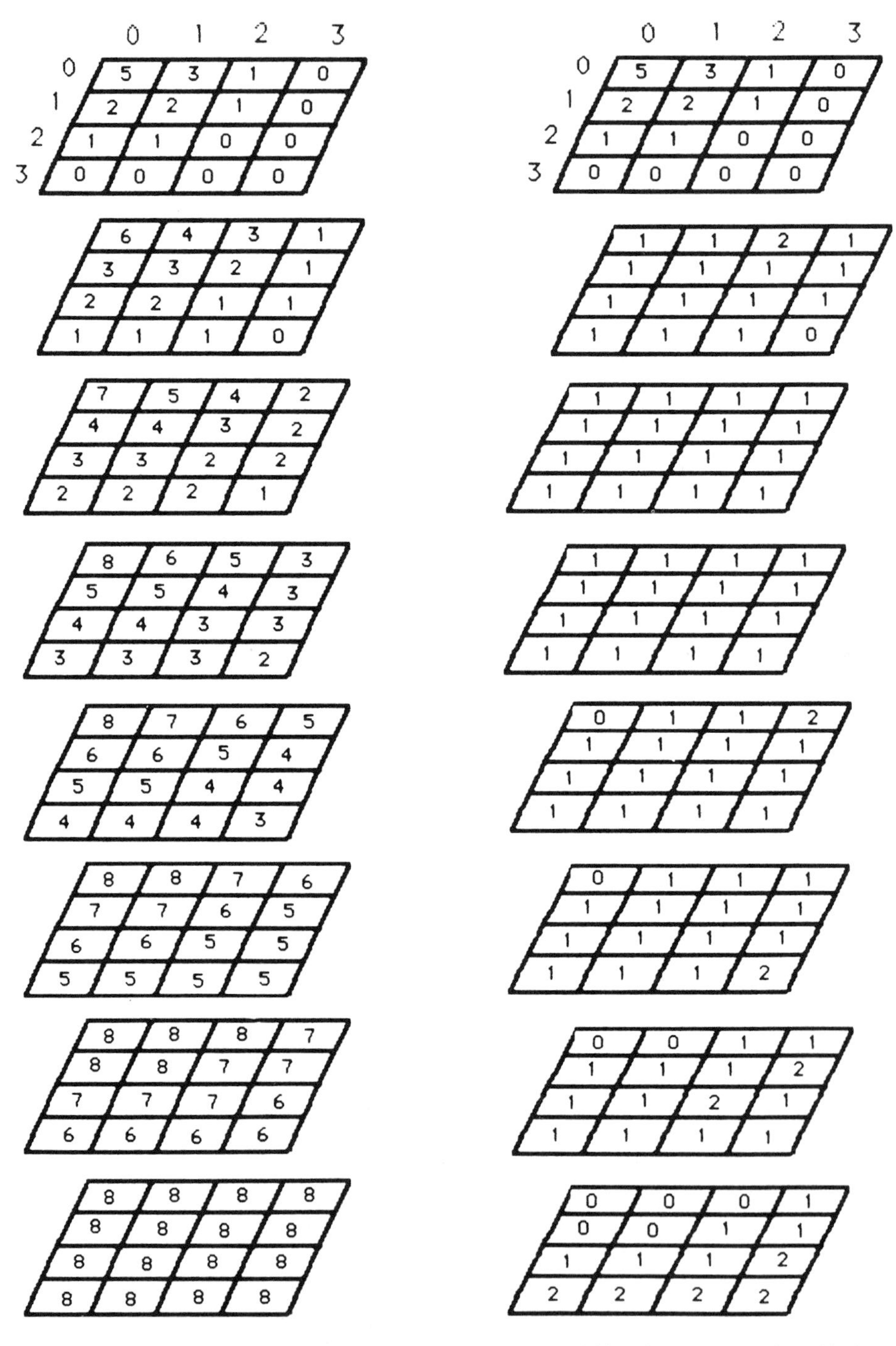

Fig. 7.176 An example of bit assignment maps and incremental bit assignment maps for a block size of 4×4 and an incremental bit rate of 1 bit/pixel-iteration [PIT-6]. (© 1986 IEEE).

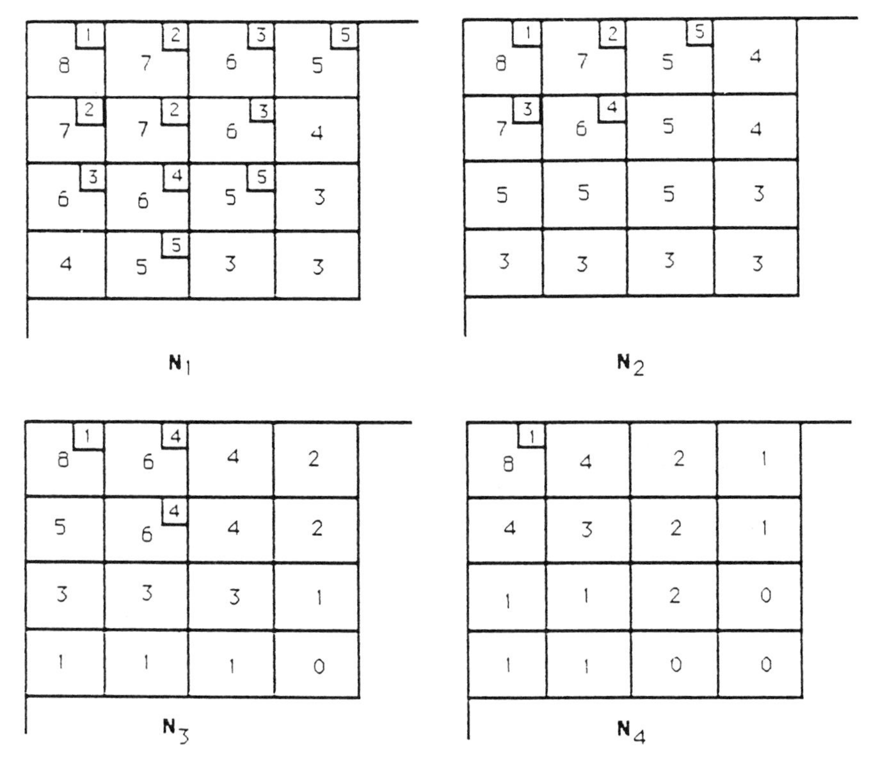

Fig. 7.177 Bit assignment matrices of four equally populated classes. (Numbers at the upper-right corners of some locations indicate the order of transmission for steps 1–5 in the progression) [PIT-8].

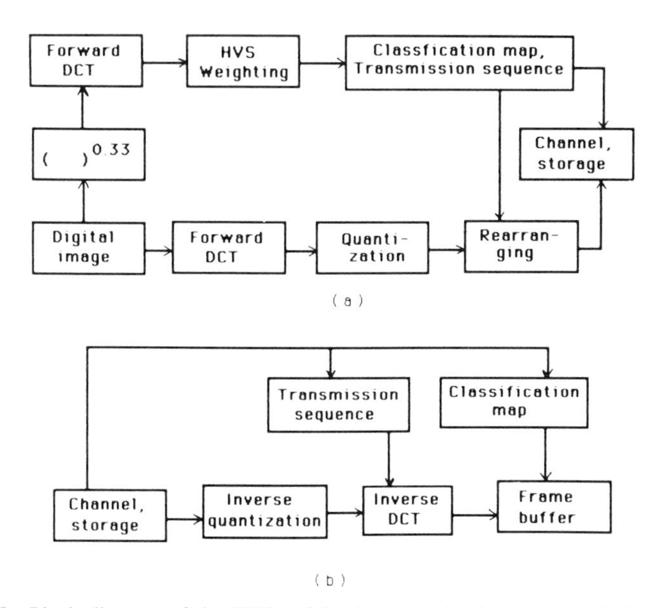

Fig. 7.178 Block diagram of the HVS weighted progressive image transmission system. The block denoted by $(\ )^{0.33}$ maps the image intensity to "brightness." (a) transmitter; (b) receiver [PIT-12, PIT-16]. (© 1990 IEEE).

CLASS 1				
	0	2	6	22
	1	9	20	
	3	17	27	
	13			
CLASS 2				
	0	4	11	26
	5	12	25	
	7	21		
	16			
CLASS 3				
	0	8	15	
	10	19		
	14			
	24			
CLASS 4				
	0	18	28	
	23			

Fig. 7.179 The ac coefficient transmission sequence map of the image "baboon," HVS weighted method. Only the first seven stages are shown [PIT-16]. (© 1990 IEEE).

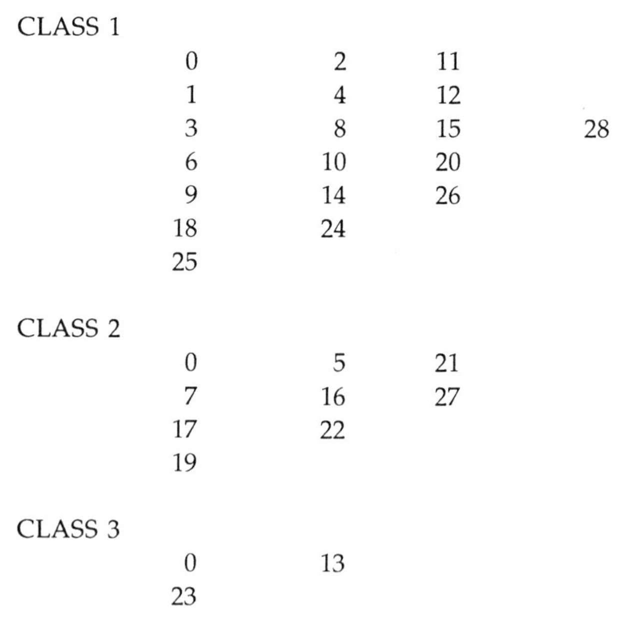

CLASS 1				
	0	2	11	
	1	4	12	
	3	8	15	28
	6	10	20	
	9	14	26	
	18	24		
	25			
CLASS 2				
	0	5	21	
	7	16	27	
	17	22		
	19			
CLASS 3				
	0	13		
	23			

Fig. 7.180 The ac coefficient transmission sequence map of the image "baboon". Only the first seven stages are shown [PIT-16]. (© 1990 IEEE).

Fig. 7.181 The original test image "baboon" [PIT-16].

number of iterations. The iterative transform coding scheme (Fig. 7.185) is further improved by applying VQ to the transform coefficients at each stage. Based on simulation of these techniques, Wang and Goldberg [PIT-11] have shown that the VQ is more efficient in removing the redundancy of the residual images compared with the scalar quantization with optimum bit allocation.

Lohscheller [PIT-2] has proposed a subjectively adaptive image transmission system in which the hierarchy of the transform coefficients is influenced by their visual perception thresholds. In this interactive scheme, the viewer can adaptively select a particular region of an image for further build-up and subsequent analysis such that a decision can be made for the entire image build-up or for aborting the image. So far the discussion on PIT has been limited to monochrome images. Several researchers have investigated various techniques for progressive transmission of color images. Specifically the Joint Photographic Experts Group (JPEG) formed from both the standards organizations, the International Telegraph and Telephone Consultative Committee (CCITT), and the International Standards Organizations (ISO) [PIT-18, PIT-20 through PIT-27, PIT-30, PIT-33 through PIT-36] has been charged with recommending the standards for transmitting facsimile, photovideotex, teletext, still-picture TV (teleconferencing, medical images, and surveillance), phototelegraphy (newspaper picture transmission), and photographic content of office documents. (A historical background towards this standardization is narrated in [PIT-24.) The

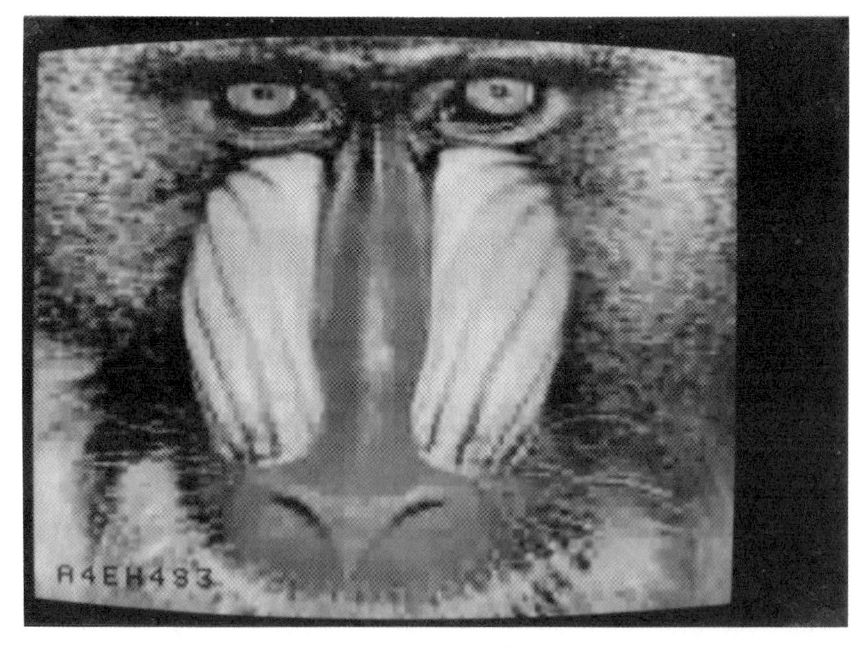

Fig. 7.182 Stage 3 of reconstruction of the "baboon" image. Unweighted method (upper) and HVS weighted method (lower). Bit rate: 1/4 bit/pel [PIT-16]. (© 1990 IEEE).

Fig. 7.183 Stage 5 of reconstruction of the "baboon" image. Unweighted method (upper) and HVS weighted method (lower). Bit rate: 3/8 bit/pel [PIT-16]. (© 1990 IEEE).

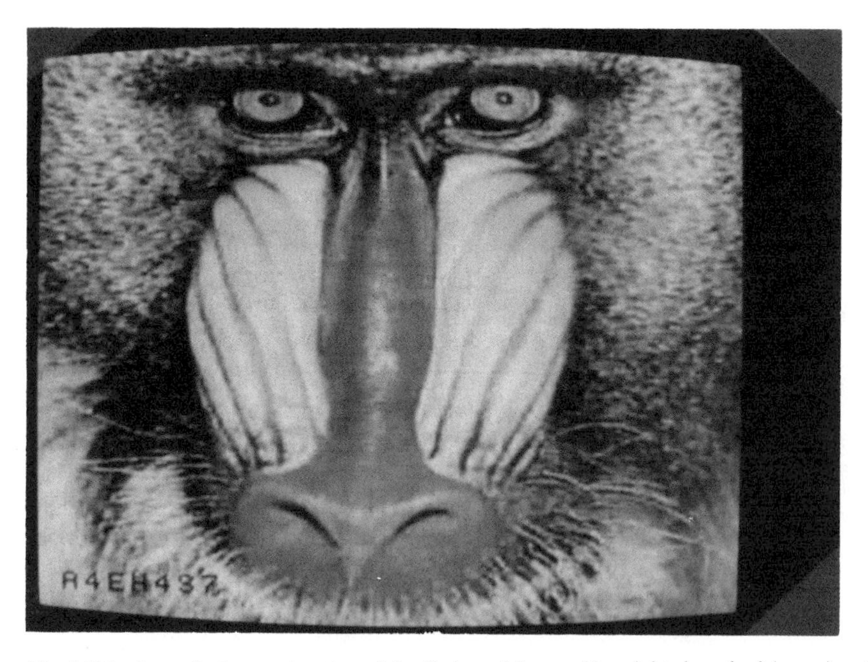

Fig. 7.184 Stage 5 of reconstruction of the "baboon" image. Unweighted method (upper) and HVS weighted method (lower). Bit rate: 1/2 bit/pel [PIT-16]. (© 1990 IEEE).

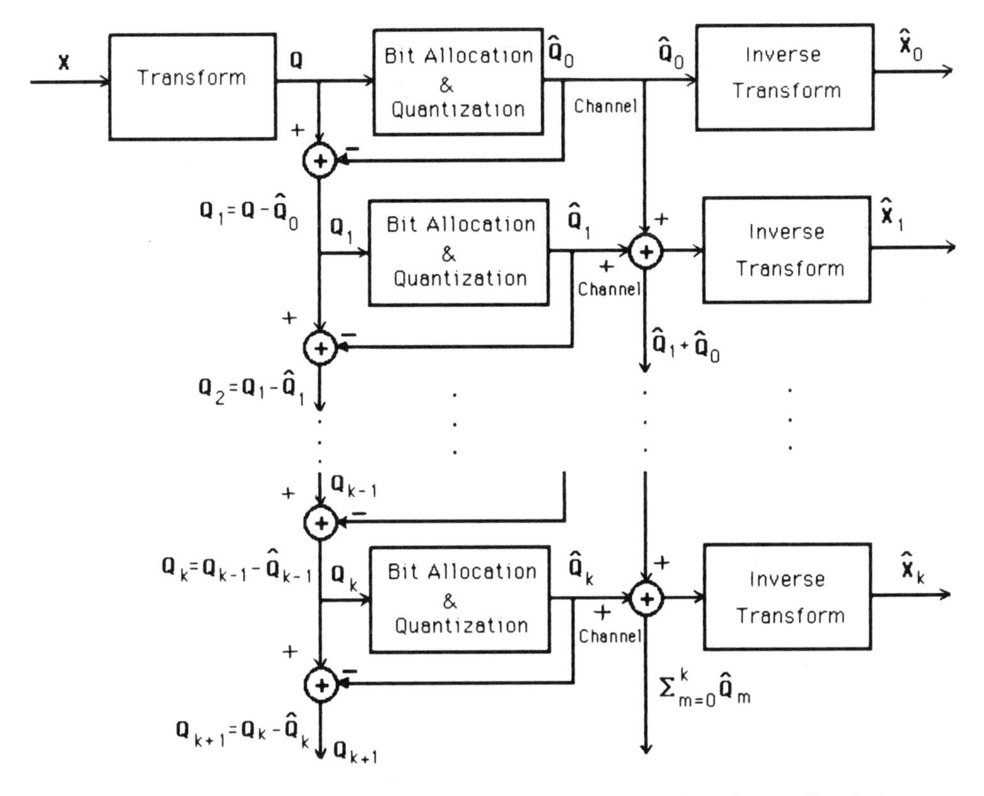

Fig. 7.185 Progressive image transform coding: As in conventional transform coding, the image X of size $N \times N$ first undergoes an orthogonal transform A; and then the transformed Q_{ij}, $i, j = 0, 1, \ldots, N-1$, are quantized. The quantized error matrix Q_1 is then considered as the input to the first stage and is quantized. The corresponding error matrix is then sent to the next stage. At the receiver, at a given stage k, the image is reconstructed by first forming the matrix $\Sigma_{m=0}^{k} \hat{Q}_m$, which sums up all the information received up to stage k. The coefficients are then inverse transformed to obtain $\hat{X}_k$, the kth approximation of the original image [PIT-7]. (© 1986 IEEE).

color image data compression algorithm being recommended by JPEG consists of three basic functional parts [PIT-35, PIT-36]: (i) a "lossy" baseline sequential system which utilizes a 2D DCT on (8×8) image blocks. This is followed by the prediction of the dc term. The dc prediction error, AC coefficients (zig-zag scanning) and run lengths of zero coefficients are Huffman coded. (ii) The baseline system is enhanced (increased precision of input and output data, and of DCT coefficients) in the extended sequential system in which an arithmetic coder (Q-coder or MEL-coder) is used. This enhancement (better compression, greater precision, and progressive coding) allows a wider range of applications such as color printers, grayscale and color scanners, and G4 facsimile machines. (iii) An independent function for providing high performance, reversible, sequential data compression which may be specially applicable to the publishing and medical fields.

An organizational chart formed from both CCITT and ISO for recommending appropriate standards for coding of still and moving images is shown in Fig. 7.186 [PIT-27]. The objective is to progressively build up multilevel monochrome and color pictures at resolutions from 256×256 through 2048×2048,

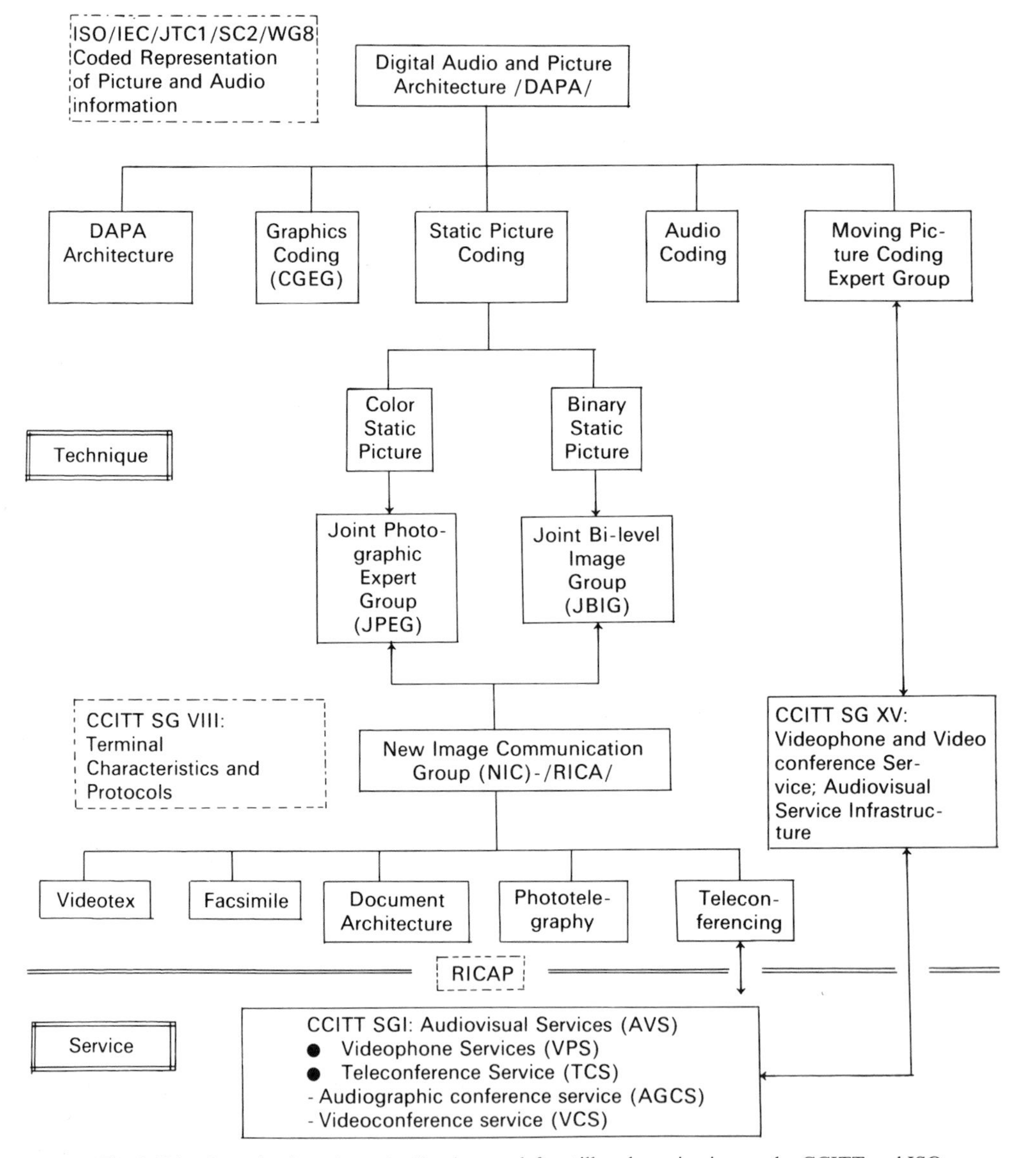

Fig. 7.186 Organization of standardization work for still and moving images by CCITT and ISO [PIT-27]. RICA = raster image communication architecture; RICAPS = raster image communication application profiles.

with the capability for eventual lossless recovery, as required in some applications such as medical imaging and newspaper and surveillance pictures. The specifications require that a recognizable picture be realized at an early stage (this is helpful for image transmission over low data-rate channels, etc., 1200 BPS on PSTN), with the indistinguishable (reversible) picture at the final stage. The codec should also have the capability for single-stage sequential build-up, which is desirable for picture transmission between computers or from a computer to a hardcopy device. Detailed requirements of the codec are described in [PIT-25]. Although several algorithms, such as adaptive block truncation coding (ABTC), recursive binary nesting (RBN), adaptive binary arithmetic coding (ABAC), ADPCM, high-correlation transform (HCT), low-correlation transform (LCT), VQ, VQ using peano scan (VQPS), color VQ (col VQ), ADCT, progressive coding scheme (PCS), and hierarchical predictive coding (HPC), have been proposed and extensively investigated, three algorithms—ADCT, ABAC, and BSPC (block-separated component progressive coding)—were selected for further study. BSPC is based on generalized BTC and PCS. These three techniques are of similar complexity and can meet all the mandatory technical features specified by the JPEG [PIT-14]. The ADCT,

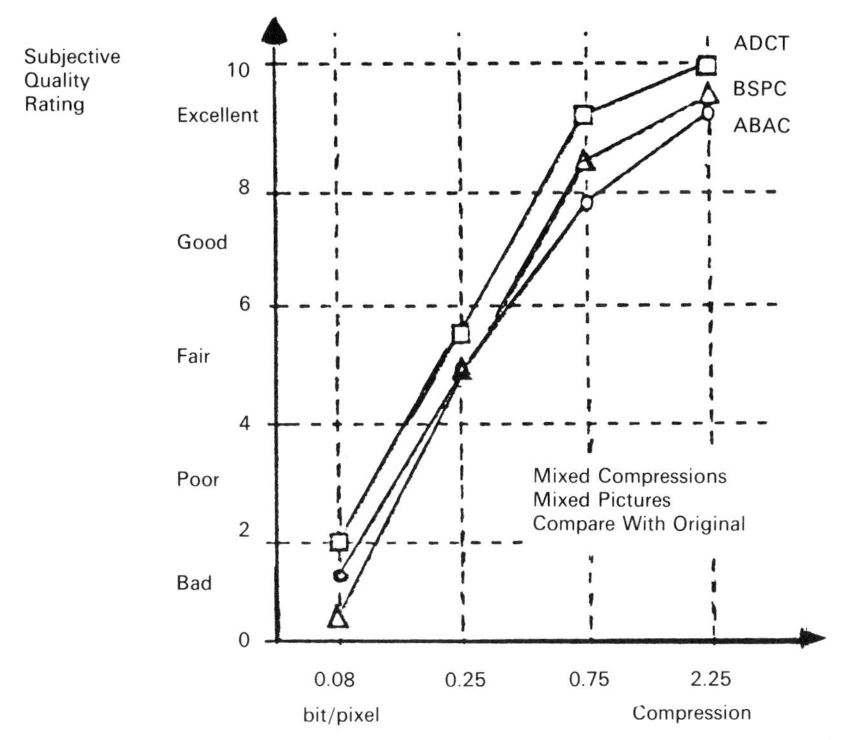

Fig. 7.187 Subjective testing results of ADCT, BSPC, and ABAC for the final selection of the JPEG algorithm [PIT-20].

however, achieved significantly higher subjective quality results compared with the other two algorithms, based on five nonstandard test pictures reconstructed at rates of 0.08, 0.25, 0.75, and 2.25 BPP (Fig. 7.187), [PIT-20]. JPEG unanimously decided to pursue further enhancement and refinement of the ADCT that meets all the requirements for the multistage progressive and sequential picture build-up. This algorithm will be the basis of a single standard capable of meeting all application requirements for color and monochrome pictures [PIT-14, PIT-20 through PIT-27, PIT-33 through PIT-36]. The ADCT algorithm is applied to the low-resolution version of the picture (2D DCT-II is applied to 8×8 blocks) obtained by subsampling the low-pass filtered image (Fig. 7.188). Preceding this operation is the decomposition of the color image into luminance (Y) and color difference signals C_R and C_B.

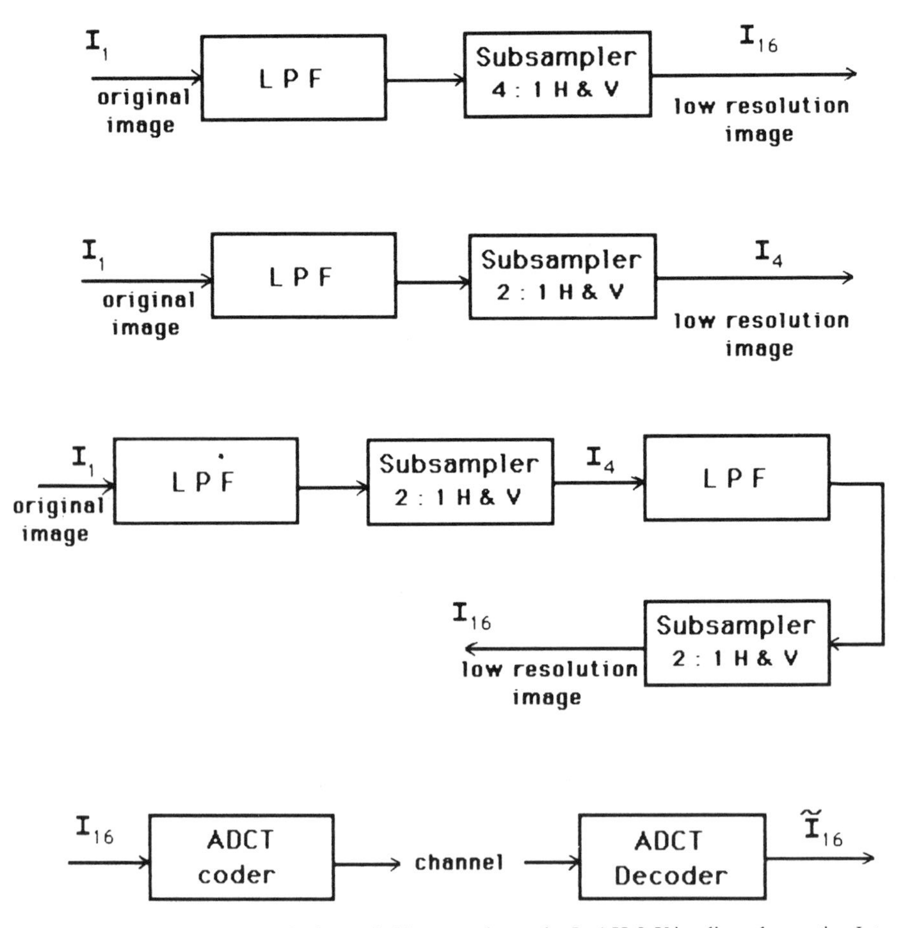

Fig. 7.188 ADCT coding of subsampled image, subsampler L : 1 H & V implies subsampling L to 1 both horizontally and vertically [PIT-13].

The decoded/subsampled image $\tilde{I}_4$ is the basis for obtaining the residual prediction error that is transform coded (Fig. 7.189). Further stages can be added to progressively build the quality and resolution of the reconstructed images, as shown in Fig. 7.190. The detailed algorithm, which involves scanning the transform coefficients, quantization, coding, and other essential operations is described in [PIT-13, PIT-18]. Various aspects of this algorithm, together with progress towards international standardization of a still-picture compression technique, are described in [PIT-20 through PIT-27, PIT-30, PIT 33 through PIT-36]. The effectiveness of this algorithm can be observed from the images reconstructed at 0.08, 0.25, and 0.75 BPP shown in Fig. 7.191. At 2.25 BPP, no visible differences can be detected between the original (8-bit PCM) and reconstructed images.

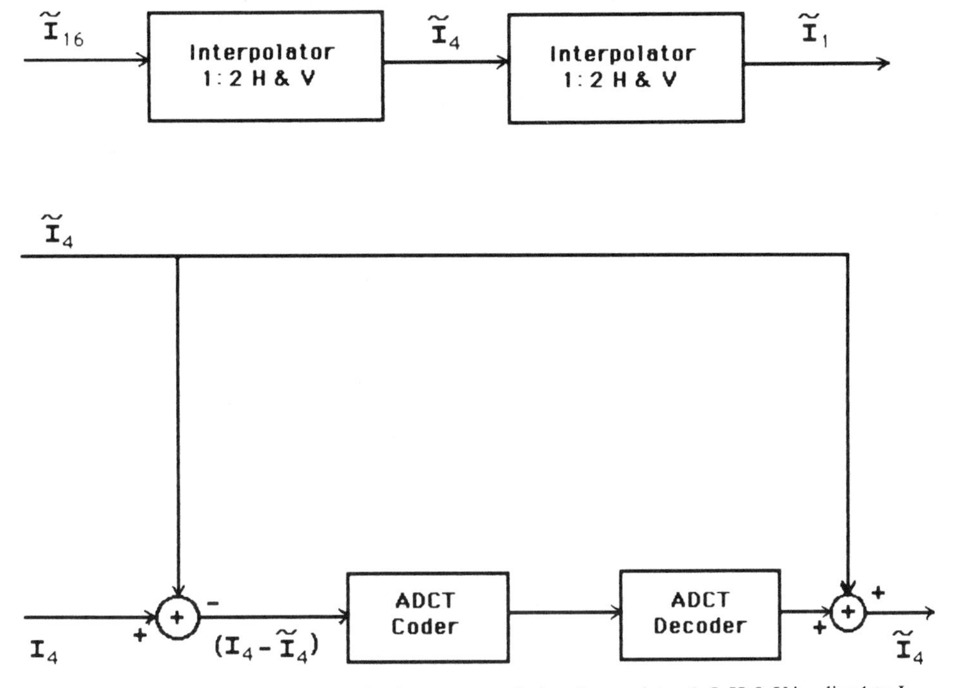

Fig. 7.189 Two stages of progressive image transmission. Interpolator 1 : L H & V implies 1 to L interpolation horizontally and vertically [PIT-13].

Another approach to PIT of color images has been proposed by Dubois and Moncet [PIT-3, PIT-5]. The compression technique is applied directly to the NTSC composite images sampled at 4 f_{SC} and 2 f_{SC}, where f_{SC} is the color subcarrier frequency. Both WHT and DCT have been utilized. A particular noteworthy feature of this technique is the scanning pattern and group segmentation of the transform coefficients. For example, the grouping and

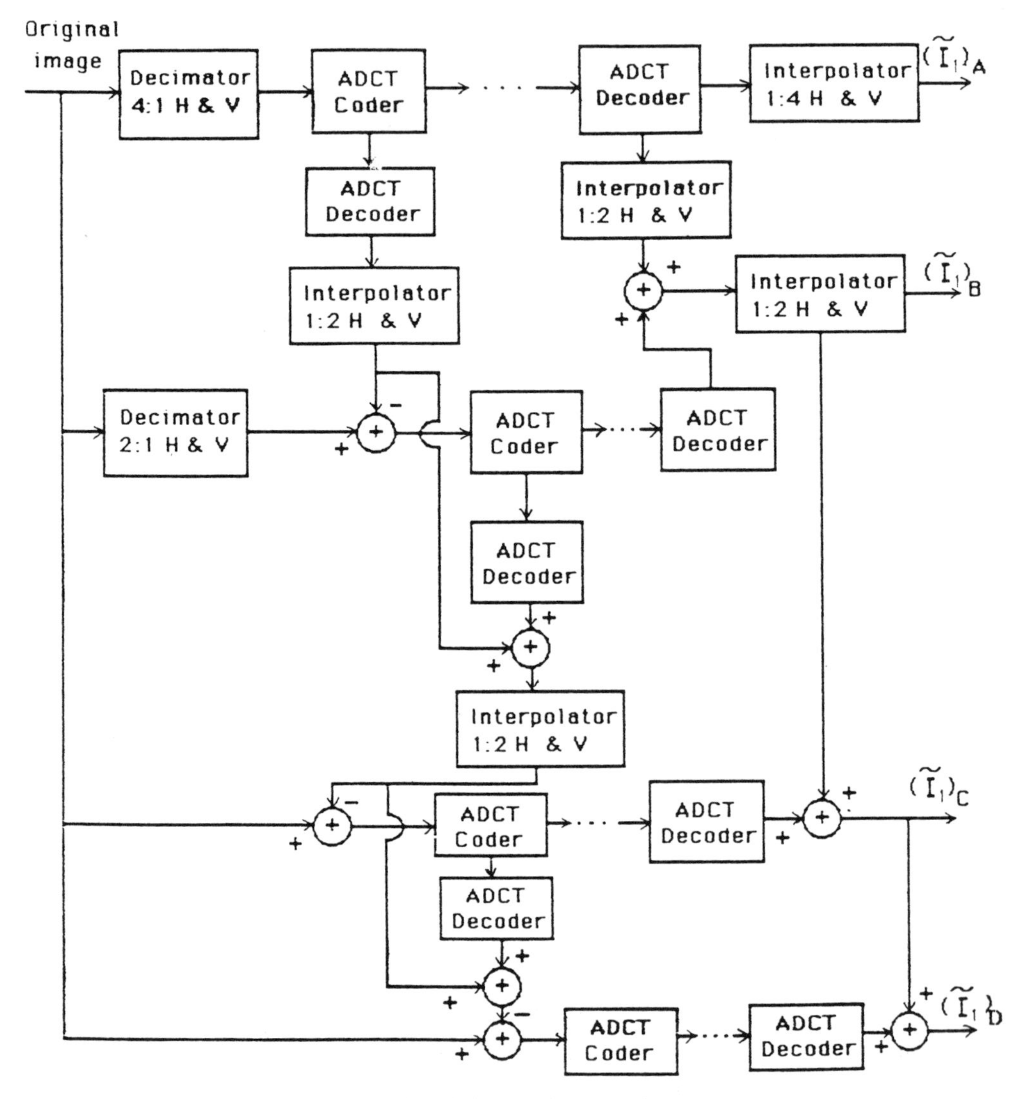

Fig. 7.190 ADCT coding scheme for still image telecommunication services. Decimator L : 1 H & V implies subsampling L to 1 both horizontally and vertically. Similarly, interpolator 1 : L H & V implies interpolation, 1 to L, both horizontally and vertically.

scanning patterns for a 16×16 block are shown in Fig. 7.192. Although no optimality has been claimed, these groups and patterns reflect the quadrature modulation of the I and Q components by the subcarrier at f_{SC}. Each stage of the image build-up is reflected by transmission of the transform coefficients controlled by the scanning pattern of a group. The hierarchy of group transmission is from 0 through 7. Several coding schemes involving various

Fig. 7.191 Reconstructed images based on the ADCT algorithm recommended by JPEG. a) 0.08 BPP; b) 0.25 BPP; c) 0.75 BPP.

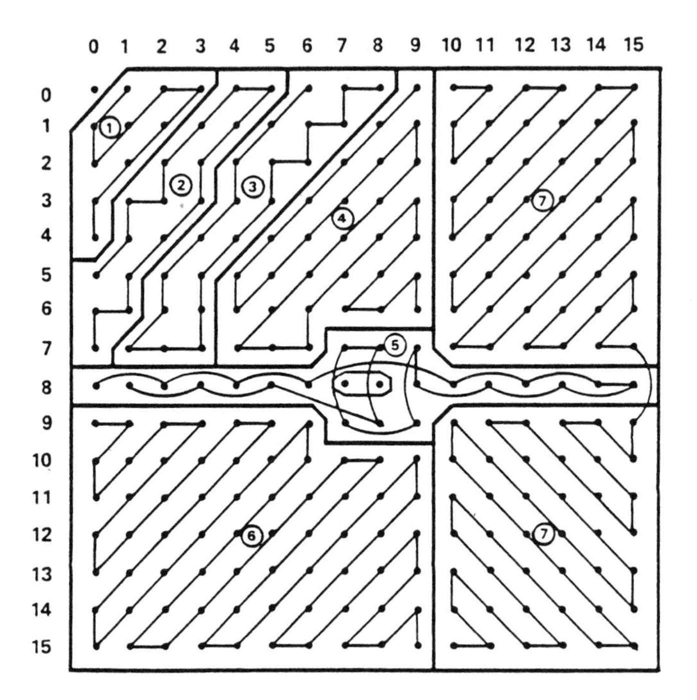

Fig. 7.192 Segmentation of the transformed block into subgroups and scanning patterns for each group. Group 0 consists of (0, 0), (8, 7), and (8, 8), representing the dc values of *Y*, *I*, and d*Q*, respectively. Group 5 consists largely of chrominance information [PIT-5]. (© 1986 IEEE).

quantizers and coders have been experimented with. The authors claim that at 4 f_{SC}, DCT gave superior results, whereas at $2 f_{SC}$ hexagonal sampling WHT was particularly appropriate.

Similar to the techniques described in Figs. 7.185 and 7.190, Burt and Adelson [PIT-32] have developed a pyramid structure for the PIT of still frames (Fig. 7.193). At each stage of the pyramid, the image is subsampled 2:1, both horizontally and vertically, and the difference between the input to that stage and the interpolated image (subsampled image is interpolated back to the original resolution) is coded and transmitted. The original image g_0 and the subsampled images at each stage $g_1, g_2, g_3, \ldots$, constitute the Gaussian pyramid, whereas the difference between the input image and the subsampled-interpolated image at each stage is part of the Laplacian pyramid. The top stage of the pyramid is a crude version of the original image, whereas the successive stages from the top to the bottom represent the gradual build-up of the image, leading to no visible degradations. Details regarding the decimation and interpolation filters, coder, decoder, bit rates at various levels, test images, etc. are furnished in [PIT-32].

Although there are various techniques applicable to the PIT, the emphasis in this section has been mainly limited to the DCT. By integrating other tools such

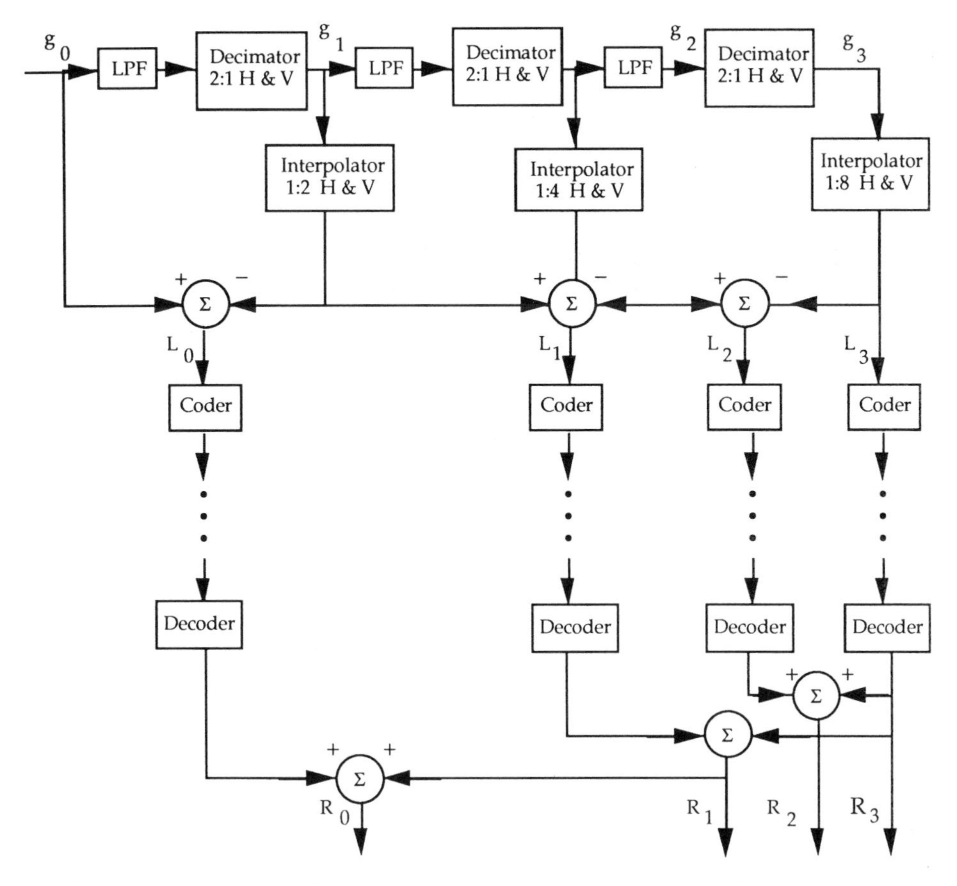

Fig. 7.193 Laplacian pyramid coding applied to progressive image transmission. $g_0, g_1, g_2, g_3, \ldots$ represent the Gaussian pyramid, L_0, L_1, L_2, L_3, ... denote the Laplacian pyramid. At the receiver, the highest level is a crude version, whereas R_0 is the final reconstructed image [PIT-32]. (© 1983 IEEE).

as subsampling, DPCM, HVS weighting, etc. with the DCT, the effectiveness of the DCT can be further improved. As with other compression schemes, the transform domain PIT involves scanning patterns, quantizers, coders, and possibly buffers. Although DCT appears to play a prominent role in interactive visual communication services, the research and development in this field is by no means complete. It is an ongoing process, and interfacing and integrating this service with desktop videophones and videoconferencing codecs offering real-time motion video can further strengthen its utility. This section has presented the transform domain (specifically DCT-II) PIT from a system/block diagram format. The reader is referred to the literature for details of the various operations, although some details are missing because of the proprietary nature of the product-based algorithms.

7.22 Printed Image Coding

In bandwidth reduction of four-color printed images [PIC-1, PIC-2], adaptive coding was implemented based on 2D DCT of 8 × 8 blocks of the four-color space yellow (Y), magenta (M), cyan (C), and black (B), followed by block classification controlled by (7.13.2) and $X^{C(2)}(0, 0)$. The classification determined the threshold (the magnitudes of the transform coefficients above the threshold are coded) (Table 7.23), step size of the uniform quantizer, and the number of high-frequency coefficients set to zero (Table 7.24). This was followed by zig-zag scan, run length of zero coefficients, and VWL coding of nonzero coefficients and the run lengths similar to [LBR-5].

Table 7.23

Classes and thresholds (no color space conversion) [PIC-2]. (© 1987 IEEE).

Class	1	2	3	4	5
$\Sigma\|AC\|$	0–100	0–100	100:600	600:1000	>1000
$\|DC\|$	<500	>500			
Y	1	4	9	13	16
M	1	4	6	10	12
C	1	4	8	12	14
B	1	4	10	14	18

Table 7.24

Number of 2D-DCT coefficients set to zero [PIC-2]. (© 1987 IEEE).

Class	1	2	3	4	5
No. set to 0	36	36	28	21	15

7.22.1 *Color Space Conversion*

A further increase in compression was obtained by applying KLT to the four color components (these are highly correlated) followed by the 2D DCT of the decorrelated components (Fig. 7.194), K_1, K_2, K_3, and K_4. As these components

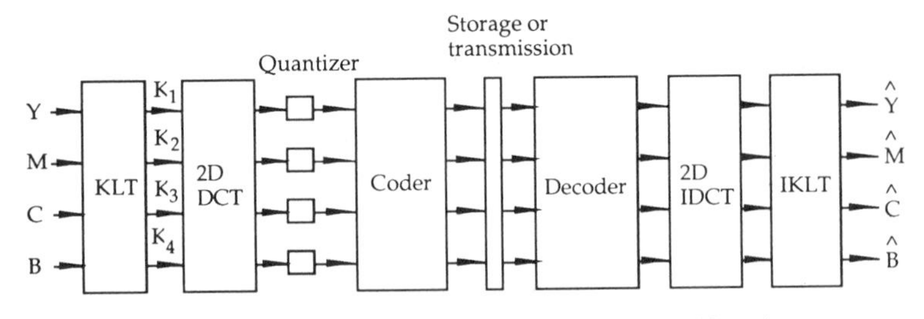

Fig. 7.194 Block diagram of four-color, printed-image coding with color-space conversion [PIC-2]. (© 1987 IEEE).

are arranged according to the decreasing eigenvalues, K_1 has the most of the structural information followed by K_2 and K_3, with K_4 contributing little information to the printed color image. Using (7.13.2), a new set of classes and thresholds are set (Table 7.25). By color space conversion, the compression ratio was nearly doubled (from 8 : 1 without the KLT to 15 : 1 with the KLT), resulting in no visual degradation of the reconstructed images.

Table 7.25

Classes and thresholds with color space conversion [PIC-2]. (© 1987 IEEE).

Class	1, 2	3	4	5
$\Sigma\|AC\|$	0–15	50–600	600 : 1000	>1000
K_1	4	8	12	16
K_2, K_3, K_4	4	15	20	25

7.22.2 *Multiple coding*

Bandwidth compression also can be applied to the storage of a large, color, printed image database. Image coding in this case, however, becomes very complex because of the editing process, i.e., combining several images into a new image. During this process the subimages may undergo translation, rotation, zooming, etc., as well as annotation of text on the new image. Gilge [PIC-2] illustrates the problems inherent in multiple image coding and indicates that several techniques are being investigated.

7.23 Packet Video

For multimedia communication and information services, the evolution of asynchronous transfer mode (ATM) networks based on packet switching represents the flexibility and freedom in maintaining the quality of these services, rather than being constrained by the constant bit rate designed for circuit switched networks. Recently, research is directed towards the development of image coding algorithms designed for constant quality at variable bit rates (VBR) [PV-4, PV-7, PV-8]. As image information representing detail, motion, etc. varies, VBR coding tailored for packet switched networks can be utilized to maintain constant image quality. Also by channel sharing among multiple video sources, transmission efficiency can be improved. The problems inherent in this scenario are packet loss and packet delay. The former can be due to random bit errors in the packet destination address. Also, the packets can be discarded during heavy traffic at a certain mode in an ATM network, due to insufficient storage. Random packet loss can be mitigated by error detection and correction codes. Also compensation techniques for packet loss such as demand refresh and retransmission are only partially effective. The impact of discarded packet loss can be minimized by introducing a priority into networks in combination with the layered structure [PV-2]. By classifying the packets into high and low priorities, a high loss rate for low-priority packets can be tolerated while preserving a low loss rate for high-priority packets. Thus the packet loss-protection-recovery is an integral part of the VBR coding. Packet delay is caused by holding the packet at any of the switching nodes until a slot is open, resulting in a differential transmission delay between packets. Buffer design must take this factor into consideration. Also, the differential delay causes problems in the timing relationship between video generation and reconstruction for display. Compression algorithms such as the transform or hybrid (DCT/DPCM, DCT/VQ, etc.) techniques compatible for packet switching must consider these problems (packet loss and packet delay) in the overall coding process. As an example, a layered structure and priority control has been developed by Nomura, Fujii, and Ohta [PV-2] for transmitting video over ATM networks based on hybrid (DPCM/DCT) coding. Similarly, error correcting protection for both random bit errors and packet losses has been incorporated by Leduc *et al.* [PV-6] in the video coding algorithm. This algorithm is designed to be universal such that it can process video with minor modifications over a wide bandwidth spectrum extending from videophone to HDTV.

7.23.1 *Layered Structure and Priority Control*

A layered structure and packet loss protection/recovery process is shown in Fig. 7.195. An optimal combination of these techniques, i.e., layers and packet loss detection and correction, can constitute a valid research topic in VBR coding. In the MC predictive transform coding [PV-1, PV-3] applied to the viceoconferencing sequence (Fig. 7.196a), the 2D DCT of 8×8 blocks is scanned in a zig-zag fashion (Fig. 7.174). Prior to the 2D DCT, the 8×8 blocks are selected by a quality estimation algorithm (quality control), which results in constant video quality. The transform coefficients along the zig-zag scan are divided into two layers based on a distortion threshold, which assigns low priority to the high-frequency layer (Fig. 7.197). The combination of layered structure and priority control has resulted in negligible degradation, even at a total packet loss rate of 10%. Simulations of layered coding are based on a non-MC algorithm (Fig. 7.196b) [PV-2]. The effects of packet loss (both random loss and layered loss) are demonstrated in Fig. 7.198 (see color insert, page I-3) for a color image sequence sampled at 480 pels/line, 512 lines/frame, 30 interlaced frames/sec, and 8-bit PCM, and in Fig. 7.199 (see color insert, page I-4) for a color still image with the same spatial resolution. The image sequence is coded at 0.7 BPP (5.1 MBPS) and the still image is coded at 2.4 BPP (18 MBPS). The effects of both the random and layered losses can be observed from the reconstructed images.

Another layered coding scheme which is robust against cell (packet) loss has been proposed by Shimamura, Hayashi and Kishino [PV-14]. In this scheme (Fig. 7.200) the quantized DCT coefficients of the interframe prediction errors of 8×8 blocks are separated into two parts i.e., most significant part (MSP) and

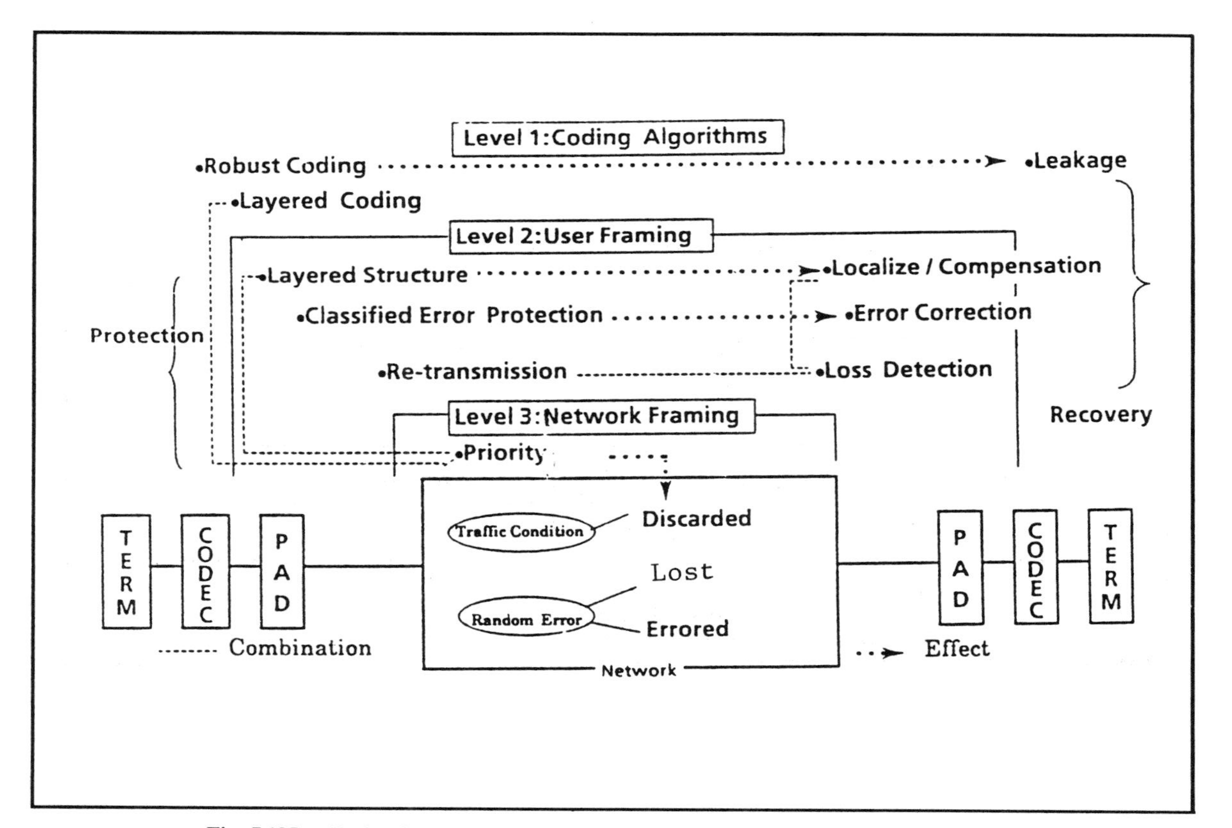

Fig. 7.195a Packet loss protection/recovery. TERM = terminal; PAD = packet assembly and disassembly [PV-2].

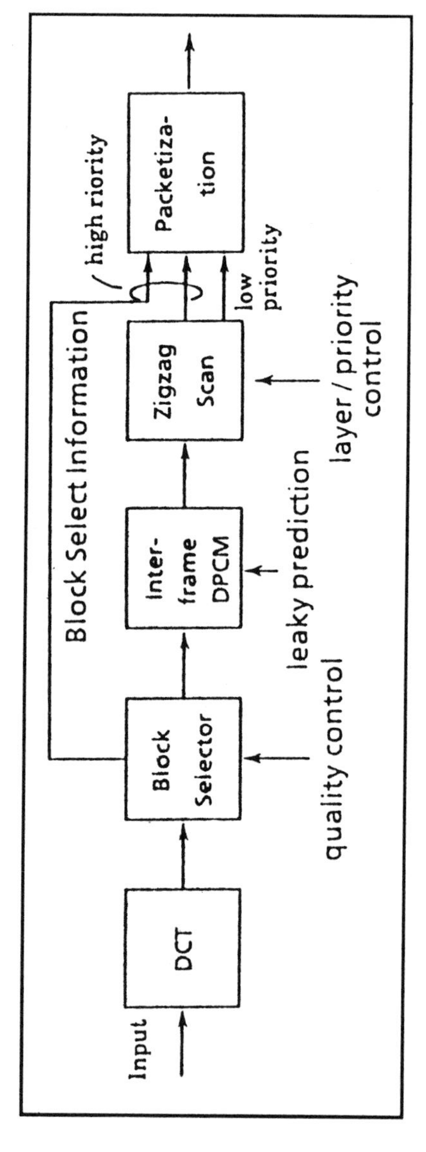

Fig. 7.195b Block diagram of coding scheme [PV-2].

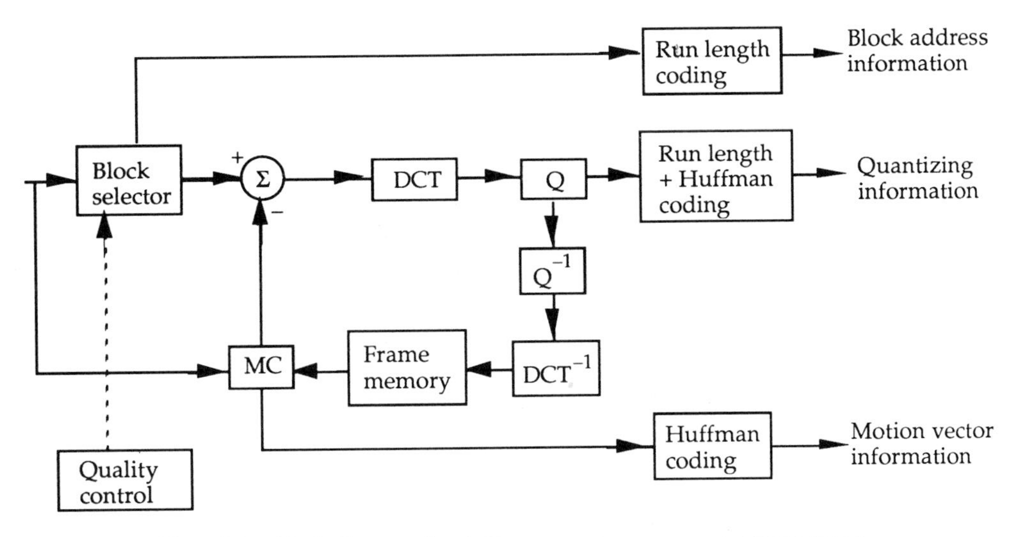

Fig. 7.196a Block diagram of variable-rate, motion-compensated DCT [PV-1].

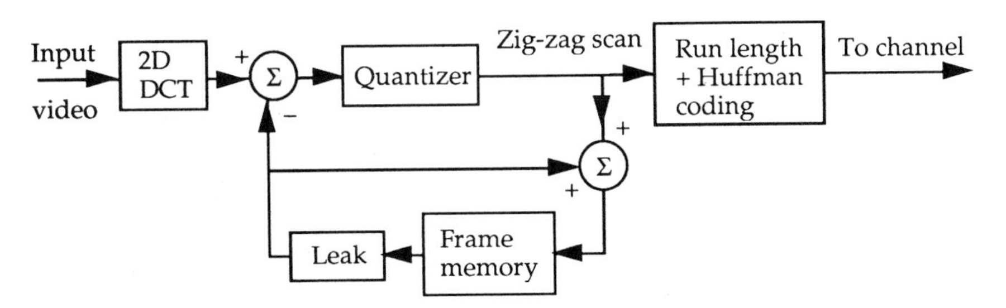

Fig. 7.196b Block diagram of variable-rate hybrid DCT [PV-2].

least significant part (LSP). In the cell multiplexer packets are created that consist of only MSPs and LSPs (Fig. 7.201). High and low priorities are allocated to MSP and LSP packets respectively. After scanning the 2D DCT coefficients (Fig. 7.202), two layering methods are chosen for simulation.

Method A

In this method the threshold for separation into MSP and LSP is based on a specified % of the number of bits required to code all the transform coefficients in the block. The coded bit length of each DCT coefficient is accumulated as per the scanning and when this accumulation reaches the threshold, all subsequent coefficients are classified as LSP. The coded bit length L_i of quantization level i is defined as

$$L_i = -\log_2 P_i$$

where P_i is the probability of occurrence of level i.

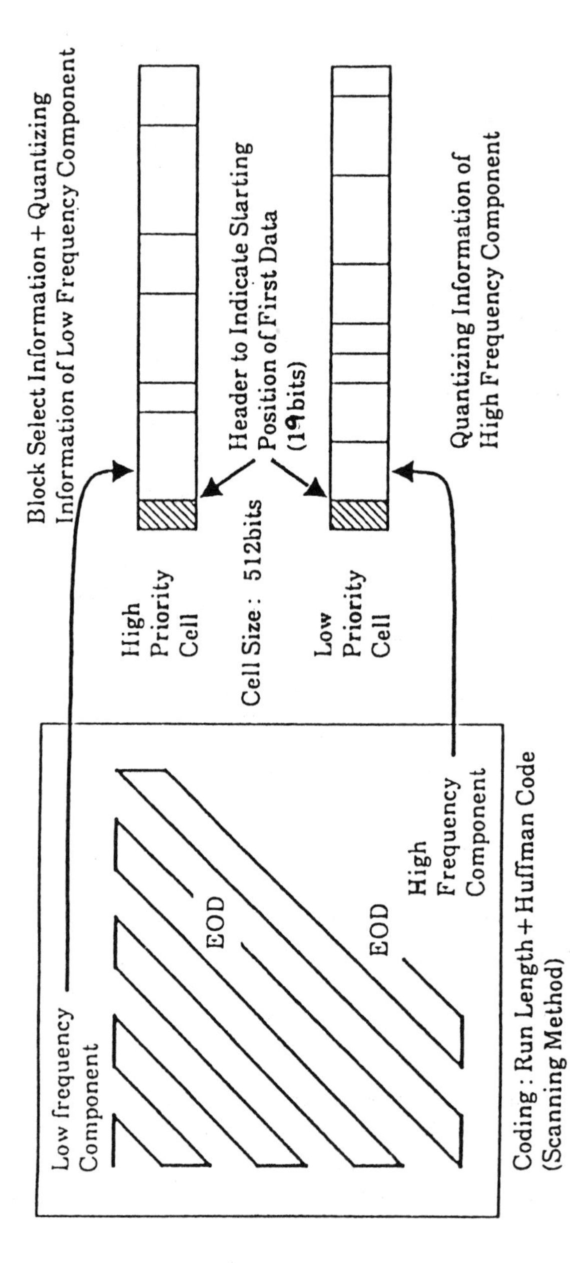

Fig. 7.197 Layer/priority control in packetization [PV-2].

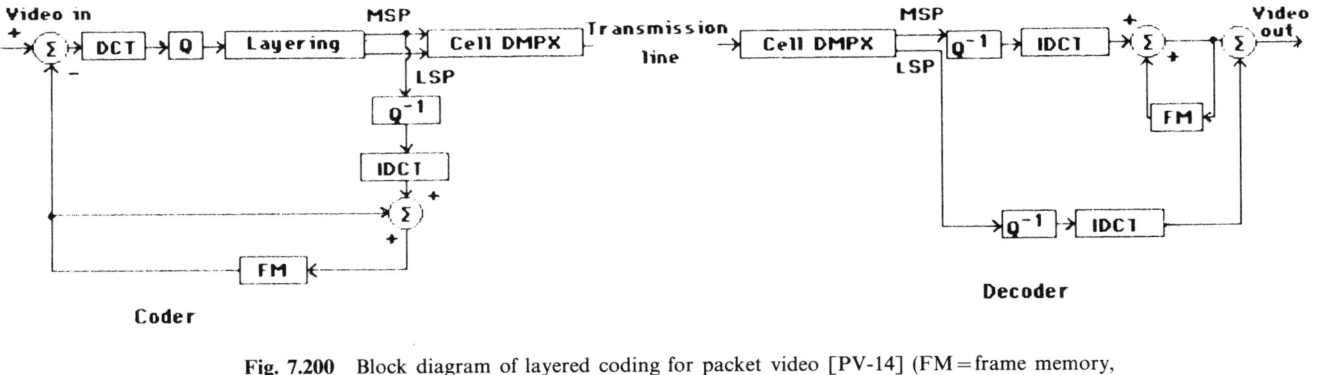

Fig. 7.200 Block diagram of layered coding for packet video [PV-14] (FM = frame memory, MPX = multiplex, DMPX = demultiplex).

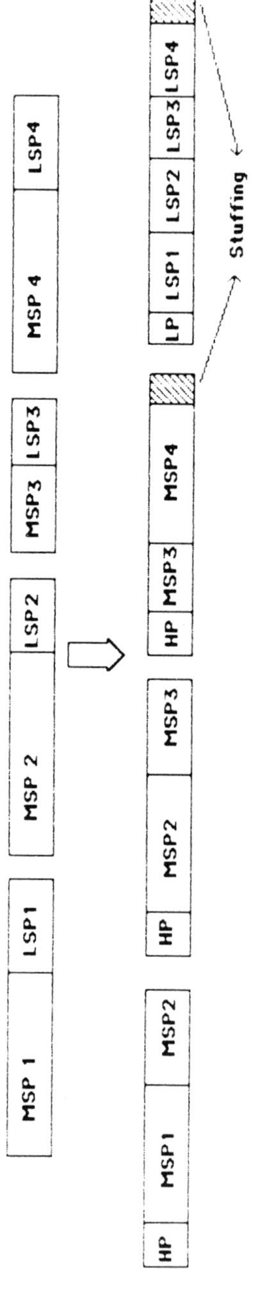

Fig. 7.201 Separation of MSPs and LSPs and formation of packets [PV-14] (HP = high priority, LP = low priority).

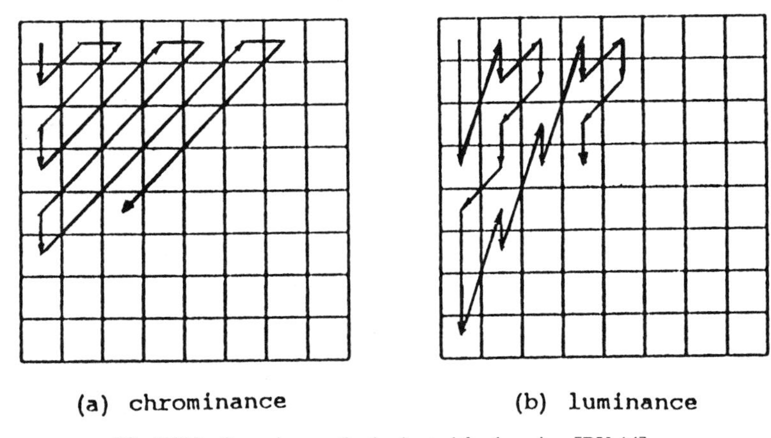

Fig. 7.202 Scanning method adopted for layering [PV-14].

Method B

The MSP/LSP separation in this method is based on the energy distribution in the DCT block. The power of each DCT coefficient is accumulated as per the scanning order. When this accumulation reaches a preset fraction of the total power in the 8×8 block, all subsequent coefficients are classified as LSP.

In simulation of test images both methods were effective in encountering cell loss. Between the two techniques, method B proved to be superior.

Another example of packet video is the universal coding algorithm developed by Leduc *et al.* [PV-6]. The standard MC DPCM/DCT of 8×8 blocks is modified by 3D coding of perceptually weighted transform coefficients. This algorithm is both universal (independent of image statistics) and versatile, i.e., for various video applications. Details as to the 3D coding, packetization, and its protection, etc. can be found in [PV-6].

Brainard and Othmer [PV-10], in a thought-provoking paper, raise various pertinent problems posed by packet networks. They suggest that the development of video compression algorithms and packet network design are mutually interactive and stress that the design changes in the codec addressing these problems need to be considered.

Video coding algorithms designed for ATM networks are still in their infancy. For standard transform and hybrid (transform/DPCM) coding, coding schemes to counter packet loss and packet delay must be considered. In component coding, the concepts of layered structure and packet prioritization need to be investigated. It is speculated that commercial ATM systems for private corporate networks will be available by the early 1990s, whereas their integration into the public networks is expected in the mid-1990s. Thus VBR video for packet switched transmission is a fertile ground for research and development.

7.24 BTC/Transform Coding

An image compression technique called block truncation coding (BTC) [BTC-1, BTC-2] is based on dividing an image into small blocks (say, 4×4) and coding the pels of each block as one of two levels. The coded pels preserve the first and second moments and hence the variance of the block. Overhead information reflecting the mean and standard deviation of the block, besides the bit plane, indicating the coded pel value (one of the two levels) has to be transmitted. Both parametric and nonparametric moment-preserving quantizers have been designed, and there are several variations to the basic BTC technique. Also, a hybrid technique in which BTC was applied to the low-frequency 2D-DCT coefficients of a block (zonal filtering) has been investigated. Some other modifications of BTC include generalized BTC (GBTC) [PIT-14] and adaptive BTC (ABTC) [PIT-18], which have been extensively investigated as possible coding schemes for still-image telecommunication services. Details regarding the performance of BTC and several of its modifications in image coding are described in these references.

7.25 Summary

A wide spectrum of the applications of the DCT covering diverse disciplines is presented. As an in-depth and detailed description of all of these applications is beyond the scope of this book, the approach, in general, has been conceptual. Block diagrams and the basic principles of bandwidth compression are provided. The extensive list of references at the end of the book is a testimony to the phenomenal popularity of the DCT in DSP in general and in image coding in particular. The reader can utilize this resource for extending the range of applications to other fields.

PROBLEMS

7.1. Chen and Fralick [IE-1] have discussed various image enhancement techniques using DCT filtering. All the processes, i.e., DCT, filtering, IDCT, etc., have been carried out on 16×16 sub-blocks. Estimate the filter matrices $H(u, v)$ used in Figs. 7.1–7.3 [IE-1]. Repeat the processes based on 8×8 blocks. Consider one or two test images having sharp edges and abrupt intensity transitions. Note that this problem requires test images and a modest image-processing facility.

7.2. Ngan and Clark [F-1] have applied low-pass filtering to images mapped into the DCT domain (Fig. 7.4). They have used a 2D Butterworth filter to obtain a modified DCT filter. Implement the DCT filtering based on second-order Tschebyshev and elliptic filters. In the postprocessing of the filtered images, 50% overlap is utilized to reduce the block edge effects. Consider 25% overlap and compare its performance with the former. Consider different block sizes and also subsampling, as adopted in [F-1].

7.3. In [PIT-12], the order in which the DCT coefficients are selected for transmission is based on HVS weighting, leading to progressive image reconstruction. If, instead, the DCT coefficients are weighted by HVS, compare the performance of these images with those in [PIT-12] for the same bit rates. Use the HVS function proposed in [PIT-12]. For simplicity, a quantizer and coder need not be incorporated in the overall processing system. As in Problems 7.1 and 7.2, selected test images need to be used.

7.4. Extend the concept of filtering, sampling rate conversion, and transform (DCT) coding developed by Adant *et al.* [F-6] to 2D signals. Carry out these operations on images both horizontally and vertically, and analyze the subsampled images. Consider various combinations of decimation factors along each dimension.

7.5. Adant *et al.* [F-6] have applied DCT transcoding for standards conversion of TV signals from 4:2:2 to 3:1:1 (Section 7.3). Extend this technique for transcoding from the 525/60 (NTSC) or 625/50 (PAL, SECAM) to the common-source intermediate format (CSIF) [LBR-23, LBR-24, LBR-37, LBR-40, LBR-41, LBR-44]. The original color signals can be in the 4:2:2 format (CCIR rec. 601) [G-21]. Apply different block sizes, i.e., 8×8, 16×8, 16×16, etc., and judge their effectiveness based on the decimated images (see [F-8]).

7.6. Suppose the standards conversion of TV signals involves interpolation, e.g., conversion, from 3:1:1 to 4:2:2. Apply the transcoding technique of Adant *et al.* [F-6] to obtain this standards conversion. Consider various block sizes, as in Problem 7.5, and comment on their effectiveness.

7.7. Repeat Problem 7.6 for conversion from the CSIF [LBR-23, LBR-37, LBR-40, LBR-41, LBR-44] to NTSC or PAL/SECAM signals (see [F-8]).

7.8. A block-type algorithm for the DFT-LMS adaptive filter has been proposed [LF-3]. Can this approach be extended to the DCT-II? If so, conduct an extensive investigation of the frequency-domain block DCT-LMS adaptive filter along lines similar to those in [LF-3].

7.9. In [SC-3] it is stated that the symmetrical DFT (SDFT) is closely related to the DCT-II. The SDFT of an $N + 1$ point sequence $x(n)$, $n = 0, 1, \ldots, N$ is defined as follows:

$$\begin{aligned} \hat{X}_F(k) &= \sum_{n=0}^{2N-1} \hat{x}(n) W_{2n}^{nk} \qquad k = 0, 1, \ldots, N \\ &= \text{SDFT of } x(n) \end{aligned}$$

where

$$\hat{x}(n) = \begin{cases} x(n) & n = 0, 1, \ldots, N, \\ x(2N - n) & n = N + 1, N + 2, \ldots, 2N - 1. \end{cases}$$

Show that the DFT of $\hat{x}(n)$ is a real, symmetrical $2N$-point sequence. Also show the close relationship between the SDFT and DCT-II.

7.10. In block quantization of transform coefficients, the optimum bit allocation scheme results in [Q-5, SC-1, Q-6, PIT-7]

$$R_i = \bar{R} + \tfrac{1}{2} \log_2 \sigma_i^2 - \frac{1}{2N} \sum_{j=1}^{N} \log_2 \sigma_j^2, \qquad \text{(P7.10.1)}$$

where R_i is the number of bits assigned to the ith transform coefficient,

$$\bar{R} = \frac{1}{N} \sum_{i=1}^{N} R_i$$

is the average bit rate without the side information, N = number of transform coefficients in the block, and σ_i^2 is the variance of the ith transform coefficient. Derive this bit assignment scheme using the constraint. Show that the average distortion is minimized by equalizing the distortion for all the transform coefficients. Note that the method of Lagrange multipliers has to be applied.

7.11. In [SC-2], a modification to the ATC scheme based on sub-blocks (Fig. 7.12a) is suggested. Divide the 128 sample block into four sub-blocks, each of length 32. Obtain the variances of the transform coefficients averaged over the sub-blocks $\hat{\sigma}_1^2, \hat{\sigma}_2^3, \ldots, \hat{\sigma}_{32}^2$. Use the technique shown in the Fig. 7.11 and send only eight variances ($M = 4$) as side information. Compare this coding scheme with that described in Fig. 7.12a at various bit rates.

7.12. Combine the TDHS of [SC-4] with the ATC of [SC-1]. Compare the performance of this scheme using DCT-II with the SBC/HS of [SC-4] in speech coding at 9.6 KBPS.

7.13. Repeat Problem 7.12 using the ATC of [SC-2]. Use the modified bit assignment scheme and the sub-block side information, as proposed in [SC-2].

7.14. An ATC/AGHMM scheme [SC-16] for speech coding has been proposed. Compare its performance with the ATC of [SC-2]. The comparison is to be based on SNR, SEGSNR, subjective quality, and complexity at various bit rates.

7.15. DFT, WHT, and rapid transform (RT) [M-23] have been applied to phonemic recognition [SR-1, SR-2]. Apply DCT to phonemic recognition and compare its performance with the DFT, WHT, and RT.

7.16. Using (7.7.2) and (7.7.3) derive (7.7.4).

7.17. Show that the DFT of a real, even sequence is real and even.

7.18. In Section 7.7 it is stated that the modified DCT (7.7.3) can be computed using an $N/4$-point complex input FFT. Derive this algorithm.

7.19. Derive (7.8.15).

7.20. The bounds for 2D-DCT coefficients are given in [LBR-5]

$$X_{(0,0)}^{C(2)} = 2x_{\max}, \tag{P7.20.1}$$

and for ac coefficients,

$$\frac{16}{\pi^2} x_{\max} \leqslant X_{(u,v)}^{C(2)} \leqslant 2x_{\max},$$

where $x_{\max}$ is the maximum value of the input signal. Prove this.

7.21. In [PIT-11], it is stated that the average reconstruction error variance is equal to the average quantization error variance in the orthogonal transform domain. Prove this.

7.22. The fast progressive reconstruction (FPR) of images based on the DFT and WHT has been developed by Takikawa [PIT-15]. This technique has been extended to DCT-II [PIT-19]. Investigate whether the FPR scheme can be extended to DST-II. Compare the performance of FPR/DST-II with the FPR based on the DFT, WHT, and DCT-II.

7.23. Dubois and Moncet [PIT-5] suggest that their PIT scheme applied to composite pictures can be improved by HVS weighting of the transform coefficients followed by nonuniform quantizers. Investigate in detail such an improved scheme and evaluate the technique based on image quality, bit rate, and system complexity during various stages of image build-up.

7.24. A HVS weighting matrix for the 8×8 2D DCT has been developed (Fig. 7.166) by Chitraprasert and Rao [PIT-16]. Develop a similar matrix for the 16×16 2D DCT based on the same sampling density.

7.25. Develop the matrix factors corresponding to the extra butterflies added to the DCT and IDCT in LOT [BSI-5, BSI-6, T-7, SC-29]. List the additional complexity involved in implementing these butterflies. The original DCT and IDCT are applied to 8×8 and 16×16 blocks.

7.26. Apply the various block distortion-minimization techniques described in Section 7.12 to test images and compare their performances based on bit rates, complexity, subjective quality, etc.

7.27. Extend the iterative transform coding scheme of Wang and Goldberg [PIT-7, PIT-11] to color images. Decompose color images to Y, C_R, and C_B or Y, I, and Q. Develop bit allocation matrices and quantizers for these components at various iterations. Note that at each stage the corresponding components have to be appropriately combined to form the composite color signal. Comment on the bit rate and image quality at each stage of progression, and compare the complexity of this technique with the ADCT scheme recommended by the JPEG [PIT-13, PIT-14, PIT-17, PIT-18, PIT-20 through PIT-27, PIT-29, PIT-30, PIT-33, PIT-34].

7.28. Mester and Frank [ACTC-2] have developed block classification based on spectral entropy-ac energy. However the ac energy used here is the absolute sum of the ac coefficients. Develop a similar classification (see Table 7.21) based on the actual ac energy as defined in (7.13.1).

7.29. Repeat Problem 7.28 by incorporating the HVS. The DCT coefficients are weighted by the HVS matrix described in Fig. 7.166.

7.30. Section 7.20 describes pattern recognition based on Mandala/DCT [PR-1]. Extend this concept to a wider range (different classes) of images and block sizes other than 8×8, defining, developing, and ranking the features and distortion measures. This research is quite involved and requires extensive simulation facilities.

7.31. In classifying VCGs [C-1], L-2 norm (7.9.1) was used. Investigate this experiment based on magnitude L-1 norm, i.e.,

$$\sum_{k=0}^{N-M-1} |X_t^{C(2)}(k) - X_{C1(i)}^{C(2)}(k)|. \tag{P7.31.1}$$

7.32. A fast implementation of LOT for $N = 8$ from a filter bank viewpoint has been developed by Malvar [T-7]. Based on [FRA-19], complete the flowgraph shown in Fig. 3 of [T-7].

7.33. Develop a detailed flowgraph for fast implementation of LOT [T-7] for $N = 16$ (see Problem 7.32).

7.34. In the classified VQ transform speech coder (Fig. 7.20), it is mentioned that the authors [SC-20] are investigating DCTs of 16 and 32 speech samples, using four to eight sub-blocks in the product code VQ and varying codevector and codebook sizes. Investigate speech coding using these combinations, with a view to reduce the bit rate and improve the speech quality.

7.35. Extend the concept of transform-based split band coder (TSBC) (Fig. 7.12b) [SC-22, SC-34] to image coding. For example, 2D DCT of image blocks of size 16×16 can be split into four 4×4 sub-blocks, which can be individually coded, similar to the TSBC. Adaptive features based on energies of the sub-blocks can be introduced for block quantization. Investigate this technique in detail and compare its performance with standard transform coding.

7.36. Guillemot and Duhamel [PC-16] have proposed a new transform for image coding. This transform has reduced complexity but the same performance as DCT when images are processed in 8×8 blocks. Investigate the performance of this transform when images are processed in 16×16 blocks.

7.37. In [PIT-12, PIT-16], HVS weighted progressive image transmission is described (see also Fig. 7.178). Replace the DCT in this system by the new transform [PC-16] and investigate its performance at various stages of image build-up.

7.38. Repeat Problem 7.37 based on the technique described in Fig. 7.185.

APPENDIX A.1

COMPUTER PROGRAMS FOR DCT-II AND IDCT-II, $N = 8$ AND 16 [FRA-3, FRA-4]

This appendix describes the computer programs for DCT-II and IDCT-II, $N = 8$ and 16 based on Lee's algorithm [FRA-3, FRA-4]. These programs are supplemented with the flowgraphs and the corresponding sparse matrix factors. Computer programs for DCT-II and IDCT-II, $N = 8$ and 16 based on (A.1.1) and (A.1.2), i.e., direct matrix multiplication, are also included. In this appendix the notation and definitions for the DCT-II and its inverse are based on [FRA-3], i.e.,

DCT-II

$$X^{C(2)}(n) = \frac{2}{N} C_n \sum_{k=0}^{N-1} x(k) \cos\left[\frac{\pi(2k+1)n}{2N}\right] \qquad n = 0, 1, \ldots, N-1 \tag{A.1.1}$$

IDCT-II

$$x(k) = \sum_{n=0}^{N-1} C_n X^{C(2)}(n) \cos\left[\frac{\pi(2k+1)n}{2N}\right] \qquad k = 0, 1, \ldots, N-1, \tag{A.1.2}$$

where

$$C_n = \frac{1}{\sqrt{2}} \qquad n = 0$$
$$= 1 \qquad n \neq 0$$

Rewrite (A.1.2) as

$$x(k) = \sum_{n=0}^{N-1} \hat{X}^{C(2)}(n) \cos\left[\frac{\pi(2k+1)n}{2N}\right], \tag{A.1.3}$$

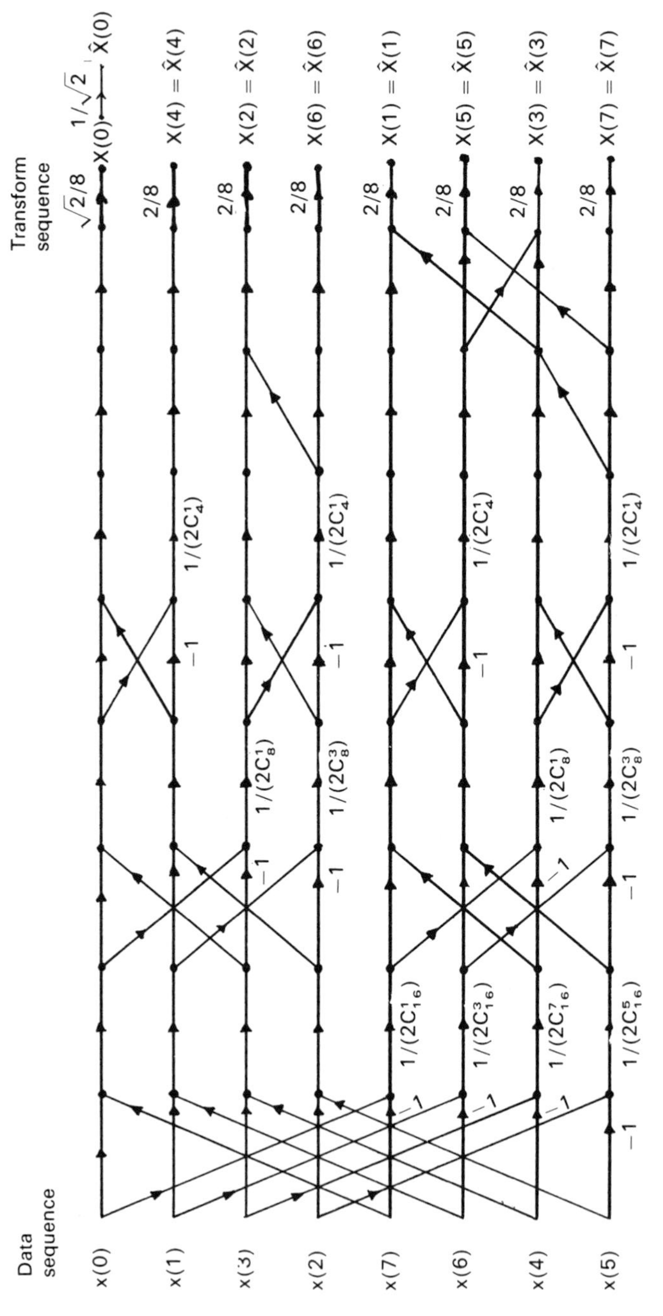
Data sequence
x(0)
x(1)
x(3)
x(2)
x(7)
x(6)
x(4)
x(5)
Transform sequence
$\sqrt{2}/8$
$1/\sqrt{2}$
$X(0) \rightarrow \hat{X}(0)$
$X(4) = \hat{X}(4)$
$X(2) = \hat{X}(2)$
$X(6) = \hat{X}(6)$
$X(1) = \hat{X}(1)$
$X(5) = \hat{X}(5)$
$X(3) = \hat{X}(3)$
$X(7) = \hat{X}(7)$
2/8
-1
$1/(2C_{16}^1)$
$1/(2C_{16}^3)$
$1/(2C_{16}^7)$
$1/(2C_{16}^5)$
$1/(2C_8^1)$
$1/(2C_8^3)$
$1/(2C_4^1)$

Fig. A.1.1

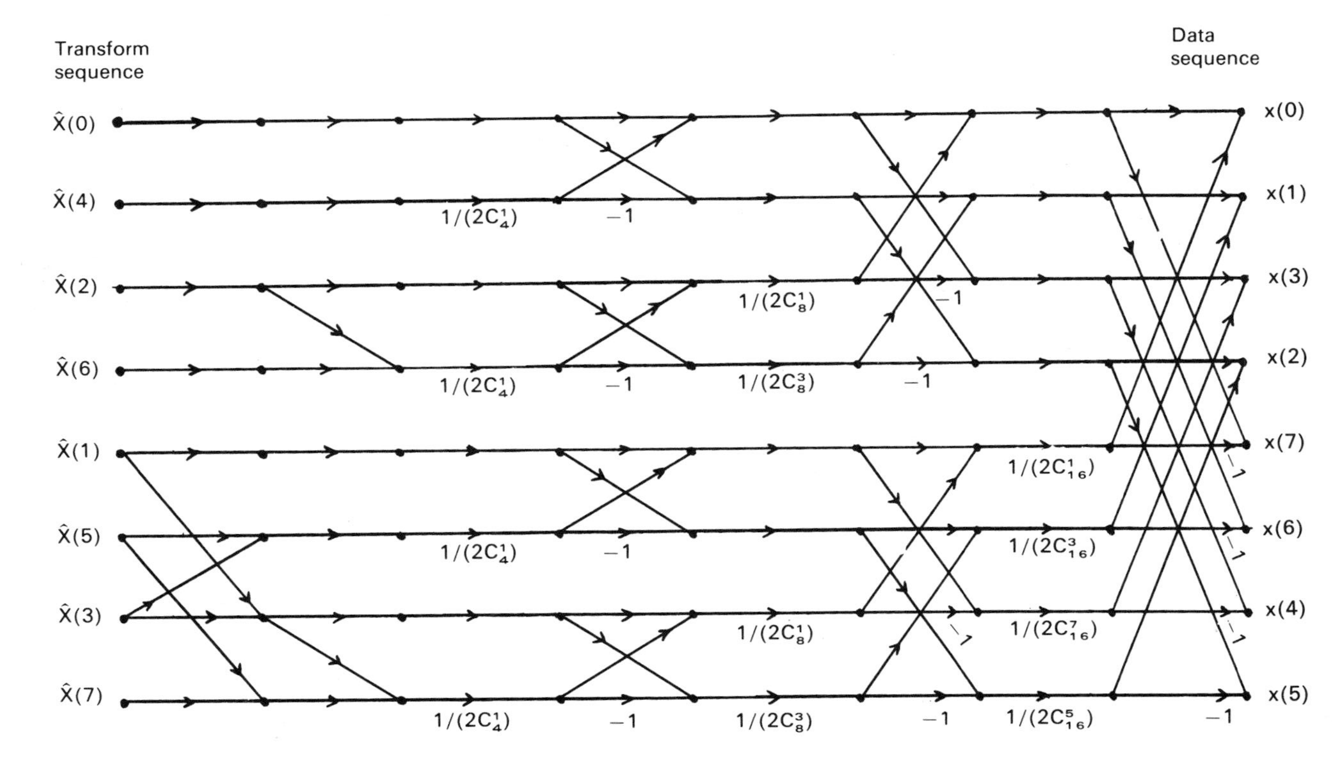
Transform sequence
X̂(0)
X̂(4)
X̂(2)
X̂(6)
X̂(1)
X̂(5)
X̂(3)
X̂(7)
Data sequence
x(0)
x(1)
x(3)
x(2)
x(7)
x(6)
x(4)
x(5)
1/(2C¹₄)
−1
1/(2C¹₈)
1/(2C³₈)
1/(2C¹₁₆)
1/(2C³₁₆)
1/(2C⁷₁₆)
1/(2C⁵₁₆)

Fig. A.1.2

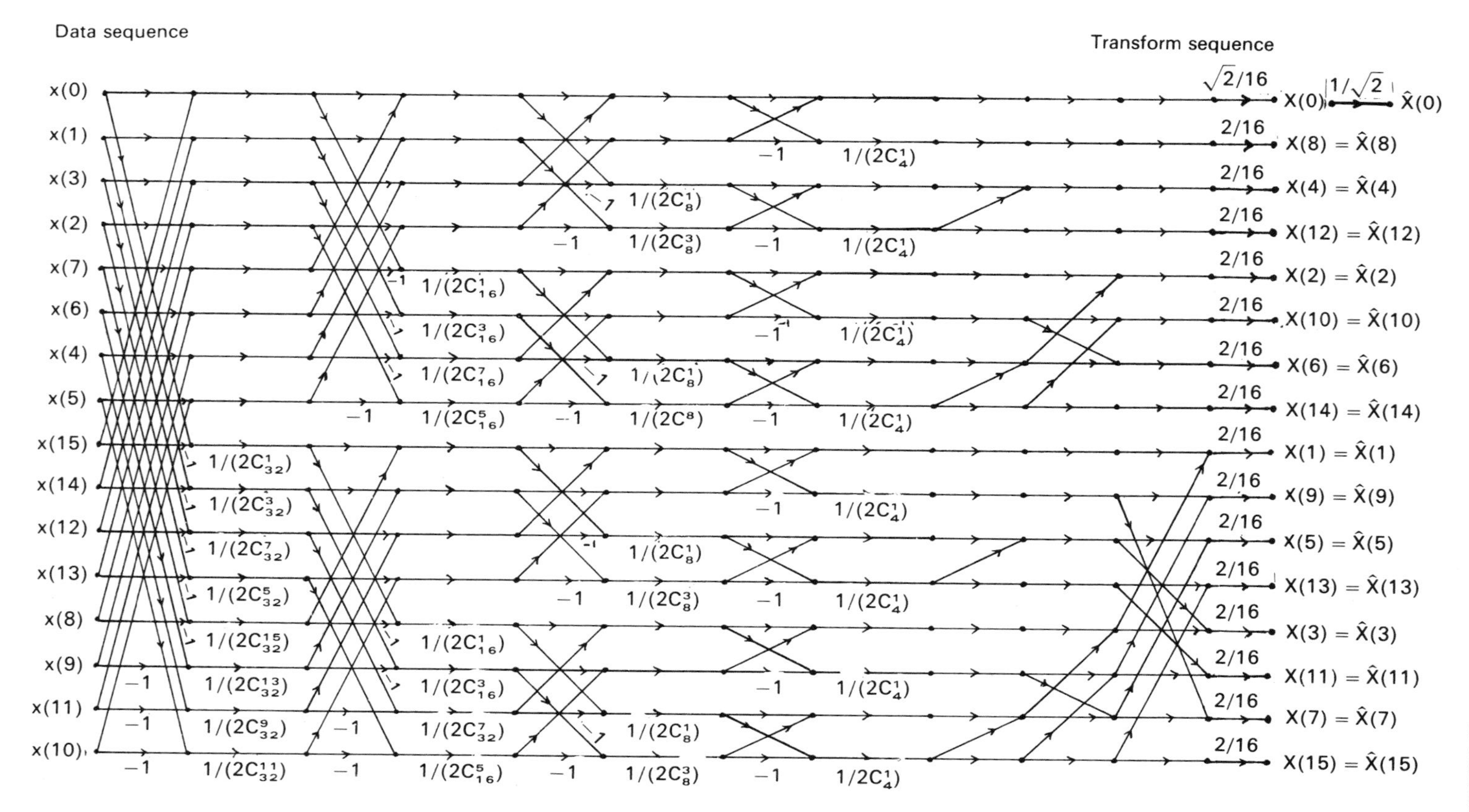
Data sequence
Transform sequence
x(0)
x(1)
x(3)
x(2)
x(7)
x(6)
x(4)
x(5)
x(15)
x(14)
x(12)
x(13)
x(8)
x(9)
x(11)
x(10)
√2/16
1/√2
X(0)
X̂(0)
X(8) = X̂(8)
X(4) = X̂(4)
X(12) = X̂(12)
X(2) = X̂(2)
X(10) = X̂(10)
X(6) = X̂(6)
X(14) = X̂(14)
X(1) = X̂(1)
X(9) = X̂(9)
X(5) = X̂(5)
X(13) = X̂(13)
X(3) = X̂(3)
X(11) = X̂(11)
X(7) = X̂(7)
X(15) = X̂(15)
2/16
1/(2C$^1_{32}$)
1/(2C$^3_{32}$)
1/(2C$^7_{32}$)
1/(2C$^5_{32}$)
1/(2C$^{15}_{32}$)
1/(2C$^{13}_{32}$)
1/(2C$^9_{32}$)
1/(2C$^{11}_{32}$)
1/(2C$^1_{16}$)
1/(2C$^3_{16}$)
1/(2C$^7_{16}$)
1/(2C$^5_{16}$)
1/(2C$^7_{32}$)
1/(2C$^1_8$)
1/(2C$^3_8$)
1/(2C^8)
1/(2C$^1_4$)
1/2C$^1_4$)
−1

Fig. A.1.3

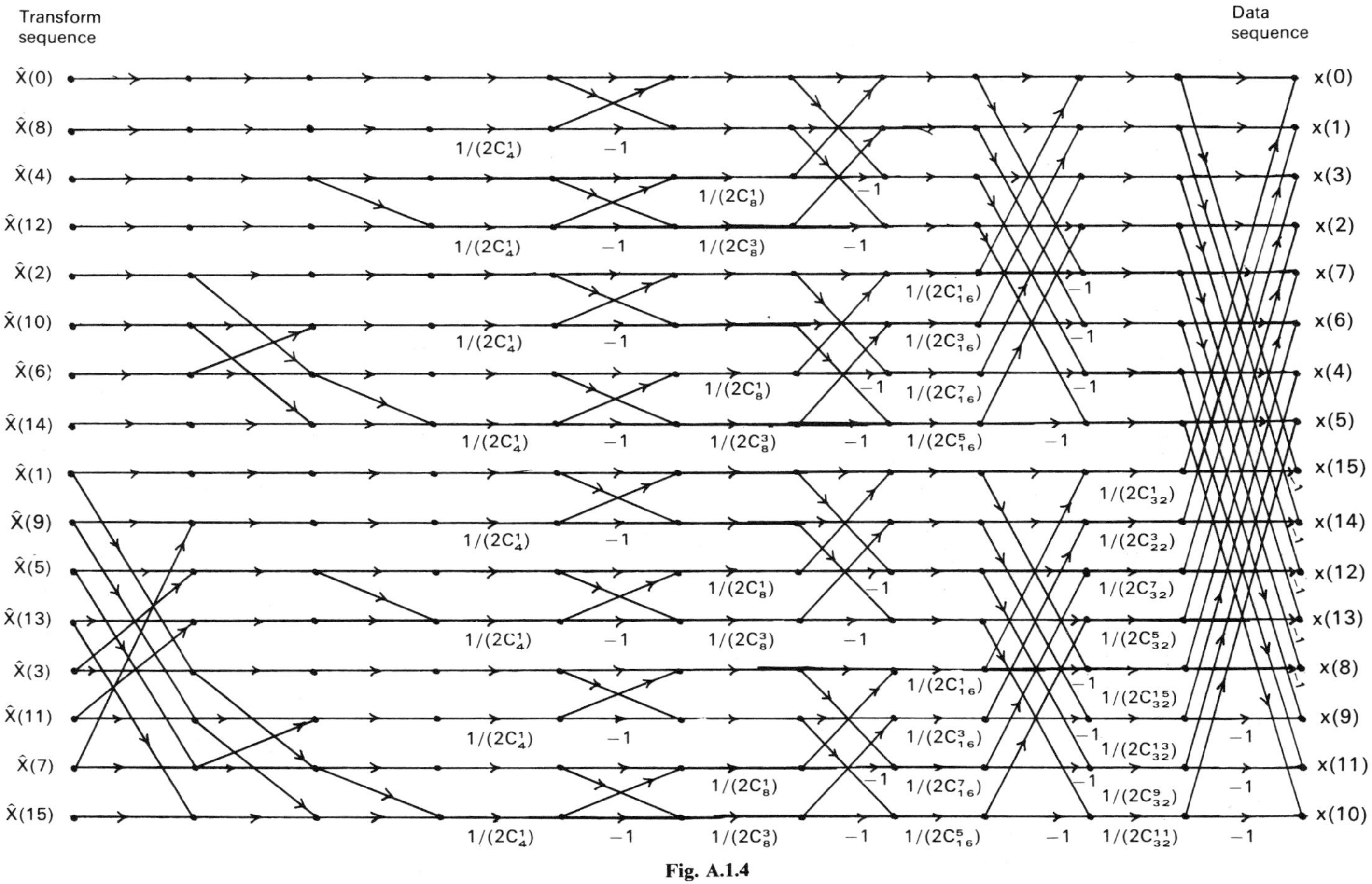
Transform sequence
$\hat{X}(0)$
$\hat{X}(8)$
$\hat{X}(4)$
$\hat{X}(12)$
$\hat{X}(2)$
$\hat{X}(10)$
$\hat{X}(6)$
$\hat{X}(14)$
$\hat{X}(1)$
$\hat{X}(9)$
$\hat{X}(5)$
$\hat{X}(13)$
$\hat{X}(3)$
$\hat{X}(11)$
$\hat{X}(7)$
$\hat{X}(15)$
$1/(2C_{4}^{1})$
$1/(2C_{8}^{1})$
$1/(2C_{8}^{3})$
$1/(2C_{16}^{1})$
$1/(2C_{16}^{3})$
$1/(2C_{16}^{7})$
$1/(2C_{16}^{5})$
$1/(2C_{32}^{1})$
$1/(2C_{22}^{3})$
$1/(2C_{32}^{7})$
$1/(2C_{32}^{5})$
$1/(2C_{32}^{15})$
$1/(2C_{32}^{13})$
$1/(2C_{32}^{9})$
$1/(2C_{32}^{11})$
-1
Data sequence
x(0)
x(1)
x(3)
x(2)
x(7)
x(6)
x(4)
x(5)
x(15)
x(14)
x(12)
x(13)
x(8)
x(9)
x(11)
x(10)

Fig. A.1.4

where

$$\hat{X}^{C(2)}(n) = C_n X^{C(2)}(n) \tag{A.1.4}$$

The flowgraphs for DCT-II and its inverse ($N = 8$) are shown in Figs. A.1.1 and A.1.2, respectively. Corresponding flowgraphs for $N = 16$ are shown in Figs. A.1.3 and A.1.4, respectively. The sparse matrix factors for IDCT-II $N = 8$ and 16 are described by (A.1.5) and (A.1.6), respectively. The corresponding sparse matrix factors for the DCT-II $N = 8$ and 16 are transpose of the matrix factors shown in (A.1.5) and (A.1.6) in reverse order. For the DCT the normalization factor $2/N$ and C_n need to be incorporated in these matrix factors. One can also write down the matrix factors from the flowgraphs. Alternately the flowgraphs can be developed from the matrix factors. Conversion from sparse matrix factors to a flowgraph and vice versa is a straightforward process. The computer programs are developed from these sparse matrix factors. Given the flowgraph for the DCT (IDCT), the corresponding flowgraph for the IDCT (DCT) can be obtained by reversing the direction of the arrows. This is valid because the DCT is an orthogonal transform. For any orthogonal transform, an efficient algorithm for its implementation is a direct result of its sparse matrix factorization. Another consequence of the orthogonality is that this efficiency is preserved in the inverse transformation.

Inspection of Fig. A.1.1 (DCT flowgraph for $N = 8$) and Fig. A.1.3 (DCT flowgraph for $N = 16$) shows that two DCT flowgraphs for $N = 8$ are embedded in the DCT flowgraph for $N = 16$. A similar structure is evident from an inspection of Fig. A.1.2 (IDCT flowgraph for $N = 8$) and Fig. A.1.4 (IDCT flowgraph for $N = 16$). This is a direct consequence of the recursivity of Lee's algorithm [FRA-3]. This property is useful in designing VSLI chips for implementing DCT of sequences of lengths 4, 8, 16, ..., [DP-28, DP-33, DP-41]

$$
\begin{aligned}
&[x(0), x(1), x(3), x(2), x(7), x(6), x(4), x(5)]^T \\
&\quad = \begin{bmatrix} I_4 & I_4 \\ I_4 & -I_4 \end{bmatrix} \operatorname{diag}\left[I_4, \frac{1}{2C_{16}^1}, \frac{1}{2C_{16}^3}, \frac{1}{2C_{16}^7}, \frac{1}{2C_{16}^5} \right] \\
&\qquad \times \left(\operatorname{diag}\left[\begin{bmatrix} I_2 & I_2 \\ I_2 & -I_2 \end{bmatrix}, \begin{bmatrix} I_2 & I_2 \\ I_2 & -I_2 \end{bmatrix} \right] \right) \\
&\qquad \times \left[\operatorname{diag}\left(I_2, \frac{1}{2C_8^1}, \frac{1}{2C_8^3}, I_2, \frac{1}{2C_8^1}, \frac{1}{2C_8^3} \right) \right] \\
&\qquad \times \left[\operatorname{diag}\left(\begin{bmatrix} 1 & 1 \\ 1 & -1 \end{bmatrix}, \begin{bmatrix} 1 & 1 \\ 1 & -1 \end{bmatrix}, \begin{bmatrix} 1 & 1 \\ 1 & -1 \end{bmatrix}, \begin{bmatrix} 1 & 1 \\ 1 & -1 \end{bmatrix} \right) \right] \\
&\qquad \times \left(\operatorname{diag}\left[1, \frac{1}{2C_4^1}, 1, \frac{1}{2C_4^1}, 1, \frac{1}{2C_4^1}, 1, \frac{1}{2C_4^1} \right] \right) \\
&\qquad \times \left[\operatorname{diag}\left(I_2, \begin{bmatrix} 1 & 0 \\ 1 & 1 \end{bmatrix}, I_2, \begin{bmatrix} 1 & 0 \\ 1 & 1 \end{bmatrix} \right) \right]
\end{aligned}
$$

$$\times \ \text{diag}\left\{ I_4, \left[\begin{array}{c|cc} I_2 & \begin{matrix} 0 & 0 \\ 1 & 0 \end{matrix} \\ \hline I_2 & I_2 \end{array}\right] \right\}\Bigg)$$

$$\times [\hat{X}^{C(2)}(0), \hat{X}^{C(2)}(4), \hat{X}^{C(2)}(2), \hat{X}^{C(2)}(6), \hat{X}^{C(2)}(1), \hat{X}^{C(2)}(5), \hat{X}^{C(2)}(3), \hat{X}^{C(2)}(7)]^T. \tag{A.1.5}$$

$$[x(0), x(1), x(3), x(2), x(7), x(6), x(4), x(5), x(15), x(14), x(12), x(13), x(8), x(9), x(11), x(10)]^T$$

$$= \begin{bmatrix} I_8 & I_8 \\ I_8 & -I_8 \end{bmatrix}$$

$$\times \left(\text{diag}\left[I_8, \frac{1}{2C_{32}^1}, \frac{1}{2C_{32}^3}, \frac{1}{2C_{32}^7}, \frac{1}{2C_{32}^5}, \frac{1}{2C_{32}^{15}}, \frac{1}{2C_{32}^{13}}, \frac{1}{2C_{32}^9}, \frac{1}{2C_{32}^{11}}, \right] \right)$$

$$\times \left[\text{diag}\left(\begin{bmatrix} I_4 & I_4 \\ I_4 & -I_4 \end{bmatrix} \begin{bmatrix} I_4 & I_4 \\ I_4 & -I_4 \end{bmatrix} \right) \right]$$

$$\times \left(\text{diag}\left[I_4, \frac{1}{2C_{16}^1}, \frac{1}{2C_{16}^3}, \frac{1}{2C_{16}^7}, \frac{1}{2C_{16}^5}, I_4, \frac{1}{2C_{16}^1}, \frac{1}{2C_{16}^3}, \frac{1}{2C_{16}^7}, \frac{1}{2C_{16}^5} \right] \right)$$

$$\times \left[\text{diag}\left(\begin{bmatrix} I_2 & I_2 \\ I_2 & -I_2 \end{bmatrix}, \begin{bmatrix} I_2 & I_2 \\ I_2 & -I_2 \end{bmatrix}, \begin{bmatrix} I_2 & I_2 \\ I_2 & -I_2 \end{bmatrix}, \begin{bmatrix} I_2 & I_2 \\ I_2 & -I_2 \end{bmatrix} \right) \right]$$

$$\times \left(\text{diag}\left[I_2, \frac{1}{2C_8^1}, \frac{1}{2C_8^3}, I_2, \frac{1}{2C_8^1}, \frac{1}{2C_8^3} \ I_2, \frac{1}{2C_8^1}, \frac{1}{2C_8^3} \ I_2, \frac{1}{2C_8^1}, \frac{1}{2C_8^3} \right] \right)$$

$$\times \left[\text{diag}\left(\begin{bmatrix} 1 & 1 \\ 1 & -1 \end{bmatrix}, \begin{bmatrix} 1 & 1 \\ 1 & -1 \end{bmatrix}, \begin{bmatrix} 1 & 1 \\ 1 & -1 \end{bmatrix}, \begin{bmatrix} 1 & 1 \\ 1 & -1 \end{bmatrix}, \begin{bmatrix} 1 & 1 \\ 1 & -1 \end{bmatrix}, \begin{bmatrix} 1 & 1 \\ 1 & -1 \end{bmatrix}, \begin{bmatrix} 1 & 1 \\ 1 & -1 \end{bmatrix}, \begin{bmatrix} 1 & 1 \\ 1 & -1 \end{bmatrix} \right) \right]$$

$$\times \left[\text{diag}\left(1, \frac{1}{2C_4^1}, 1, \frac{1}{2C_4^1}, 1, \frac{1}{2C_4^1}, 1, \frac{1}{2C_4^1}, 1, \frac{1}{2C_4^1}, 1, \frac{1}{2C_4^1}, 1, \frac{1}{2C_4^1}, 1, \frac{1}{2C_4^1} \right) \right]$$

$$\times \left[\text{diag}\left(I_2, \begin{bmatrix} 1 & 0 \\ 1 & 1 \end{bmatrix}, I_2, \begin{bmatrix} 1 & 0 \\ 1 & 1 \end{bmatrix}, I_2, \begin{bmatrix} 1 & 0 \\ 1 & 1 \end{bmatrix}, I_2, \begin{bmatrix} 1 & 0 \\ 1 & 1 \end{bmatrix} \right) \right]$$

$$\times \left[\text{diag}\left(I_4, \left[\begin{array}{c|cc} I_2 & \begin{matrix} 0 & 0 \\ 1 & 0 \end{matrix} \\ \hline I_2 & I_2 \end{array}\right], I_4, \left[\begin{array}{c|cc} I_2 & \begin{matrix} 0 & 0 \\ 1 & 0 \end{matrix} \\ \hline I_2 & I_2 \end{array}\right] \right) \right]$$

$$\times \left[\operatorname{diag} \left(I_8, \left[\begin{array}{c|c} I_4 & \begin{array}{c|cc} 0 & 0 & 0 \\ & 1 & 0 \\ \hline I_2 & & 0 \end{array} \\ \hline \begin{array}{c|cc} I_2 & & 0 \\ \hline 0 & 0 & 0 \\ & 0 & 1 \end{array} & I_4 \end{array} \right] \right) \right]$$

$$\times [\hat{X}^{C(2)}(0), \hat{X}^{C(2)}(8), \hat{X}^{C(2)}(4), \hat{X}^{C(2)}(12), \hat{X}^{C(2)}(2), \hat{X}^{C(2)}(10), \hat{X}^{C(2)}(6), \hat{X}^{C(2)}(14), \hat{X}^{C(2)}(1), \hat{X}^{C(2)}(9), \hat{X}^{C(2)}(5), \hat{X}^{C(2)}(13), \hat{X}^{C(2)}(3), \hat{X}^{C(2)}(11), \hat{X}^{C(2)}(7), \hat{X}^{C(2)}(15)]. \quad (A.1.6)$$

```
C.............................................................
C
C    PROGRAM FOR FAST DISCRETE COSINE TRANSFORM
C    (N = 8) BASED ON B.G. LEE ALGORITHM
C    REFERENCE: IEEE TRANS. ON ACOUSTICS, SPEECH, AND
C               SIGNAL PROCESSING, VOL. ASSP-32,
C               PP.1243-1245, DEC. 1984.
C               IMPLEMENTED BY J.Y. NAM
C
C.............................................................
        REAL X(8),Z(8)
        N=8
C   INPUT DATA X    ///////////////////////////////////////////
        OPEN (UNIT=3,FILE='DCT8.DAT',FORM='FORMATTED',
     1        STATUS='NEW')
        DO 10 I=1,N
                WRITE(*,44) I
 44     FORMAT(2X,'PLEASE INPUT DATA X',I1)
                READ(*,*) X(I)
                WRITE(*,55) I,X(I)
                WRITE(3,55) I,X(I)
 55     FORMAT(2X,'X',I1,'=',F9.4)
 10     CONTINUE
C CALL FAST DCT SUBROUTINE    //////////////////////////////
        CALL DCT8(X,Z,N)
C OUTPUT THE DCT DATA Z   ///////////////////////////////////
        WRITE(*,*) ' THE DCT DATA'
        WRITE(3,*) ' THE DCT DATA'
        DO 20 I=1,N
                WRITE(*,66) I,Z(I)
                WRITE(3,66) I,Z(I)
 66     FORMAT(2X,'Z',I1,'=',F9.4)
 20     CONTINUE
        CLOSE(3)
        END
```

```
C
C.......................................
C   SUBROUTINE FOR FAST DCT (N = 8)
C   BASED ON  B.G. LEE'S ALGORITHM
C.......................................
C
      SUBROUTINE DCT8(X,Z,N)
C
      REAL X(8),Z(8),G1(8),G2(8),G3(8),G4(8),G5(8)
      REAL G6(8),G7(8),G8(8)
      N1=N/2
      N2=N/4
      PI=3.141592654
      NM1=N-1
      NV2=N/2
C
C CHANGE NATURAL ORDER TO B.G.LEE'S ORDER  ////////////////
C
      DO 5 I=1,2
      G1(I)=X(I)
      G1(2+I)=X(5-I)
      G1(4+I)=X(9-I)
      G1(6+I)=X(4+I)
  5   CONTINUE
C
C IMPLEMENT FAST DCT USING THE FLOW GRAPH OF N = 8  ///////
      DO 10 I=1,N1
      G2(I)=G1(I)+G1(N1+I)
      G2(N1+I)=G1(I)-G1(N1+I)
 10   CONTINUE
C..................................
      DO 20 I=1,N2
      G3(I)=G2(I)
      G3(N2+I)=G2(N2+I)
      G3(N1+I)=G2(N1+I)*(1/(2*COS((2*I-1)*PI/(2*N))))
      G3(N1+N2+I)=G2(N1+N2+I)*(1/(2*COS((N+1-2*I)*PI
     1                         /(2*N))))
 20   CONTINUE
C..................................
      DO 30 I=1,N2
      DO 30 J=1,N2
      G4(4*(I-1)+J)=G3(4*(I-1)+J)+G3(4*(I-1)+N2+J)
      G4(4*(I-1)+N2+J)=G3(4*(I-1)+J)-G3(4*(I-1)+N2+J)
 30   CONTINUE
C..................................
      DO 40 I=1,N2
      DO 40 J=1,N2
      G5(4*(I-1)+J)=G4(4*(I-1)+J)
      G5(4*(I-1)+N2+J)=G4(4*(I-1)+N2+J)*(1/(2*COS((2*J-1)
     1                                   *2*PI/(2*N))))
 40   CONTINUE
C..................................
      DO 50 I=1,N1
      G6(2*I-1)=G5(2*I-1)+G5(2*I)
      G6(2*I)=G5(2*I-1)-G5(2*I)
 50   CONTINUE
```

```
C.................................
        DO 60 I=1,N1
        G7(2*(I-1)+1)=G6(2*(I-1)+1)
        G7(2*I)=G6(2*I)*(1/(2*COS((4*PI)/(2*N))))
  60    CONTINUE
C.................................
        DO 70 I=1,N2
        G7(4*(I-1)+N2+1)=G7(4*(I-1)+N2+1)+G7(4*(I-1)+N1)
  70    CONTINUE
C.................................
        DO 80 I=1,N2
        G8(I)=G7(I)
        G8(N2+I)=G7(N2+I)
        G8(N1+I)=G7(N1+I)+G7(N1+N2+I)
  80    CONTINUE
        G8(N1+N2+1)=G7(N1+N2+1)+G7(N1+N2)
        G8(N)=G7(N)
C.................................
C
C REORDERING THE BIT REVERSE ORDER DATA INTO NATURAL ORDER /////
        J=1
        DO 13 I=1,NM1
           IF(I.GE.J) GO TO 1
           T=G8(J)
           G8(J)=G8(I)
           G8(I)=T
   1       K=NV2
   2    IF(K.GE.J) GO TO 13
           J=J-K
           K=K/2
           GO TO 2
  13    J=J+K
C
C NORMALIZATION COEFFICIENTS  ///////////////////////////////////
        DO 17 I=1,N
        Z(I)=2./FLOAT(N)*G8(I)
        IF(I.EQ.1) Z(I)=2./FLOAT(N)*(1./SQRT(2.))*G8(I)
  17    CONTINUE
C
        RETURN
        END
```

```
C.......................................................
C
C   PROGRAM FOR FAST INVERSE DISCRETE COSINE TRANSFORM
C   (N = 8) BASED ON B.G. LEE ALGORITHM
C   REFERENCE: IEEE TRANS. ON ACOUSTICS, SPEECH, AND
C              SIGNAL PROCESSING, VOL. ASSP-32,
C              PP.1243-1245, DEC. 1984.
C              IMPLEMENTED BY J.Y. NAM
C
C.......................................................
      REAL X(8),Z(8)
C
C INPUT DATA Z  //////////////////////////////////////////
      N=8
      OPEN (UNIT=3,FILE='IDCT8.DAT',FORM='FORMATTED',
   1        STATUS='NEW')
      DO 10 I=1,N
            WRITE(*,40) I
 40   FORMAT(2X,'PLEASE INPUT DATA Z',I1)
            READ(*,*) Z(I)
            WRITE(*,50) I,Z(I)
            WRITE(3,50) I,Z(I)
 50   FORMAT(2X,'Z',I1,'=',F9.4)
 10   CONTINUE
C CALL FAST INVERSE DCT SUBROUTINE   /////////////////////
      CALL IDCT8(Z,X,N)
C OUTPUT THE IDCT DATA X   //////////////////////////////
      WRITE(*,*) ' THE IDCT DATA'
      WRITE(3,*) ' THE IDCT DATA'
      DO 25 I=1,N
            WRITE(*,60) I,X(I)
            WRITE(3,60) I,X(I)
 60   FORMAT (2X,'X',I1,'=',F9.4)
 25   CONTINUE
      CLOSE(3)
      STOP
      END
C
C.......................................
C   SUBROUTINE FOR INVERSE DCT (N = 8)
C   BASED ON  B.G. LEE'S ALGORITHM
C.......................................
C
      SUBROUTINE IDCT8(Z,X,N)
C
      REAL Z(8),X(8),G(8),G1(8),G2(8),G3(8),G4(8),G5(8)
      REAL G6(8),G7(8),G8(8)
      N1=N/2
      N2=N/4
      PI=3.141592654
      NM1=N-1
      NV2=N/2
C
C  NORMALIZE COEFFICIENTS     ///////////////////////////
C       _
```

```
C       X(N) = e(N)*X(N)
C
C                         , where  e(N) = 1./SQRT(2.)  ,FOR N=1
C                                       = 1.           ,FOR N=2,...7
C       X̄(N) = Z(N)
C ......................................................
C
        DO 19 I=1,N
        Z(I)=Z(I)
        IF(I.EQ.1) Z(I)=(1./SQRT(2.))*Z(I)
  19    CONTINUE
C
C  REORDERING THE INPUT DATA ( BRO )   /////////////////////
C
        J=1
        DO 13 I=1,NM1
           IF(I.GE.J) GO TO 1
                T=Z(J)
                Z(J)=Z(I)
                Z(I)=T
  1             K=NV2
  2        IF(K.GE.J) GO TO 13
                J=J-K
                K=K/2
                GO TO 2
  13    J=J+K
C
C IMPLEMENT IDCT USING THE FLOW GRAPH OF N = 8   //////////
C
        DO 20 I=1,N1+1
        G1(I)=Z(I)
  20    CONTINUE
        G1(N1+N2)=Z(N1+N2)+Z(N1+N2+1)
        DO 30 I=1,N2
        G1(N1+N2+I)=Z(N1+I)+Z(N1+N2+I)
  30    CONTINUE
C.....................................
        DO 40 I=1,N2
        G1(4*(I-1)+N1)=G1(4*(I-1)+N1)+G1(4*(I-1)+N1-1)
  40    CONTINUE
C.....................................
        DO 50 I=1,N1
        G2(2*(I-1)+1)=G1(2*(I-1)+1)
        G2(2*I)=G1(2*I)*(1/(2*COS((4*PI)/(2*N))))
  50    CONTINUE
C.....................................
        DO 60 I=1,N1
        G3(2*I-1)=G2(2*I-1)+G2(2*I)
        G3(2*I)=G2(2*I-1)-G2(2*I)
  60    CONTINUE
C.....................................
        DO 70 I=1,N2
        DO 70 J=1,N2
        G4(4*(I-1)+J)=G3(4*(I-1)+J)
        G4(4*(I-1)+N2+J)=G3(4*(I-1)+N2+J)*(1/(2*COS((2*J-1)
```

```
     1                        *2*PI/(2*N))))
  70    CONTINUE
C..................................
        DO 75 I=1,N2
        DO 75 J=1,N2
        G5(4*(I-1)+J)=G4(4*(I-1)+J)+G4(4*(I-1)+N2+J)
        G5(4*(I-1)+N2+J)=G4(4*(I-1)+J)-G4(4*(I-1)+N2+J)
  75    CONTINUE
C..................................
        DO 80 I=1,N2
        G6(I)=G5(I)
        G6(N2+I)=G5(N2+I)
        G6(N1+I)=G5(N1+I)*(1/(2*COS((2*I-1)*PI/(2*N))))
        G6(N1+N2+I)=G5(N1+N2+I)*(1/(2*COS((N+1-2*I)*PI/(2*N))))
  80    CONTINUE
C..................................
        DO 90 I=1,N1
        G7(I)=G6(I)+G6(N1+I)
        G7(N1+I)=G6(I)-G6(N1+I)
  90    CONTINUE
C..................................
C
C CHANGE  B.G. LEE'S ORDER TO NATURAL ORDER   /////////////
        DO 100 I=1,2
        X(I)=G7(I)
        X(5-I)=G7(2+I)
        X(9-I)=G7(4+I)
        X(4+I)=G7(6+I)
 100    CONTINUE
C
C
        RETURN
        END
```

```
C.......................................................
C
C    PROGRAM FOR FAST DISCRETE COSINE TRANSFORM
C    (N = 16) BASED ON  B.G. LEE ALGORITHM
C    REFERENCE: IEEE TRANS. ON ACOUSTICS, SPEECH, AND
C               SIGNAL PROCESSING, VOL. ASSP-32,
C               PP.1243-1245, DEC. 1984.
C               IMPLEMENTED BY J.Y. NAM
C
C.......................................................

        REAL X(16),Z(16)
        N=16
C    INPUT DATA X   ////////////////////////////////////////
        OPEN(UNIT=3,FILE='DCT16.DAT',FORM='FORMATTED',
     1       STATUS='NEW')
        DO 10 I=1,N
                WRITE(*,44) I
  44    FORMAT(2X,'PLEASE INPUT DATA X ',I2)
                READ(*,*) X(I)
                WRITE(*,55) I,X(I)
                WRITE(3,55) I,X(I)
  55    FORMAT(2X,'X',I2,'=',F9.4)
  10    CONTINUE
C    CALL FAST DCT SUBROUTINE   /////////////////////////////
        CALL DCT16(X,Z,N)
C    OUTPUT THE DCT DATA Z    //////////////////////////////
        WRITE(*,*) ' THE DCT DATA '
        WRITE(3,*) ' THE DCT DATA '
C
        DO 20 I=1,N
                WRITE(*,66) I,Z(I)
                WRITE(3,66) I,Z(I)
  66    FORMAT(2X,'Z',I2,'=',F9.4)
  20    CONTINUE
        CLOSE(3)
C
        STOP
        END
C
C............................................
C    SUBROUTINE FOR FAST DCT (N = 16)
C    BASED ON  B.G. LEE'S ALGORITHM
C............................................
C
        SUBROUTINE DCT16(X,Z,N)
        REAL G(16),X(16),Z(16),G1(16),G2(16),G3(16),G4(16)
        REAL G5(16),G6(16),G7(16),G8(16),G9(16),G10(16)
        N1=N/2
        N2=N/4
        N3=N/8
        PI=3.141592654
        NM1=N-1
        NV2=N/2
C
```

```
C CHANGE NATURAL ORDER TO  B.G. LEE'S ORDER   /////////////
        DO 11 I=1,N3
        G(I)=X(I)
        G(2+I)=X(5-I)
        G(4+I)=X(9-I)
        G(6+I)=X(4+I)
        G(8+I)=X(17-I)
        G(10+I)=X(12+I)
        G(12+I)=X(8+I)
        G(14+I)=X(13-I)
  11    CONTINUE
C
C IMPLEMENT DCT USING THE FLOW GRAPH OF N = 16   //////////
C
        DO 10 I=1,N1
        G1(I)=G(I)+G(N1+I)
        G1(N1+I)=G(I)-G(N1+I)
  10    CONTINUE
C......................................
        DO 22 I=1,N1
        G2(I)=G1(I)
  22    CONTINUE
        DO 20 I=1,N3
        G2(N1+I)=G1(N1+I)*(1/(2*COS((2*I-1)*PI/(2*N))))
        G2(N1+N3+I)=G1(N1+N3+I)*(1/(2*COS((N1+1-2*I)*
     1                              PI/(2*N))))
        G2(N1+N2+I)=G1(N1+N2+I)*(1/(2*COS((N+1-2*I)*
     1                              PI/(2*N))))
        G2(N1+N2+N3+I)=G1(N1+N2+N3+I)*(1/(2*COS((N1-1+2*I)*
     1                              PI/(2*N))))
  20    CONTINUE
C......................................
        DO 30 I=1,N3
        DO 30 J=1,N2
        G3(8*(I-1)+J)=G2(8*(I-1)+J)+G2(8*(I-1)+N2+J)
        G3(8*(I-1)+N2+J)=G2(8*(I-1)+J)-G2(8*(I-1)+N2+J)
  30    CONTINUE
C......................................
        DO 44 I=1,N3
        DO 44 J=1,N2
        G4(8*(I-1)+J)=G3(8*(I-1)+J)
  44    CONTINUE
        DO 40 I=1,N3
        DO 40 J=1,N3
        G4(8*(I-1)+N2+J)=G3(8*(I-1)+N2+J)*(1/(2*COS((2*J-1)
     1                              *2*PI/(2*N))))
        G4(8*(I-1)+N2+N3+J)=G3(8*(I-1)+N2+N3+J)*(1/(2*COS
     1                    ((N1+1-2*J)*2*PI/(2*N))))
  40    CONTINUE
C......................................
        DO 50 I=1,N2
        DO 50 J=1,N3
        G5(4*(I-1)+J)=G4(4*(I-1)+J)+G4(4*(I-1)+N3+J)
        G5(4*(I-1)+N3+J)=G4(4*(I-1)+J)-G4(4*(I-1)+N3+J)
  50    CONTINUE
```

```
C..................................
      DO 60 I=1,N2
      DO 60 J=1,N3
      G6(4*(I-1)+J)=G5(4*(I-1)+J)
      G6(4*(I-1)+N3+J)=G5(4*(I-1)+N3+J)*(1/(2*COS((2*J-1)
     1                                 *4*PI/(2*N))))
 60   CONTINUE
C..................................
      DO 70 I=1,N1
      G7(2*I-1)=G6(2*I-1)+G6(2*I)
      G7(2*I)=G6(2*I-1)-G6(2*I)
 70   CONTINUE
C..................................
      DO 80 I=1,N1
      G8(2*I-1)=G7(2*I-1)
      G8(2*I)=G7(2*I)*(1/(2*COS((8*PI)/(2*N))))
 80   CONTINUE
C..................................
      DO 90 I=1,N2
      G8(4*(I-1)+N3+1)=G8(4*(I-1)+N3+1)+G8(4*(I-1)+N2)
 90   CONTINUE
C..................................
      DO 99 I=1,N3
      DO 99 J=1,N2
      G9(8*(I-1)+J)=G8(8*(I-1)+J)
 99   CONTINUE
      DO 100 I=1,N3
      DO 101 J=1,N3
      G9(8*(I-1)+N2+J)=G8(8*(I-1)+N2+J)+G8(8*(I-1)+N2+N3+J)
101   CONTINUE
      G9(8*(I-1)+N2+N3+1)=G8(8*(I-1)+N2+N3+1)
     1                    +G8(8*(I-1)+N2+N3)
      G9(8*(I-1)+N1)=G9(8*(I-1)+N1)
100   CONTINUE
C..................................
      DO 122 I=1,N1
      G10(I)=G9(I)
122   CONTINUE
      DO 120 I=1,N2
      G10(N1+I)=G9(N1+I)+G9(N1+N2+I)
120   CONTINUE
      DO 130 I=1,N3
      G10(N1+N2+I)=G9(N1+N2+I)+G9(N1+N3+I)
130   CONTINUE
      G10(N1+N2+N3+1)=G9(N1+N2+N3+1)+G9(N1+N3)
      G10(N)=G9(N)
C..................................
C
C  REORDER THE BRO DATA INTO NATURAL ORDER  ///////////////
      J=1
      DO 13 I=1,NM1
              IF(I.GE.J) GO TO 1
                      T=G10(J)
                      G10(J)=G10(I)
                      G10(I)=T
```

```
  1                   K=NV2
  2             IF(K.GE.J) GO TO 13
                      J=J-K
                      K=K/2
                      GO TO 2
  13    J=J+K
C
C   NORMALIZE COEFFICIENTS   ////////////////////////////
        DO 17 I=1,N
        Z(I)=2./FLOAT(N)*G10(I)
        IF(I.EQ.1) Z(I)=2./FLOAT(N)*(1./SQRT(2.))*G10(I)
  17    CONTINUE
C
C
        RETURN
        END
```

```
C.............................................................
C
C   PROGRAM FOR FAST INVERSE DISCRETE COSINE TRANSFORM
C   (N = 16) BASED ON  B.G. LEE ALGORITHM
C   REFERENCE: IEEE TRANS. ON ACOUSTICS, SPEECH, AND
C              SIGNAL PROCESSING, VOL. ASSP-32,
C              PP.1243-1245, DEC. 1984.
C              IMPLEMENTED BY J.Y. NAM
C
C.............................................................
C
      REAL X(16),Z(16)
      N=16
C   INPUT DATA Z   ///////////////////////////////////////
      OPEN(UNIT=3,FILE='IDCT16.DAT',FORM='FORMATTED',
     1     STATUS='NEW')
      DO 10 I=1,N
            WRITE(*,44) I
 44   FORMAT(2X,'PLEASE INPUT DATA Z',I2)
            READ(*,*) Z(I)
            WRITE(*,55) I,Z(I)
            WRITE(3,55) I,Z(I)
 55   FORMAT(2X,'Z',I2,'=',F9.4)
 10   CONTINUE
C   CALL FAST IDCT SUBROUTINE   ///////////////////////////
      CALL IDCT16(Z,X,N)
C   OUTPUT THE IDCT DATA X   ///////////////////////////////
      WRITE(*,*) ' THE IDCT DATA '
      WRITE(3,*) ' THE IDCT DATA '
C
      DO 20 I=1,N
            WRITE(*,66) I,X(I)
            WRITE(3,66) I,X(I)
 66   FORMAT(2X,'X',I2,'=',F9.4)
 20   CONTINUE
      CLOSE(3)
C
      STOP
      END
C
C.....................................................
C   SUBROUTINE FOR INVERSE DCT (N = 16)
C   BASED ON B.G. LEE'S ALGORITHM
C.....................................................
C
      SUBROUTINE IDCT16(Z,X,N)
      REAL X(16),Z(16),G(16),G1(16),G2(16),G3(16),G4(16)
      REAL G5(16),G6(16),G7(16),G8(16),G9(16),G10(16)
      N1=N/2
      N2=N/4
      N3=N/8
      PI=3.141592654
      NM1=N-1
      NV2=N/2
C
```

```
C
C  NORMALIZE COEFFICIENTS    ////////////////////////////////
C
C       X̄(N) = e(N)*X(N)
C                       ,where  e(N) = 1./SQRT(2.) ,FOR N=1
C                                    = 1.          ,FOR N=2,...,16
C
C       X̄(N) = Z(N)
C.....................................................
C
       DO 19 I=1,N
       Z(I)=Z(I)
       IF(I.EQ.1) Z(I)=(1./SQRT(2.))*Z(I)
  19   CONTINUE
C
C  REORDERING THE INPUT DATA ( BRO ) !!!!
C
       J=1
       DO 13 I=1,NM1
          IF(I.GE.J) GO TO 1
               T=Z(J)
               Z(J)=Z(I)
               Z(I)=T
  1            K=NV2
  2       IF(K.GE.J) GO TO 13
               J=J-K
               K=K/2
               GO TO 2
  13   J=J+K
C
C IMPLEMENT IDCT USING THE FLOW GRAPH OF N = 16  //////////
C
       DO 20 I=1,N1+1
       G1(I)=Z(I)
  20   CONTINUE
       G1(N1+N3)=Z(N1+N3)+Z(N1+N2+N3+1)
       DO 30 I=1,N3
       G1(N1+N3+I)=Z(N1+N3+I)+Z(N1+N2+I)
  30   CONTINUE
       DO 40 I=1,N2
       G1(N1+N2+I)=Z(N1+I)+Z(N1+N2+I)
  40   CONTINUE
C
C.................................
C
       DO 55 I=1,N3
       DO 55 J=1,N2+1
       G2(8*(I-1)+J)=G1(8*(I-1)+J)
  55   CONTINUE
       DO 50 I=1,N3
       DO 60 J=1,N3
       G2(8*(I-1)+N2+N3+J)=G1(8*(I-1)+N2+J)+
    1                          G1(8*(I-1)+N2+N3+J)
  60   CONTINUE
       G2(8*(I-1)+N2+N3)=G1(8*(I-1)+N2+N3)+
```

```
     1                      G1(8*(I-1)+N2+N3+1)
  50    CONTINUE
C..................................
        DO 70 I=1,N2
        G2(4*(I-1)+N2)=G2(4*(I-1)+N2)+G2(4*(I-1)+N2-1)
  70    CONTINUE
C..................................
        DO 80 I=1,N1
        G3(2*I-1)=G2(2*I-1)
        G3(2*I)=G2(2*I)*(1/(2*COS((8*PI)/(2*N))))
  80    CONTINUE
C..................................
        DO 90 I=1,N1
        G4(2*I-1)=G3(2*I-1)+G3(2*I)
        G4(2*I)=G3(2*I-1)-G3(2*I)
  90    CONTINUE
C..................................
        DO 100 I=1,N2
        DO 100 J=1,N3
        G5(4*(I-1)+J)=G4(4*(I-1)+J)
        G5(4*(I-1)+N3+J)=G4(4*(I-1)+N3+J)*(1/(2*COS((2*J-1)*4
     1                                 *PI/(2*N))))
 100    CONTINUE
C..................................
        DO 110 I=1,N2
        DO 110 J=1,N3
        G6(4*(I-1)+J)=G5(4*(I-1)+J)+G5(4*(I-1)+N3+J)
        G6(4*(I-1)+N3+J)=G5(4*(I-1)+J)-G5(4*(I-1)+N3+J)
 110    CONTINUE
C..................................
        DO 122 I=1,N3
        DO 122 J=1,N2
        G7(8*(I-1)+J)=G6(8*(I-1)+J)
 122    CONTINUE
        DO 120 I=1,N3
        DO 120 J=1,N3
        G7(8*(I-1)+N2+J)=G6(8*(I-1)+N2+J)*(1/(2*COS((2*J-1)*
     1                                 2*PI/(2*N))))
        G7(8*(I-1)+N2+N3+J)=G6(8*(I-1)+N2+N3+J)*(1/(2*COS((N1
     1                              +1-2*J)*2*PI/(2*N))))
 120    CONTINUE
C..................................
        DO 130 I=1,N3
        DO 130 J=1,N2
        G8(8*(I-1)+J)=G7(8*(I-1)+J)+G7(8*(I-1)+N2+J)
        G8(8*(I-1)+N2+J)=G7(8*(I-1)+J)-G7(8*(I-1)+N2+J)
 130    CONTINUE
C..................................
        DO 144 I=1,N1
        G9(I)=G8(I)
 144    CONTINUE
        DO 140 I=1,N3
        G9(N1+I)=G8(N1+I)*(1/(2*COS((2*I-1)*PI/(2*N))))
        G9(N1+N3+I)=G8(N1+N3+I)*(1/(2*COS((N1+1-2*I)*PI
     1                        /(2*N))))
```

```
      G9(N1+N2+I)=G8(N1+N2+I)*(1/(2*COS((N+1-2*I)*PI
     1                             /(2*N))))
      G9(N1+N2+N3+I)=G8(N1+N2+N3+I)*(1/(2*COS((N1-1+2*I)
     1                                    *PI/(2*N))))
 140  CONTINUE
C
C....................................
      DO 150 I=1,N1
      G10(I)=G9(I)+G9(N1+I)
      G10(N1+I)=G9(I)-G9(N1+I)
 150  CONTINUE
C....................................
C
C  CHANGE B.G. LEE'S ORDER TO NOTURAL ORDER    //////////////
      DO 22 I=1,N3
      X(I)=G10(I)
      X(5-I)=G10(2+I)
      X(9-I)=G10(4+I)
      X(4+I)=G10(6+I)
      X(17-I)=G10(8+I)
      X(12+I)=G10(10+I)
      X(8+I)=G10(12+I)
      X(13-I)=G10(14+I)
 22   CONTINUE
C
      RETURN
      END
```

APPENDIX A.2

COMPUTER PROGRAMS FOR DCT-II AND IDCT-II, $N = 8$ AND 16 [FD-1, FRA-12]

This appendix describes the computer programs for DCT-II and IDCT-II, $N = 8$ and 16 based on the algorithms developed by Wang [FD-1], and Suehiro and Hatori [FRA-12]. As in Appendix A.1 these programs are supplemented with flowgraphs and corresponding sparse matrix factors.

In this appendix the definitions for DCT-II and its inverse are based on [FD-1] and [FRA-12], i.e.,

DCT-II

$$X^{C(2)}(n) = \sqrt{\frac{2}{N}}\, C(n) \sum_{k=0}^{N-1} x(k) \cos\left[\frac{\pi(2k+1)n}{2N}\right]$$

$$n = 0, 1, \ldots, N-1 \qquad \text{(A.2.1)}$$

IDCT-II

$$x(k) = \sqrt{\frac{2}{N}} \sum_{n=0}^{N-1} C(n) X^{C(2)}(n) \cos\left[\frac{\pi(2k+1)n}{2N}\right]$$

$$k = 0, 1, \ldots, N-1, \qquad \text{(A.2.2)}$$

where

$$C(n) = 1/\sqrt{2} \qquad n = 0$$
$$= 1 \qquad n \neq 0.$$

The flowgraphs for DCT-II and its inverse ($N = 8$) are shown in Figs. A.2.1 and A.2.2, respectively. Corresponding flowgraphs for $N = 16$ are shown in Figs. A.2.3 and A.2.4, respectively. The sparse matrix factors for DCT-II, $N = 8$ and 16 are described in (A.2.3) and (A.2.4), respectively. Corresponding matrix factors for the IDCT-II, $N = 8$ and 16 can be easily obtained. As described in Appendix A.1, conversion from flowgraph to matrix factors or vice versa is a straightforward process. By reversing the direction of arrows in the flowgraph for DCT (IDCT), the corresponding flowgraph for IDCT (DCT) can be obtained.

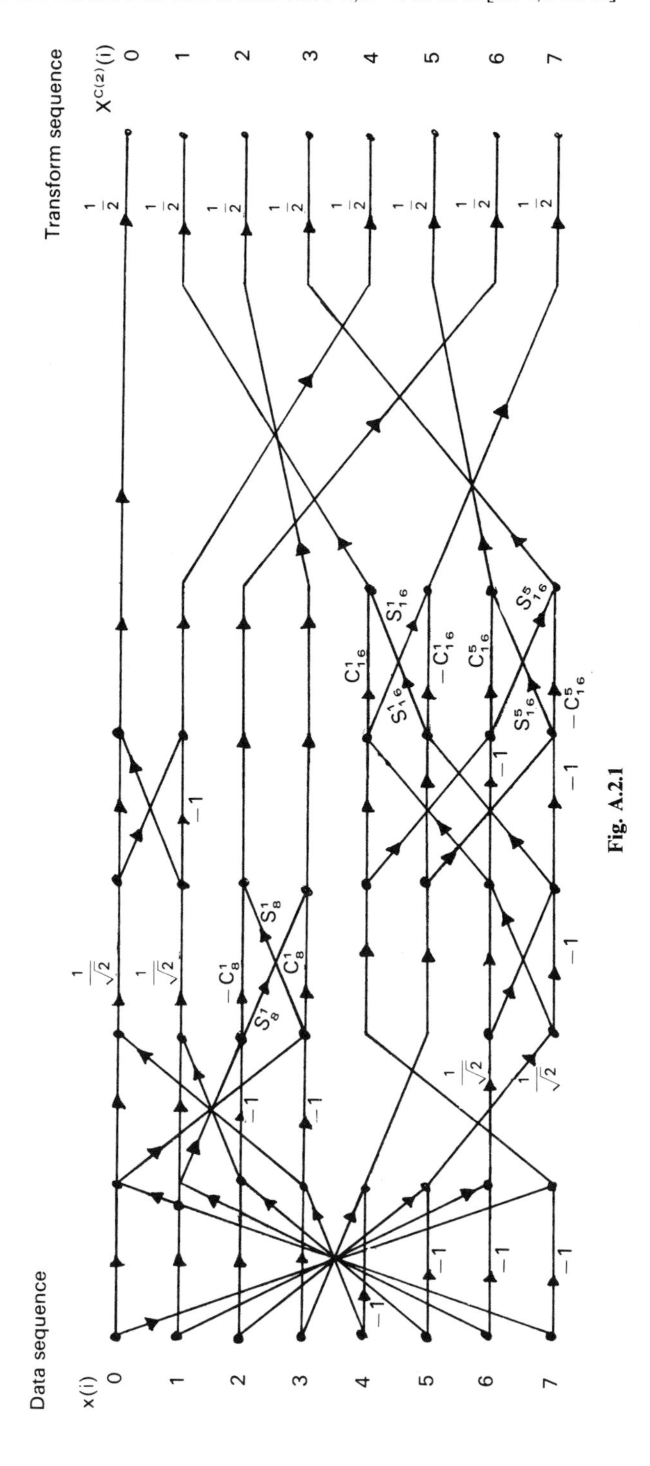

Fig. A.2.1

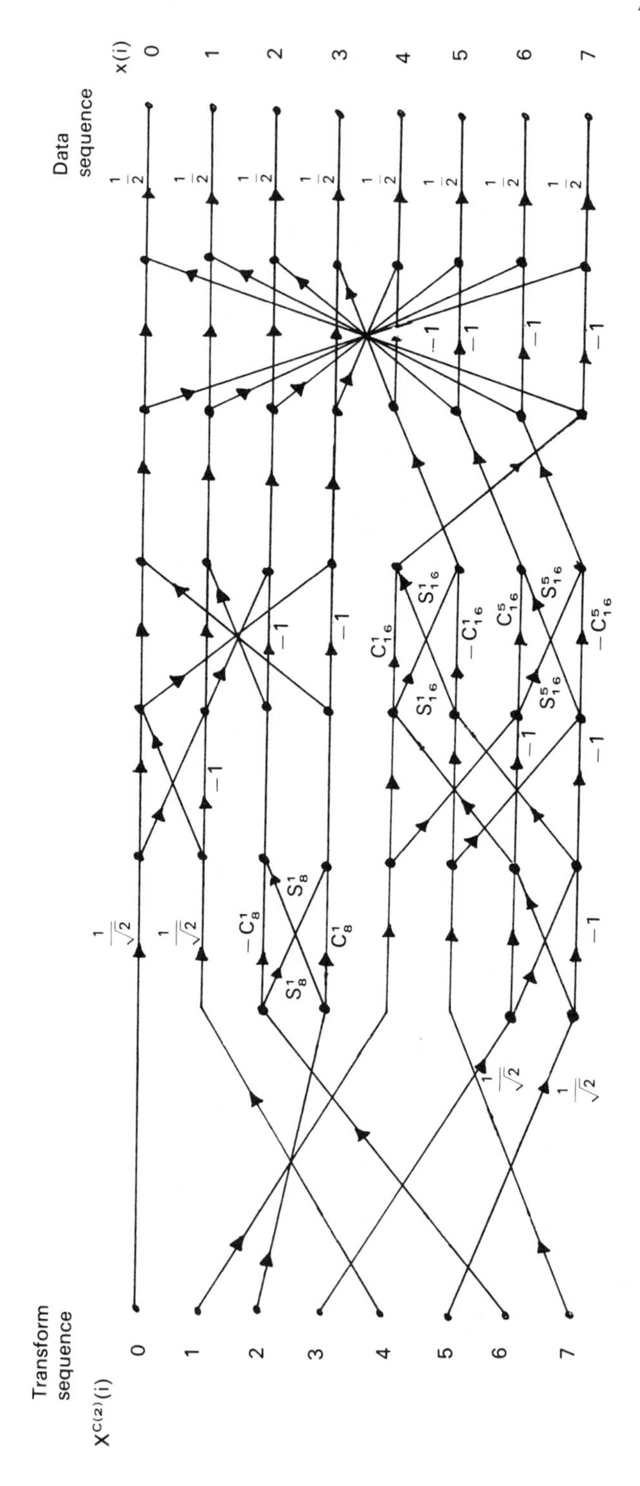
Data sequence
x(i)
0
1
2
3
4
5
6
7
Transform sequence
$X^{C(2)}(i)$
0
1
2
3
4
5
6
7
$\frac{1}{2}$
$\frac{1}{\sqrt{2}}$
$-C_8^1$
C_8^1
S_8^1
C_{16}^1
$-C_{16}^1$
C_{16}^5
$-C_{16}^5$
S_{16}^1
S_{16}^5
-1

Fig. A.2.2

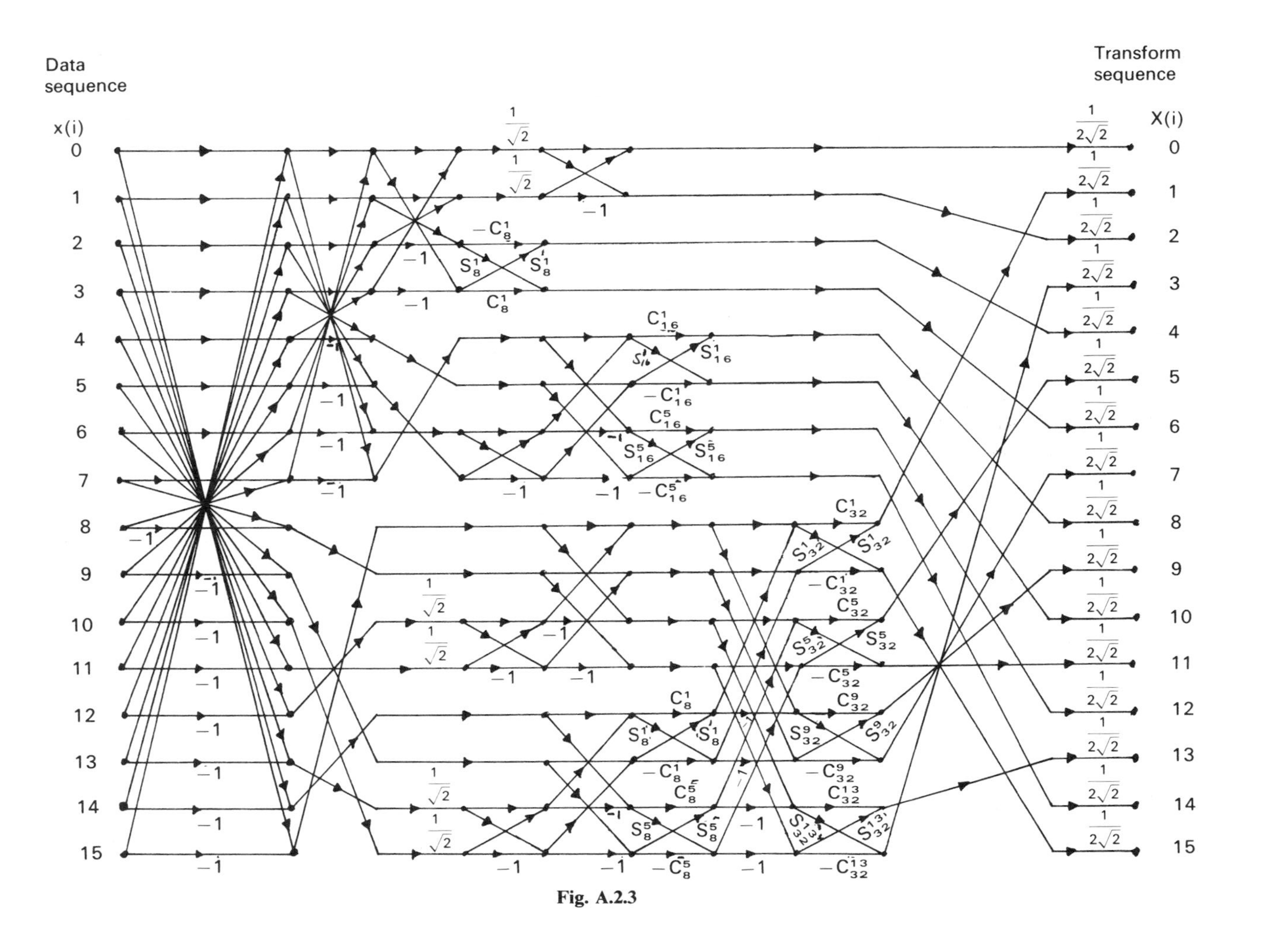

Fig. A.2.3

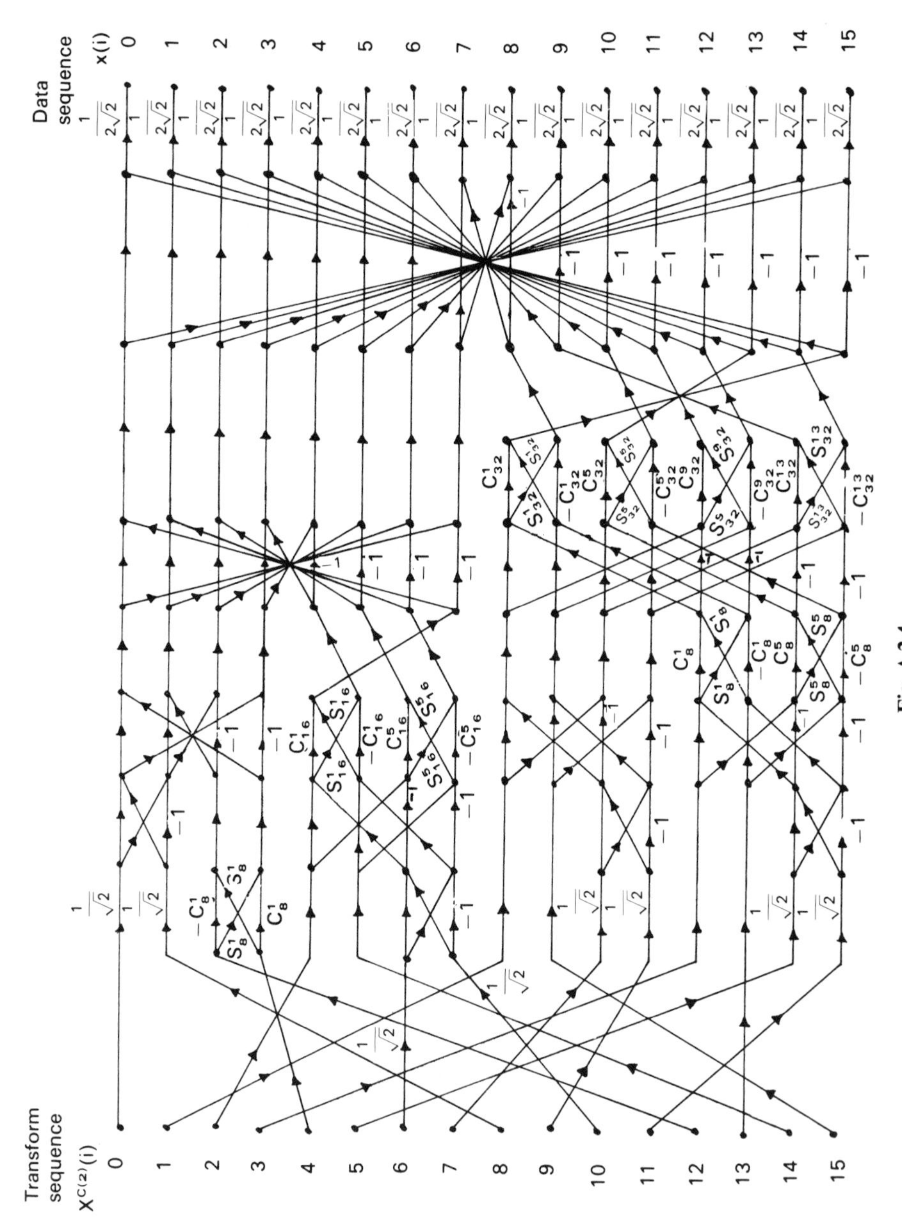
Transform sequence
$X^{C(2)}(i)$
Data sequence
$x(i)$
$\frac{1}{\sqrt{2}}$
$\frac{1}{2\sqrt{2}}$
-1
C_8^1
$-C_8^1$
S_8^1
C_8^5
$-C_8^5$
S_8^5
C_{16}^1
$-C_{16}^1$
S_{16}^1
C_{16}^5
$-C_{16}^5$
S_{16}^5
C_{32}^1
$-C_{32}^1$
S_{32}^1
C_{32}^5
$-C_{32}^5$
S_{32}^5
C_{32}^9
$-C_{32}^9$
S_{32}^9
C_{32}^{13}
$-C_{32}^{13}$
S_{32}^{13}

Fig. A.2.4

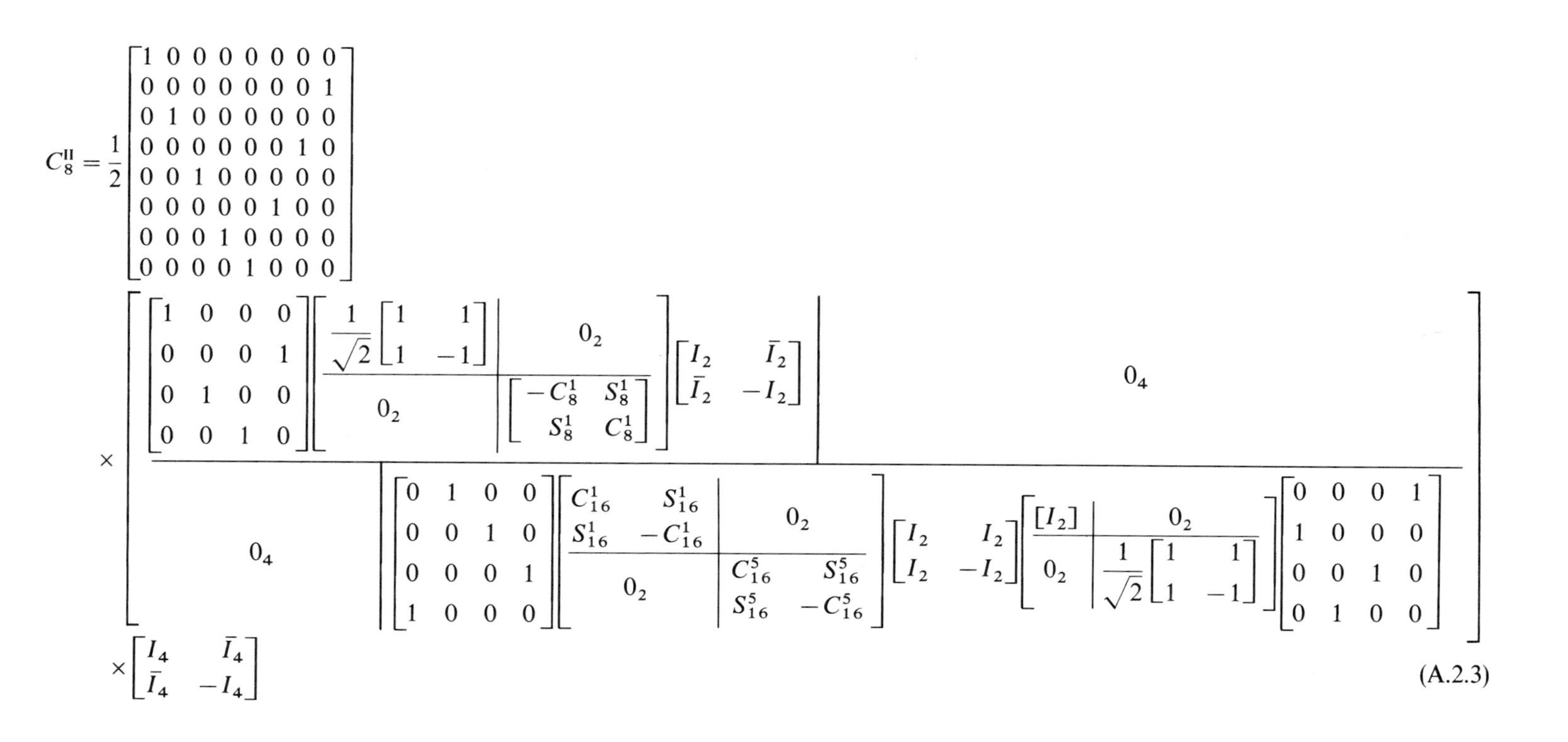

$$C_8^{\mathrm{II}} = \frac{1}{2}\begin{bmatrix} 1&0&0&0&0&0&0&0\\ 0&0&0&0&0&0&0&1\\ 0&1&0&0&0&0&0&0\\ 0&0&0&0&0&0&1&0\\ 0&0&1&0&0&0&0&0\\ 0&0&0&0&0&1&0&0\\ 0&0&0&1&0&0&0&0\\ 0&0&0&0&1&0&0&0 \end{bmatrix}$$

$$\times \left[\begin{array}{c|c} \begin{bmatrix}1&0&0&0\\0&0&0&1\\0&1&0&0\\0&0&1&0\end{bmatrix}\left[\begin{array}{c|c} \frac{1}{\sqrt{2}}\begin{bmatrix}1&1\\1&-1\end{bmatrix} & 0_2 \\ \hline 0_2 & \begin{bmatrix}-C_8^1 & S_8^1\\ S_8^1 & C_8^1\end{bmatrix}\end{array}\right]\begin{bmatrix}I_2 & \bar{I}_2\\ \bar{I}_2 & -I_2\end{bmatrix} & 0_4 \\ \hline 0_4 & \begin{bmatrix}0&1&0&0\\0&0&1&0\\0&0&0&1\\1&0&0&0\end{bmatrix}\left[\begin{array}{c|c} \begin{matrix}C_{16}^1 & S_{16}^1\\ S_{16}^1 & -C_{16}^1\end{matrix} & 0_2 \\ \hline 0_2 & \begin{matrix}C_{16}^5 & S_{16}^5\\ S_{16}^5 & -C_{16}^5\end{matrix}\end{array}\right]\begin{bmatrix}I_2 & I_2\\ I_2 & -I_2\end{bmatrix}\left[\begin{array}{c|c} [I_2] & 0_2 \\ \hline 0_2 & \frac{1}{\sqrt{2}}\begin{bmatrix}1&1\\1&-1\end{bmatrix}\end{array}\right]\begin{bmatrix}0&0&0&1\\1&0&0&0\\0&0&1&0\\0&1&0&0\end{bmatrix} \end{array}\right]$$

$$\times \begin{bmatrix}I_4 & \bar{I}_4\\ \bar{I}_4 & -I_4\end{bmatrix} \tag{A.2.3}$$

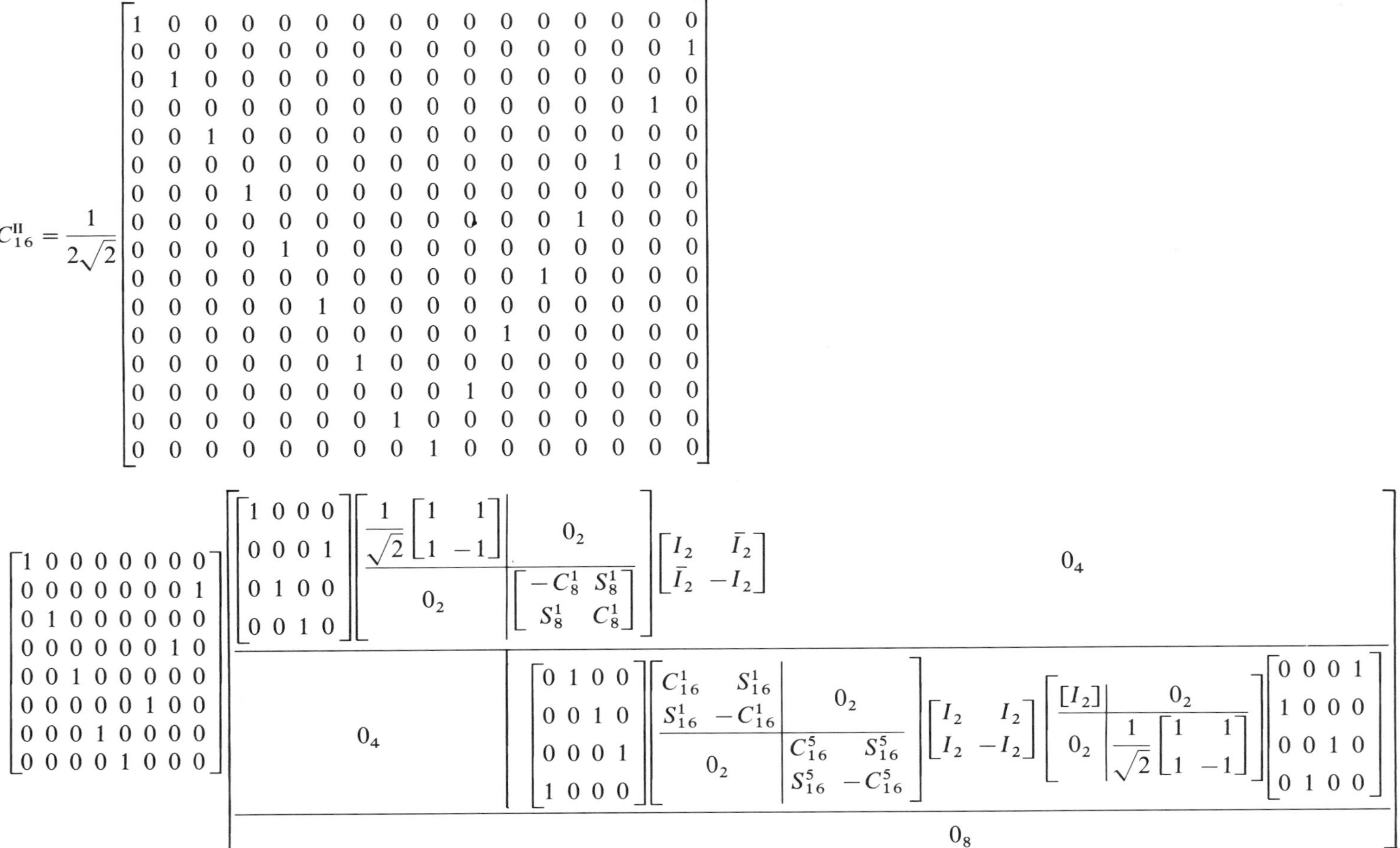

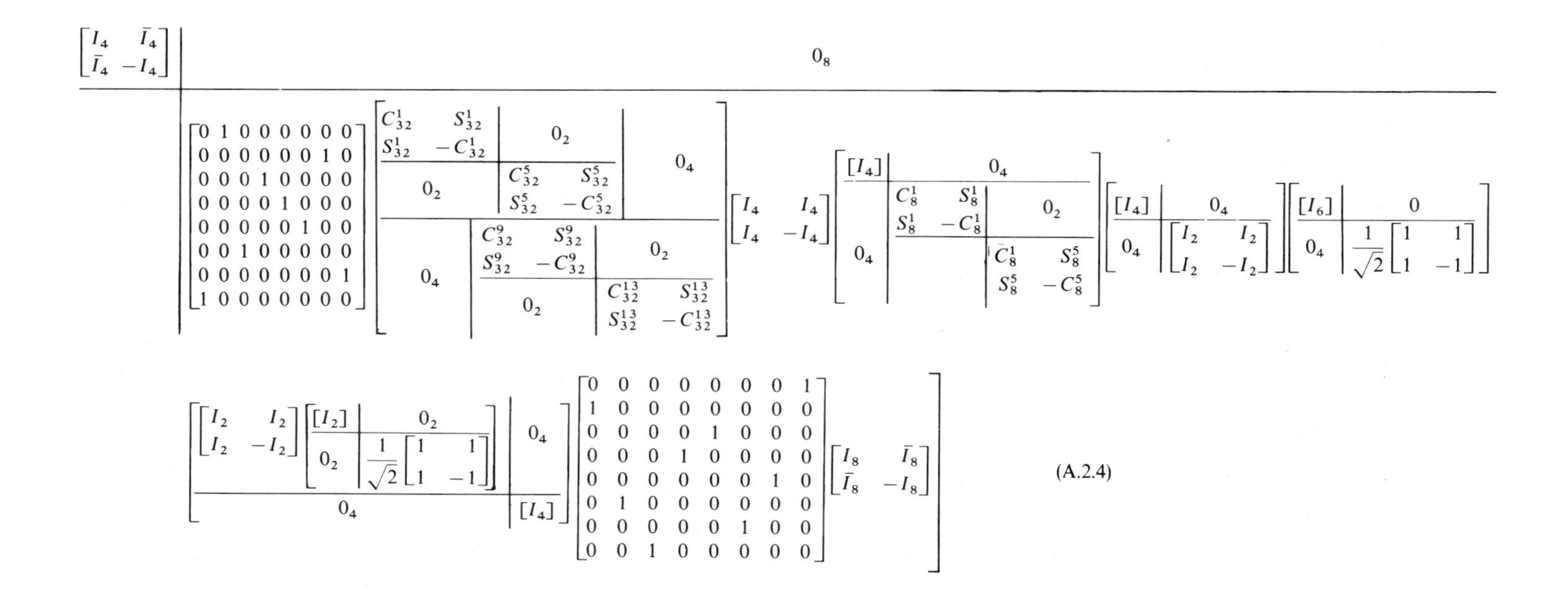

(A.2.4)

Program for DCT-II, N=8 based on [FD-1, FRA-12]

```
C*****************************************************
C      8 POINT FDCT-II
C*****************************************************
        PROGRAM FDCT-II
C*****************************************************
        REAL X1(8),X2(8),X3(4),X4(4),X5(4),X7(4),X8(4)
        REAL X9(4),X10(4),X12(4),Y(8)
        REAL PI,C1I4,S1I4,C1I8,S1I8,C1I4B,S1I4B,V1
        REAL C1I16,S1I16,C5I16,S5I16

C*****************************************************
C       INPUT DATA X
        OPEN (UNIT=6,FILE='FDCT2.DAT',STATUS='UNKNOWN')
        WRITE(6,*) 'THE INPUT DATA '
        DO 5 K=1,8
           WRITE (*,40) K
 40     FORMAT(2X,'PLEASE INPUT DATA X ',I1)
           READ *,X1(K)
           WRITE(*,50) K,X1(K)
           WRITE(6,50) K,X1(K)
 50     FORMAT (2X,'X',I1,'=',F9.4)
 5      CONTINUE

C******************************************************
C       X2=MA8II*X1
C  --------------------
        X2(1)=X1(1)+X1(8)
        X2(2)=X1(2)+X1(7)
        X2(3)=X1(3)+X1(6)
        X2(4)=X1(4)+X1(5)
        X2(5)=X1(4)-X1(5)
        X2(6)=X1(3)-X1(6)
        X2(7)=X1(2)-X1(7)
        X2(8)=X1(1)-X1(8)

C  ----------------------
C       X3=MA4II*X2(1-4)
C  ----------------------
        X3(1)=X2(1)+X2(4)
        X3(2)=X2(2)+X2(3)
        X3(3)=X2(2)-X2(3)
        X3(4)=X2(1)-X2(4)
C  ----------------------
C       X4=MC4V*X3
C  ----------------------
        PI=3.141592658
        C1I4=(COS(PI/4))*0.5
        S1I4=C1I4
        C1I8=(COS(PI/8))*0.5
        S1I8=(SIN(PI/8))*0.5
        X4(1)=(X3(1)+X3(2))*C1I4
        X4(2)=(X3(1)-X3(2))*C1I4
        X4(3)=(X3(3)*(-C1I8))+(X3(4)*S1I8)
        X4(4)=(X3(3)*S1I8)+(X3(4)*C1I8)
C  ----------------------
```

```
C        X5=MP4*X4
C  ----------------------
         X5(1)=X4(1)
         X5(2)=X4(4)
         X5(3)=X4(2)
         X5(4)=X4(3)
C  -----------------------
C        X7=MH4*X6=MH4*MII4*X2(5-8)
C  -----------------------
         X7(1)=X2(8)
         X7(2)=X2(5)
         X7(3)=X2(7)
         X7(4)=X2(6)
C  -----------------------
C        X8=MX4*X7
C  -----------------------
         C1I4B=(COS(PI/4))**2
         S1I4B=C1I4B
         V1=1/(SQRT(2.0))
         X8(1)=X7(1)*V1
         X8(2)=X7(2)*V1
         X8(3)=(X7(3)+X7(4))*C1I4B
         X8(4)=(X7(3)-X7(4))*C1I4B
C  -------------------------
C        X9=MR4*X8
C  -------------------------
         X9(1)=(X8(1)+X8(3))*V1
         X9(2)=(X8(2)+X8(4))*V1
         X9(3)=(X8(1)-X8(3))*V1
         X9(4)=(X8(2)-X8(4))*V1
C  --------------------------
C        X10=MV4*X9
C  --------------------------
         C1I16=COS(PI/16)
         S1I16=SIN(PI/16)
         C5I16=COS(PI*5/16)
         S5I16=SIN(PI*5/16)
         X10(1)=X9(1)*C1I16+X9(2)*S1I16
         X10(2)=X9(1)*S1I16-X9(2)*C1I16
         X10(3)=X9(3)*C5I16+X9(4)*S5I16
         X10(4)=X9(3)*S5I16-X9(4)*C5I16
C  --------------------------
C        X12=MII4*X11=MII4*MO4*X10
C  --------------------------
         X12(1)=X10(2)
         X12(2)=X10(3)
         X12(3)=X10(4)
         X12(4)=X10(1)
C  --------------------------
C        Y=MP8*[X5,X12]
         Y(1)=X5(1)
         Y(2)=X12(4)
         Y(3)=X5(2)
         Y(4)=X12(3)
         Y(5)=X5(3)
```

```
      Y(6)=X12(2)
      Y(7)=X5(4)
      Y(8)=X12(1)
C***********************************************************
C     OUTPUT THE DCT DATA Y
      WRITE(*,*) 'THE DCT DATA'
      WRITE(6,*) 'THE DCT DATA'
      DO 10 K=1,8
         WRITE(*,60) K,Y(K)
         WRITE(6,60) K,Y(K)
 60      FORMAT(2X,'Y',I1,'=',F9.4)
 10   CONTINUE
      CLOSE(6)
      STOP
      END
```

Program for IDCT-II, N=8 based on [FD-1, FRA-12]

```
C*****************************************************
C       8 POINT IFDCT-II
C*****************************************************
        PROGRAM IFDCT-II
C*****************************************************
        REAL X1(8),X2(8),X3(4),X4(4),X5(4),X7(4),X8(4)
        REAL X9(4),X10(4),X12(4),Y(8)
        REAL PI,C1I4,S1I4,C1I8,S1I8,C1I4B,S1I4B,V1
        REAL C1I16,S1I16,C5I16,S5I16
C*****************************************************
C       INPUT DATA X
        OPEN (UNIT=6,FILE='IFDCT2.DAT',STATUS='UNKNOWN')
        WRITE(6,*) 'THE INPUT DATA '
        DO 5 K=1,8
           WRITE (*,40) K
 40     FORMAT(2X,'PLEASE INPUT DATA X ',I1)
           READ *,X1(K)
           WRITE(*,50) K,X1(K)
           WRITE(6,50) K,X1(K)
 50     FORMAT (2X,'X',I1,'=',F9.4)
 5      CONTINUE

C*****************************************************
C       X2=MP8T*X1
C  -------------------
        X2(1)=X1(1)
        X2(2)=X1(3)
        X2(3)=X1(5)
        X2(4)=X1(7)
        X2(5)=X1(8)
        X2(6)=X1(6)
        X2(7)=X1(4)
        X2(8)=X1(2)
C  ---------------------
C       X3=MP4T*X2(1-4)
C  ---------------------
        X3(1)=X2(1)
        X3(2)=X2(3)
        X3(3)=X2(4)
        X3(4)=X2(2)
C  ---------------------
C       X4=MC4V*X3
C  ---------------------
        PI=3.141592658
        C1I4=(COS(PI/4))*0.5
        S1I4=C1I4
        C1I8=(COS(PI/8))*0.5
        S1I8=(SIN(PI/8))*0.5
        X4(1)=(X3(1)+X3(2))*C1I4
        X4(2)=(X3(1)-X3(2))*C1I4
        X4(3)=(X3(3)*(-C1I8))+(X3(4)*S1I8)
        X4(4)=(X3(3)*S1I8)+(X3(4)*C1I8)
C  ----------------------
C       X5=MA4II*X4
```

```
C  -----------------------
         X5(1)=X4(1)+X4(4)
         X5(2)=X4(2)+X4(3)
         X5(3)=X4(2)-X4(3)
         X5(4)=X4(1)-X4(4)
C  ------------------------
C        X7=MH4*X6=MH4*MII4*X2(5-8)
C  ------------------------
         X7(1)=X2(8)
         X7(2)=X2(5)
         X7(3)=X2(7)
         X7(4)=X2(6)
C  ------------------------
C        X8=MX4*X7
C  ------------------------
         C1I4B=(COS(PI/4))**2
         S1I4B=C1I4B
         V1=1/(SQRT(2.0))
         X8(1)=X7(1)*V1
         X8(2)=X7(2)*V1
         X8(3)=(X7(3)+X7(4))*C1I4B
         X8(4)=(X7(3)-X7(4))*C1I4B
C  --------------------------
C        X9=MR4*X8
C  --------------------------
         X9(1)=(X8(1)+X8(3))*V1
         X9(2)=(X8(2)+X8(4))*V1
         X9(3)=(X8(1)-X8(3))*V1
         X9(4)=(X8(2)-X8(4))*V1
C  ---------------------------
C        X10=MV4*X9
C  ---------------------------
         C1I16=COS(PI/16)
         S1I16=SIN(PI/16)
         C5I16=COS(PI*5/16)
         S5I16=SIN(PI*5/16)
         X10(1)=X9(1)*C1I16+X9(2)*S1I16
         X10(2)=X9(1)*S1I16-X9(2)*C1I16
         X10(3)=X9(3)*C5I16+X9(4)*S5I16
         X10(4)=X9(3)*S5I16-X9(4)*C5I16
C  ---------------------------
C        X12=MII4*X11=MII4*MO4*X10
C  ---------------------------
         X12(1)=X10(2)
         X12(2)=X10(3)
         X12(3)=X10(4)
         X12(4)=X10(1)
C  ---------------------------
C        Y=MA8II*[X5,X12]
         Y(1)=X5(1)+X12(4)
         Y(2)=X5(2)+X12(3)
         Y(3)=X5(3)+X12(2)
         Y(4)=X5(4)+X12(1)
         Y(5)=X5(4)-X12(1)
         Y(6)=X5(3)-X12(2)
```

```
      Y(7)=X5(2)-X12(3)
      Y(8)=X5(1)-X12(4)
C***********************************************************
C     OUTPUT THE IDCT DATA Y
      WRITE(*,*) 'THE IDCT DATA'
      WRITE(6,*) 'THE IDCT DATA'
      DO 10 K=1,8
         WRITE(*,60) K,Y(K)
         WRITE(6,60) K,Y(K)
 60      FORMAT(2X,'Y',I1,'=',F9.4)
 10   CONTINUE
      CLOSE(6)
      STOP
      END
```

Program for DCT-II, N=16 based on [FD-1, FRA-12]

```
C*****************************************************************
C       16 POINT FDCT-II
C*****************************************************************
        PROGRAM FDCT16-II
C*****************************************************************
        REAL X0(16),X1(16),X2(8),X3(4),X4(4),X5(4),X7(4),X8(4)
        REAL X9(4),X10(4),X12(4),X13(8),X14(8),X15(8),X16(4)
        REAL X17(4),X18(4),X19(8),X20(8),X21(8),X22(8),X23(8)
        REAL X24(8),X25(8),Y(16)
        REAL PI,C1I4A,S1I4A,C1I8A,S1I8A,C1I4B,S1I4B,V1
        REAL C1I16,S1I16,C5I16,S5I16,C1I4,S1I4,C1I8,S1I8
        REAL C5IB,S5IB,C1I32,S1I32,C5I32,S5I32,C9I32
        REAL S9I32,C13I32,S13I32
C*****************************************************************
C       INPUT DATA X
        OPEN (UNIT=6,FILE='FDCT16-I.DAT',STATUS='UNKNOWN')
        WRITE(6,*) 'THE INPUT DATA '
        DO 5 K=1,16
           WRITE (*,40) K
 40     FORMAT(2X,'PLEASE INPUT DATA X ',I2)
           READ *,X0(K)
           WRITE(*,50) K,X0(K)
           WRITE(6,50) K,X0(K)
 50     FORMAT (2X,'X',I2,'=',F9.4)
 5      CONTINUE
C*****************************************************
        V1=1/(SQRT(2.0))
        X1(1)=(X0(1)+X0(16))*V1
        X1(2)=(X0(2)+X0(15))*V1
        X1(3)=(X0(3)+X0(14))*V1
        X1(4)=(X0(4)+X0(13))*V1
        X1(5)=(X0(5)+X0(12))*V1
        X1(6)=(X0(6)+X0(11))*V1
        X1(7)=(X0(7)+X0(10))*V1
        X1(8)=(X0(8)+X0(9))*V1
        X1(9)=(X0(8)-X0(9))*V1
        X1(10)=(X0(7)-X0(10))*V1
        X1(11)=(X0(6)-X0(11))*V1
        X1(12)=(X0(5)-X0(12))*V1
        X1(13)=(X0(4)-X0(13))*V1
        X1(14)=(X0(3)-X0(14))*V1
        X1(15)=(X0(2)-X0(15))*V1
        X1(16)=(X0(1)-X0(16))*V1
C  ---------------------------------
        X2(1)=X1(1)+X1(8)
        X2(2)=X1(2)+X1(7)
        X2(3)=X1(3)+X1(6)
        X2(4)=X1(4)+X1(5)
        X2(5)=X1(4)-X1(5)
        X2(6)=X1(3)-X1(6)
        X2(7)=X1(2)-X1(7)
        X2(8)=X1(1)-X1(8)
C  ---------------------------------
```

```
      X3(1)=X2(1)+X2(4)
      X3(2)=X2(2)+X2(3)
      X3(3)=X2(2)-X2(3)
      X3(4)=X2(1)-X2(4)
C     ----------------------
      PI=3.141592658
      C1I4A=(COS(PI/4))*0.5
      S1I4A=C1I4A
      C1I8A=(COS(PI/8))*0.5
      S1I8A=(SIN(PI/8))*0.5
      X4(1)=(X3(1)+X3(2))*C1I4A
      X4(2)=(X3(1)-X3(2))*C1I4A
      X4(3)=(X3(3)*(-C1I8A))+(X3(4)*S1I8A)
      X4(4)=(X3(3)*S1I8A)+(X3(4)*C1I8A)
C     -----------------------
      X5(1)=X4(1)
      X5(2)=X4(4)
      X5(3)=X4(2)
      X5(4)=X4(3)
C     ------------------------
      X7(1)=X2(8)
      X7(2)=X2(5)
      X7(3)=X2(7)
      X7(4)=X2(6)
C     ------------------------
      C1I4B=(COS(PI/4))**2
      S1I4B=C1I4B
      V1=1/(SQRT(2.0))
      X8(1)=X7(1)*V1
      X8(2)=X7(2)*V1
      X8(3)=(X7(3)+X7(4))*C1I4B
      X8(4)=(X7(3)-X7(4))*C1I4B
C     --------------------------
      X9(1)=(X8(1)+X8(3))*V1
      X9(2)=(X8(2)+X8(4))*V1
      X9(3)=(X8(1)-X8(3))*V1
      X9(4)=(X8(2)-X8(4))*V1
C     ---------------------------
      C1I16=COS(PI/16)
      S1I16=SIN(PI/16)
      C5I16=COS(PI*5/16)
      S5I16=SIN(PI*5/16)
      X10(1)=X9(1)*C1I16+X9(2)*S1I16
      X10(2)=X9(1)*S1I16-X9(2)*C1I16
      X10(3)=X9(3)*C5I16+X9(4)*S5I16
      X10(4)=X9(3)*S5I16-X9(4)*C5I16
C     ---------------------------
      X12(1)=X10(2)
      X12(2)=X10(3)
      X12(3)=X10(4)
      X12(4)=X10(1)
C     ---------------------------
      X13(1)=X5(1)
      X13(2)=X12(4)
      X13(3)=X5(2)
```

```
      X13(4)=X12(3)
      X13(5)=X5(3)
      X13(6)=X12(2)
      X13(7)=X5(4)
      X13(8)=X12(1)
C  -----------------------------
      X15(1)=X1(16)
      X15(2)=X1(9)
      X15(3)=X1(13)
      X15(4)=X1(12)
      X15(5)=X1(15)
      X15(6)=X1(10)
      X15(7)=X1(14)
      X15(8)=X1(11)
C  -----------------------------
      C1I4=COS(PI/4)
      S1I4=SIN(PI/4)
      X16(1)=X15(1)
      X16(2)=X15(2)
      X16(3)=X15(3)*C1I4+X15(4)*S1I4
      X16(4)=X15(3)*S1I4-X15(4)*C1I4
C  -------------------------------
      X17(1)=(X16(1)+X16(3))*V1
      X17(2)=(X16(2)+X16(4))*V1
      X17(3)=(X16(1)-X16(3))*V1
      X17(4)=(X16(2)-X16(4))*V1
C  ---------------------------------
      X18(1)=X15(5)
      X18(2)=X15(6)
      X18(3)=X15(7)
      X18(4)=X15(8)
C  ---------------------------------
      C1I8=COS(PI/8)
      S1I8=SIN(PI/8)
      X19(1)=X17(1)
      X19(2)=X17(2)
      X19(3)=X17(3)
      X19(4)=X17(4)
      X19(5)=X18(1)
      X19(6)=X18(2)
      X19(7)=X18(3)*C1I4+X18(4)*S1I4
      X19(8)=X18(3)*S1I4-X18(4)*C1I4
C  ---------------------------------
      X20(1)=X19(1)
      X20(2)=X19(2)
      X20(3)=X19(3)
      X20(4)=X19(4)
      X20(5)=(X19(5)+X19(7))*V1
      X20(6)=(X19(6)+X19(8))*V1
      X20(7)=(X19(5)-X19(7))*V1
      X20(8)=(X19(6)-X19(8))*V1
C  ----------------------------------
      C5I8=COS(PI*5/8)
      S5I8=SIN(PI*5/8)
      X21(1)=X20(1)
```

```
      X21(2)=X20(2)
      X21(3)=X20(3)
      X21(4)=X20(4)
      X21(5)=X20(5)*C1IB+X20(6)*S1IB
      X21(6)=X20(5)*S1I8-X20(6)*C1I8
      X21(7)=X20(7)*C5I8+X20(8)*S5I8
      X21(8)=X20(7)*S5IB-X20(8)*C5I8
C  ------------------------------
      X22(1)=(X21(1)+X21(5))*V1
      X22(2)=(X21(2)+X21(6))*V1
      X22(3)=(X21(3)+X21(7))*V1
      X22(4)=(X21(4)+X21(8))*V1
      X22(5)=(X21(1)-X21(5))*V1
      X22(6)=(X21(2)-X21(6))*V1
      X22(7)=(X21(3)-X21(7))*V1
      X22(8)=(X21(4)-X21(8))*V1
C  -------------------------------
      C1I32=COS(PI/32)
      S1I32=SIN(PI/32)
      C5I32=COS(PI*5/32)
      S5I32=SIN(PI*5/32)
      C9I32=COS(PI*9/32)
      S9I32=SIN(PI*9/32)
      C13I32=COS(PI*13/32)
      S13I32=SIN(PI*13/32)
      X23(1)=X22(1)*C1I32+X22(2)*S1I32
      X23(2)=X22(1)*S1I32-X22(2)*C1I32
      X23(3)=X22(3)*C5I32+X22(4)*S5I32
      X23(4)=X22(3)*S5I32-X22(4)*C5I32
      X23(5)=X22(5)*C9I32+X22(6)*S9I32
      X23(6)=X22(5)*S9I32-X22(6)*C9I32
      X23(7)=X22(7)*C13I32+X22(8)*S13I32
      X23(8)=X22(7)*S13I32-X22(8)*C13I32
C  -----------------------------------
      X25(1)=X23(2)
      X25(2)=X23(7)
      X25(3)=X23(4)
      X25(4)=X23(5)
      X25(5)=X23(6)
      X25(6)=X23(3)
      X25(7)=X23(8)
      X25(8)=X23(1)
C  -----------------------------------
      Y(1)=X13(1)
      Y(2)=X25(8)
      Y(3)=X13(2)
      Y(4)=X25(7)
      Y(5)=X13(3)
      Y(6)=X25(6)
      Y(7)=X13(4)
      Y(8)=X25(5)
      Y(9)=X13(5)
      Y(10)=X25(4)
      Y(11)=X13(6)
      Y(12)=X25(3)
```

```
      Y(13)=X13(7)
      Y(14)=X25(2)
      Y(15)=X13(8)
      Y(16)=X25(1)
C**********************************************************
C     OUTPUT THE DCT DATA Y
      WRITE(*,*) 'THE DCT DATA'
      WRITE(6,*) 'THE DCT DATA'
      DO 10 K=1,16
         WRITE(*,60) K,Y(K)
         WRITE(6,60) K,Y(K)
 60      FORMAT(2X,'Y',I2,'=',F9.4)
 10   CONTINUE
      CLOSE(6)
      STOP
      END
```

Program for IDCT-II, N=16 based on [FD-1, FRA-12]

```
C*************************************************************
C       16 POINT IFDCT-I
C*************************************************************
        PROGRAM IFDCT16-I
C*************************************************************
        REAL X0(16),X1(16),X2(8),X3(4),X4(4),X5(4),X7(4),X8(4)
        REAL X9(4),X10(4),X12(4),X13(8),X14(8),X15(8),X16(4)
        REAL X17(4),X18(4),X19(8),X20(8),X21(8),X22(8),X23(8)
        REAL X24(8),X25(8),Y(16)
        REAL PI,C1I4A,S1I4A,C1I8A,S1I8A,C1I4B,S1I4B,V1
        REAL C1I16,S1I16,C5I16,S5I16,C1I4,S1I4,C1I8,S1I8
        REAL C5IB,S5IB,C1I32,S1I32,C5I32,S5I32,C9I32
        REAL S9I32,C13I32,S13I32
C*************************************************************
C       INPUT DATA Y
        OPEN (UNIT=6,FILE='IFDCT16-I.DAT',STATUS='UNKNOWN')
        WRITE(6,*) 'THE INPUT DATA '
        DO 5 K=1,16
           WRITE (*,40) K
 40     FORMAT(2X,'PLEASE INPUT DATA Y ',I2)
           READ *,X0(K)
           WRITE(*,50) K,X0(K)
           WRITE(6,50) K,X0(K)
 50     FORMAT (2X,'Y',I2,'=',F9.4)
 5      CONTINUE
C****************************************************
        V1=1/(SQRT(2.0))
        X1(1)=X0(1)
        X1(2)=X0(3)
        X1(3)=X0(5)
        X1(4)=X0(7)
        X1(5)=X0(9)
        X1(6)=X0(11)
        X1(7)=X0(13)
        X1(8)=X0(15)
        X1(9)=X0(16)
        X1(10)=X0(14)
        X1(11)=X0(12)
        X1(12)=X0(10)
        X1(13)=X0(8)
        X1(14)=X0(6)
        X1(15)=X0(4)
        X1(16)=X0(2)
C  ---------------------------------
        X2(1)=X1(1)
        X2(2)=X1(3)
        X2(3)=X1(5)
        X2(4)=X1(7)
        X2(5)=X1(8)
        X2(6)=X1(6)
        X2(7)=X1(4)
        X2(8)=X1(2)
C  ---------------------------------
```

```
      X3(1)=X2(1)
      X3(2)=X2(3)
      X3(3)=X2(4)
      X3(4)=X2(2)
C  ---------------------
      PI=3.141592658
      C1I4=(COS(PI/4))*0.5
      S1I4=C1I4
      C1I8=(COS(PI/8))*0.5
      S1I8=(SIN(PI/8))*0.5
      X4(1)=(X3(1)+X3(2))*C1I4
      X4(2)=(X3(1)-X3(2))*C1I4
      X4(3)=(X3(3)*(-C1I8))+(X3(4)*S1I8)
      X4(4)=(X3(3)*S1I8)+(X3(4)*C1I8)
C  ----------------------
      X5(1)=X4(1)+X4(4)
      X5(2)=X4(2)+X4(3)
      X5(3)=X4(2)-X4(3)
      X5(4)=X4(1)-X4(4)
C  -----------------------
      X7(1)=X2(8)
      X7(2)=X2(5)
      X7(3)=X2(7)
      X7(4)=X2(6)
C  -----------------------
      C1I4B=(COS(PI/4))**2
      S1I4B=C1I4B
      V1=1/(SQRT(2.0))
      X8(1)=X7(1)*V1
      X8(2)=X7(2)*V1
      X8(3)=(X7(3)+X7(4))*C1I4B
      X8(4)=(X7(3)-X7(4))*C1I4B
C  -------------------------
      X9(1)=(X8(1)+X8(3))*V1
      X9(2)=(X8(2)+X8(4))*V1
      X9(3)=(X8(1)-X8(3))*V1
      X9(4)=(X8(2)-X8(4))*V1
C  --------------------------
      C1I16=COS(PI/16)
      S1I16=SIN(PI/16)
      C5I16=COS(PI*5/16)
      S5I16=SIN(PI*5/16)
      X10(1)=X9(1)*C1I16+X9(2)*S1I16
      X10(2)=X9(1)*S1I16-X9(2)*C1I16
      X10(3)=X9(3)*C5I16+X9(4)*S5I16
      X10(4)=X9(3)*S5I16-X9(4)*C5I16
C  --------------------------
      X12(1)=X10(2)
      X12(2)=X10(3)
      X12(3)=X10(4)
      X12(4)=X10(1)
C  --------------------------
      X13(1)=X5(1)+X12(4)
      X13(2)=X5(2)+X12(3)
      X13(3)=X5(3)+X12(2)
```

```
      X13(4)=X5(4)+X12(1)
      X13(5)=X5(4)-X12(1)
      X13(6)=X5(3)-X12(2)
      X13(7)=X5(2)-X12(3)
      X13(8)=X5(1)-X12(4)
C ----------------------------
      X15(1)=X1(16)
      X15(2)=X1(9)
      X15(3)=X1(13)
      X15(4)=X1(12)
      X15(5)=X1(15)
      X15(6)=X1(10)
      X15(7)=X1(14)
      X15(8)=X1(11)
C ----------------------------
      C1I4=COS(PI/4)
      S1I4=SIN(PI/4)
      X16(1)=X15(1)
      X16(2)=X15(2)
      X16(3)=X15(3)*C1I4+X15(4)*S1I4
      X16(4)=X15(3)*S1I4-X15(4)*C1I4
C ------------------------------
      X17(1)=(X16(1)+X16(3))*V1
      X17(2)=(X16(2)+X16(4))*V1
      X17(3)=(X16(1)-X16(3))*V1
      X17(4)=(X16(2)-X16(4))*V1
C -------------------------------
      X18(1)=X15(5)
      X18(2)=X15(6)
      X18(3)=X15(7)
      X18(4)=X15(8)
C -------------------------------
      C1I8=COS(PI/8)
      S1I8=SIN(PI/8)
      X19(1)=X17(1)
      X19(2)=X17(2)
      X19(3)=X17(3)
      X19(4)=X17(4)
      X19(5)=X18(1)
      X19(6)=X18(2)
      X19(7)=X18(3)*C1I4+X18(4)*S1I4
      X19(8)=X18(3)*S1I4-X18(4)*C1I4
C --------------------------------
      X20(1)=X19(1)
      X20(2)=X19(2)
      X20(3)=X19(3)
      X20(4)=X19(4)
      X20(5)=(X19(5)+X19(7))*V1
      X20(6)=(X19(6)+X19(8))*V1
      X20(7)=(X19(5)-X19(7))*V1
      X20(8)=(X19(6)-X19(8))*V1
C ---------------------------------
      C5I8=COS(PI*5/8)
      S5I8=SIN(PI*5/8)
      X21(1)=X20(1)
```

```
      X21(2)=X20(2)
      X21(3)=X20(3)
      X21(4)=X20(4)
      X21(5)=X20(5)*C1IB+X20(6)*S1IB
      X21(6)=X20(5)*S1I8-X20(6)*C1I8
      X21(7)=X20(7)*C5I8+X20(8)*S5I8
      X21(8)=X20(7)*S5IB-X20(8)*C5I8
C     -------------------------------
      X22(1)=(X21(1)+X21(5))*V1
      X22(2)=(X21(2)+X21(6))*V1
      X22(3)=(X21(3)+X21(7))*V1
      X22(4)=(X21(4)+X21(8))*V1
      X22(5)=(X21(1)-X21(5))*V1
      X22(6)=(X21(2)-X21(6))*V1
      X22(7)=(X21(3)-X21(7))*V1
      X22(8)=(X21(4)-X21(8))*V1
C     --------------------------------
      C1I32=COS(PI/32)
      S1I32=SIN(PI/32)
      C5I32=COS(PI*5/32)
      S5I32=SIN(PI*5/32)
      C9I32=COS(PI*9/32)
      S9I32=SIN(PI*9/32)
      C13I32=COS(PI*13/32)
      S13I32=SIN(PI*13/32)
      X23(1)=X22(1)*C1I32+X22(2)*S1I32
      X23(2)=X22(1)*S1I32-X22(2)*C1I32
      X23(3)=X22(3)*C5I32+X22(4)*S5I32
      X23(4)=X22(3)*S5I32-X22(4)*C5I32
      X23(5)=X22(5)*C9I32+X22(6)*S9I32
      X23(6)=X22(5)*S9I32-X22(6)*C9I32
      X23(7)=X22(7)*C13I32+X22(8)*S13I32
      X23(8)=X22(7)*S13I32-X22(8)*C13I32
C     -------------------------------------
      X25(1)=X23(2)
      X25(2)=X23(7)
      X25(3)=X23(4)
      X25(4)=X23(5)
      X25(5)=X23(6)
      X25(6)=X23(3)
      X25(7)=X23(8)
      X25(8)=X23(1)
C     -------------------------------------
      Y(1)=(X13(1)+X25(8))*V1
      Y(2)=(X13(2)+X25(7))*V1
      Y(3)=(X13(3)+X25(6))*V1
      Y(4)=(X13(4)+X25(5))*V1
      Y(5)=(X13(5)+X25(4))*V1
      Y(6)=(X13(6)+X25(3))*V1
      Y(7)=(X13(7)+X25(2))*V1
      Y(8)=(X13(8)+X25(1))*V1
      Y(9)=(X13(8)-X25(1))*V1
      Y(10)=(X13(7)-X25(2))*V1
      Y(11)=(X13(6)-X25(3))*V1
      Y(12)=(X13(5)-X25(4))*V1
```

```
        Y(13)=(X13(4)-X25(5))*V1
        Y(14)=(X13(3)-X25(6))*V1
        Y(15)=(X13(2)-X25(7))*V1
        Y(16)=(X13(1)-X25(8))*V1
C*********************************************************
C       OUTPUT THE IDCT DATA Y
        WRITE(*,*) 'THE IDCT DATA'
        WRITE(6,*) 'THE IDCT DATA'
        DO 10 K=1,16
           WRITE(*,60) K,Y(K)
           WRITE(6,60) K,Y(K)
 60        FORMAT(2X,'Y',I2,'=',F9.4)
 10     CONTINUE
        CLOSE(6)
        STOP
        END
```

APPENDIX A.3

COMPUTER PROGRAM FOR 2D 16×16 DCT AND IDCT

This appendix describes the computer programs [HTC-6] for implementing 2D 16 × 16 DCT and its inverse using the row-column approach (Fig. 5.1) based on the fast algorithm developed by Chen, Smith, and Fralick [FRA-1]. The 2D DCT of a 2D array $x(m, n)$ is defined as

$$X^{C(2)}(u, v) = \frac{4}{MN} C(u)C(v) \sum_{m=0}^{M-1} \sum_{n=0}^{N-1} x(m, n) \cos\left[\frac{(2m+1)u\pi}{2M}\right] \cos\left[\frac{(2n+1)v\pi}{2N}\right]$$

$$u = 0, 1, \ldots, M-1, v = 0, 1, \ldots, N-1, \tag{A.3.1}$$

where $X^{C(2)}(u, v)$ is the 2D-DCT coefficient.

The original 2D array $x(m, n)$ can be recovered from the 2D IDCT, i.e.,

$$x(m, n) = \sum_{u=0}^{M-1} \sum_{v=0}^{N-1} C(u)C(v)X^{C(2)}(u, v) \cos\left[\frac{(2m+1)u\pi}{2M}\right] \cos\left[\frac{(2n+1)v\pi}{2N}\right]$$

$$m = 0, 1, \ldots, M-1, n = 0, 1, \ldots, N-1. \tag{A.3.2}$$

where

$$C(m) = \frac{1}{\sqrt{2}} \qquad m = 0$$

$$= 1 \qquad m \neq 0$$

```
                  SUBROUTINE     CHEN16(X,IDIR)
C
C     THIS SUBROUTINE COMPUTES THE TWO-DIMENSIONAL FAST COSINE
C     TRANSFORM USING FAST ALGORITHM BASED ON [FRA-1]
C
C     X = INPUT ON ENTRY, OUTPUT ON EXIT
C
C     IDIR = DIRECTION OF TRANSFORMATION
```

```
C         0 : FORWARD TRANSFORM
C         1 : INVERSE TRANSFORM
C
          DIMENSION X(16,16), Y(16)
C     FIRST DIMENSION
          DO 3 I1 = 1, 16
          DO 1 I2 = 1, 16
1         Y(I2) = X(I2,I1)
          IF(IDIR .EQ. 0) CALL FDCT16(Y)
          IF(IDIR .EQ. 1) CALL IDCT16(Y)
          DO 2 I2 = 1, 16
2         X(I2,I1) = Y(I2)
3         CONTINUE
C         SECOND DIMENSION
          DO 6 I1 = 1, 16
          DO 4 I2 = 1, 16
4         Y(I2) = X(I1,I2)
          IF(IDIR .EQ. 0) CALL FDCT16(Y)
          IF(IDIR .EQ. 1) CALL IDCT16(Y)
          DO 5 I2 = 1, 16
5         X(I1,I2) = Y(I2)
6         CONTINUE
          RETURN
          END
C
```

```
      SUBROUTINE FDCT16(Z)
C
C     THIS SUBROUTINE COMPUTES THE FORWARD FAST COSINE TRANSFORM
C     FOR A ONE-DIMENSIONAL DATA VECTOR OF DIMENSION 16.
C
C     Z = INPUT ON ENTRY, OUTPUT ON EXIT
C
      DIMENSION Y(16), Z(16)
      PARAMETER PI = 3.1416
C     FIRST BUTTERFLY LOOP
      DO 100 I1 = 1, 8
      J1 = 17 - I1
      Y(I1) = Z(I1) + Z(J1)
100   Y(J1) = Z(I1) - Z(J1)
C     SECOND BUTTERFLY LOOP
      A1 = Y(1) + Y(8)
      A2 = Y(2) + Y(7)
      A3 = Y(3) + Y(6)
      A4 = Y(4) + Y(5)
      A5 = Y(4) - Y(5)
      A6 = Y(3) - Y(6)
      A7 = Y(2) - Y(7)
      A8 = Y(1) - Y(8)
      A9 = Y(9)
      A10 = Y(10)
      A11 = (Y(14) - Y(11)) * COS( PI/4.0 )
      A12 = (Y(13) - Y(12)) * COS( PI/4.0 )
      A13 = (Y(12) + Y(13)) * COS( PI/4.0 )
      A14 = (Y(11) + Y(14)) * COS( PI/4.0 )
      A15 = Y(15)
      A16 = Y(16)
C     THIRD BUTTERFLY LOOP
      B1 = A1 + A4
      B2 = A2 + A3
      B3 = A2 - A3
      B4 = A1 - A4
      B5 = A5
      B6 = (A7 - A6) * COS( PI/4.0 )
      B7 = (A6 + A7) * COS( PI/4.0 )
      B8 = A8
      B9 = A9 + A12
      B10 = A10 + A11
      B11 = A10 - A11
      B12 = A9 - A12
      B13 = A16 - A13
      B14 = A15 - A14
      B15 = A14 + A15
      B16 = A13 + A16
C     FOURTH BUTTERFLY LOOP
      C1 = (B1 + B2) * COS( PI/4.0)
      C2 = (B1 - B2) * COS( PI/4.0 )
      C3 = B3 * SIN( PI/8.0 ) - B4 * COS( PI / 8.0)
      C4 = B4 * COS( 3 * PI / 8.0 ) - B3 * SIN(3 * PI / 8.0)
      C5 = B5 + B6
      C6 = B5 - B6
```

```
          C7 = B8 - B7
          C8 = B7 + B8
          C9 = B9
          C10 = B15 * SIN( PI/8.0 ) - B10 * COS( PI/8.0 )
          C11 = -B11 * SIN( PI/8.0 ) - B14 * COS( PI/8.0)
          C12 = B12
          C13 = B13
          C14 = B14 * COS( 3*PI/8.0 ) - B11 * SIN( 3*PI/8.0 )
          C15 = B10 * COS( 3*PI/8.0 ) + B15 * SIN( 3*PI/8.0 )
          C16 = B16
C     FIFTH BUTTERFLY LOOP
          D5 = C5 * SIN( PI/16.0 ) + C8 * COS( PI/16.0 )
          D6 = C6 * SIN( 5*PI/16.0 ) + C7 * COS( 5*PI/16.0 )
          D7 = C7 * COS( 3*PI/16.0 ) - C6 * SIN( 3*PI/16.0 )
          D8 = C8 * COS( 7*PI/16.0 ) - C5 * SIN( 7*PI/16.0 )
          D9 = C9 + C10
          D10 = C9 - C10
          D11 = C12 - C11
          D12 = C11 + C12
          D13 = C13 + C14
          D14 = C13 - C14
          D15 = C16 - C15
          D16 = C15 + C16
C     SIXTH BUTTERFLY LOOP
      E9 = D9 * SIN( PI/32.0 ) + D16 * COS( PI/32.0 )
      E10 = D10 * SIN( 9*PI/32.0 ) + D15 * COS( 9*PI/32.0 )
      E11 = D11 * SIN( 5*PI/32.0 ) + D14 * COS( 5*PI/32.0 )
      E12 = D12 * SIN( 13*PI/32.0 ) + D13 * COS( 13*PI/32.0 )
      E13 = D13 * COS( 3*PI/32.0 ) - D12 * SIN( 3*PI/32.0 )
      E14 = D14 * COS( 11*PI/32.0 ) - D11 * SIN( 11*PI/32.0 )
      E15 = D15 * COS( 7*PI/32.0 ) - D10 * SIN( 7*PI/32.0 )
      E16 = D16 * COS( 15*PI/32.0 ) - D9 * SIN( 15*PI/32.0)
C     NORMALIZED FORWARD TRANSFORM COEFFICIENTS
      Z(1)   = C1 / 8.0
      Z(9)   = C2 / 8.0
      Z(5)   = C3 / 8.0
      Z(13)  = C4 / 8.0
      Z(3)   = D5 / 8.0
      Z(11)  = D6 / 8.0
      Z(7)   = D7 / 8.0
      Z(15)  = D8 / 8.0
      Z(2)   = E9 / 8.0
      Z(10) = E10 / 8.0
      Z(6)  = E11 / 8.0
      Z(14) = E12 / 8.0
      Z(4)  = E13 / 8.0
      Z(12) = E14 / 8.0
      Z(8)  = E15 / 8.0
      Z(16) = E16 / 8.0
          RETURN
          END
C
```

```
      SUBROUTINE IDCT16(Z)
C
C     THIS SUBROUTINE COMPUTES THE INVERSE FAST COSINE TRANSFORM
C     FOR A ONE-DIMENSIONAL DATA VECTOR OF DIMENSION 16.
C
C     Z = INPUT ON ENTRY, OUTPUT ON EXIT
C
      DIMENSION Y(16), Z(16)
      PARAMETER PI = 3.1416
C     FIRST INVERSE BUTTERFLY LOOP
      E9  = Z(2) * SIN( PI/32.0 ) - Z(16) * SIN( 15*PI/32.0 )
      E10 = Z(10) * SIN( 9*PI/32.0 ) - Z(8) * SIN( 7*PI/32.0 )
      E11 = Z(6) * SIN( 5*PI/32.0 ) - Z(12) * SIN( 11*PI/32.0 )
      E12 = Z(14) * SIN( 13*PI/32.0 ) - Z(4) * SIN( 3*PI/32.0 )
      E13 = Z(4) * COS( 3*PI/32.0 ) + Z(14) * COS( 13*PI/32.0 )
      E14 = Z(12) * COS( 11*PI/32.0 ) + Z(6) * COS( 5*PI/32.0 )
      E15 = Z(8) * COS( 7*PI/32.0 ) + Z(10) * COS( 9*PI/32.0 )
      E16 = Z(16) * COS( 15*PI/32.0 ) + Z(2) * COS( PI/32.0 )
C     SECOND INVERSE BUTTERFLY LOOP
      D5 = Z(3) * SIN( PI/16.0 ) - Z(15) * SIN( 7*PI/16.0 )
      D6 = Z(11) * SIN( 5*PI/16.0 ) - Z(7) * SIN( 3*PI/16.0 )
      D7 = Z(7) * COS( 3*PI/16.0 ) + Z(11) * COS( 5*PI/16.0 )
      D8 = Z(15) * COS( 7*PI/16.0 ) + Z(3) * COS( PI/16.0 )
      D9 = E9 + E10
      D10 = E9 - E10
      D11 = E12 - E11
      D12 = E11 + E12
      D13 = E13 + E14
      D14 = E13 - E14
      D15 = E16 - E15
      D16 = E15 + E16
C     THIRD INVERSE BUTTERFLY LOOP
      C1 = (Z(1) + Z(9)) * COS( PI/4.0 )
      C2 = (Z(1) - Z(9)) * COS( PI/4.0 )
      C3 = Z(5) * SIN( PI/8.0 ) - Z(13) * SIN( 3*PI/8.0 )
      C4 = Z(13) * COS( 3*PI/8.0 ) + Z(5) * COS( PI/8.0 )
      C5 = D5 + D6
      C6 = D5 - D6
      C7 = D8 - D7
      C8 = D7 + D8
      C9 = D9
      C10 = D15 * COS( 3*PI/8.0 ) - D10 * COS( PI/8.0 )
      C11 = -D11 * SIN( PI/8.0 ) - D14 * SIN( 3*PI/8.0 )
      C12 = D12
      C13 = D13
      C14 = D14 * COS( 3*PI/8.0 ) - D11 * COS( PI/8.0 )
      C15 = D10 * SIN( PI/8.0 ) + D15 * SIN( 3*PI/8.0 )
      C16 = D16
C     FOURTH INVERSE BUTTERFLY LOOP
      B1 = C1 + C4
      B2 = C2 + C3
      B3 = C2 - C3
      B4 = C1 - C4
      B5 = C5
```

```
      B6 = (C7 - C6) * COS( PI/4.0 )
      B7 = (C6 + C7) * COS( PI/4.0 )
      B8 = C8
      B9 = C9 + C12
      B10 = C10 + C11
      B11 = C10 - C11
      B12 = C9 - C12
      B13 = C16 - C13
      B14 = C15 - C14
      B15 = C14 + C15
      B16 = C13 + C16
C     FOURTH INVERSE BUTTERFLY LOOP
      Y(1) = B1 + B8
      Y(2) = B2 + B7
      Y(3) = B3 + B6
      Y(4) = B4 + B5
      Y(5) = B4 - B5
      Y(6) = B3 - B6
      Y(7) = B2 - B7
      Y(8) = B1 - B8
      Y(9) = B9
      Y(10) = B10
      Y(11) = (B14 - B11) * COS( PI/4.0 )
      Y(12) = (B13 - B12) * COS( PI/4.0 )
      Y(13) = (B12 + B13) * COS( PI/4.0 )
      Y(14) = (B11 + B14) * COS( PI/4.0 )
      Y(15) = B15
      Y(16) = B16
C     FIFTH INVERSE BUTTERFLY LOOP
      DO 100 I1 = 1, 8
      J1 = 17 - I1
      Z(I1) = (Y(I1) + Y(J1)) / 8.0
100   Z(J1) = (Y(I1) - Y(J1)) / 8.0
      RETURN
      END
C
```

APPENDIX A.4

COMPUTER PROGRAM FOR GENERATING LLOYD–MAX QUANTIZER

This appendix describes the computer program [HTC-6] for generating the Lloyd–Max quantizer [Q-1, Q-3] based on minimizing the mean square quantization error. This program is based on a fast convergence method developed by Ngan, Leong, and Singh [Q-4]. The midriser quantizer (Fig. 7.47) is designed for Gaussian, Laplacian, and Rayleigh distributions. A 256 level Laplacian symmetrical quantizer generated from this program is shown in Table A.4.1.

Table A.4.1

An optimum 256-level Laplacian quantizer table [HTC-6].

Decision levels		Reconstruction levels	
0.00000E + 00	1.4589	0.82416E-02	1.4753
0.16548E-01	1.4920	0.24854E-01	1.5086
0.33225E-01	1.5256	0.41597E-01	1.5425
0.50035E-01	1.5597	0.58474E-01	1.5769
0.66980E-01	1.5944	0.57486E-01	1.6119
0.84061E-01	1.6297	0.92636E-01	1.6475
0.10128	1.6655	0.10993	1.6836
0.11864	1.7020	0.12736	1.7204
0.13615	1.7391	0.14494	1.7578
0.15380	1.7769	0.16266	1.7960
0.17160	1.8154	0.18053	1.8348
0.18955	1.8545	0.19856	1.8743
0.20765	1.8944	0.21674	1.9146
0.22590	1.9351	0.23507	1.9556
0.24432	1.9766	0.25357	1.9975

Table A.4.1

Decision levels		Reconstruction levels	
0.26290	2.0188	0.27223	2.0402
0.28164	2.0620	0.29105	2.0838
0.30055	2.1060	0.31004	2.1283
0.31962	2.1510	0.32930	2.1737
0.33887	2.1969	0.34854	2.2202
0.35830	2.2439	0.36806	2.2677
0.37791	2.2919	0.38776	2.3162
0.39769	2.3411	0.40763	2.3659
0.41766	2.3913	0.42770	2.4168
0.43783	2.4428	0.44798	2.4689
0.45819	2.4956	0.46842	2.5223
0.47875	2.5497	0.48908	2.5772
0.49952	2.6053	0.50995	2.6334
0.52048	2.6623	0.53101	2.6912
0.54165	2.7209	0.55229	2.7505
0.56303	2.7811	0.57377	2.8116
0.58462	2.8430	0.59548	2.8745
0.60644	2.9069	0.61741	2.9392
0.62850	2.9726	0.63958	3.0061
0.65078	3.0406	0.66198	3.0750
0.67330	3.1107	0.68462	3.1463
0.69606	3.1832	0.70750	3.2201
0.71906	3.2584	0.73062	3.2966
0.74231	3.3363	0.75400	3.3760
0.76582	3.4171	0.77764	3.4583
0.78960	3.5011	0.80156	3.5440
0.81365	3.5886	0.82074	3.6332
0.83797	3.6797	0.85020	3.7263
0.86257	3.7750	0.87494	3.8238
0.88746	3.8748	0.89997	3.9259
0.91264	3.9795	0.92531	4.0331
0.93814	4.0896	0.95096	4.1461
0.96394	4.2058	0.97692	4.2655
0.99005	4.3287	1.0032	4.3920
1.0165	4.4593	1.0298	4.5266
1.0433	4.5984	1.0567	4.6703
1.0704	4.7473	1.0841	4.8244
1.0979	4.9076	1.1117	4.9907
1.1258	5.0809	1.1398	5.1711
1.1540	5.2697	1.1682	5.3683
1.1826	5.4771	1.1970	5.5858
1.2116	5.7070	1.2262	5.8282
1.2409	5.9650	1.2557	6.1018
1.2707	6.2589	1.2857	6.4160
1.3009	6.6005	1.3162	6.7851
1.3316	7.0087	1.3471	7.2323
1.3627	7.5160	1.3784	7.7997
1.3943	8.1885	1.4103	8.5772
1.4264	9.1987	1.4425	9.8202

1) Fortran program for fast generation of Lloyd-Max optimum quantizer [Q-1,Q-3]

Main Program : PROGRAM MAXZ

Subroutine Call :

AREA * Total area under segment

AINFIN * Area between y(N) and infinity

BISECT * Bisection method for initial y(1)

BIST * Bisection method for segment centroid

TRAPEZ * Trapezoidal method for finding area under curve

Function Call :

Gaussian p.d.f. GF
GDF
GDSTF
GDSTDS
GSF

Laplacian p.d.f. PLF
PLDF
PLDSTF
PLDSTDF
PLSF

Rayleigh p.d.f. RF
RDF
RDSTF
RDSTDF
RSF

```
      PROGRAM MAXZ
C
C     This program calculates the Max's optimum quantizers with
C     even levels by bisection method with successive approximation.
C     Formulas for the Max's quantizer are:
C     1.   X(I) = (Y(I) + Y(I+1)) / 2
C     2.   Integral of (X - YI) * P(X) * DX between X = X(I)
C          and X = X(I+1) is equal to 0.
C
C     The number of levels in quantizer is 2 ** (N+1) where N lies
      between N1 and N2 (user input numbers).
C     Quantizer ID   1    :    Gaussian
C                    2    :    Laplacian
C                    3    :    Rayleigh
C
C     The total distortion and the entropy of the quantizer is C
      also given.
C
      COMMON/QN/X(256),Y(256)
      DIMENSION GXY(2,256)
      INTEGER N, K, N1, N2, IERR
      REAL X, Y, TOL, Y1, YL, YR, YM
      REAL A1, AL, AR, AM
      EXTERNAL AREA, GF, GDF, GSF, GDSTF, GDSTDF, GSDF
      EXTERNAL PLF, PLDF, PLSF, PLDSTF, PLSDF, PDSTDF
      EXTERNAL RF, RDF, RSF, RDSTF, RDSTDF, RSDF
      DATA IERR, TOL/O,1.OE-5/
C
      TYPE *, 'OPTIMUM QUANTIZATION LEVEL FOR EVEN NUMBER OF O/P'
C     User input quantizer ID and levels required.
      TYPE 110
      TYPE 111
      ACCEPT *, ID
      TYPE 115
      ACCEPT *, N1, N2
      TME = SECNDS( 0.0 )
      DO 80 K = N1, N2
      N = 2**K
      TYPE 102, 2*N
C     Assume initial Y(1) values.
      A1 = 0.0
      Y1 = 0.05/N
C     Y1 = 0.008
      YL = Y1
C     Calculate the quantizer with the subroutine BISECT.
      GOTO( 51,52,53 ) ID
51    CALL BISECT(AREA, Y1, YL, A1, N, GF, GDF, TOL, IERR)
      GOTO 55
52    CALL BISECT(AREA, Y1, YL, A1, N, PLF, PLDF, TOL, IERR)
      GOTO 55
53    CALL BISECT(AREA,Y1,YL,A1,N,RF,RDF,TOL,IERR)
55    IF( IERR .GT. 0 ) GOTO 70    *if error then print error.
      DA = 0.0
      S = 0.0
      X(N+1) = 100.0
C
C     Print quantizer table
      DO 60 II = 1,N
      TYPE 103, X(II), Y(II)
      GXY(1,N-1+II) = X(II)
      GXY(2,N-1+II) = Y(II)
C     Calculate total distortion and entropy.
      GOTO(91,92,93) ID
91    CALL TRAPEZ( X(II), X(II+1), Y(II), DSA, GDSTF, GDSTDF, TOL)
      CALL TRAPEZ( X(II), X(II+1), Y(II), SA, GSF, GSDF, TOL)
      GOTO 95
```

```
92        CALL TRAPEZ( X(II),X(II+1),Y(II), DSA, PLDSTF,PDSTDF,TOL)
          CALL TRAPEZ( X(II), X(II+1), Y(II), SA, PLSF, PLSDF, TOL)
          GOTO 95
93        CALL TRAPEZ( X(II), X(II+1), Y(II), DSA, RDSTF, RDSTDF, TOL)
          CALL TRAPEZ( X(II), X(II+1), Y(II), SA, RSF, RSDF, TOL)
95        DA = DA + 2 * DSA
          S = S - 2 * SA * LOG( SA ) / LOG(2.0)
60        CONTINUE
          TYPE 108, DA              * Total distortion
          TYPE 109, S               * Total entropy
80        CONTINUE
C         Store quantizer on file
          OPEN(UNIT=1, NAME='GX.DAT', TYPE='NEW',
     /    ACCESS='DIRECT', RECORDSIZE=128)
          WRITE(1'1) (GXY(1,I), I=1,128)
          WRITE(1'2) (GXY(2,I), I=1,128)
          WRITE(1'3) (GXY(1,I), I=129, 256)
          WRITE(1'4) (GXY(2,I), I=129, 256)
          CLOSE(UNIT=1)
          GOTO 999
70        TYPE 101, K, IERR
999       TYPE *, 'TIME TAKEN = ', SECNDS( TME )
          STOP
101       FORMAT('$ ERROR OR NO ROOT FOR ORDER = ', I2,2X,
     *    ' NO. OF ITERATION ERROR = ', I5/)
102       FORMAT(/'$ THE QUANTIZER LEVELS = ', I5/)
103       FORMAT(2(5X,G12.5))
107       FORMAT(19X, G12.5)
108       FORMAT(/'$ THE DISTORTION = ', G12.5)
109       FORMAT(/'$ THE ENTROPY = ', F9.5/)
110       FORMAT(/' GAUSSIAN = 1    LAPLACIAN = 2     RAYLEIGH = 3 ')
111       FORMAT(/'$ THE DISTRIBUTION REQUIRED (1/2/3) = ')
115       FORMAT(/'$ THE QUANTIZATION ORDER REQUIRED N1 TO N2 = ')
          END
```

```
      SUBROUTINE AREA(XX,Y1,YY,ADIF,N,F,DF,TOL,IERR)
C
C     Obtain the centroid of each segment by equating the +ve and -ve
C     area.
C
      COMMON /QN/X(256), Y(256)
      INTEGER N, I, IERR
      REAL XL, Y1, XX, YY, TOL
      REAL A1, AN, AINF, ADIF
      DATA X,Y/256*0.0,256*0.0/
C
C     Assign initial values to X1 and Y1
      Y(1) = Y1
C     X(1) = Y1 * 0.5
      IF( N .EQ. 1 ) GOTO 60
      DO 50 I = 1, N-1
C     Calculate the partial area between XI and YI.
C     Estimate X(I+1) and calculate the area involved.
      CALL TRAPEZ( X(I), Y(I), Y(I), A1, F, DF, TOL )
      XL = 2.0 * Y(I) - X(I)
      CALL BIST( TRAPEZ,Y(I),XL, A1,F,DF,TOL,IERR )
      IF( IERR .GT. 0 ) GOTO 90
      IF( A1 .GT. 80.0 ) GOTO 88
40    X(I+1) = XL
C     Calculate Y(I+1).
      Y(I+1) = 2 * X(I+1) - Y(I)
50    CONTINUE
C     Calculate the partial area between XN and YN.
60    CALL TRAPEZ(X(N), Y(N), Y(N), AN, F, DF, TOL)
C     Calculate the partial area between YN and infinity.
      CALL AINFIN( Y(N), AINF, F, DF, TOL )
      ADIF = AINF + AN
C     To be consistence with subroutine BISECT.
      ADIF =  -ADIF
C     TYPE 103, X(N), Y(N), AINF, AN
      GOTO 90
88    ADIF = 90.0
90    RETURN
103   FORMAT('$X(N), Y(N), AINF, AN = ', 4(5X,F9.5)/)
      END
```

```
      SUBROUTINE AINFIN(YN, AREA, F, DF, TOL)
C
C     Calculate the area between centroid in segment N and infinity.
      INTEGER I
      REAL YN, A, AREA, LOWER, UPPER, TOL
      EXTERNAL TRAPEZ, F, DF
      AREA = 0.0
      UPPER = YN
            DO 10 I=1,100
      LOWER = UPPER
      UPPER = UPPER + 10.0
      CALL TRAPEZ(LOWER,UPPER, YN, A, F, DF, TOL)
      AREA = AREA + A
      IF (ABS(A) .LT. ABS(AREA*TOL)) GOTO 50
10    CONTINUE
50    RETURN
      END
```

```
        SUBROUTINE BISECT( SUBR, YI, YL, A1,N,F,DF,TOL, IERR)
C       The subroutine calculates the quantizer table with bisection
C            method.
C       With initial y(1), y(N) is check if it is the centroid of the
C       last area segment.  If not, then a better approx. of y(1) is
C       chosen by bisection method.
C
        INTEGER IERR, N
        REAL YI, ST, TOL, YL, YR, YM
        REAL A1, A2, AL, AR, AM, NF
        EXTERNAL SUBR, F, DF
        NF = 1.0
        ST = YI
        CALL SUBR( ST,YL,YI,A2, N, F, DF, TOL, IERR)
        AL = A2 + A1
        IF( ABS( AL ) .LT. TOL ) GOTO 999
        IF( AL .GT. 0.0 ) NF = -1.0
C       Locate left and right bisection limit.
C       The initial values of y(1) is improved after each bisection.
        YR = YL
        DO 10 J = 1, 20
        YR = YR + 0.1 * NF
        CALL SUBR( ST, YR, YI, A2, N, F, DF, TOL, IERR )
        AR = A1 + A2
        IF( YR .LT. 0.0 ) GOTO 15
        IF( ABS( AR ) .LT. ABS(AL*TOL) ) GOTO 999
        IF( (AL * AR) .LT. 0.0 ) GOTO 20
        YL = YR
        AL = AR
10      CONTINUE
15      TYPE 106
        IERR = IERR + 1
        GOTO 999
C       Successive bisection method.
20      DO 50 I=1, 1000
        YM = ( YL+YR ) * 0.5
        CALL SUBR( ST, YM, YI, A2, N, F, DF, TOL, IERR )
        AM = A1 + A2
C       If approx. within limit then exit else repeat bisection.
        IF( ABS( YL - YM ) .LT. ABS(YL*TOL) ) GOTO 999
        IF( ABS( AM ) .LT. ABS(AL*TOL) ) GOTO 999
        IF( (AL * AM) .LT. 0.0 ) GOTO 40
        YL = YM
        AL = AM
        GOTO 50
40      YR = YM
        AR = AM
50      CONTINUE
        IERR = IERR + 1
999     IF( AM .GE. 80.0 ) CALL SUBR(ST,YL,YI,AL,N,F,DF,TOL,IERR)
        RETURN
104     FORMAT(/'$YL, AL, YR, AR = ', 2(F9.5, 3X,E12.5,3X))
105     FORMAT(/'$YM, AM, YR, AR = ', 2(F9.5, 3X,E12.5,3X))
106     FORMAT(/'Bisection limit less than 0.0'/)
        END
```

```
      SUBROUTINE BIST( SUBR, YI, XL, A1,F,DF, TOL,IERR )
C
C     This subroutine calculates the decision points of each segment.
C     The decision points are found by bisection method such that the
C     reconstruction level is the centroid of the decision points.
      INTEGER N, IERR
      REAL YI, ST, TOL, XL, XR, XM
      REAL A1, AL, AR, AM, NF
      EXTERNAL SUBR, F, DF
      NF = 1.0
      ST = YI
      CALL SUBR( ST, XL, YI, A2, F, DF, TOL)
      AL = A1 + A2
      IF( ABS( AL ) .LT. ABS(A1*TOL) ) GOTO 999
      IF ( AL .GT. 0.0 ) NF = -1.0
C     Locate left and right bisection limit.
      XR = XL
      DO 10 J = 1, 20
      XR = XR + (XR - YI) * 0.1 * NF
      CALL SUBR( ST, XR, YI, A2, F, DF, TOL )
      AR = A1 + A2
      IF( XR .LT. YI ) GOTO 11
      IF( ABS( AR ) .LT. ABS(A1*TOL) ) GOTO 997
      IF( (AL * AR) .LT. 0.0 ) GOTO 20
      XL = XR
      AL = AR
10    CONTINUE
11    AR = 99.0
      A1 = 99.0
      GOTO 999
C     Successive bisection method.
20    DO 50 I = 1, 1000
      XM = ( XL + XR ) * 0.5
      CALL SUBR( ST,XM,YI,A2,F,DF,TOL )
      AM = A1 + A2
      IF( ABS( XL - XM ) .LT. ABS(XL*TOL) ) GOTO 998
      IF( ABS( AM ) .LT. ABS( A1*TOL ) ) GOTO 998
      IF( (AL * AM) .LT. 0.0 ) GOTO 40
      XL = XM
      AL = AM
      GOTO 50
40    XR = XM•
      AR = AM
50    CONTINUE
      IERR = IERR + 1
997   XL = XR
      GOTO 999
998   XL = XM
999   RETURN
104   FORMAT(/'$XL, AL, XR, AR = ', 2(F9.5,3X,E12.5,3X))
      END
```

```
      SUBROUTINE TRAPEZ(LOWER, UPPER, YI, SUM, F, DF, TOL)
C
C     Numerical integration by the trapezoid method.
C
      INTEGER PIECES, I, P2, N
      EXTERNAL F, DF
      REAL X, DELTA, SUM, UPPER, LOWER, TOL
      REAL ENDSUM, MIDSUM, SUM1, ENDCOR
C
      PEICES = 1
      DELTA = (UPPER - LOWER) / PIECES
      ENDSUM = F(UPPER, YI) + F(LOWER, YI)
      ENDCOR = (DF(UPPER, YI) - DF(LOWER,YI)) / 12.0
      SUM = ENDSUM * DELTA/2.0
      MIDSUM = 0.0
5     PIECES = PIECES * 2
      P2 = PIECES / 2
      SUM1 = SUM
      DELTA = (UPPER - LOWER) / PIECES
       DO 10 I = 1, P2
          X = LOWER + DELTA * (2 * I - 1)
          MIDSUM = MIDSUM + F(X,YI)
10    CONTINUE
      SUM = (ENDSUM + 2.0 * MIDSUM) * DELTA * 0.5
     * - DELTA * DELTA * ENDCOR
C     TYPE *, PIECES, SUM
      IF( P2 .GT. 10000 ) GOTO 50
      IF( ABS(SUM - SUM1) .GT. ABS(SUM1*TOL) ) GOTO 5
50    RETURN
      END
```

```
      FUNCTION GF(X,YI)
C     Quantizer with Gaussian p.d.f.
C     f(X) = (X-Y(I)) * EXP(-X**2/2) / SQRT(2*PI)
C     df(X) = (1+X*Y(I)-X**2) * EXP(-X**2/2) / SQRT(2*PI)
C     1 / SQRT( 2*PI ) = 0.3989422
      SQRP2 = 0.3989422
      GF = (X-YI) * EXP(-X**2*0.5) * SQRP2
      RETURN
      ENTRY GDF(X,YI)
      GDF = (1+YI*X-X**2) * EXP(-X**2*0.5) * SQRP2
      RETURN
C
      ENTRY GDSTF(X,YI)
C
C     Distortion Measure
C     f(X) = (X-YI) ** 2 * EXP( -X**2/2 ) / SQRT( 2*PI )
C     df(X) = (2+X*YI-X**2) * (X-YI) * EXP(-X**2/2)/SQRT(2*PI)
C     SQRP2 = 1 / SQRT( 2*PI ) = 0.3989422
      GDSTF = (X-YI) ** 2 * EXP(-X**2*0.5) * SQRP 2
      RETURN
      ENTRY GDSTDF(X,YI)
      GDSTDF = (X-YI) * (2+YI*X-X**2) * EXP(-X**2*0.5) * SQRP2
      RETURN
C
      ENTRY GSF(X,YI)
C
C     Gaussian Distribution
C     f(X) = EXP(-X**2/2) / SQRT(2*PI)
C     df(X) = -X * EXP(-X**2/2)/SQRT(2*PI)
C     SQRP2 = 1 / SQRT( 2*PI ) = 0.3989422
      GSF = EXP(-X**2*0.5) * SQRP2
      RETURN
      ENTRY GSDF(X,YI)
      GSDF = -X * EXP(-X**2*0.5) * SQRP2
      RETURN
      END
```

```
          FUNCTION PLF(X,YI)
C
C         Quantizer for Laplacian Distribution
C         f(X) = (X-Y(I)) * EXP(-X*SQRT(2)) / SQRT(2)
C         Df(X) = (1/SQRT(2) - X + Y(I)) * EXP( -X*SQRT(2))
          PLF = (X-YI) * EXP(-X*1.4142135) * 0.7071068
          RETURN
          ENTRY PLDF(X,YI)
          PLDF = (0.7071068 - X + YI) * EXP(-X*1.1412135)
          RETURN
C
          ENTRY PLDSTF(X,YI)
C
C         Distortion Measure.
C         Laplacian Distribution
C         f(X) = (X-YI) * EXP(-X * SQRT(2)) / SQRT(2)
C         df(X) = (1/SQRT(2) - X+Y(I)) * EXP(-X*SQRT(2))

C         f(X) = (X-YI) ** 2 * EXP(-X*SQRT(2)) / SQRT(2)
C         df(X) = (SQRT(2)-X+YI) * (X-YI) * EXP(-X * SQRT(2))
          PLDSTF = (X-YI) ** 2 * EXP(-X*1.4142135 ) * 0.7071068
          RETURN
          ENTRY PDSTDF(X,YI)
          PDSTDF = (X-YI)*(1.4142135+YI-X) * EXP(-X*1.4142135)
          RETURN
C
          ENTRY PLSF(X,YI)
C
C         Laplacian Distribution
C         f(X) = EXP(-X*SQRT(2)) / SQRT(2)
C         df(X) = - * EXP(-X*SQRT(2))
          PLSF = EXP(-X*1.4142135) * 0.7071068
          RETURN
          ENTRY PLSDF(X,YI)
          PLSDF = -EXP(-X*1.4142135)
          RETURN
          END
```

```
          FUNCTION RF(X,YI)
C
C         Rayleigh Distribution
C         f(X) = (X-Y(I)) * X * EXP(-X ** 2)
C         df(X) = (-2X**3 + 2YI*X**2 + 2X - YI) * EXP(-X**2)
          RF = (X-YI) * X * EXP(-X ** 2)
          RETURN
          ENTRY RDF(X,YI)
          RDF = (-2*X**3 + 2*YI*X**2 + 2*X - YI) * EXP(-X**2)
          RETURN
C
          ENTRY RDSTF(X,YI)
C
C         Distortion Measure
C         f(X) = (X-YI) ** 2 * X * EXP(-X**2)
C         df(X) = (X - YI) * (3*X - YI + 2*YI*X**2 - 2*X**3)
C         * EXP(-X ** 2)
          RDSTF = (X-YI) ** 2 * X EXP(-X ** 2)
          RETURN
          ENTRY RDSTDF(X,YI)
          RDSTDF = (X - YI) * (3*X - YI + 2*YI*X**2 - 2*X**3)
          *EXP(-X ** 2)
          RETURN
C
          ENTRY RSF(X,YI)
C
C         f(X) = X * EXP(-X ** 2)
C         df(X) = (1 - 2 * X**2) * EXP(-X ** 2)
          RETURN
          END
```

APPENDIX A.5

COMPUTER PROGRAM FOR BLOCK QUANTIZATION

This appendix describes the computer program for generating bit allocation (P7.10.1) in block transform image coding. This program has been developed by Tzou and is based on [Q-6].

Computer program for block quantization [Q-6]

```
C      *****************************************************************
C      *                                                               *
C      *                         BQ_DESIGN.FOR                         *
C      *                                                               *
C      *    This Subroutine will find the optimal bit map assignment   *
C      *    of transform image coding based upon the quantization      *
C      *    error.  The DC term is assigned 8 bits, but it should be   *
C      *    taken care of in the main program.                         *
C      *                                                               *
C      *         QERR(I) : The normalized quantization error for       *
C      *                   I bits output levels                        *
C      *         NORD(I) : The ITH element of RVAR is NORD(I)-TH       *
C      *                   largest                                     *
C      *         INORD(I): The ITH largest element is from the         *
C      *                   INORD(I)-TH element of RVAR                 *
C      *         TOTBIT  : The total number of bits used for BITMAP    *
C      *                   assignment                                  *
C      *         ITYPE   : type of quantizer used                      *
C      *                                                               *
C      *              1 - Nonuniform Max Quantizer for Gaussian        *
C      *              2 - Suboptimal Nonuniform Q. for Gaussian        *
C      *              3 - Optimal Uniform Quantizer for Gaussian       *
C      *                                                               *
C      *              4 - Nonuniform Max Quantizer for Laplacian       *
C      *              5 - Suboptimal Nonuniform Q. for Laplacian       *
C      *              6 - Optimal Uniform Quantizer for Laplacian      *
C      *                                                               *
C      *              7 - Nonuniform Max Quantizer for Gamma           *
C      *              8 - Suboptimal Nonuniform Q. for Gamma           *
C      *              9 - Optimal Uniform Quantizer for Gamma          *
C      *                                                               *
C      *****************************************************************
C
C
      SUBROUTINE BQ_DESIGN(RVAR, NORD, BITMAP, NSIZE, TOTBIT, ITYPE)
C
C
C         PARAMETER PASSED: TOTBIT: INTEGER, TOTAL NUMBER OF BITS
C                                          USED TO ENCODE A DCT BLOCK
C
C                             RVAR: NSIZE X NSIZE ARRAY FOR DCT
C                                          VARIANCE
C                             BITMAP: NSIZE X NSIZE INTEGER MATRIX
C                                          FOR BLOCK CODE, I.E. BIT MAP
C
C         BEFORE CALLED: TOTBIT: INPUT TO SPECIFY NO. OF BITS
C                         NORD : THIS DATA MUST BE GENERATED BY
C                                   CALLING QKSORT SUBROUTINE TO SORT
C                                   THE ORDER OF RVAR
C
C         UPON RETURN   : THE OPTIMAL BITMAP IS DESIGNED
C
C   @@@ WARNING: IF BLOCK SIZE IS LARGER THAN 64*64, INCREASE THE DIMENSION
C                FOR INORD(NSIZE*NSIZE)
C
```

```
          IMPLICIT        NONE
          PARAMETER       NBITMX = 8
          INTEGER*4       ITYPE, NSIZE, NSQ,I,II,K,TOTBIT,MINPNT,ORDER,OPTPNT
          INTEGER*4       PU(0:NBITMX), PL(0:NBITMX)
          INTEGER*4       BITMAP(NSIZE*NSIZE), NORD(NSIZE*NSIZE), INORD(4096)
          REAL            QERR(0:NBITMX), DLTERR(0:NBITMX)
          REAL            RVAR(NSIZE*NSIZE), DLTMAX, REF

          DATA            PU /NBITMX * 0, 0/
          DATA            PL /NBITMX * 0, 0/

          QERR(0) = 1.0
          GOTO TO (501,502,503,504,505,506,507,508,509) ITYPE

501       QERR(1) = .3634
          QERR(2) = .1175
          QERR(3) = .03455
          QERR(4) = .009501
          QERR(5) = .002505
          QERR(6) = .0006442
          QERR(7) = .0001635
          QERR(8) = .00004117
cc        WRITE(6,511)
511       FORMAT(/,3X,'Bit map based on nonuinform',
          1        'Max quantizer for Gaussian ',/)
          GO TO 550

502       QERR(1) = .3634
          QERR(2) = .1203
          QERR(3) = .03485
          QERR(4) = .009501
          QERR(5) = .002524
          QERR(6) = .0006596
          QERR(7) = .0001705
          QERR(8) = .00004374
cc        WRITE(6,512)
512       FORMAT(/,3X,'Bit map based on suboptimal',
          1        'Max quantizer for Gaussian ',/)
          GO TO 550

503       QERR(1) = .3634
          QERR(2) = .1188
          QERR(3) = .03744
          QERR(4) = .01154
          QERR(5) = .003495
          QERR(6) = .001041
          QERR(7) = .0003035
          QERR(8) = .00008714
cc        WRITE(6,513)
513       FORMAT(/,3X,
     1    'Bit map based on optimal uniform ',
          1    'quantizer for Gaussian ',/)
          GO TO 550

504       QERR(1) = .5000
          QERR(2) = .1762
          QERR(3) = .05448
          QERR(4) = .01537
          QERR(5) = .004102
          QERR(6) = .001061
          QERR(7) = .0002699
          QERR(8) = .00006806
cc        WRITE(6,514)
514       FORMAT(/,3X,
     1    'Bit map based on nonuniform Max',
          1    'quantizer for Laplacian',/)
          GO TO 550
```

```
505       QERR(1) = .5000
          QERR(2) = .1826
          QERR(3) = .05516
          QERR(4) = .01537
          QERR(5) = .004146
          QERR(6) = .001097
          QERR(7) = .0002863
          QERR(8) = .00007403
cc        WRITE(6,515)
515       FORMAT(/,3X,
     1    'Bit map based on suboptimal Max',
          1    'quantizer for Laplacian',/)
          GO TO 550

506       QERR(1) = .5000
          QERR(2) = .1963
          QERR(3) = .07175
          QERR(4) = .02535
          QERR(5) = .008713
          QERR(6) = .002913
          QERR(7) = .0009486
          QERR(8) = .0003014
cc        WRITE(6,516)
516       FORMAT(/,3X,'Bit map based on optimal',
          1    'uniform quantizer for Laplacian',/)
          GO TO 550

507       QERR(1) = .6680
          QERR(2) = .2318
          QERR(3) = .07047
          QERR(4) = .01961
          QERR(5) = .005185
          QERR(6) = .001334
          QERR(7) = .0003382
          QERR(8) = .00008516
cc        WRITE(6,517)
517       FORMAT(/,3X,'Bit map based on nonuniform ',
          1       'Max quantizer for Gamma ',/)
          GO TO 550

508       QERR(1) = .6680
          QERR(2) = .2445
          QERR(3) = .07165
          QERR(4) = .01961
          QERR(5) = .005250
          QERR(6) = .001386
          QERR(7) = .00036175
          QERR(8) = .00009360
cc        WRITE(6,518)

518       FORMAT(/,3X,'Bit map based on suboptimal ',
          1       'Max quantizer for Gamma',/)
          GO TO 550

509       QERR(1) = .6680
          QERR(2) = .3200
          QERR(3) = .1323
          QERR(4) = .0501
          QERR(5) = .01784
          QERR(6) = .006073
          QERR(7) = .001996
          QERR(8) = .0006379
cc        WRITE(6,519)
519       FORMAT(/,3X,
     1    'Bit map based on optimal uniform',
          1    'quantizer for Gamma',/)
          GO TO 550
```

```
550       CONTINUE

          NSQ=NSIZE*NSIZE
          DO 10 I=1,NSQ
          ORDER = NORD(I)
          INORD(ORDER)=I
10        CONTINUE
C
          DO 20 I=0, NBITMX-1
               DLTERR(I) = QERR(I) - QERR(I+1)
20        CONTINUE
C
C         DO 30 I=1,NSQ
          BITMAP(I) = 0
30        CONTINUE
C
C         INITIALIZATION
C
          ORDER = INORD(1)
          BITMAP(ORDER) = 1
          PU(1) = 1
          PL(1) = 1
          PU(0) = 2
          PL(0) = NSQ
          MINPNT = 0
C
          DO 200 K = 1, TOTBIT-1
               OPTPNT = MINPNT
80        ORDER = INORD(PU(MINPNT))
               DLTMAX = DLTERR(MINPNT) * RVAR(ORDER)
               DO 100 I = MINPNT+1, NBITMX-1
                    IF(PU(I) .LE. 0) GOTO 100
                    ORDER = INORD(PU(I))
                    REF = DLTERR(I)*RVAR(ORDER)
                    IF (DLMAX .GT. REF) GOTO 100
                    DLTMAX = REF
                    OPTPNT = I
100            CONTINUE
               ORDER = INORD(PU(OPTPNT))
               BITMAP (ORDER) = BITMAP(ORDER) + 1
101            IF (PU(OPTPNT+1)) 110, 120, 130
110            PU(OPTPNT+1) = PU(OPTPNT)
               PL(OPTPNT+1) = PU(OPTPNT)
               GOTO 150
120            PU(OPTPNT + 1) = 1
               PL(OPTPNT + 1) = 1

               GOTO 150
130            PL(OPTPNT + 1) = PL(OPTPNT + 1) + 1
150            CONTINUE
               IF (PU(OPTPNT) - PL(OPTPNT)) 160, 170, 170
160            PU(OPTPNT) = PU(OPTPNT) + 1
               GOTO 180
170            PU(OPTPNT) = -1
               PL(OPTPNT) = -1
180            CONTINUE
          DO 190 II = 1, NBITMX
          IF (PU(II-1) .GT. 0) GOTO 195
          MINPNT = II
190       CONTINUE
195       CONTINUE
200       CONTINUE
C         WRITE(6,91) BITMAP
91        FORMAT(3X,'DESIGNED BITMAP IS:',/,16(16I4,/))
          RETURN
          END
```

APPENDIX A.6

COMPUTER PROGRAM FOR DHT

```
 /* ------------------------------------------------------------- *
  *                                                               *
  *   Module:    Orthonormal Hartley Transform                    *
  *                                                               *
  *   Authors:   Francisco Assis O. Nascimento and                *
  *              Henrique S. Malvar                               *
  *                                                               *
  *   Date:      December 27, 1988                                *
  *                                                               *
  *   Algorithm: Split-radix FHT                                  *
  *                                                               *
  *   Reference:  H.V. Sorensen et al.,                           *
  *               "On Computing the Discrete Hartley              *
  *               Transform" IEEE Trans. A.S.S.P.,                *
  *               Oct. 1985, pp. 1231-1245 [M-29].                *
  *                                                               *
  *   This is a translation to "C" of the FORTRAN program         *
  *   presented in the above paper.  Array references are made    *
  *   heavily through pointers, for good code generation.  The    *
  *   table of sines and cosines is generated only when the       *
  *   length of the transform is changed.                         *
  *                                                               *
  *   Note that the Hartley transform is symmetrical.  Thus, the  *
  *   inverse transform is identical to the direct transform.     *
  *                                                               *
  * ------------------------------------------------------------- */

/* ------- Prototype, to be included in the calling program -------- */

void hartley(
   float *x,      /* input/output vector                              */
   int   m,       /* log_2(vector length).  E.g. length=256 ->  m=8   */
   float *tab);   /* table of sines and cossines, must have space for
                     n/2  numbers                                     */

/* ------- Includes ------- */

#include <math.h>

/* ------- Defines -------- */

#define  PI       3.141592653589793     /* pi                 */
#define  SQRT_2   1.414213562373095     /* square root of 2   */
#define  SQH      0.707106781186547     /* square root of 1/2 */

/* --- Local variables ---- */

static float      x11,x12,x13,x14,x21,x22,x23,x24;
```

```
/* --------------------- Beginning of module ----------------------- */
void hartley(
   float *x,
   int   m,
   float *tab)
{
   int         n, im, i, j, k, pont;
   int         n2, n4, n8, n_tab, is, id;
   float       *px, *py, *pw, *pz, *px1, *py1, *pw1, *pz1;
   float       *pcc1, *pcc3, *pss1, *pss3;
   float       a, a3, temp;

   /* Compute:  n     = length of transform
                n2    = n/2
                n_tab = how many angles in the table */
   n  = 1 << m;
   n2 = n << 1;
   n_tab = n/8 - 1;

   /* Direct code for the trivial cases  n = 1, 2, and 4  */
   if (n == 1) return;

   if (n == 2) {
      temp   = SQH*(*x + *(x+1));
      *(x+1) = SQH*(*x - *(x+1));
      *x     = temp;
      return;
   }

   if (n == 4) {
      temp = 0.5*(*x + *(x+2));
      *(x+2) = 0.5*(*x - *(x+2));
      *x     = temp;
      temp = 0.5*(*(x+1) + *(x+3));
      *(x+3) = 0.5*(*(x+1) - *(x+3));
      *(x+1) = temp;

      temp = *x - *(x+1);
      *x   = *x + *(x+1);
      *(x+1) = *(x+2) + *(x+3);
      *(x+3) = *(x+2) - *(x+3);
      *(x+2) = temp;
      return;
   }

   /*
      Generate table of sines and cosines.  The angles start at 2*PI/n,
      increasing in steps of 2*PI/n radians.  The no. of different angle
      used is  n/8-1.
      A new table is generated only if the program is called with a
      new value of m.  The first entry in the table holds the value of
      n for which the table is built.
   */

   if( (int) (*tab) != n) {

       *tab = n;
       pcc1 = tab + 1;
       pss1 = pcc1 + n_tab;
       pcc3 = pss1 + n_tab;
       pss3 = pcc3 + n_tab;
       temp = (2*PI)/n;
       a    = temp;
```

```
      for(j = 0; j < n_tab; j++) {
         a3 = 3*a;
         *pcc1++ = (float) cos( (double) a );
         *pss1++ = (float) sin( (double) a );
         *pcc3++ = (float) cos( (double) a3 );
         *pss3++ = (float) sin( (double) a3 );
         a += temp;
      }
   }

/* Start computation of the transform.  Code translated from
   the reference */

pont = 1;
for(im = 0; im < m-1; im++) {
   is = 1;
   id = n2;
   n2 = n2 >> 1;
   n4 = n2 >> 2;
   n8 = n2 >> 3;
   do {
      for (i = is - 1; i < n; i += id) {
         px  = x + i;
         py  = px + n4;
         pw  = py + n4;
         pz  = pw + n4;
         x11 = *px;
         *px = *pw + x11;
         x12 = *py;
         *py = *py + *pz;
         x13 = *pw;
         *pw = x11 - x13 + x12 - *pz;
         *pz = x11 - x13 - x12 + *pz;
         if(n4 > 1) {
            px  = px + n8;
            py  = px + n4;
            pw  = py + n4;
            pz  = pw + n4;
            x11 = *px;
            *px = *pw + x11;
            x12 = *py;
            *py = *pz + x12;
            *pw = (x11 - *pw)*SQRT_2;
            *pz = (x12 - *pz)*SQRT_2;
            pcc1  = tab + pont;
            pss1  = pcc1 + n_tab;
            pcc3  = pss1 + n_tab;
            pss3  = pcc3 + n_tab;
           for (j = 1; j < n8; j++) {
              px  = x + i + j;
              py  = px + n4;
              pw  = py + n4;
              pz  = pw + n4;
              x11 = *px;
              x12 = *py;
              x13 = *pw;
              x14 = *pz;
              px1 = x + i + n4 - j;
              py1 = px1 + n4;
              pw1 = py1 + n4;
              pz1 = pw1 + n4;
              x21 = *px1;
```

```
                x22 = *py1;
                x23 = *pw1;
                x24 = *pz1;

                *px  = x11 + x13;
                *py  = x12 + x14;
                *pw  = (x11-x13+x21-x23)*(*pcc1) +
                       (x14-x12+x22-x24)*(*pss1);
                *pz  = (x11-x13-x21+x23)*(*pcc3) -
                       (x14-x12-x22+x24)*(*pss3);
                *px1 = x21 + x23;
                *py1 = x22 + x24;
                *pw1 = (x11-x13+x21-x23)*(*pss1) -
                       (x14-x12+x22-x24)*(*pcc1);
                *pz1 = (x11-x13-x21+x23)*(*pss3) +
                       (x14-x12-x22+x24)*(*pcc3);
                pcc1 += pont;
                pcc3 += pont;
                pss1 += pont;
                pss3 += pont;
             }
          }
       }
       is = 2 * id - n2 + 1;
       id = id << 2;
    } while(is < n);

    pont <<= 1;

    /* Normalize  butterflies */
    px = x;
    for(i = 0; i < n; i++) {
       *px /= SQRT_2;
       px++;
    }
}

/* Compute last stage of transform */

is = 1;
id = 4;
do {
   for (i = is - 1; i < n; i += id) {
      px      = x + i;
      x11     = *px;
      *px     = x11 + *(px+1);
      *(px+1) = x11 - *(px+1);
   }
   is = 2 * id - 1;
   id = id << 2;
} while(is < n);

/* Normalize last stage butterflies */
px = x;
for(i = 0; i < n; i++) {
   *px /= SQRT_2;
   px++;
}

/* Bit-reverse unshuffling of output sequence */
j = 1;
py = x;
```

```
   for(i = 1; i < n; i++) {
      if(i < j) {
         px  = x + j - 1;
         x11 = *px;
         *px = *py;
         *py = x11;
      }
         py ++;
      k = n >> 1;
      while (k < j) {
         j = j - k;
         k = k >> 1;
      }
      j = j + k;
   }
}

/* ----------------------- End of module --------------------------- */
```

APPENDIX A.7

COMPUTER PROGRAM FOR DCT VIA DHT

```
/* -------------------------------------------------------------- *
 *                                                                *
 *    Module:     Orthonormal Discrete Cosine Transform           *
 *                                                                *
 *    Author:     Henrique S. Malvar                              *
 *                                                                *
 *    Date:       December 27, 1988                               *
 *                                                                *
 *    Algorithm: DCT via Hartley transform                        *
 *                                                                *
 *    Reference: H.S. Malvar, "Fast computation of the discrete   *
 *                cosine transform through fast Hartley           *
 *                transform," Electron. Lett., vol. 22, no. 7,    *
 *                pp. 352-353, 27 March 1986 [FRA-19].            *
 *                                                                *
 * -------------------------------------------------------------- */

/* ------- Prototypes, to be included in the calling program ------- */

void dct(          /* Direct transform                                  */
   float *x,       /* input/output vector, length  n                    */
   float *y,       /* working vector, length n                          */
   int   m,        /* log_2(vector length).  E.g. length=256 ->  m=8    */
   float *tab);    /* table of sines and cossines, must have space for
                      3n/2  numbers                                     */

void idct(         /* Inverse transform                                 */
   float *x,       /* input/output vector, length  n                    */
   float *y,       /* working vector, length n                          */
   int   m,        /* log_2(vector length).  E.g. length=256 ->  m=8    */
   float *tab);    /* table of sines and cossines, must have space for
                      3n/2  numbers                                     */

void hartley(
   float *x,       /* input/output vector                               */
   int   m,        /* log_2(vector length).  E.g. length=256 ->  m=8    */
   float *tab);    /* table of sines and cossines, must have space for
                      n/2  numbers                                      */

/* ------- Includes ------- */

#include <math.h>

/* ------- Defines -------- */

#define  PI      3.141592653589793      /* pi                   */
#define  SQH     0.707106781186547      /* square root of 1/2 */
```

```
/* -------------- Beginning of direct DCT module ----------------- */
void dct(
   float *x,
   float *y,
   int   m,
   float *tab)
{
   int      n, n2, i;
   float    *cp1, *cp2, *xp1, *xp2, *yp1, *yp2;
   double   arg, delta_w;

   /* Compute:  n     = length of transform
                n2    = n/2  */
   n  = 1 << m;
   n2 = n >> 1;

   /*
      Generate table of sines and cosines.
      A new table is generated only if the program is called with a
      new value of m.  The first entry in the table holds the value of
      n for which the table is built.
      The first  n/2  elements of the table are used by the fhartley
      module.  The next  n  elements is a table of the  cas(.)
      function used in the hartley-to-dct mapping.
   */
   if( (int) (*tab) != n) {
      delta_w = PI / (2 * n);
      cp1 = tab + n2;
      cp2 = cp1 + n2;
      arg = 0;
      for ( i = 0; i < n2; i++ ) {
         *(cp1++) = (float) ((cos(arg) - sin(arg)) * SQH);
         *(cp2++) = (float) ((cos(arg) + sin(arg)) * SQH);
         arg += delta_w;
      }
   }

   /* Compute  y  sequence (see reference) */
   xp1 = x;
   xp2 = x + (n-1);
   yp1 = y;
   yp2 = y + n2;
   for ( i = 0; i < n2; i++ ) {
      *(yp1++) = *xp1;
      *(yp2++) = *xp2;
      xp1 += 2;
      xp2 -= 2;
   }

   hartley(y, m, tab);

   /* Rotation back to  x */

   *x      = *y;
   *(x+n2) = *(y+n2);
   cp1 = tab + n2 + 1;
   cp2 = cp1 + n2;
   xp1 = x + 1;
   xp2 = x + (n-1);
   yp1 = y + 1;
   yp2 = y + (n-1);
   for ( i = 1; i < n2; i++ ) {
     *(xp1++) = (*cp1) * (*yp1)     + (*cp2) * (*yp2);
     *(xp2--) = (*cp2++) * (*yp1++) - (*cp1++) * (*yp2--);
   }
}

/* ----------------- End of direct DCT module -------------------- */
```

```
/* -------------- Beginning of inverse DCT module ---------------- */

void idct(
   float *x,
   float *y,
   int   m,
   float *tab)

   int      n, n2, i;
   float    *cp1, *cp2, *xp1, *xp2, *yp1, *yp2;
   double   arg, delta_w;

   /* Compute:  n     = length of transform
                n2    = n/2  */
   n  = 1 << m;
   n2 = n >> 1;

   /* Update table of sines & cossines if  n  has chaged */

   if( (int) (*tab) != n) {
      delta_w = PI / (2 * n);
      cp1 = tab + n2;
      cp2 = cp1 + n2;
      arg = 0;
      for ( i = 0; i < n2; i++ ) {
         *(cp1++) = (float) ((cos(arg) - sin(arg)) * SQH);
         *(cp2++) = (float) ((cos(arg) + sin(arg)) * SQH);
         arg += delta_w;
      }
   }

   /* Rotation from  x  to  y */

   *y      = *x;
   *(y+n2) = *(x+n2);
   cp1 = tab + n2 + 1;
   cp2 = cp1 + n2;
   xp1 = x + 1;
   xp2 = x + (n-1);
   yp1 = y + 1;
   yp2 = y + (n-1);
   for ( i = 1; i < n2; i++ ) {
      *(yp1++) = (*cp1) * (*xp1)     + (*cp2) * (*xp2);
      *(yp2--) = (*cp2++) * (*xp1++) - (*cp1++) * (*xp2--);
   }

   hartley(y, m, tab);

   /* Back to  x  sequence */

   xp1 = x;
   xp2 = x + (n-1);
   yp1 = y;
   yp2 = y + n2;
   for ( i = 0; i < n2; i++ ) {
      *xp1 = *(yp1++);
      *xp2 = *(yp2++);
      xp1 += 2;
      xp2 -= 2;
   }

}

/* ---------------- End of inverse DCT module ------------------ */
```

APPENDIX A.8

COMPUTER PROGRAM FOR DCT-IV

```
  /* ---------------------------------------------------------------- *
   *                                                                  *
   *    Module:     Orthonormal Type-IV Discrete Cosine Transform     *
   *                                                                  *
   *    Author:     Henrique S. Malvar                                *
   *                                                                  *
   *    Date:       December 27, 1988                                 *
   *                                                                  *
   *    Algorithm: DCT-IV via DCT                                     *
   *                                                                  *
   *    Reference: Z. Wang, "On computing the discrete Fourier        *
   *               and cosine transforms," IEEE Trans. ASSP,          *
   *               vol. 33, pp. 1341-1344, Oct. 1985 [FRA11].         *
   *                                                                  *
   *    Note that the DCT-IV matrix is symmetric.  Thus, the          *
   *    direct and inverse transforms are the same.                   *
   *                                                                  *
   * ---------------------------------------------------------------- */

/* ------- Prototypes, to be included in the calling program ------- */

void dctiv(        /* Direct or inverse transform                          */
   float *x,       /* input/output vector, length  n                       */
   float *y,       /* working vector, length n                             */
   int   m,        /* log_2(vector length).  E.g. length=256 ->  m=8       */
   float *tab);    /* table of sines and cosines, must have space for
                      7n/4  numbers                                        */

void dct(          /* Direct transform                                     */
   float *x,       /* input/output vector, length  n                       */
   float *y,       /* working vector, length n                             */
   int   m,        /* log_2(vector length).  E.g. length=256 ->  m=8       */
   float *tab);    /* table of sines and cosines, must have space for
                      3n/2  numbers                                        */

/* ------- Includes ------- */

#include <math.h>

/* ------- Defines -------- */

#define  PI      3.141592653589793      /* pi                 */
#define  SQH     0.707106781186547      /* square root of 1/2 */
#define  FC2     0.92387953251129       /* constant for  n=2  */
#define  FS2     0.38268343236509       /* constant for  n=2  */
```

```
/* ---------------- Beginning of DCT-IV module ------------------- */

void dctiv(
   float *x,
   float *y,
   int   m,
   float *tab)
{
   int      n, n2, n4, i;
   float    *cp1, *cp2, *xp1, *xp2, *yp1, *yp2, temp;
   double   arg, delta_w;


   /* Compute:  n     = length of transform
                n2    = n/2
                n4    = n/4  */
   n  = 1 << m;
   n2 = n >> 1;
   n4 = n2 >> 1;

   /* Direct code for the trivial cases  n=1  and  n=2  */

   if (n == 1) return;

   if (n == 2) {
      temp   = FC2 * (*x) + FS2 * (*(x+1));
      *(x+1) = FS2 * (*x) - FC2 * (*(x+1));
      *x     = temp;
      return;
   }

   /*
      Generate table of sines and cosines.
      A new table is generated only if the program is called with a
      new value of m.  The first entry in the table holds the value of
      n for which the table is built.
      The first  n/2  elements of the table are used by the fhartley
      module.  The next  n  elements is a table of the  cas(.)
      function used in the hartley-to-dct mapping.  The last  n/4
      elements are for the  TN  matrix  (see reference).


      Note that this module will call a DCT and a DST of half length.
      Thus,  tab[0]  must contain  n/2.
   */

   if( (int) (*tab) != n2) {
      delta_w = PI / (2 * n);
      cp1 = tab + (3 * n) / 4;
      cp2 = cp1 + n2;
      arg = PI / ( 4 * n);
      for ( i = 0; i < n2; i++ ) {
         *(cp1++) = (float) cos(arg);
         *(cp2++) = (float) sin(arg);
         arg += delta_w;
      }
   }

   /* Multiplication by the  TN  matrix (see reference) */

      cp1 = tab + (3 * n) / 4;
      cp2 = cp1 + n2;
      xp1 = x;
      xp2 = x + (n-1);
      for ( i = 0; i < n2; i++ ) {
```

```
      temp = (*cp1) * (*xp1) + (*cp2) * (*xp2);
      *xp2 = (*cp2) * (*xp1) - (*cp1) * (*xp2);
      *xp1 = temp;
      cp1++; cp2++;
      xp1++; xp2--;
   }

   /* Reordering of last  n/2  samples */

   xp1 = x + n2;
   xp2 = x + (n-1);
   for (i = 0; i < n4; i++ ) {
      temp = *xp1;
      *xp1 = *xp2;
      *xp2 = temp;
      xp1++; xp2--;
   }

   /* Change sign of alternating samples */

   xp1 = x + n2 + 1;
   for (i = 0; i < n4; i++ ) {
      *xp1 = - *xp1;
      xp1 += 2;
   }

   /* Get DCT of first half of vector  x  */

   dct(&x[0], y, m-1, tab);

   /* Get DCT of second half of vector  x  */

   dct(&x[n2], y, m-1, tab);

   /* Copy  x  into  y, reordering */

   xp1 = x;
   xp2 = x + (n-1);
   yp1 = y;
   yp2 = y + 1;
   for ( i = 0; i < n2; i++ ) {
      *yp1 = *xp1++;
      *yp2 = *xp2--;
      yp1 += 2;
      yp2 += 2;
   }

   /* Last stage of butterflies, output back in   x  */

   x[0]   = y[0];
   x[n-1] = y[n-1];
   xp1 = x + 1;
   xp2 = x + 2;
   yp1 = y + 1;
   yp2 = y + 2;
   for ( i = 1; i < n2; i++ ) {
      *xp1 = SQH * (*yp1 + *yp2);
      *xp2 = SQH * (*yp2 - *yp1);
      xp1 += 2; xp2 += 2;
      yp1 += 2; yp2 += 2;
   }
}

/* -------------------- End of DCT-IV module ----------------------- */
```

APPENDIX A.9

COMPUTER PROGRAM FOR DST

```
/* ------------------------------------------------------------ *
 *                                                              *
 *    Module:     Orthonormal Discrete Sine Transform           *
 *                                                              *
 *    Author:     Henrique S. Malvar                            *
 *                                                              *
 *    Date:       December 27, 1988                             *
 *                                                              *
 *    Algorithm: DST via DCT                                    *
 *                                                              *
 *    Reference: Z. Wang, "A fast algorithm for the discrete    *
 *               sine transform implemented through the fast    *
 *               cosine transform," IEEE Trans. ASSP, vol. 30,  *
 *               pp. 814-815, 1982 [DDD-4].                     *
 *                                                              *
 * ------------------------------------------------------------ */

/* ------- Prototypes, to be included in the calling program ------- */

void dst(           /* Direct transform                                 */
   float *x,        /* input/output vector, length  n                   */
   float *y,        /* working vector, length n                         */
   int   m,         /* log_2(vector length).  e.g. length=256 ->  m=8   */
   float *tab);     /* table of sines and cosines, must have space for
                       3n/2  numbers                                    */

void idst(          /* Inverse transform                                */
   float *x,        /* input/output vector, length  n                   */
   float *y,        /* working vector, length n                         */
   int   m,         /* log_2(vector length).  e.g. length=256 ->  m=8   */
   float *tab);     /* table of sines and cosines, must have space for
                       3n/2  numbers                                    */

void dct(           /* Direct transform                                 */
   float *x,        /* input/output vector, length  n                   */
   float *y,        /* working vector, length n                         */
   int   m,         /* log_2(vector length).  e.g. length=256 ->  m=8   */
   float *tab);     /* table of sines and cosines, must have space for
                       3n/2  numbers                                    */

void idct(          /* Inverse transform                                */
   float *x,        /* input/output vector, length  n                   */
   float *y,        /* working vector, length n                         */
   int   m,         /* log_2(vector length).  e.g. length=256 ->  m=8   */
   float *tab);     /* table of sines and cosines, must have space for
                       3n/2  numbers                                    */
```

```
/* ------- Includes ------- */

#include <math.h>

/* --------------- Beginning of direct DST module ------------------ */

void dst(
   float *x,
   float *y,
   int   m,
   float *tab)
{
   int      n, n2, i;
   float    *xp1, *xp2;
   float    temp;

   /* Compute:  n     = length of transform
                n2    = n/2  */
   n  = 1 << m;
   n2 = n >> 1;

   /* Multiply data by sequence of alternating +1's and -1's */

   xp1 = x + 1;
   for (i = 0; i < n2; i++ ) {
      *xp1 = - *xp1;
      xp1 += 2;
   }

   /* Get DCT */

   dct(x, y, m, tab);

   /* Reverse order of transformed coefficients */

   xp1 = x;
   xp2 = x + n - 1;
   for (i = 0; i < n2; i++ ) {
      temp = *xp1;
      *xp1 = *xp2;
      *xp2 = temp;
      xp1++;
      xp2--;
   }
}

/* ------------------ End of direct DST module --------------------- */

/* --------------- Beginning of inverse DST module ----------------- */

void idst(
   float *x,
   float *y,
   int   m,
   float *tab)
{
   int      n, n2, i;
   float    *xp1, *xp2;
   float    temp;

   /* Compute:  n     = length of transform
                n2    = n/2  */
   n  = 1 << m;
   n2 = n >> 1;
```

```
   /* Reverse order of transformed coefficients */

   xp1 = x;
   xp2 = x + n - 1;
   for (i = 0; i < n2; i++ ) {
      temp = *xp1;
      *xp1 = *xp2;
      *xp2 = temp;
      xp1++;
      xp2--;
   }

   /* Get IDCT */

   idct(x, y, m, tab);

   /* Multiply data by sequence of alternating +1's and -1's */

   xp1 = x + 1;
   for (i = 0; i < n2; i++ ) {
      *xp1 = - *xp1;
      xp1 += 2;
   }
}

/* ----------------- End of inverse DST module ------------------- */
```

APPENDIX A.10

COMPUTER PROGRAM FOR DST-IV

```
/* ------------------------------------------------------------ *
 *                                                              *
 *    Module:     Orthonormal Type-IV Discrete Sine Transform   *
 *                                                              *
 *    Author:     Henrique S. Malvar                            *
 *                                                              *
 *    Date:       December 27, 1988                             *
 *                                                              *
 *    Algorithm: DST-IV via DCT                                 *
 *                                                              *
 *    Reference: Z. Wang, "On computing the discrete Fourier    *
 *               and cosine transforms," IEEE Trans. ASSP,      *
 *               vol. 33, pp. 1341-1344, Oct. 1985 [FRA-11].    *
 *                                                              *
 *    Note that the DST-IV matrix is symmetric.  Thus, the      *
 *    direct and inverse transforms are the same.               *
 *                                                              *
 * ------------------------------------------------------------ */


/* ------- Prototypes, to be included in the calling program ------- */

void dstiv(       /* Direct or inverse transform                         */
   float *x,      /* input/output vector, length  n                      */
   float *y,      /* working vector, length n                            */
   int   m,       /* log_2(vector length).  E.g. length=256 ->  m=8      */
   float *tab);   /* table of sines and cosines, must have space for
                     7n/4  numbers                                       */

void dct(         /* Direct transform                                    */
   float *x,      /* input/output vector, length  n                      */
   float *y,      /* working vector, length n                            */
   int   m,       /* log_2(vector length).  E.g. length=256 ->  m=8      */
   float *tab);   /* table of sines and cosines, must have space for
                     3n/2  numbers                                       */

/* ------- Includes ------- */

#include <math.h>

/* ------- Defines -------- */

#define  PI       3.141592653589793     /* pi                 */
#define  SQH      0.707106781186547     /* square root of 1/2 */
#define  FC2      0.92387953251129      /* constant for  n=2  */
#define  FS2      0.38268343236509      /* constant for  n=2  */
```

```
/* ---------------- Beginning of DST-IV module -------------------- */

void dstiv(
   float *x,
   float *y,
   int   m,
   float *tab)
{
   int      n, n2, n4, i;
   float    *cp1, *cp2, *xp1, *xp2, *yp1, *yp2, temp;
   double   arg, delta_w;

   /* Compute:  n     = length of transform
                n2    = n/2
                n4    = n/4   */
   n  = 1 << m;
   n2 = n >> 1;
   n4 = n2 >> 1;

   /* Direct code for the trivial cases  n=1  and  n=2  */

   if (n == 1) return;

   if (n == 2) {
      temp   = FS2 * (*x) + FC2 * (*(x+1));
      *(x+1) = FC2 * (*x) - FS2 * (*(x+1));
      *x     = temp;
      return;
   }

   /*
      Generate table of sines and cosines.
      A new table is generated only if the program is called with a
      new value of m.  The first entry in the table holds the value of
      n for which the table is built.
      The first  n/2  elements of the table are used by the fhartley
      module.  The next  n  elements is a table of the  cas(.)
      function used in the hartley-to-dct mapping.  The last  n/4
      elements are for the  TN  matrix  (see reference).

      Note that this module will call a DCT and a DST of half length.
      Thus,  tab[0]  must contain  n/2.
   */

   if( (int) (*tab) != n2) {
      delta_w = PI / (2 * n);
      cp1 = tab + (3 * n) / 4;
      cp2 = cp1 + n2;
      arg = PI / ( 4 * n);
      for ( i = 0; i < n2; i++ ) {
         *(cp1++) = (float) cos(arg);
         *(cp2++) = (float) sin(arg);
         arg += delta_w;
      }
   }

   /* Multiplication by the matrix  J * TN (see reference) */

      cp1 = tab + (3 * n) / 4;
      cp2 = cp1 + n2;
      xp1 = x;
      xp2 = x + (n-1);
      for ( i = 0; i < n2; i++ ) {
         temp = (*cp2) * (*xp1) + (*cp1) * (*xp2);
         *xp2 = (*cp2) * (*xp2) - (*cp1) * (*xp1);
```

```
        *xp1 = temp;
        cp1++; cp2++;
        xp1++; xp2--;
    }

    /* Reordering of last  n/2  samples */

    xp1 = x + n2;
    xp2 = x + (n-1);
    for (i = 0; i < n4; i++ ) {
        temp = *xp1;
        *xp1 = *xp2;
        *xp2 = temp;
        xp1++; xp2--;
    }

    /* Change sign of alternating samples */

    xp1 = x + n2 + 1;
    for (i = 0; i < n4; i++ ) {
        *xp1 = - *xp1;
        xp1 += 2;
    }

    /* Get DCT of first half of vector  x  */

    dct(&x[0], y, m-1, tab);

    /* Get DCT of second half of vector  x  */

    dct(&x[n2], y, m-1, tab);

    /* Copy  x  into  y, reordering */

    xp1 = x;
    xp2 = x + (n-1);
    yp1 = y;
    yp2 = y + 1;
    for ( i = 0; i < n2; i++ ) {
        *yp1 = *xp1++;
        *yp2 = *xp2--;
        yp1 += 2;
        yp2 += 2;
    }

    /* Last stage of butterflies, and multiplication by  DN,
       output back in  x  */
       x[0]   =  y[0];
       x[n-1] = -y[n-1];
       xp1 = x + 1;
       xp2 = x + 2;
       yp1 = y + 1;
       yp2 = y + 2;
       for ( i = 1; i < n2; i++ ) {
           *xp1 = -SQH * (*yp1 + *yp2);
           *xp2 =  SQH * (*yp2 - *yp1);
           xp1 += 2; xp2 += 2;
           yp1 += 2; yp2 += 2;
       }
    }

    /* -------------------- End of DST-IV module ----------------------- */
```

APPENDIX A.11

COMPUTER PROGRAM FOR LOT

```
/* --------------------------------------------------------------- *
 *                                                                 *
 *    Module:     Lapped Orthogonal Transform                      *
 *                                                                 *
 *    Author:     Henrique S. Malvar                               *
 *                                                                 *
 *    Date:       December 27, 1988                                *
 *                                                                 *
 *    Algorithm: LOT via DCT & DST-IV                              *
 *                                                                 *
 *    Reference: H.S. Malvar and R. Duarte, "Transform/subband     *
 *               coding of speech with the lapped orthogonal       *
 *               transform," Proc. 1989 International Symposium    *
 *               on Circuits and Systems, PP,1268-1271,Portland,   *
 *               May 1989 [SC-29].                                 *
 *                                                                 *
 *    Unlike the DCT, DHT, and DST-IV modules, the output of the   *
 *    LOT  and  ILOT  modules are not time aligned with the        *
 *    input, because of the  50%  overlap of the basis functions. *
 *    Thus, there is an extra delay of  n/2  samples in the  LOT   *
 *    and  ILOT  modules.  Thus a call to  lot(...)   followed     *
 *    by a call to  ilot(...)  recovers the original signal        *
 *    back, delayed by  n  samples.                                *
 *                                                                 *
 *    Note 1:  The transform size was called  M  in the            *
 *    reference.  Here, we refer to the transform size as  n,      *
 *    and  m  is its logarithm (base 2).                           *
 *                                                                 *
 *    Note 2:  Because the samples stored in the  z^(-M)  delay    *
 *    units are stored in the last  n/2  elements of the working   *
 *    vector  y  below,  a program that calls both the  lot(...)   *
 *    and  ilot(...)  modules SHOULD NOT use the same working      *
 *    vector for both modules.  The table of sines and cosines     *
 *    can be the same for both modules.                            *
 *                                                                 *
 *     Example of calling code:                                    *
 *                                                                 *
 *     float  x[64], y1[96], y2[96], tab[200];                     *
 *                                                                 *
 *     main() {                                                    *
 *        .                                                        *
 *        .                                                        *
 *        for ( i = 0; i < (n/2); i++ ) {     <-- Clear internal   *
 *           y1[n + i] = 0;                   storage area for     *
 *           y2[n + i] = 0;                   lot(..) and ilot(..) *
 *        }                                   modules.             *
 *        .                                                        *
 *        for ( frame = 0; frame < no_of_frames; frames++ ) {      *
```

```
 *          .                                                   *
 *          read x form input buffer or file                    *
 *          .                                                   *
 *          lot( x, y1, 6, tab );   <--  direct  LOT            *
 *          .                                                   *
 *          .                                                   *
 *          ilot( x, y2, 6, tab );  <--  inverse LOT            *
 *          .                                                   *
 *          copy x to output buffer or file                     *
 *          .                                                   *
 *       }                                                      *
 *       .                                                      *
 *       .                                                      *
 *    }                                                         *
 *                                                              *
 *    Note 3: the overlap-add of the inverse  LOT  is performed *
 *    internally by the  ilot(..)  module, so that the calling  *
 *    program can work just with the  n  samples of  x, as      *
 *    indicated above.                                          *
 *                                                              *
 * ------------------------------------------------------------ */

/* ------- Prototypes, to be included in the calling program ------- */

void lot(          /* Direct LOT                                      */
   float *x,       /* input/output vector, length  n                  */
   float *y,       /* working vector, length 3n/2                     */
   int   m,        /* log_2(vector length).  e.g. length=256 ->  m=8  */
   float *tab);    /* table of sines and cosines, must have space for
                      25n/8  numbers                                  */

void ilot(         /* Inverse LOT                                     */
   float *x,       /* input/output vector, length  n                  */
   float *y,       /* working vector, length 3n/2                     */
   int   m,        /* log_2(vector length).  e.g. length=256 ->  m=8  */
   float *tab);    /* table of sines and cosines, must have space for
                      25n/8  numbers                                  */

void dstiv(        /* Direct or inverse transform                     */
   float *x,       /* input/output vector, length  n                  */
   float *y,       /* working vector, length n                        */
   int   m,        /* log_2(vector length).  e.g. length=256 ->  m=8  */
   float *tab);    /* table of sines and cosines, must have space for
                      7n/4  numbers                                   */

void dct(          /* Direct transform                                */
   float *x,       /* input/output vector, length  n                  */
   float *y,       /* working vector, length n                        */
   int   m,        /* log_2(vector length).  e.g. length=256 ->  m=8  */
   float *tab);    /* table of sines and cosines, must have space for
                      3n/2  numbers                                   */

void idct(         /* Inverse transform                               */
   float *x,       /* input/output vector, length  n                  */
   float *y,       /* working vector, length n                        */
   int   m,        /* log_2(vector length).  e.g. length=256 ->  m=8  */
   float *tab);    /* table of sines and cosines, must have space for

                      3n/2  numbers                                   */

/* ------- Includes ------- */

#include <math.h>

/* ------- Defines -------- */
```

```
#define  PI      3.141592653589793      /* pi                   */
#define  SQH     0.707106781186547      /* square root of 1/2 */

/* --------------- Beginning of direct LOT module ------------------ */

void lot(
   float *x,
   float *y,
   int   m,
   float *tab)
{
   int      n, n2, ntb1, ntb2, i;
   float    *xp1, *xp2, *yp1, *yp2;

   /* Compute:  n     = length of transform
                n2    = n/2  */
   n  = 1 << m;
   n2 = n >> 1;
   ntb1 = 3 * n2;
   ntb2 = 9 * (n >> 2);

   /* Compute DCT */

   dct(x, y, m, tab);

   /* Copy  even-indexed  x's  on  y[0]  and
      odd-indexed  x's on  y[n/2] */

   xp1 = x;
   yp1 = y;
   yp2 = y + n2;
   for ( i = 0; i < n2; i++ ) {
      *(yp1++) = *(xp1++);
      *(yp2++) = *(xp1++);
   }

   /* First butterflies with  +1/-1  factors, with  1/2  factor,
      output in  x  */

   xp1 = x;
   xp2 = x + n2;
   yp1 = y;
   yp2 = y + n2;
   for ( i = 0; i < n2; i++ ) {
      *(xp1++) = 0.5 * ( *yp1     + *yp2 );
      *(xp2++) = 0.5 * ( *(yp1++) - *(yp2++) );
   }
/* This piece of code corresponds to the  z^(-M)  delays indicated in
   the reference.  The stored values are in the last  n/2  samples
   of  y  */
yp1 = y;
yp2 = y + n;
for ( i = 0; i < n2; i++ ) {
   *(yp1++) = *(yp2++);
}
xp2 = x + n2;
yp2 = y + n;
for ( i = 0; i < n2; i++ ) {
   *(yp2++) = *(xp1++);
}

/* Copy first  n/2  coefficients of  y  in  x[n/2]  */

yp1 = y;
xp2 = x + n2;
```

```
    for ( i = 0; i < n2; i++ ) {
       *(xp2++) = *(yp1++);
    }

    /* Second stage of  +1/-1  butterflies, output in  y */

    xp1 = x;
    xp2 = x + n2;
    yp1 = y;
    yp2 = y + n2;
    for ( i = 0; i < n2; i++ ) {
       *(yp1++) = *xp1 + *xp2;
       *(yp2++) = *xp1 - *xp2;
       xp1++; xp2++;
    }

    /* Length-(n/2)  IDCT  */

    idct(&y[n2], x, m-1, &tab[ntb1]);

    /* Length-(n/2)  DST-IV  */

    dstiv(&y[n2], x, m-1, &tab[ntb2]);

    /* Even/odd re-indexing, output in  x  */

    xp1 = x;
    yp1 = y;
    yp2 = y + n2;
    for ( i = 0; i < n2; i++ ) {
       *(xp1++) = *(yp1++);
       *(xp1++) = *(yp2++);
    }

}
/* ------------------ End of direct LOT module --------------------- */

/* --------------- Beginning of inverse LOT module ----------------- */

void ilot(
   float *x,
   float *y,
   int   m,
   float *tab)
{
   int      n, n2, ntb1, ntb2, i;
   float    *xp1, *xp2, *yp1, *yp2;

   /* Compute:  n     = length of transform
                n2    = n/2  */
   n  = 1 << m;
   n2 = n >> 1;
   ntb1 = 3 * n2;
   ntb2 = 9 * (n >> 2);

   /* Even/odd re-indexing, output in  y  */

   xp1 = x;
   yp1 = y;
   yp2 = y + n2;
   for ( i = 0; i < n2; i++ ) {
      *(yp1++) = *(xp1++);
      *(yp2++) = *(xp1++);
   }

   /* Length-(n/2)  DST-IV  */

   dstiv(&y[n2], x, m-1, &tab[ntb2]);
```

```
/* Length-(n/2)  DCT  */

dct(&y[n2], x, m-1, &tab[ntbl]);

/* First butterflies with  +1/-1  factors, with  1/2  factor,
   output in  x  */

xp1 = x;
xp2 = x + n2;
yp1 = y;
yp2 = y + n2;
for ( i = 0; i < n2; i++ ) {
   *(xp1++) = 0.5 * ( *yp1     + *yp2 );
   *(xp2++) = 0.5 * ( *(yp1++) - *(yp2++) );
}

/* This piece of code corresponds to the  z^(-M)  delays indicated in
   the reference.  The stored values are in the last  n/2  samples
   of  y .  Note that the  z^(-M)  delays are now in the branches
   marked by asterisks in the LOT flowgraph (see reference). */

   yp1 = y;
   yp2 = y + n;
   for ( i = 0; i < n2; i++ ) {
      *(yp1++) = *(yp2++);
   }
   xp1 = x;
   yp2 = y + n;
   for ( i = 0; i < n2; i++ ) {
      *(yp2++) = *(xp1++);
   }

   /* Copy first  n/2  coefficients of  y  in  x[0]  */

   yp1 = y;
   xp1 = x;
   for ( i = 0; i < n2; i++ ) {
      *(xp1++) = *(yp1++);
   }

   /* Second stage of  +1/-1  butterflies, output in  y */

   xp1 = x;
   xp2 = x + n2;
   yp1 = y;
   yp2 = y + n2;
   for ( i = 0; i < n2; i++ ) {
      *(yp1++) = *xp1 + *xp2;
      *(yp2++) = *xp1 - *xp2;
      xp1++; xp2++;
   }

   /* Copy  y[0]  on  even-indexed  x's  and
      y[n/2]  on odd-indexed  x's */

   xp1 = x;
   yp1 = y;
   yp2 = y + n2;
   for ( i = 0; i < n2; i++ ) {
      *(xp1++) = *(yp1++);
      *(xp1++) = *(yp2++);
   }

   /* Compute IDCT */

   idct(x, y, m, tab);
}

/* ----------------- End of inverse LOT module ------------------- */
```

APPENDIX B.1

DCT VLSI CHIP MANUFACTURERS

This appendix provides a partial list of DCT chip developers/manufacturers (Table B.1.1). Some of these, such as Universities and Research Institutes, have designed and developed the DCT chip purely from a research and development viewpoint (Technology transfer to industry is a possibility), whereas the industry's objective is based on marketing strategy including applications in consumer electronics such as videophone, videoconferencing, photovideotex, video disk, CD ROM, etc. Some of these activities are limited to design and architecture, prototype development, etc. With simple modifications, the chip can implement DCT of various lengths as well as the IDCT. In general, the architecture is based on implementing the 2D DCT using the separability property. Some of the design takes advantage of the recursive property of the fast algorithms such that the same chip can implement 1D DCT of different lengths and/or 2D DCT of different block sizes. Another development is to integrate other operations such as zig-zag scanning, quantization, coding, etc. with the DCT on the same chip. It is conceivable that the entire coding process may be integrated into a single chip. Another flexibility is implementation of both forward and inverse DCTs by the same chip. This is simplified because the forward and inverse DCT matrices are transpose (within a scale factor) of each other. The current trend is to increase the speed (throughput rate) and/or processing of various block sizes. The DCT chips developed by some manufacturers are for their internal use only. VLSI of DCT/IDCT has gained momentum, in part, due to the standards being recommended by various international organizations such as CCITT, ISO, CMTT, T1.Y1., etc. for video coding systems (videophone, videoconferencing, freeze frame image transmission, broadcast quality video codec, etc.).

Table B.1.1

DCT VLSI chip manufacturers

Address	DCT	Status	Throughput (Speed)	Reference
Lincoln Lab MIT, Lexington, MA 02173 USA	1D 16-point	CCD device has been fabricated, tested and is working.	10 MHz 0.1 μsec	[DP-31]
	2D (16 × 16) or (8 × 8)	Design and architecture	2D (16 × 16) DCT in 8.5 μsec	[DP-26]
Dept. of Electrical Engineering University of Linköping S-58183 Linköping Sweden	1D 8-point	Design and architecture	0.75×10^6 DCTs/sec	[DP-13, DP-23, DP-24]
	1D 16-point 2D (8 × 8) or (16 × 16)	Design and architecture	46000 2D DCTs/sec	[DP-18]
Electronic & Computer Engineering Dept. Univ. of California Davis, CA 95616 USA	1D 8-point	Unified DCT/IDCT architecture Design and layout	10 MHz	[DP-35]
AT&T Bell Labs Crawfords Corner Rd. Holmdel, NJ 07733 USA	2D (8 × 8)	Chip has been fabricated	2D (8 × 8) DCT in 10 μsec	[DP-27, DP-22, FRA-24]
Bellcore Inc. 331 Newman Springs Rd. Red Bank, NJ 07701 USA	2D (16 × 16)	Chip is functional. Technology licensing is available.	14.3 MHz	[DP-29, DP-34, DP-37,DP-39, DP-49]
	2D (8 × 8)	Design architecture and layout	27 MHz	
VLSI Research Center Toshiba R&D Center Kawasaki 210, Japan		DCT chip architecture is being developed for CCITT standard	Video rate	
Inmos Ltd. 1000 Aztec West Almondbury Bristol BS 124SQ UK	IMS A121 2D (8 × 8) DCT or IDCT including on chip addition/subtraction	Chip is functional. It can also function as a 2D linear filter or perform matrix transposition. Chip is available on the market.	20 MHz	IMS A121 2D DCT image processor data sheet

SGS-Thomson Microelectronics 17 Ave des Martyrs, BP 217 33019 Grenoble, France	STV3200 2D (4 × 4), (4 × 8), (8 × 4), (8 × 8), (8 × 16), (16 × 8), (16 × 16)	DCT and IDCT chip has been fabricated and tested. Chip is available on the market.	13.5 MHz	STV 3200 Central advanced datasheet [DP-33, DP-28, DP-41]
	2D (8 × 8)	DCT/IDCT, CCITT compatibility, on chip zig-zag conversion of coefficient scanning. Chip is available in the market.	0 to 27 MHz	Advanced information
CCETT, 4, Rue du Clos Courtel, BP59 35512 Cesson Sevigne France	2D (8 × 8)	Design and architecture. Chip has been fabricated and tested. Chip is functional	0 to 27 MHz	[DP-44]
IMT UNI NE Institut de Microtechnique Universite de Neuchatel CH-2000 Neuchatel, Switzerland	1D, 2D (16 × 16), (16 × 8), (8 × 8) (8 × 4), (4 × 4)	Design, architecture and layout. Chip is to be fabricated.	16.7 MHz	[DP-45]
Telettra via Trento, 30 20059 Vimercate Italy 039/66551	2D (8 × 8)	Chip is functional and is utilized in 34 MBPS and DS-3 video codecs.	11.25 MHz	LSI Logic, UK has manufactured the chip for Telettra. [IC-53, IC-67, IC-73]
Matsushita Elec. Ind. Co. Ltd. Semiconductor Research Center Moriguchi Osaka, 570 Japan	MN8520 2D(8 × 8), (16 × 16)	Chip is functional. It can also function as 1D or 2D filter	20 MHz	[DP-54]
ANT Nachrichtentechnik GmbH Postfach 1120, D-7150 Backnang W. Germany		Semicustom chip has been fabricated		
Phillips Research Labs P.O. Box 80000-5600 JA Eindhoven, The Netherlands	2D (8 × 8) DCT	Chip is functional	13.5 MHz	[DP-46]

Table B.1.1 (*continued*)

Address	DCT	Status	Throughput (Speed)	Reference
Electrical Engrg. Dept. Stanford University Stanford, CA 94305	2D (8 × 8) DCT	Design and architecture. Chip has been fabricated.	13.5 MHz	[DP-47]
Siemens AG Otto-Hahn-Ring 6 D-8000 Munich 83 W. Germany	2D (8 × 8) or (16 × 16)	Test chip being designed	30 MHz (8 × 8) 15 MHz (16 × 16)	[DP-52]
HHI GmbH Berlin Einsteinufer 37 D-1000 Berlin 10 West Germany	2D (8 × 8)	Design, architecture	80 MHz and layout	[DP-53]
Fujitsu Limited 1015 Kamikodanaka Nakahara-ku Kawasaki 211, Japan	2D (8 × 8) 1D (2 × 2), (4 × 4), (8 × 8), (16 × 16)	Chip is functional. It has 2D (8 × 8) DCT, loop filter and zig-zag scan converter	15 MHz 2D (8 × 8) 1D (16 × 16) 30 MHz 1D (2 × 2), (4 × 4, (8 × 8)	Private correspondence
TRW LSI Products Inc. 4243 Campus Point Ct.	2D (8 × 8) including on chip auxiliary adder/subtractor 1D 8-point	DCT/IDCT. CCITT compatible. Chip can also be used in other waveform coding techniques such as LPC and DPCM. Available in the market.	15 MHz	TMC 2311 data sheet
Graphics Communication Technologies Column-M.A. 7–1–5 Minami-Aoyama, Minato-ku, Tokyo 107 Japan	2D (8 × 8) DCT	Zig-zag scanning on chip addition/ subtraction. Chip is being fabricated.	27 MHz	Private corre-spondence

APPENDIX B.2

IMAGE COMPRESSION BOARDS

The appendix describes image compression/decompression boards based on DCT (Table B.2.1) for increasing the image storage in magnetic or optical media or for reducing transmission time over local or wide-area networks. These boards are in general designed for IBM PC, PC/AT, PC/XT, or compatible systems. Together with frame grabber and digitizer boards, they can perform a number of operations such as RGB $\Leftrightarrow$ YIQ transformation, 2D DCT, quantization, and coding (Fig. B.2.1). They are supplemented with a software library designed for a number of DSP operations, such as split screen, windowing, filtering, overlay, zoom, scaling, graphics, text, etc. They also have user selectable compression ratios, a library of software packages that enable different image processing operations, on screen menu, etc. These boards can process images generated from any video source (video camera, graphics camera, VCR, video disc, image files stored on magnetic or optical media). A basic desktop image processing system (Fig. B.2.2) can be set up for about $10,000.00 to $15,000.00. Such a system has the following capabilities:

Generating image data base (from video camera or scanner)
User selectable compression ratios
Store or transmit images in compressed form
Retrieve (decompress) and display images
Perform various DSP operations
Hard copy output

Using several boards based on INTEL 80386 microprocessor for the CPU and 16 TMS 320C25 DSP chips for the video processing subsystem, an IBM PC compatible desktop workstation has been developed by VideoTelecom Corp. [LBR-47]. The video compression is based on the 2D 16×16 DCT. The multifunction/multimedia workstation processes video and audio in real time for real time communication over digital networks. During videophone calls, the

Table B.2.1

Image compression boards

Address	DCT	Compression	References
Zoran Corp 3450 Central Expressway Santa Clara, CA 95051 USA	RGB ⇔ YIQ 2D (8×8) DCT, Quantization and Huffman coding	Can compress RGB ($512 \times 480 \times 24$) image file in 20 secs. up to 30:1 compression (user selectable)	Zoran ZR 73650 Imagineering™ Board April 1989 [DP-40]
Telephoto Communications, Inc. 11722 Sorrento Valley Rd. Suite D San Diego, CA 92121 USA	RGB ⇔ YIQ, 2D (16×16) DCT, truncation	Up to 14:1 compression (user selectable) Alice Type-120 hardware/software package	[IC-68] Telephoto product news Feb. 1989
Discrete Time Systems, Inc. 511 West Gold Road Arlington Heights, IL 60005 USA	To our knowledge, DCT is not used for compression. This is included, however, for the sake of completeness.	Up to 30:1 compression (user selectable) Image capture, compression, storage, transmission, retrieval and display	CVP-200 color video- processor family (boards) CVP-2000™, Hardware product specifications.
Optivision, Inc. 744 San Antonio Road Suite 10 Palo Alto, CA 94303 USA	DCT for controlled quality compression. DPCM and VLC for lossless coding. CCITT algorithms for binary images. Can update/enhance with new compression algorithms	Up to 115:1 compression (user selectable) with various modes of operation	Optipac™ brochure. Optivision press release, Aug. 4, 1989
Video Telecom Corp 1908 Kramer Lane Austin, TX 78758 USA	2D (16×16) DCT	Real time image compression (both sequences and still images) storage, retrieval and transmission	[LBR-47]

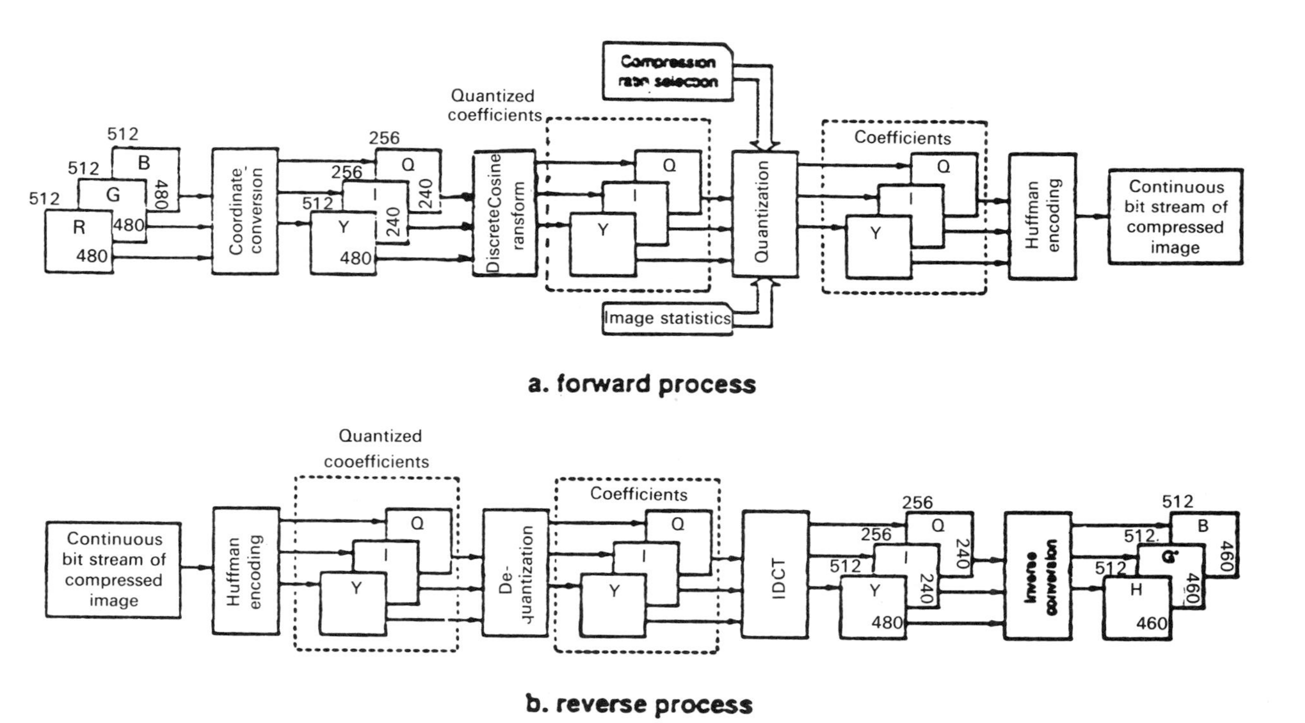

Fig. B.2.1 Image compression algorithm based on Zoran ZR 73650 Imagineering™ board [DP-40].

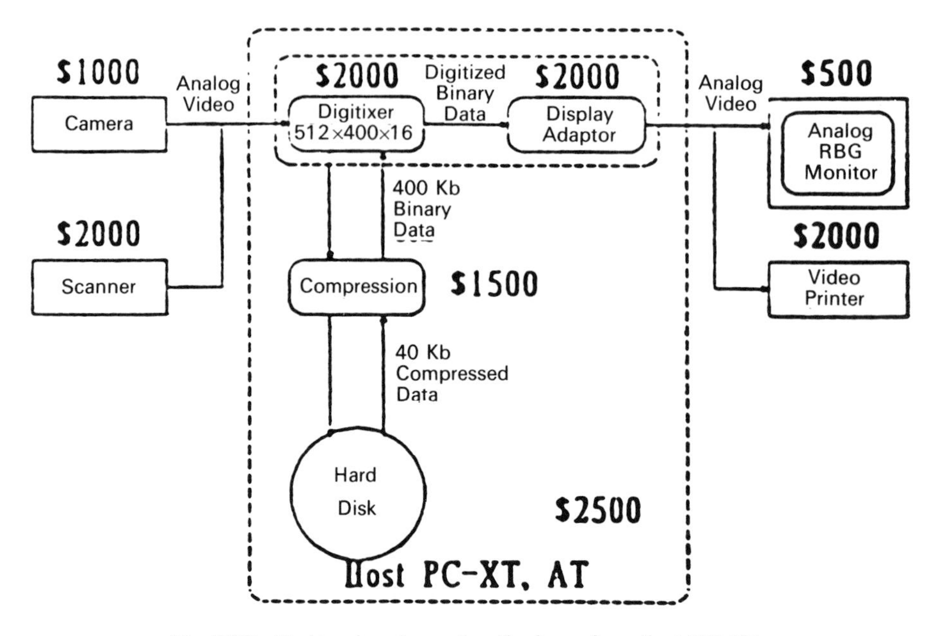

Fig. B.2.2 Desktop imaging system (basic configuration) [IC-68].

user can transfer computer files, execute programs, send snapshots of computer screens, send FAX messages, and transfer data through other devices. Other capabilities are storage and retrieval of video sequences with sound and still images on magnetic and optical discs. Another advantage is handsfree operation as audio communication is over a full duplex echo cancelling speakerphone. Through an RS-449/422 interface, the workstation is connected to the digital network communication equipment and operates at speeds of 9.6 KBPS through 768 KBPS. The system, however, is optimized at 384 KBPS.

APPENDIX B.3

MOTION ESTIMATION VLSI CHIP MANUFACTURERS

This appendix compiles a partial list of VLSI chip manufacturers for implementing motion estimation useful for image coding. The motion estimation of a block in an image frame (see Fig. 7.108) is based on full (exhaustive) search using, in general, MAE (7.10.3) as the distortion criterion. Even though several efficient algorithms (BMA) have been developed for block motion estimation (see Section 7.10, also Figs. 7.109 through 7.120), their implementation by VLSI is inhibited due to large control overhead and irregular data flow. Some of the chips are designed for fractional pel/line resolution of the motion vector, and the range of motion vector can be varied based on video sampling rates. Also, motion vector range can be increased by cascading the VLSI chips. Other options may include random access to any distortion within the motion vector range for implementing more elaborate algorithms at the system level. For example, fractional pel/line resolution of the motion vector requires interpolation between distortions. As motion compensation plays a significant part in predictive and/or hybrid (DPCM/transform) coding in terms of increasing the compression ratio at the same fidelity or improving the fidelity at the same compression ratio, availability of these chips further contributes to the codec development for various visual communication services (videophone, teleconferencing, broadcast TV, HDTV, etc.). Design, architecture, and layout for pel-by-pel motion estimators are described in [MC-38, MC-44].

Table B.3.1

Motion estimation VLSI chip.

Address	Details	Status	Throughput (Speed)	Reference
Bellcore 331 Newman Springs Rd. Red Bank, NJ 07701 and 445 South St. Morristown, NJ 07960	Full search BMA. Various block sizes (8×8, 16×16, 32×32, ...). Motion vector range -8 to $+7$ pels-lines/frame. Same chips can be cascaded for larger tracking ranges. Fractional pel precision.	Design architecture and layout	30 MHz	[MC-30, MC-33, MC-40, MC-41, MC-42]
Plessy Research Caswell Ltd. UK				[MC-39]
Telecom Paris University 46 rue Barrault 75634 Paris, France	Full search BMA. Block size and motion vector range are variable. Maximum motion vector range is up to ± 15 pel-line/frame. Same chips can be cascaded for larger tracking ranges.	Design and architecture	Up to 18 MHz	[MC-31]
University of Hannover Appelstr. 9A D-3000, Hannover 1, W. Germany	Full search BMA. Block size (8×8). Motion vector range ± 8 pel-line per frame. Systolic array processors.	Design and architecture		[MC-32, MC-43]
Siemens AG Corporate R&D Otto-Hahn-Ring, 6 D-8000 Munich 83, W. Germany	Full search BMA (line scan mode and block scan mode)	Architecture	Long range of data rates	[MC-37]

Table B.3.1

Motion estimation VLSI chip.

Address	Details	Status	Throughput (Speed)	Reference
CSTS-CNR CEFRIEL and ITALTEL SIT Milano, Italy	Pel-by-pel motion estimation	Design architecture and layout		[MC-38]
SGS-Thomson Microelectronics 17 Avenue des Martyrs BP 217 38019 Grenoble France	Full search BMA. 8×8, 8×16, 16×16 block sizes. Motion vector range -8 to $+7$ pels-lines/frame. Extension to -16 to $+15$ pels-lines/frame with a single chip for lower video rates or with multiple chips for higher video rates	Chip is functional	13.5 MHz	STV 3220
Graphics Communication Technologies Column-M.A. 7–1–5 Minami-Aoyama, Minato-ku, Tokyo 107 Japan	Full search BMA 8×8, 16×16, 32×32 block sizes. Motion vector range -8 to $+7$ pels-lines/frame This range can be extended.	Chip is being fabricated.	30 MHz	Private correspondence

REFERENCES

Activity Classification in Transform Coding

ACTC-1 J. K. Wu and R. E. Burge, "Adaptive bit allocation for image compression," Comput. Graphics and Image Process, vol. 19, pp. 392–400, 1982.

ACTC-2 R. Mester and U. Franke, "Spectral entropy-activity classification in adaptive transform coding," IEEE Trans. Commun. (under review).

ACTC-3 W. H. Chen and C. H. Smith, "Adaptive coding of color images using cosine transform," ICC 76, Intl. Conf. on Commun., pp. 47-7 through 47-13, Philadelphia, PA, June 1976.

ACTC-4 W. H. Chen and C. H. Smith, "Adaptive coding of monochrome and color images," IEEE Trans. Commun., vol. COM-25, pp. 1285–1292, Nov. 1977.

ACTC-5 J. I. Gimlett, "Use of 'activity' classes in adaptive transform image coding," IEEE Trans. Commun., vol. COM-23, pp. 785–786, July 1975.

ACTC-6 O. Franceschi, Y. Shtarkov and R. Forchheimer, "An adaptive source coding method for still images," PCS 88, Picture Coding Symp., pp. 6.5-1 through 6.5-2, Torrino, Italy, Sept. 1988.

Books

B-1 N. Ahmed and K. R. Rao, "Orthogonal transforms for digital signal processing," New York, NY: Springer, 1975.

B-2 D. F. Elliott and K. R. Rao, "Fast transforms: Algorithms, analyses and applications," New York, NY: Academic Press, 1982.

B-3 K. R. Rao (ed.), "Discrete transforms and their applications," New York, NY: Van Nostrand Reinhold, 1985.

B-4 N. S. Jayant and P. Noll, "Digital coding of waveforms," Englewood Cliffs, NJ: Prentice–Hall, 1984.

B-5 H. C. Andrews and B. R. Hunt, "Digital image restoration," Englewood Cliffs, NJ: Prentice–Hall, 1977.

B-6 W. K. Pratt, "Digital image processing," (Chapter 10: "Two-dimensional unitary transforms," New York, NY: Wiley–Interscience, 1978.

B-7 K. G. Beauchamp, "Applications of Walsh and related functions," New York, NY: Academic Press, 1984.
B-8 A. Rosenfeld and A. C. Kak, "Digital picture processing," New York, NY: Academic Press, vol. 1, 1982.
B-9 K. R. Rao and R. Srinivasan (eds.), "Teleconferencing," New York, NY: Van Nostrand Reinhold, 1985.
B-10 R. J. Clarke, "Transform coding of images," Orlando, FL: Academic Press, 1985.
B-11 R. H. Stafford, "Digital television," (Chapter 4), New York, NY: Wiley–Interscience, 1980.
B-12 P. Yip and K. R. Rao, "Discrete transforms," chapter in Handbook on Digital Signal Processing, edited by D. F. Elliott, Orlando, FL: Academic Press, 1987.
B-13 A. N. Netravali and B. G. Haskell, "Digital pictures, representation and compression," New York, NY: Plenum Press, 1988.
B-14 J. G. Wade, "Signal coding and processing: An introduction based on video systems," Ellis Horwood Limited, England, 1987.
B-15 H. C. Andrews, "Two dimensional transforms," chapter in "Picture processing and digital filtering," Topics in Applied Physics, vol. 6, edited by T. S. Huang, New York, NY: Springer, 1975.
B-16 A. K. Jain, "Fundamentals of digital image processing," Englewood Cliffs, NJ: Prentice–Hall, 1989.
B-17 A. Gersho and R. M. Gray, "Vector quantization and signal compression," Norwell, MA: Kluwer Academic Publishers, 1990.
B-18 J. S. Lim, "Two-dimensional signal and image processing," Englewood Cliffs, NJ: Prentice–Hall, 1989.

Block Structure in Image Coding

BSI-1 H. C. Reeve and J. S. Lim, "Reduction of blocking effect in image coding," ICASSP 83, Intl. Conf. on Acoust., Speech, and Signal Process., pp. 1212–1215, Boston, MA, April 1983.
BSI-2 A. K. Jain and P. M. Farrelle, "Recursive block coding", 16th Asilomar Conf. on Circuits, Systems and Computers, pp. 431–436, Pacific Grove, CA, Nov. 1982.
BSI-3 M. Schlichte, "Block-overlap transform coding of image signals," Siemens Forsch.-u. Entwickl.-Ber. Bd. 13, pp. 100–104, Springer-Verlag, 1984.
BSI-4 P. M. Farrelle and A. K. Jain, "Recursive block coding—a new approach to transform coding," IEEE Trans. Commun., vol. COM-34, pp. 112–117, Feb. 1986.
BSI-5 H. S. Malvar and D. H. Staelin, "Reduction of blocking effects in image coding with a lapped orthogonal transform," ICASSP 88, Int. Conf. on Acoust., Speech, and Signal Process., pp. 781–784, New York, NY, April 1988.
BSI-6 H. S. Malvar and D. H. Staelin, "The LOT: Transform coding without blocking effects," IEEE Trans. Acoust., Speech, and Signal Process., vol. ASSP-37, pp. 553–559, April 1989.
BSI-7 M. Miyahara and K. Kotani, "Block distortion in orthogonal transform coding—analysis, minimization and distortion measure," IEEE Trans. Commun., vol. COM-33, pp. 90–96, Jan. 1985.
BSI-8 L. T. Watson, R. M. Haralick and O. A. Zuniga, "Constrained transform coding and surface fitting," IEEE Trans. Commun., vol. COM-31, pp. 717–726, May 1983.
BSI-9 N. M. Nasrabadi and R. A. King, "Computationally efficient adaptive block-transform coding," Signal Process. II, Theories and Applications, EURASIP, pp. 729–733, Erlangen, W. Germany, Sept. 12–16, 1983.
BSI-10 H. M. Hang and B. G. Haskell, "Interpolative vector quantization of color images," IEEE Trans. Commun., vol. COM-36, pp. 465–470, April 1988.
BSI-11 K. H. Tzou, "Post-filtering of transform coded images," SPIE, Advances in Image Processing, vol. 974, pp. 121–126, San Diego, CA, Aug. 1988.
BSI-12 K. Nemoto and T. Omachi, "Average separation method for DCT coding using smooth interpolation", PCS 88, Picture Coding Symp., pp. 8.12-1 through 8.12-2, Torino, Italy, Sept. 1988.
BSI-13 P. M. Cassereau, D. H. Staelin and G. D. Jager, "Encoding of images based on lapped orthogonal transform," IEEE Trans. Commun., vol. COM-37, pp. 189–193, Feb. 1989.

BTC/Transform Coding

BTC-1 E. J. Delp and O. R. Mitchell, “Image compression using block truncation coding,” IEEE Trans. Commun., vol. COM-27, pp. 1335–1342, Sept. 1979.

BTC-2 O. R. Mitchell, E. J. Delp and S. G. Carlton, “Block truncation coding: A new approach to image compression,” ICC 78, Intl. Conf. on Commun., Toronto, Canada, pp. 12B.1.1–12B.1.4, June 1978.

Classification

C-1 S. A. Dyer, N. Ahmed, and D. R. Hummels, “Classification of vectorcardiograms using Walsh and cosine orthogonal transforms,” IEEE Trans., Electromag., Compata., vol. EMC-27, pp. 35–40, Feb. 1985.

Cepstral Analysis

CA-1 H. Hassanein and M. Rudko, “On the use of discrete cosine transform in cepstral analysis,” IEEE Trans. Acoust., Speech, Signal, and Process., vol. ASSP-32, pp. 922–925, Aug. 1984.

Data Compression

DC-1 N. Ahmed, P. J. Milne and S. G. Harris, “Electrocardiographic data compression via orthogonal transforms,” IEEE Trans. Biomed. Engrg., vol. BME-22, pp. 484–487, Nov. 1975.

DC-2 S. A. Dyer, N. Ahmed, and D. R. Hummels, “Vectorcardiographic data compression via Walsh and cosine transforms,” IEEE Trans., Electromag., Compata., vol. EMC-27, pp. 24–34, Feb. 1985.

DFT-DST-DCT

DDD-1 O. Ersoy, “On relating discrete Fourier, sine and symmetric cosine transforms,” IEEE Trans. Acoust., Speech, and Signal Process., vol. ASSP-33, pp. 219–222, Feb. 1985.

DDD-2 O. Ersoy, “Real discrete Fourier transform,” IEEE Trans. Acoust., Speech, and Signal Process., vol. ASSP-33, pp. 880–882, Aug. 1985.

DDD-3 O. Ersoy, and N. C. Hu, “A unified approach to the fast computation of all discrete trigonometric transforms,” Intl. Conf. on Acoust., Speech, and Signal Process., pp. 1843–1846, Dallas, TX, April 6–9, 1987.

DDD-4 Z. Wang, “A fast algorithm for the discrete sine transform implemented by the fast cosine transform,” IEEE Trans. Acoust., Speech, and Signal Process., vol. ASSP-30, pp. 814–815, Oct. 1982.

DCT Processors

DP-1 E. Arnould and J. P. Dugre, “Real time discrete cosine transform: An original architecture,” Intl. Conf. on Acoust, Speech, and Signal Process., pp. 48.6.1–48.6.4, San Diego, CA, March 1984.

DP-2 J. S. Ward and B. J. Stanier, “Fast discrete cosine transform algorithm for systolic arrays,” Electronics Lett., vol. 19, pp. 58–60, Jan. 20, 1983.

DP-3 G. Bertocci, B. W. Schoenbern and D. G. Messerschmitt, "An approach to the implementation of a discrete cosine transform," IEEE Trans. Commun., vol. COM-30, pp. 635–641, April 1982.

DP-4 A. Jalali and K. R. Rao, "A high speed processor for real time processing of NTSC color TV signal," IEEE Trans. Electromag. Compata., vol. EMC-24, pp. 278–286, May 1982.

DP-5 H. Whitehouse, E. Wrench, A. Weber, G. Claffie, J. Richards, J. Rudnick, W. Schaming, and J. Schanne, "A digital real time intraframe video bandwidth compression system," Proc. of SPIE Conf. on Applications of Digital Image Processing, vol. 119, pp. 64–78, San Diego, CA, Aug. 22–26, 1977.

DP-6 G. G. Murray, "Microprocessor system for TV imagery compression," Proc. SPIE Conf. on Applications of Digital Image Processing, vol. 119, pp. 121–129, San Diego, CA, Aug. 22–26, 1977.

DP-7 L. W. Martinson, "A 10 MHz image bandwidth compression model," IEEE Conf. on Pattern Recognition and Image Processing, pp. 132–136, Chicago, IL, 1978.

DP-8 R. A. Belt, R. V. Keele and G. G. Murray, "Digital TV microprocessor systems," National Telecommun., Conf., pp. 10:6-1 through 10:6-6, Los Angeles, CA, Dec. 3–5, 1977.

DP-9 S. Srinivasan, A. K. Jain and T. M. Chin, "Cosine transform block codec for images using TMS 32010," ISCAS 86, Intl. Symp. Circuits and Systems, pp. 299–302, San Jose, CA, May 5–7, 1986.

DP-10 H. Harasaki, I. Tamitami, Y. Endo, T. Nishitani, M. Yamashina, T. Enomoto, and N. Suzuki, "Real time video signal processor module," ICASSP 87, Intl. Conf. Acoust., Speech, and Signal Process., pp. 1961–1964, Dallas, TX, April 6–9, 1987

DP-11 A. Leger, J. P. Duhamel, J. L. Sicre, G. Madec, and J. M. Knoepfli, "Distributed arithmetic implementation of the DCT for real time photovideotex on ISDN," PCS 87, Picture Coding Symp., pp. 175–176, Stockholm, Sweden, June 9–11, 1987.

DP-12 G. Madec, "A comparison between several fast-DCT algorithms for hardware implementation," PCS 87, Picture Coding Symp., pp. 177–178, Stockholm, Sweden, June 9–11, 1987.

DP-13 S. Matsumura, B. Sikström, U. Sjöström, and L. Wanhammar, "LSI implementation of an 8 point DCT," Intl. Conf. on Computers, Systems and Signal Process., pp. 1473–1477, Bangalore, India, Dec. 1984.

DP-14 A. Leger, "Implementations of fast DCT for full color videotex services and terminals," GLOBECOM 84, Global Telecommun., Conf., pp. 333–337, Atlanta, GA, Nov. 1984.

DP-15 M. Vetterli and A. Ligtenberg, "A discrete Fourier-cosine transform chip," IEEE Journal on Selected Areas in Commun., vol. SAC-4, pp. 49–61, Jan. 1986.

DP-16 P. P. N. Yang and M. J. Narasimha, "Prime Factor decomposition of the DCT, and its hardware realization," ICASSP 85, Intl. Conf. Acoust., Speech, and Signal Process., pp. 20.5.1–20.5.4, Tampa, FL, March 1985.

DP-17 J. Francis, N. Demassieux, C. Gilles, G. Jacques, and C. Eric, "A single chip video rate 16×16 DCT," ICASSP 86, Intl. Conf. Acoust., Speech, and Signal Process., pp. 15.8.1–15.8.4, Tokyo, Japan, Apr. 1986.

DP-18 M. Afghahi, S. Matsumura, J. Pencz, B. Sikström, U. Sjöström, and L. Wanhammar, "An array processor for 2-D discrete cosine transforms," EUSIPCO 86, European Signal Process., Conf., pp. 1283–1286, Hague, The Netherlands, Sept. 1986.

DP-19 T. Mochizuki and S. Mizuno, "DCT of image data by using DSPs," PCS 86, Picture Coding Symp., pp. 182–183, Tokyo, Japan, Apr. 1986.

DP-20 N. Demassieux, D. Jean-Pierre, and J. Francis, "An optimized VLSI architecture for a multiformat discrete cosine transform," ICASSP 87, Intl. Conf. Acoust, Speech, and Signal Process., pp. 547–550, Dallas, TX, June 6–9, 1987.

DP-21 P. Duhamel and H. H'Mida, "New 2^n DCT algorithms suitable for VLSI implementation," ICASSP 87, Intl. Conf. Acoust., Speech, and Signal Process., pp. 1805–1808, Dallas, TX, June 6–9, 1987.

DP-22 A. Ligtenberg and J. H. O'Neill, "A single chip solution for an 8 by 8 two dimensional DCT," ISCAS 87, Intl. Symp. on Circuits and Systems, pp. 1128–1131, Philadelphia, PA, May, 5–7, 1987.

DP-23 T. Kronander, S. Matsumura, B. Sikström, U. Sjöstrom, and L. Wanhammer, "VLSI implementation of the discrete cosine transform," Nordic Symp. on VLSI in Computers and Commun., Tampere, Finland, June 13–16, 1984.

DP-24 S. Matsumura, "Discrete cosine transforms—theory and LSI implementation," M.S. Thesis, Dept. of Elect. Engrg., Linkoping Univ., Linkoping, Sweden, 1985.

DP-25 M. Gilge and W. Guse, "A multi-digital-signal-processor system for image processing," PCS 87, Picture Coding Symp., pp. 207–208, Stockholm, Sweden, June 9–11, 1987.

DP-26 A. M. Chiang, "A video-rate CCD two-dimensional cosine transform processor," Visual Commun. and Image Process. II, SPIE, vol. 845, pp. 2–5, Cambridge, MA, Oct. 1987.

DP-27 A. Ligtenberg, R. H. Wright, and J. H. O'Neill, "A VLSI orthogonal transform chip for real-time image compression," Visual Commun. and Image Process. II, SPIE, Cambridge, MA, Oct. 1987.

DP-28 F. Jutand, N. Demassieux, M. Dana, J-P. Durandeau, G. Concordel, A. Artieri, E. Mackowiack, and L. Bergher, "A 13.5 MHz single chip multiformat discrete cosine transform," Visual Commun. and Image Process. II, SPIE, Vol. 845, pp. 6–12, Cambridge, MA, Oct. 1987.

DP-29 M. T. Sun, T. C. Chen, A. Gottlieb, L. Wu, and M. L. Liou, "A 16 × 16 discrete cosine transform chip," Visual Commun. and Image Process. II, SPIE, Vol. 845, pp. 13–18, Cambridge, MA, Oct. 1987.

DP-30 T. R. Hsing and M. L. Liou, "The challenge of VLSI technology to video communications," SPIE Annual Technical Symposium, vol. 829, pp. 80–84, San Diego, CA, Aug. 1987.

DP-31 A. M. Chiang, P. C. Bennett, B. B. Kosincki, and R. W. Mountain, "A 100-ns CCD 16-point cosine transform processor," 1987 ISSCC, Digest of Technical Papers, pp. 306–307, New York, NY, Feb. 1987.

DP-32 D. Davies, S. Eidson and M. Stauffer, "Improvements in image compression," Advanced Imaging, pp. A25–A28, Jan./Feb. 1988.

DP-33 A. Artieri, S. Kritter, F. Jutand, and N. Demassieux, "A one chip VLSI for real time two-dimensional discrete cosine transform," 1988 Intl. Symp. on Circuits and Systems, pp. 701–704, Helsinki, Finland, June 1988.

DP-34 T. C. Chen, A. Gottlieb and M. T. Sun, "VLSI implementation of a 16 × 16 DCT," ICASSP 88, Intl. Conf. Acoust., Speech, and Signal Process, pp. 1973–1976, New York, NY, April 11–14, 1988

DP-35 J. R. Parkhurst, K. W. Current, A. K. Jain, and J. E. Grishaw, "A unified DCT/IDCT architecture for VLSI implementation," ICASSP 88, Intl. Conf. Acoust., Speech, and Signal Process, pp. 1993–1996, New York, NY, April 11–14, 1988.

DP-36 Z. H. Gu, J. R. Leger and S. H. Lee, "Optical computations of cosine transforms," Optics Commun., vol. 39, pp. 137–142, Oct. 1981.

DP-37 M. L. Liou and J. A. Bellisio, "VLSI implementation of discrete cosine transform for visual communication," Intl. Conf. on Commun. Technology, World Scientific, pp. 666–669, Nov. 9–11, 1987.

DP-38 B. Sikström, L. Wanhammar, M. Afghahi, and J. Pencz, "A high speed 2-D discrete cosine transform chip," Integration the VLSI Journal, vol. 5, pp. 159–169, June 1987.

DP-39 M. T. Sun, L. Wu, and M. L. Liou, "A concurrent architecture for VLSI implementation of discrete cosine transform," IEEE Trans. Circuits and Systems, vol. CAS-34, pp. 992–994, Aug. 1987.

DP-40 B. D. Kulp, "An FCT based image compression board," Electronic Imaging 88, Boston, MA, Oct. 1988.

DP-41 A. Artieri, N. Demassieux and F. Jutand, "A VLSI for DCT coding at video rate," PCS 88, Picture Coding Symp., pp. 5.4-1 through 5.4-2, Torino, Italy, Sept. 1988.

DP-42 C. S. Choy, W. K. Cham and L. Lee, "An integer cosine transform chip using ASIC technology," PCS 88, Picture Coding Symp., pp. 5.5-1 through 5.5-2, Torino, Italy, Sept. 1988.

DP-43 C. S. Choy, W. K. Cham and L. Lee, "A LSI implementation of integer cosine transform," Intl. Conf. on Commun., Systems, pp. 17.5-1 through 17.5-5, Singapore, Nov. 1988.

DP-44 J. C. Carlach, P. Penard and J. L. Sicre, "TCAD: A 27 MHz 8 × 8 discrete cosine transform chip," Intl. Conf. on Acoust., Speech, Signal Process., pp. 2429–2432, Glasgow, Scotland, May 23–26, 1989.

DP-45 I. Defilippis, U. Sjöström, M. Ansorge, and F. Pellandini, "A 2 dimensional 16 point discrete cosine transform chip for real time video applications," Douzieme colloque sur le traitement du signal et des images, pp. 813–816, Juan-Les-Pins, France, June 12–16, 1989.

DP-46 L. Matterne, D. Chong, B. McSweeney and R. Woudsma, "A flexible high performance 2-D discrete cosine transform IC," ISCAS 89, Intl. Symp. Circuits and Systems, pp. 618–621, Portland, OR, May 9–11, 1989.

DP-47 V. Rampa and G. De Micheli, "Computer-aided synthesis of a bidimensional discrete cosine transform chip," ISCAS 89, Intl. Symp. Circuits and Systems, pp. 220–224, Portland, OR, May 9–11, 1989.

DP-48 T. Denayer, E. Vanzieleghem, and P. Jespers, "A 50 MIPS multiprocessor chip for image processing," IEEE 1988 Custom Integrated Circuits Conf., pp. 8.3.1–8.3.4, Rochester, NY, May 16–19, 1988.

DP-49 A. M. Goltlieb, M. T. Sun, and T. C. Chen, "A video rate 16 × 16 discrete cosine transform IC," IEEE 1988 Custom Integrated Circuits Conf., pp. 8.2.1–8.2.4, Rochester, NY, May 16–19, 1988.

DP-50 W. Li, I. L. Sidharta and H. S. Fetterman, "A computation engine for the FFT/FCT/FHT using AT&T DSP32," Intl. Symp. Circuits and Systems, pp. 1272–1275, Portland, OR, May 9–11, 1989.

DP-51 N. I. Cho and S. U. Lee, "DCT algorithms for VLSI implementations," IEEE Trans. Acoust., Speech and Signal Process., vol. ASSP-38, pp. 121–127, Jan. 1990.

DP-52 U. Totzek and F. Matthiesen, "Two dimensional discrete cosine transform with linear systolic arrays," Intl. Conf. on Systolic Arrays, pp. 387–397, Killarney, Co. Kerry, Ireland, 1989. (Publishers: Prentice Hall, Englewood Cliffs, NJ).

DP-53 W. Liebsch, "Parallel architecture for VLSI implementation of a 2-dimensional discrete cosine transform," IEE Third Intl. Conf. on Image Processing and its Applications, pp. 609–612, Warwick, Coventry, UK, July 18–20, 1989.

DP-54 K. Aono, M. Toyokura and T. Arak, "A 30 ns (600 MOPS) image processor with a reconfigurable pipeline architecture," IEEE Custom Integrated Circuits Conf. pp. 24.4.1–24.4.4, San Diego, CA, May 1989.

DP-55 H. R. Wu and F. J. Paoloni, "Implementation of 2-D DCT for image coding using FDP™ A41102," Conference of "Image Processing and the Impact of New Technologies," Canberra, Australia, Dec. 18–20, 1989.

DP-56 M. T. Sun, T-C. Chen and A. M. Gottlieb, "VLSI implementation of a 16 × 16 discrete cosine transform," IEEE Trans. Circuits and Systems, vol. CAS-36, pp. 610–617, April 1989.

Dynamic Range of DCT Coefficients

DRD-1 R. J. Clarke, "On the dynamic range of coefficients generated in transform processing of digitised image data," IEE Proc., vol. 132, Pt. F, pp. 107–110, April 1985.

DCT/VQ

DV-1 T. Saito, H. Takeo, K. Aizawa, H. Harashima, and H. Miyakawa, "Adaptive discrete cosine transform image coding using gain/shape vector quantizers," ICASSP 86, Intl. Conf. on Acoust. Speech, and Signal Process., pp. 129–132, Tokyo, Japan, April 7–11, 1986.

DV-2 S. Venkataraman and K. R. Rao, "Applications of vector quantizers and BTC in image coding," GLOBECOM' 85, IEEE Global Commun. Conf., pp. 602–608, New Orleans, LA, Dec. 2–5, 1985.

DV-3 M. Gotze, "Adaptive vector quantization of images in the discrete cosine transform domain," PCS 86, Picture Coding Symp., pp. 142–143, Tokyo, Japan, April 2–4, 1986.

DV-4 T. Saito, H. Takeo, K. Aizawa, H. Harashima, and H. Miyakawa, "Discrete cosine transform coding system using gain/shape vector quantizers and its application to color image coding," PCS 86, Picture Coding Symp., pp. 225–226, Tokyo, Japan, April 2–4, 1986.

DV-5 K. Aizawa, H. Harashima and H. Miyakawa, "Adaptive discrete cosine transform coding with vector quantization," PCS 86, Picture Coding Symp., pp. 239–240, Tokyo, Japan, April 2–4, 1986.

DV-6 M. Goldberg, P. R. Boucher and S. Shlien, "Image compression using adaptive vector quantization," IEEE Trans. Commun., vol. COM-34, pp. 180–187, Feb. 1986.

DV-7 Y. Kato, N. Mukawa and H. Watanabe," Hybrid vector-and-scalar quantization for image coding," PCS' 86, Picture Coding Symp., pp. 78–79, Tokyo, Japan, April 2–4, 1986.

DV-8 J. P. Marescq and C. Labit, "Image sequence coding by vector quantization in a transformed domain," PCS 86, Picture Coding Symp., pp. 241–242, Tokyo, Japan, April 2–4, 1986.

DV-9 J. P. Marescq and C. Labit, "Vector quantization in transformed image coding," ICASSP 86, Intl. Conf. on Acoust., Speech, and Signal Process., pp. 145–147, Tokyo, Japan, April 7–11, 1986.

DV-10 A. A. Abdelwahab and S. C. Kwatra, "Image data compression with vector quantization in the transform domain," ICC '86, IEEE Intl. Conf. on Commun., pp. 1286–1289, Toronto, Canada, June 22–25, 1986.

DV-11 B. Ramamurthi and A. Gersho, "Classified vector quantization of images," IEEE Trans. Commun., vol. COM-34, pp. 1105–1115, Nov. 1986.

DV-12 Y. Kato, N. Mukawa, S. Okubo, H. Hashimoto, and H. Yasuda, "Discrete cosine transform coding for video conferencing using vector and scalar quantization," GLOBECOM 86, IEEE Global Telecommun., Conf., pp. 266–270, Houston, TX, Dec. 1–4, 1986.

DV-13 M. E. Blain and T. R. Fischer, "Optimum rate allocation in pyramid vector quantizer transform coding of imagery," ICASSP 87, Intl. Conf. on Acoust., Speech, and Signal Process., pp. 729–732, Dallas, TX, April 6–9, 1987.

DV-14 H. C. Tseng and T. R. Fischer, "Transform and hybrid transform/DPCM coding of images using pyramid vector quantization," IEEE Trans. Commun., vol. COM-35, pp. 79–86, Jan. 1987.

DV-15 D. J. Vaisey and A. Gersho, "Variable block-size image coding," ICASSP 87, Intl. Conf. Acoust., Speech, and Signal Process., pp. 1051–1054, Dallas, TX, April 6–9, 1987.

DV-16 G. Conte, L. Corgnier and M. Guglielmo, "Optimization of the scanning paths in 2-D transform coding by VQ technique," PCS 87, Picture Coding Symp., pp. 122–123, Stockholm, Sweden, June 9–11, 1987.

DV-17 T. Omachi, Y. Takashima, and H. Okada, "DCT-VQ coding scheme using categorization with adaptive band partition," PCS 87, Picture Coding Symp., pp. 161–162, Stockholm, Sweden, June 9–11, 1987.

DV-18 G. Tu, G. Verbeek, J. Rommelaere, and A. Oosterlinck, "Adaptive coding in transform domain using vector quantization and scalar correction," IEEE Region 10 Conf., pp. 418–422, Seoul, Korea, Aug. 1987.

DV-19 G. Tu, G. Verbeek, J. Rommelaere, and A. Oosterlinck, "Hybrid coding scheme for still images: adaptive VQ with correction information in DCT domain," SPIE's 31st Annual Intl. Tech. Symp. on Optical & Optoelectronic Applied Sci. & Engrg., vol. 829, pp. 119–124, San Diego, CA, Aug. 16–21, 1987.

DV-20 Y. H. Ang and T. S. Durrani, "Adaptive transform coding using vector quantization," II Intl. Conf. on Image Process., Imperial College of Science and Technology, pp. 11–16, London, UK, June 1986.

DV-21 M. Breeuwer, "Adaptive transform coding using cascaded vector quantization for coefficient quantization," PCS 87, Picture Coding Symp., pp. 159–160, Stockholm, Sweden, June 9–11, 1987.

DV-22 S. U. Lee and D. S. Kim, "Image vector quantizer based on a classification in the DCT domain," IEEE Region 10 Conf., pp. 413–417, Seoul, Korea, Aug. 1987.

DV-23 A. A. Abdelwahab and S. C. Kwatra, "Component coding of color video signal with vector quantization in the transform domain," IEEE Region 10 Conf., pp. 433–437, Seoul, Korea, Aug. 1987.

DV-24 K. Aizawa, H. Harashima, and H. Miyakawa, "Adaptive discrete cosine transform coding with vector quantization for color images," ICASSP 86, Intl. Conf. on Acoust. Speech Signal Process., pp. 985–988, Tokyo, Japan, April 7–11, 1986.

DV-25 K. Aizawa, H. Harashima and H. Miyakawa, "Vector quantization of picture signals using discrete cosine transforms (DCT-VQ)," J. of Inst. of TV Engineers of Japan, vol. 39, pp. 920–925, Oct. 1985.

DV-26 M. Breeuwer, "Transform coding of images using directionally adaptive vector quantization," ICASSP 88, Intl. Conf. on Acoust., Speech, and Signal Process., pp. 788–791, New York, NY, April 11–14, 1988.

DV-27 S. E. Budge, T. G. Stockham Jr., D. M. Chabries, R. W. Christiansen, "Vector quantization of color digital images within a color visual model," ICASSP 88, Intl. Conf. on Acoust., Speech, and Signal Process., pp. 816–819, New York, NY, April 11–14, 1988.

DV-28 C. Fong and M. Baraniecki, "Adaptive transform coding with multi-band vector quantization for video signals," ICASSP 88, Intl. Conf. on Acoust., Speech, and Signal Process., pp. 1184–1187, New York, NY, April 11–14, 1988.

DV-29 R. M. Gray, "Vector quantization," IEEE ASSP Magazine, vol. 1, pp. 4–29, April 1984.

DV-30 Y. Linde, A. Buzo, and R. M. Gray, "An algorithm for vector quantizer design," IEEE Trans. Commun., vol. COM-28, pp. 84–95, Jan. 1980.

DV-31 W. H. Equitz, "Fast algorithms for vector quantization picture coding," M.Sc. Thesis, Dept. of Elect. Engrg. and Comput. Sci., M.I.T., MA, June 1984. Also in Proc. IEEE Intl. Conf. Acoust., Speech, and Signal Process., pp. 725–728, Dallas, TX, April 1987. Also vol. ASSP-37, pp. 1568–1575, Oct. 1989.

DV-32 V. Cuperman and A. Gersho, "Vector predictive coding of speech at 16 Kbits/s," IEEE Trans. Commun., vol. COM-33, pp. 685–696, July 1985.

DV-33 C. Labit and J. P. Marescq, "Temporal adaptive vector quantization for image sequence coding," SPIE, Advances in Image Process., vol. 804, pp. 371–378, The Hague, Netherlands, March 31–April 3, 1987.

DV-34 C. H. Yim and J. K. Kim, "A simple DCT-CVQ based on two DCT coefficients," PCS 88, Picture Coding Symp., pp. 8.11-1 through 8.11-2, Torino, Italy, Sept. 1988.

DV-35 N. M. Nasrabadi and R. A. King, "Image coding using vector quantization: A review," IEEE Trans. Commun., vol. COM-36, pp. 957–971, Aug. 1988.

DV-36 G. Tu, L. V. Eycken, and A. Oosterlinck, "Hybrid coding of image sequences using vector quantization and motion compensation techniques," PCS 88, Picture Coding Symp., pp. 13.12-1 through 13.12-2, Torino, Italy, Sept. 1988.

DV-37 M. Breewer, "Transform coding of images using adaptive tree searched vector quantization," PCS 88, Picture Coding Symp., pp. 10.6-1 through 10.6-2, Torino, Italy, Sept. 1988.

DV-38 Y. Du, "Application of vector quantization to threshold coding of image signals," PCS 88, Picture Coding Symp., pp. 10.2-1 through 10.2-2, Torino, Italy, Sept. 1988.

DV-39 F. Bellifemine and R. Picco, "2D-DCT coding with pyramidal vector quantization," PCS 88, Picture Coding Symp., pp. 13.2-1 through 13.2-2, Torino, Italy, Sept. 1988.

DV-40 G. R. Davidson, P. R. Cappello, and A. Gersho, "Systolic architectures for vector quantization," IEEE Trans. Acoust., Speech, and Signal Process., vol. ASSP-36, pp. 1651–1664, Oct. 1988.

DV-41 J. W. Kim and S. U. Lee, "Discrete cosine transform classified VQ technique for image coding," Intl. Conf. on Acoust., Speech, and Signal Process., pp. 1831–1834, Glasgow, Scotland, May 23–26, 1989.

DV-42 Y. S. Ho and A. Gersho, "Classified transform coding of images using vector quantization," Intl. Conf. on Acoust., Speech, and Signal Process., pp. 1890–1893, Glasgow, Scotland, May 23–26, 1989.

DV-43 H. S. Wu and H. B. Chen, "A DCT gain-shape vector quantizer for image coding," ICC 88, Intl. Conf. on Commun., pp. 1245–1247, Philadelphia, PA, June 1988.

DV-44 A. Gersho and Y. Shoham, "Hierarchical vector quantization of speech with dynamic codebook allocation," Intl. Conf. on Acoust., Speech, and Signal Process., pp. 18.9.1 through 18.9.4, San Diego, CA, March 1984.

DV-45 G. Benelli and L. Alparone, "An integrated coding procedure using cosine transform and vector quantization for high resolution digital image compression," Douzieme Colloque sur le traitement du signal et des images, Juan-Les-Pins, France, June 12–16, 1989.

DV-46 M. Götze and Y. Du, "Adaptive transform coding of images using vector quantization," EUSIPCO 86, European Signal Process. Conf., pp. 789–792, The Hague, Netherlands, Elsevier Science Publishers, North Holland, Sept. 1986.

DV-47 Q. A. Qureshi and T. R. Fischer, "A hardware processor for implementing the pyramid vector quantizer," IEEE Trans. Acoust., Speech, and Signal Process., vol. ASSP-37, pp. 1135–1142, July 1989.

DV-48 T. Moriya and H. Suda, "An 8 Kbit/s transform coder for noisy channels," Intl. Conf. on Acoust., Speech, and Signal Process., pp. 196–199, Glasgow, Scotland, May 23–26, 1989.

DCT via WHT

DW-1 S. Venkataraman *et al.*, "Discrete transforms via the Walsh-Hadamard transform," 26th Midwest Symp. Circuits and Systems, Puebla, Mexico, Proc. pp. 74–78, Aug. 1983.

DW-2 D. Hein and N. Ahmed, "On a real-time Walsh-Hadamard cosine transform image processor," IEEE Trans. Electromag. Compat., vol. EMC-20, pp. 453–457, Aug. 1978.

DW-3 H. W. Jones, D. N. Hein, and S. C. Knauer, "The Karhunen-Loeve, discrete cosine and related transforms via the Hadamard transform," Intl. Telemetering Conf., pp. 87–98, Los Angeles, CA, Nov. 14–16, 1978.

DW-4 B. G. Kashef and A. Habibi, "Direct computation of higher-order DCT coefficients from lower-order DCT coefficients," SPIE's 28th Annual Intl., Tech. Symp., pp. 425–431, San Diego, CA, Aug. 19–24, 1984.

DW-5 S. Venkataraman, V. R. Kanchan, K. R. Rao and M. Mohanty, "Discrete transforms via the Walsh-Hadamard transform," Signal Process., vol. 14, pp. 371–382, 1988.

Filtering

F-1 K. N. Ngan and R. J. Clarke, "Lowpass filtering in the cosine transform domain," Proc. Intrnl. Conf. Commun., pp. 31.7.1–31.7.5, Seattle, WA, June 1980.

F-2 H. J. Nussbaumer, "Polynomial transform implementation of digital filter banks," IEEE Trans. Acoust., Speech, and Signal Process., vol. ASSP-31, pp. 616–622, June 1983.

F-3 K. N. Ngan, "Two-dimensional transform domain decimation techniques," ICASSP 86, Intl. Conf. on Acoust., Speech, and Signal Process., pp. 1001–1004, Tokyo, Japan, April 7–11, 1986.

F-4 K. N. Ngan, "Experiments on two-dimensional decimation in time and orthogonal transform domains," Signal Process., vol. 11, pp. 249–263, 1986.

F-5 J. M. Adant, P. Delogne, B. Macq, and L. Vandendorpe, "Multidimensional block filtering," PCS 87, Picture Coding Symp., pp. 40, Stockholm, Sweden, June 9–11, 1987.

F-6 J. M. Adant, P. Delogne, E. Lasker, B. Macq, L. Stroobants, and L. Vandendorpe, "Block operations in digital signal processing with application to TV coding," Signal Process., vol. 13, pp. 385–397, Dec. 1987.

F-7 B. Chitprasert and K. R. Rao, "Discrete cosine transform filtering" Signal Process., vol. 16, pp. 000–000.

F-8 B. V. Caillie, P. Delogne, L. Vandendorpe, and B. Macq, "Multidimensional block processing for CIF conversion in digital videotelephonie," Intl. Workshop on 64 Kbit/s Coding of Moving Video, pp. 3–5, Hannover, W. Germany, June 14–16, 1988.
F-9 C. L. Gundel, "Filter bank interpretation of DFT and DCT," EUSIPCO 88, European Signal Process. Conf. pp. 943–946, Grenoble, France, Sept. 1988.
F-10 O. Privat and L. Paris, "Design of digital filters for video circuits," IEEE Journal of Solid-State Circuits, vol. SC-21, pp. 441–445, June 1986.
F-11 K. Ramachandran, D. F. Daly, and R. R. Cordell, "A CMOS Video FIR filter and correlator," pp. 273–280, in VLSI Signal Processing II, IEEE Press Book, 1986.
F-12 P. Pirsch, T. Micke, and H. Bao, "Digital filters for video codecs with oversampled ADC and DAC," Intl. Symp. Circuits and Systems, pp. 217–220, Philadelphia, PA, May 1987.
F-13 W. Demmer and J. Speidel, "Programmable digital filter for high speed applications," Intl. Symp. Circuits and Systems, pp. 1285–1288, Helsinki, Finland, June 1988.
F-14 G. Chiappano and D. Raveglia, "Anti-aliasing VLSI digital filters for video signal coding," Intl. Symp. Circuits and Systems, pp. 709–713, Helsinki, Finland, June 1988.

Family of DCTs

FD-1 Z. Wang, "Fast algorithms for the discrete W transform and for the discrete Fourier transform," IEEE Trans. Acoust., Speech, and Signal Process., vol. ASSP-32, pp. 803–816, Aug. 1984.
FD-2 P. M. Farrelle, S. Srinivasan, and A. K. Jain, "A unified transform architecture," ICASSP 86, Intl. Conf. on Acoust., Speech, and Signal Process., pp. 293–296, Tokyo, Japan, April 7–11, 1986.

FDCT via FFT

FF-1 M. J. Narasimha and A. M. Peterson, "On the computation of the discrete cosine transform," IEEE Trans. Commun., vol. COM-26, pp. 934–946, June 1978.
FF-2 B. D. Tseng and W. C. Miller, "On computing the discrete cosine transform," IEEE Trans. Comput., Vol. C-27, pp. 966–968, Oct. 1978.
FF-3 R. M. Haralick, "A storage efficient way to implement the discrete cosine transform," IEEE Trans. Comput., vol. C-25, pp. 764–765, June 1976.
FF-4 M. Vetterli and H. Nussbaumer, "Simple FFT and DCT algorithms with reduced number of operations," Signal Processing, vol. 6, pp. 267–278, Aug. 1984.
FF-5 J. Makhoul, "A Fast cosine transform in one and two dimensions," IEEE Trans. Acoust. Speech, and Signal Process., vol. ASSP-28, pp. 27–34, Feb. 1980.
FF-6 P. Duhamel, "Implementation of 'Split-Radix' FFT algorithms for complex, real and real-symmetric data," IEEE Trans. Acoust., Speech, and Signal Process., vol. ASSP-34, pp. 285–295, April 1986.

Fast Implementation of 2D-DCT

FID-1 F. A. Kamangar, and K. R. Rao, "Fast algorithms for the 2D-discrete cosine transform," IEEE Trans. Comput., vol. C-31, pp. 899–906, Sept. 1982.
FID-2 H. J. Nussbaumer, "Fast polynomial transform computation of the 2-D DCT," in Proc. Intl. Conf. on Digital Processing, pp. 276–283, Florence, Italy, Sept. 2–5, 1981.
FID-3 M. Vetterli, "Fast 2-D discrete cosine transforms," Intl. Conf. on Acoust., Speech, and Signal Process., pp. 1538–1541, Tampa, FL, March 26–29, 1985.
FID-4 N. Nasrabadi, and R. King, "Computationally efficient discrete cosine transform algorithm," Electronics Lett., vol. 19, pp. 24–25, Jan. 6, 1983.

FID-5 S. C. Pei, and E. F. Huang, "Improved 2D discrete cosine transforms using generalized polynomial transforms and DFTs" Intl. Conf. on Commun., pp. 242–244, Amsterdam, The Netherlands, May 14–17, 1984.

FID-6 M. L. Haque, "A two-dimensional fast cosine transform," IEEE Trans. Acoust., Speech, and Signal Process., vol. ASSP-33, pp. 1532–1538, Dec. 1985.

FID-7 Y. Morikawa, H. Hamada, and K. Watabu, "Fast computation algorithm of the two-dimensional cosine transform using Chebyshev polynomial transform," Electron. Commun., Japan, Part 1, vol. 69, pp. 33–44, Nov. 1986.

FID-8 M. Vetterli, P. Duhamel, and C. Guillemot, "Trade-off's in the computation of mono—and multi—dimensional DCTs," Intl. Conf. Acoust., Speech, and Signal Process., pp. 999–1002, Glasgow, Scotland, May 23–26, 1989.

FID-9 C. Ma, "A fast recursive two dimensional cosine transform," SPIE, Intelligent Robots and Computer Vision, vol. 1002, pp. 541–548, Cambridge, MA, Nov. 7–11, 1988.

FDCT by Real Arithmetic

FRA-1 W. H. Chen, C. H. Smith, and S. C. Fralick, "A fast computational algorithm for the discrete cosine transform," IEEE Trans. Commun., vol. COM-25, pp. 1004–1009, Sept. 1977.

FRA-2 Z. Wang, "Reconsideration of a fast computational algorithm for the discrete cosine transform," IEEE Trans. Commun., vol. COM-31, pp. 121–123, Jan. 1983.

FRA-3 B. G. Lee, "A new algorithm for the discrete cosine transform," IEEE Trans. Acoust., Speech, and Signal Process., vol. ASSP-32, pp. 1243–1245, Dec. 1984.

FRA-4 B. G. Lee, "FCT—a fast cosine transform," Intl. Conf. on Acoust., Speech, and Signal Process., pp. 28A.3.1–28A.3.3, San Diego, CA, March 1984.

FRA-5 S. A. Dyer, N. Ahmed, and D. R. Hummels, "Computation of the discrete cosine transform via the arcsine transform," Intl. Conf. on Acoust., Speech, and Signal Process., pp. 231–234, Denver, CO, April 1980.

FRA-6 M. S. Corrington, "Implementation of fast cosine transforms using real arithmetic," NAECON, Dayton, OH, May 16–18, 1978.

FRA-7 M. D. Wagh and H. Ganesh, "A new algorithm for the discrete cosine transform of arbitrary number of points," IEEE Trans. Comput., vol. C-29, pp. 269–277, April 1980.

FRA-8 P. Yip and K. R. Rao, "Fast DIT algorithms for DST's and DCT's," Circuits, Systems and Signal Process., vol. 3, No. 4, pp. 387–408, 1984.

FRA-9 P. Yip and K. R. Rao, "DIF algorithms for DCT and DST," Intl., Conf. on Acoust., Speech, and Signal Process., pp. 776–779 Tampa, FL, March 26–29, 1985.

FRA-10 P. P. N. Yang and M. J. Narasimha, "Prime factor decomposition of the discrete cosine transform," Intl. Conf. on Acoust., Speech, and Signal Process., pp. 772–775, Tampa, FL, March 26–29, 1985.

FRA-11 Z. Wang, "On computing the discrete Fourier and cosine transforms," IEEE Trans. Acoust. Speech, and Signal Process., vol. ASSP-33, pp. 1341–1344, Oct. 1985.

FRA-12 N. Suehiro and M. Hatori, "Fast algorithms for the DFT and other sinusoidal transforms," IEEE Trans. Acoust., Speech, and Signal Process., vol. ASSP-34, pp. 642–644, June 1986.

FRA-13 H. S. Hou, "A fast recursive algorithm for computing the discrete cosine transform," IEEE Trans. Acoust., Speech, and Signal Process., vol. ASSP-35, pp. 1455–1461, Oct. 1987.

FRA-14 P. Yip and K. R. Rao, "The decimation-in-frequency algorithms for a family of discrete sine and cosine transforms," Circuits, Systems, and Signal Process., pp. 4–19, 1988.

FRA-15 Y. Morikawa, H. Hamada, and N. Yamane, "A fast algorithm for the cosine transform based on successive order reduction of the Tchebycheff polynomial," IECE Trans. (in Japanese), vol. J68-A, pp. 173–180, 1985.

FRA-16 H. S. Malvar, "Fast computation of the discrete cosine transform and the discrete Hartley transform," IEEE Trans. Acoust., Speech, and Signal Process., vol. ASSP-35, pp. 1484–1485, Oct. 1987.

FRA-17 W. Kou and T. Fjallbrant, "A fast algorithm for computing DCT coefficients of a signal block which consists partly of one block and partly of the next consecutive block," IEEE Trans. Acoust., Speech, and Signal Process., vol. ASSP-38, pp. 000–000, 1990.

FRA-18 B. G. Lee, "Input and output index mappings for a prime-factor-decomposed computation of discrete cosine transform," IEEE Trans. Acoust., Speech, and Signal Process., vol. ASSP-37, pp. 237–244, Feb. 1989.

FRA-19 H. S. Malvar, "Fast computation of the discrete cosine transform through the fast Hartley transform," Electronics Letters, vol. 22, pp. 352–353, Mar. 27, 1986.

FRA-20 Y. Morikawa, H. Hamada, and N. Yamane, "A fast algorithm for the cosine transform based on successive order reduction of the Tchebycheff polynomial," Electron Commun., Japan, part 1, vol. 69, pp. 45–54, March 1986.

FRA-21 H. S. Malvar, Corrections to, "Fast computation of the discrete cosine transform and the discrete Hartley transform," IEEE Trans. Acoust., Speech, and Signal Process., vol. ASSP-36, pp. 610, April 1988.

FRA-22 H. S. Hou, "The fast Hartley transform algorithm," IEEE Trans. Comput., vol. C-36, pp. 147–156, Feb. 1987.

FRA-23 C. Loeffler, A. Ligtenberg, and G. S. Moschytz, "Algorithm-architecture mapping for custom DSP chips," Intl., Symp. on Circuits and Systems, pp. 1953–1956, Espoo, Finland, June 1988.

FRA-24 Y. Arai, T. Agui, and M. Nakazima, "Fast DCT algorithm for picture coding scheme," PCS 88, Picture Coding Symp., pp. 8.1-1 through 8.1-2, Torino, Italy, Sept. 1988.

FRA-25 Z. Wang, "Pruning the fast discrete cosine transform," Intl. Conf. on Commun. Systems, pp. 26.6.1 through 26.6.5, Singapore, Nov. 1988.

FRA-26 C. Loeffler, A. Ligtenberg, and G. S. Moschytz, "Practical fast 1-D DCT algorithms with 11 multiplications," Intl. Conf. on Acoust., Speech, and Signal Process., pp. 988–991, Glasgow, Scotland, May 23–26, 1989.

FRA-28 V. Nagesha, "Comments on fast computation of the discrete cosine transform and discrete Hartley transform," IEEE Trans. Acoust., Speech, and Signal Process., vol. ASSP-37, pp. 439–440, March 1989.

FRA-29 C. Sun and P. Yip, "Split radix algorithms for DCT and DST," Asilomar Conf. on Signals, Systems, and Computers, pp. 508–512, Pacific Grove, CA, Nov. 1989.

General

G-1 N. Ahmed, T. Natarajan, and K. R. Rao, "Discrete cosine transform," IEEE Trans. Comput., vol. C-23, pp. 90–93, Jan. 1974.

G-2 K. S. Shanmugam, "Comments on discrete cosine transform," IEEE Trans. Comput., vol. C-24, pp. 759, July 1975.

G-3 M. D. Flickner and N. Ahmed, "A derivation for the discrete cosine transform," Proc. IEEE, vol. 70, pp. 1132–1134, Sept. 1982.

G-4 N. Ahmed and M. D. Flickner, "Some considerations of the discrete cosine transform," 16th Asilomar Conf. on Circuits, Systems and Computers, Pacific Grove, CA, pp. 295–299, Nov. 8–10, 1982.

G-5 R. J. Clarke, "Relation between the Karhunen–Loeve and cosine transforms," Proc. IEE, Part F, vol. 128, pp. 359–360, Nov. 1981.

G-6 R. J. Clarke, "Spectral response of the discrete cosine and Walsh–Hadamard transforms," IEE Proc., vol. 130, pp. 309–313, June 1983.

G-7 M. Chelehmal and K. R. Rao, "Fast computational algorithms for the discrete cosine transform," 18th Asilomar Conf. on Circuits, Systems, and Computers, Pacific Grove, CA, pp. 273–280, Nov. 6–8, 1985.

G-8 K. R. Rao, "Theory and the applications of the discrete cosine transform," Jordan Intl. Elect., and Electronic Engrg. Conf., pp. 259–264, Amman, Jordan, April 28–May 1, 1985.

G-9 A. G. Tescher and J. A. Saghri, "Adaptive transform coding and image quality," Optical Engineering, vol. 25, pp. 979–983, Aug. 1986.

G-10 A. K. Jain, "Image data compression: a review," Proc. IEEE, vol. 69, pp. 349–389, March 1981.
G-11 A. N. Netravali and J. O. Limb, "Picture coding: a review," Proc. IEEE, vol. 68, pp. 366–406, March 1980.
G-12 T. A. Soame, "Bandwidth compression using transform techniques for image transmission systems," Marconi Review, vol. XLIII, pp. 228–240, 1980.
G-13 K. R. Rao, "Discrete transforms," IMACS, Math. and Computers in Simulation, vol. 27, pp. 421–430, 1985.
G-14 H. G. Musmann, P. Pirsch, and H. J. Grallert, "Advances in picture coding," Proc. IEEE, vol. 73, pp. 523–548, April 1985.
G-15 P. A. Wintz, "Transform picture coding," Proc. IEEE, vol. 60, pp. 809–820, July 1972.
G-16 P. Camana, "Video-bandwidth compression: a study in tradeoffs," IEEE Spectrum, vol. 16, pp. 24–29, June 1979.
G-17 A. Habibi, "Survey of adaptive image coding techniques," IEEE Trans. Commun., vol. COM-25, pp. 1275–1284, Nov. 1977.
G-18 J. O. Limb, C. B. Rubinstein, and J. E. Thompson, "Digital coding of color video signals—a review," IEEE Trans. Commun., vol. COM-25, pp. 1349–1385, Nov. 1977.
G-19 A. K. Jain, P. M. Farrelle, and V. R. Algazi, "Image data compression," Chapter 5 from the book, "Digital image processing techniques" edited by M. P. Ekstrom, New York, NY: Academic Press, 1984.
G-20 H. Gaggioni, and D. L. Gall, "Digital video transmission and coding for the broadband ISDN," IEEE Trans. Consumer Electronics, vol. CE-34, pp. 16–35, Feb. 1988.
G-21 CCIR Rec. 601, "Encoding parameters of digital television for studios," CCIR Tec. and Rep., Intl. Telecommun. Union, vol. XI—part 1, Plenary Assembly, Geneva, Switzerland, 1982.

Finite Wordlength Effects

FWE-1 A. Jalali and K. R. Rao, "Limited wordlength and FDCT processing accuracy," ICASSP, Intl. Conf. on Acoust., Speech, and Signal Process, pp. 1180–1183, Atlanta, GA, March 30–April 1, 1981.
FWE-2 A. D'Souza, "Spectral and finite register length aspects of the discrete cosine transform," M.S. Thesis, Univ. of Texas at Arlington, Arlington, TX., Aug, 1984.
FWE-3 M. Guglielmo, "An analysis of error behavior in the implementation of 2-D orthogonal transformations," IEEE Trans. Commun., vol. COM-34, pp. 973–975, Sept. 1986.
FWE-4 G. Madec, "A comparison between several fast-DCT algorithms for hardware implementation," PCS 87, Picture Coding Symp., pp. 177–178, Stockholm, Sweden, June 9–11, 1987.

HDTV Image Coding

HDTV-1 F. Bellifemine, H. Ferreyra, R. Picco, L. Sobrero, "Vector quantization for HDTV-DCT coding," II Intl. Workshop on Signal Processing of HDTV, L'Aquila, Italy, Elsevier Science Publishers, North Holland, pp. 239–247, Feb.–March, 1988.
HDTV-2 R. Kutka and K. Waidhas, "Optimizing the quantization levels in high resolution image coding," PCS 88, Picture Coding Symp., pp. 6.6–1 through 6.6-2, Torino, Italy, Sept. 1988.
HDTV-3 G. Schamel, "Coding of HDTV-signals with bit-rates between 140 and 280 Mbit/s," PCS 88, Picture Coding Symp., pp. 13.10-1 through 13.10-2, Torino, Italy, Sept. 1988.
HDTV-4 K. H. Tzou and T. C. Chen, "An embedded HDTV coding scheme using the decimation in frequency domain method," PCS 88, Picture Coding Symp., pp. 14.4-1 through 14.4-2, Torino, Italy, Sept. 1988.
HDTV-5 M. Barbero and G. F. Barbieri, "Consideration on the feasibility of codecs based on DCT for HDTV applications," First EURASIP Workshop in Signal Processing of HDTV, vol. 1, L'Aquila, Italy, Nov. 1986.

HDTV-6 "Special issue on high definition television," IEEE Trans. Broadcasting, vol. BC-33, Dec. 1987.

HDTV-7 "Special issue on advanced television systems," IEEE Trans. Consumer Electronics, vol. CE-34, Feb. 1988.

HDTV-8 K. H. Tzou, "Compatible HDTV Coding for broadband ISDN," GLOBECOM 88, Global Commun., Conf., pp. 743–749, Hollywood, FL, Nov. 1988.

HDTV-9 M. Barbero, S. Cucchi, and M. Stroppiana, "Coding strategies based on DCT for the transmission of HDTV," II Intl. Workshop on Signal Processing of HDTV, L'Aquila, Italy, Elsevier Science Publishers, North Holland, pp. 503–508, Feb–March, 1988.

HDTV-10 O. Chantelou and C. Remus, "Adaptive transform coding of HDTV pictures," II Intl. Workshop on Signal Processing of HDTV, L'Aquila, Italy, Elsevier Science Publishers, North Holland, pp. 231–238, Feb.–March, 1988.

HDTV-11 R. Kutka and K. Waidhas, "A formalism for calculating the optimal quantization of DCT coefficients for HDTV coding," II Intl. Workshop on Signal Processing of HDTV, L'Aquila, Italy, Elsevier Science Publishers, North Holland, pp. 249–258, Feb.–March, 1988.

HDTV-12 D. L. Gall, H. Gaggioni, and C. T. Chen, "Transmission of HDTV signals under 140 Mbits/s using a sub-band decomposition and discrete cosine transform coding," II Intl. Workshop on Signal Processing of HDTV, L'Aquila, Italy, Elsevier Science Publishers, North Holland, pp. 287–293, Feb.–March, 1988.

HDTV-13 Third international workshop on HDTV, Proceedings, Turin, Italy, Aug. 30–Sept. 1, 1989.

HDTV-14 R. K. Jurgen, "Consumer electronics," IEEE Spectrum, vol. 26, pp. 59–61, Jan. 1989.

HDTV-15 R. Hopkins, "Advanced television systems," IEEE Trans. Consumer Electronics, vol. CE-34, pp. 1–15, Feb. 1988.

HDTV-16 A. G. Toth, M. Tsinberg, and C. W. Rhodes, "NTSC compatible high definition television emission system," IEEE Trans. Consumer Electronics, vol. CE-34, pp. 40–47, Feb. 1988.

HDTV-17 G. J. Tonge, "Image processing for higher definition television," IEEE Trans. Circuits and Systems, vol. CAS-34, pp. 1385–1398, Nov. 1987.

HDTV-18 G. Barbieri, F. Molo, and J. L. Tejerina, "A modular and flexible video codec architecture for application to TV and HDTV," Intl. TV Symp. and Techl., Exhibition, Broadcast session, pp. 410–420, Montreux, Switzerland, June 17–22, 1989.

HDTV-19 J. Sabatier and D. Nasse, "Standardization activities in HDTV broadcasting," Signal Process: Image Commun., vol. 1, pp. 17–28, June 1989.

HDTV-20 R. K. Jurgen, "Chasing Japan in the HDTV race", IEEE Spectrum," vol. 26, pp. 26–30, Oct. 1989.

HVS in Transform Coding

HTC-1 J. L. Mannos and D. J. Sakrison, "The effect of a visual fidelity criterion on the encoding of images," IEEE Trans. Inform. Theory, vol. IT-20, pp. 525–536, July 1974.

HTC-2 N. C. Griswold, "Perceptual coding in the cosine transform domain," Optical Engrg., vol. 19, pp. 306–311, May–June 1980.

HTC-3 S. Ericsson, "Frequency weighted interframe hybrid coding," Report No. TRITA-TTT-8401, Telecommun. Theory, The Royal Inst. of Tech., Stockholm, Sweden, Jan. 1984.

HTC-4 N. B. Nill, "A visual model weighted cosine transform for image compression and quality assessment," IEEE Trans. Commun., vol. COM-33, pp. 551–557, June 1985.

HTC-5 K. N. Ngan, K. S. Leong, and H. Singh, "Cosine transform coding incorporating human visual system model," SPIE Fiber 86, vol. 707, pp. 165–171, Cambridge, MA, Sept. 14–20, 1986.

HTC-6 K. S. Leong, "Adaptive cosine transform image coding incorporating human visual system model," M.S. Thesis, National Univ. of Singapore, Singapore, Dec. 1987.

HTC-7 K. N. Ngan, K. S. Leong, and H. Singh, "A HVS-weighted transform coding scheme with adaptive quantization," SPIE Visual Commun. and Image Process., vol. 1001, pp. 702–708, Cambridge, MA, Nov. 9–11, 1988.

HTC-8 J. Ameye, J. Bursens, S. Desmet, K. Vanhoof, G. Tu, J. Rommelaere, and A. Oostevlinck, "Image coding using the human visual system," Intl. Workshop on image coding, The Korea Inst. of Commun. Sciences, pp. 229–308, Seoul, Korea, Aug. 1987.

HTC-9 V. Thomas, J. L. Blin, and M. Hias, "Experiments on visibility thresholds of DCT Coefficients," PCS 88, Picture Coding Symp., pp. 3.1-1 through 3.1-2, Torino, Italy, Sept. 1988.

HTC-10 K. N. Ngan, K. S. Leong, and H. Singh, "Adaptive cosine transform coding of images in perceptual domain," IEEE Trans., Acoust., Speech, and Signal Process., vol. ASSP-37, pp. 1743–1750, Nov. 1989.

Image Coding

IC-1 J. W. Modestino, N. Farvardin, and M. R. Ogrinc, "Performance of block cosine image coding with adaptive quantization," IEEE Trans. Commun., vol. COM-33, pp. 210–217, Mar. 1985.

IC-2 J. W. Modestino, D. G. Daut, and A. L. Vickers, "Combined source channel coding of images using the block cosine transform," IEEE Trans. Commun., vol. COM-29, pp. 1261–1274, Sept. 1981.

IC-3 R. C. Reininger and J. D. Gibson, "Distribution of the two dimensional DCT coefficients for images," IEEE Trans. Commun., vol. COM-31, pp. 835–839, June 1983.

IC-4 K. N. Ngan, "Adaptive transform coding of video signals," IEE Proc., vol. 129, pp. 28–40, Feb. 1982.

IC-5 D. R. Comstock and J. D. Gibson, "Hamming coding of DCT-compressed images over noisy channels," IEEE Trans. Commun., vol. COM-32, pp. 856–861, July 1984.

IC-6 A. Ploysongsang and K. R. Rao, "DCT/DPCM processing of NTSC composite signal," IEEE Trans. Commun., Vol. COM-30, pp. 541–549, March 1982.

IC-7 F. A. Kamangar and K. R. Rao, "Interframe hybrid coding of NTSC component video signal," IEEE Trans. Commun., vol. COM-29, pp. 1740–1753, Dec. 1981.

IC-8 J. A. Roese, W. K. Pratt, and G. S. Robinson, "Interframe cosine transform image coding," IEEE Trans. Commun., Vol. COM-25, pp. 1329–1338, Nov. 1977.

IC-9 J. M. Schumpert and R. J. Jenkins, "A two-component image coding scheme based on two dimensional interpolation and the discrete cosine transform," Intl. Conf. on Acoust., Speech, and Signal Process., pp. 1232–1235, Boston, MA, April 14–16, 1983.

IC-10 J. B. O'Neal, Jr. and T. R. Natarajan, "Coding isotropic images," IEEE Trans. Inform. Theory, vol. IT-23, pp. 679–707, Nov. 1977.

IC-11 R. Srinivasan and K. R. Rao, "CMT and hybrid coding of the component color TV signal," IEEE Trans. Syst. Man., and Cybern., vol. SMC-14, pp. 506–510, May/June 1984.

IC-12 S. C. Kwatra and C. Ekambaram, "A new architecture for adaptive intrafield transform compression of NTSC composite video signal," IEEE Trans. Commun., vol. COM-32, pp. 1349–1351, Dec. 1984.

IC-13 W. A. Pearlman and P. Jakatdar, "The effectiveness and efficiency of hybrid transform/DPCM interframe image coding," IEEE Trans. Commun., vol. COM-32, pp. 832–838, July 1984.

IC-14 H. Murakami and H. Yamamoto, "Performance of transform coding for carrier chrominance signals," IEEE Trans. Commun., vol. COM-32, pp. 324–327, March 1984.

IC-15 W. K. Pratt and W. H. Chen, "Slant transform image coding," IEEE Trans. Commun., vol. COM-22, pp. 1075–1093, Aug. 1974.

IC-16 R. C. Reininger and J. D. Gibson, "Soft decision demodulation and transform coding of images," IEEE Trans. Commun., vol. COM-31, pp. 572–577, April 1983.

IC-17 S. C. Kwatra and H. Fatmi, "NTSC composite video at 1.6 bits/pel," Intl. Conf. on Commun., pp. 458–462, Boston, MA, June 1983.

IC-18 T. R. Natarajan and N. Ahmed, "On interframe transform coding," IEEE Trans. Commun., vol. COM-25, pp. 1323–1329, Nov. 1977.
IC-19 A. G. Tescher, "A dual transform coding algorithm," National Telecommun. Conf., pp. 53.4.1–53.4.4, Washington, D.C., Nov. 27–29, 1979.
IC-20 A. G. Tescher and R. V. Cox, "An adaptive transform coding algorithm," Intl. Conf. on Commun., pp. 47-20 through 47-23, Philadelphia, PA, July 14–16, 1976.
IC-21 N. Garguir, "Comparative performance of SVD and adaptive cosine transform in coding images," IEEE Trans. Commun., vol. COM-27, pp. 1230–1234, Aug, 1979.
IC-22 R. J. Clarke, "Hybrid intraframe transform coding of image data," IEE Proc., vol. 131, part. F, pp. 2–6, Feb. 1984.
IC-23 W. K. Cham and R. J. Clarke, "DC coefficient restoration in transform image coding," IEE Proc., vol. 131, Part F, pp. 709–712, Dec. 1984.
IC-24 H. Gharavi and A. Jalali, "DCT/conditional-entropy coding of images," PCS 86, Picture Coding Symp., pp. 238, Tokyo, Japan, April 2–4, 1986.
IC-25 M. Ghanbari and D. E. Pearson, "Fast cosine transform implementation for television signals," IEE Proc., part F, vol. 129, pp. 59–68, Feb. 1982.
IC-26 S. Ericsson, "Fixed and adaptive predictor for hybrid predictive transform coding," Royal Inst. of Technology, TRI TA-TTA-8305, Stockholm, Sweden, Dec. 1983.
IC-27 P. Strobach, H. Holzlwimmer, and R. Kutka, "Two methods for reduction of typical DCT-transform-coding errors," PCS 87, Picture Coding Symp., pp. 43–44, Stockholm, Sweden, June 9–11, 1987.
IC-28 M. R. Civanlar, S. A. Rajala, and T. K. Miller, "Implementation of a projection onto convex sets iteration based image coder," SPIE, Visual Commun., and Image Processing, Vol. 707, pp. 98–103, Cambridge, MA, Sept. 1986.
IC-29 H. Gharavi and A. Jalali, "Entropy coding of cosine transformed images," GLOBECOM 86, IEEE Global Telecommun., Conf., pp. 1165–1169, Houston, TX, Dec. 1–4, 1986.
IC-30 R. Mester and U. Franke, "On 'optimal' thresholds in the context of transform coding with stabilized quality," PCS 87, Picture Coding Symp., pp. 98–99, Stockholm, Sweden, June 9–11, 1987.
IC-31 M. Gilge, "Block boundary detection in transform coded images," PCS 87, Picture Coding Symp., pp. 100–101, Stockholm, Sweden, June 9–11, 1987.
IC-32 J. M. Adant, P. Delogne, B. Macq, and L. Vandendorpe, "Perceptual transforms," PCS 87, Picture Coding Symp., pp. 102–103, Stockholm, Sweden, June 9–11, 1987.
IC-33 L. Bengtsson, and P. Weiss, "A comparison between DPCM and transform/hybrid coding techniques for transmission of broadcast signals at 30.0 Mbit/s." PCS 87, Picture Coding Symp., pp. 136, Stockholm, Sweden, June 9–11, 1987.
IC-34 S. Acharya, "Design of a freeze-frame coder," O-F Lase-8 Optoelectronics and Laser Applications in Science and Engineering and Electro-Optic Imaging System & Devices '87, SPIE, vol. 757, pp. 23–29, Los Angeles, CA, Jan. 1987.
IC-35 O. R. Mitchell and A. J. Tabatabai, "Channel error recovery for transform image coding," IEEE Trans. Commun., vol. COM-29, pp. 1754–1762, Dec. 1981.
IC-36 K. Rose, A. Heiman and I. Dinstein, "DCT/DST alternate transform image coding," GLOBECOM 87 Global Telecommun. Conf., pp. 426–430, Tokyo, Japan, Nov. 1987.
IC-37 D. G. Daut, "Two-dimensional hybrid image coding and transmission," SPIE, Visual Communications and Image Processing II, vol. 845, pp. 24–31, Cambridge, MA, Oct. 1987.
IC-38 E. Dubois, Y. Rahmonni, and F. Lortie, "Experiments on image coding with distortion below visual threshold," SPIE, Visual Communications and Image Processing II, vol. 845, pp. 126–131, Cambridge, MA, Oct. 1987.
IC-39 A. Saghri, A. G. Tescher, and A. Habibi, "Block size considerations for adaptive image coding," National Telesystem, Conf., pp. E1.2.1–E1.2.4, Galveston, TX, Nov. 1982.
IC-40 J. D. Eggerton and M. D. Srinath, "Statistical distributions of image DCT coefficients," Comput. and Elecl. Engrg., vol. 12, pp. 137–145, 1986.
IC-41 J. D. Eggerton and M. D. Srinath, "Design of optimal quantizers with an entropy constraint," Comput. and Elecl. Engrg., vol. 12, pp. 147–153, 1986.
IC-42 W. C. Wong and R. Steele, "Adaptive discrete cosine transformation of pictures using an

energy distribution logarithmic model," The Radio and Electronic Engineer, vol. 51, pp. 571–578, Nov./Dec. 1981.

IC-43 W. A. Pearlman, "Variable block rate and blockwise spectral adaptation in cosine transform image coding," ICASSP, Intl. Conf. Acoust., Speech, and Signal Process., pp. 773–776, New York, NY, April 11–14, 1988.

IC-44 G. W. Wornell and D. H. Staelin, "Transform image coding with a new family of models," ICASSP, Intl. Conf. Acoust., Speech, and Signal Process., pp. 777–780, New York, NY, April 11–14, 1988.

IC-45 B. G. Haskell, "Interpolative, predictive and pyramid transform coding of color images," ICASSP, Intl. Conf. Acoust., Speech, and Signal Process., pp. 785–787, New York, NY, April 11–14, 1988.

IC-46 H. B. Kekre and S. A. Aleem, "Adaptive linear quantization and reconstruction scheme in the discrete cosine transform domain," Intl. J. Electronics, vol. 54, pp. 31–45, Jan. 1983.

IC-47 O. Telese and G. Zarone, "Video hybrid coding by picture domain segmentation," Alta Frequenza, vol. L, pp. 248–253, Sept.–Oct. 1981.

IC-48 A. Baskurt and R. Goutte, "3-dimensional image compression by discrete cosine transform," EUSIPCO-88, European Signal Process. Conf., pp. 79–82, Grenoble, France, Sept. 5–8, 1988.

IC-49 A. Sanz, C. Munoz, and N. Garcia, "Statistical analysis of predictive hierarchical image encoding," EUSIPCO-88, European Signal Process. Conf., pp. 1649–1652, Grenoble, France, Sept. 5–8, 1988.

IC-50 B. Zoran, "Analysis of output signal-to-noise ratio in hybrid DCT/DPCM image coding system," EUSIPCO-88, European Signal Process. Conf., pp. 1661–1664, Grenoble, France, Sept. 5–8, 1988.

IC-51 T Ohira, M. Hayakawa, and K. Matsumoto, "Orthogonal transform coding system for NTSC color television signals," IEEE Trans. Commun., vol. COM-26, pp. 1454–1463, Oct. 1978.

IC-52 A. Jalali and K. R. Rao, "An architecture for hybrid coding of NTSC TV signals," Comput. and Elec. Engrg., vol. 9, pp. 45–51, 1982.

IC-53 Telettra, "DCT algorithm for broadcast quality encoding of NTSC television for transmission at 44.736 Megabits/second (DS3)," Document number: T1.Y1.1/88-035, Aug. 2, 1988.

IC-54 J. A. Roese and W. K. Pratt, "Theoretical performance models for interframe transform and hybrid transform/DPCM coders," SPIE, Advances in Image Transmission Techniques, vol. 87, pp. 172–179, San Diego, CA, Aug. 1976.

IC-55 N. Ahmed and T. Natarajan, "Some aspects of adaptive transform coding of multispectral data," 16th Asilomar Conf. on Circuits, Systems and Computers, pp. 583–587, Pacific Grove, CA, Nov. 22–24, 1976.

IC-56 A. Habibi, "Hybrid coding of pictorial data," IEEE Trans. Commun., vol. COM-22, pp. 614–624, May 1974.

IC-57 W. H. Chen, "Slant transform image coding," Univ. of Southern California, Image Process. Inst., USCEE Report 441, Los Angeles, CA, May 1973.

IC-58 W. A. Pearlman, "Variable rate, adaptive transform tree coding of images," SPIE, Advances in image coding, vol. 1001, pp. 1026–1037, Cambridge, MA, Nov. 1988.

IC-59 N. Doi, H. Hanyu, M. Izumita, S. Mita, Y. Eto, and H. Imai, "Adaptive DCT coding of the home digital VTR," GLOBECOM 88, Global Commun, Conf., pp. 1073–1079, Hollywood, FL, Nov.–Dec. 1988.

IC-60 V. Thomas, "DCT coding for TV contributions applications," PCS 88, Picture Coding Symp., pp. 13.14-1 through 13.14-2, Torino, Italy, Sept. 1988.

IC-61 G. Madec, "A 15 Mbit/s codec for 4:2:2 TV signals," PCS 88, Picture Coding Symp., pp. 14.7-1, Torino, Italy, Sept. 1988.

IC-62 C. Perron, "Behaviour of a hardware DCT television codec faced to transmission errors," PCS 88, Picture Coding Symp., pp. 15.6-1 through 15.6-2, Torino, Italy, Sept. 1988.

IC-63 E. Peters, "A 15 Mbit/s codec for component video signals," PCS 88, Picture Coding Symp., pp. 15.7-1 through 15.7-2, Torino, Italy, Sept. 1988.

IC-64 P. Pirsch, "Design of DPCM quantizers for video signals using subjective tests," IEEE Trans. Commun., vol. COM-29, pp. 990–1000, July 1981.
IC-65 R. C. Brainard and J. H. Othmer, "VLSI implementation of a DPCM compression algorithm for digital TV," IEEE Trans. Commun., vol. COM-35, pp. 854–856, Aug. 1987.
IC-66 C. T. Chen, "Adaptive transform coding via quadtree-based variable blocksize DCT," Intl. Conf. on Acoust., Speech, and Signal Proces., pp. 1854–1857, Glasgow, Scotland, May 23–26, 1989.
IC-67 S. Cucchi and F. Molo, "DCT based television codec for DS3 digital transmission," 130th SMPTE Technical Conf., Preprint No. 130-12, New York, NY, Oct. 1988.
IC-68 D. R. Ahlgren, J. Crosbie, and D. Eriqat, "Compression of digitized images for transmission and storage applications," SPIE, Image Processing, Analysis, Measurement and Quality, vol. 901, pp. 105–113, Los Angeles, CA, Jan. 1988.
IC-69 P. Pirsch, "Design of a DPCM codec for VLSI realization in CMOS technology," Proc. IEEE, vol. 73, pp. 592–598, April 1985.
IC-70 A. Rothermel, "Realization of a DPCM coder for 13.5 MHz sampling rate in CMOS technology," IEEE Journal of Solid State Circuits, vol. SC-22, pp. 1196–1197, Dec. 1987.
IC-71 B. Zehner et al., "Video chip set for data rate compression by filtering and DPCM coding," Intl. Symp. Circuits and Systems, pp. 697–700, Helsinki, Finland, June 1988.
IC-72 P. Pirsch, "VLSI design of DPCM codecs for video signals," IEEE press book "Visual Communications Systems," Feb. 1989.
IC-73 F. Cesa, M. Modena, and G. L. Sicuranza, "2-D DCT intrafield codec for high quality TV signals," Digital Signal Process., '87, Proc. of the Intl. Conf., pp. 512–516, Florence, Italy, Sept. 7–10, 1987.
IC-74 H. D. Ferreyra, R. Picco, and L. A. Sobrero, "Analysis and project considerations on 2D-DCT adaptive coding," Digital Signal Processing-87, V. Cappelini and A. G. Constantinides (eds.), Elsevier Science Publishers, North Holland, 1987.
IC-75 M. Muratori and M. Stroppiana, "Bit rate reduction techniques for coding standard and high definition TV signals," Young Researchers Seminar, MIT, Boston, MA, Oct. 1988.
IC-76 CMTT/2-DCT Group chairman report, "Digital transmission of component-coded television signals at 30–34 Mbit/s and 45 Mbit/s using discrete cosine transform," CMTT/2-66, July 1988.
IC-77 L. Stenger, "Digital coding of television signals—CCIR activities for standardization," Signal Process: Image Commun., vol. 1, pp. 29–43, June 1989.
IC-78 M. Barbero and M. Stroppiana, "Coding of digital TV signal: systems for redundancy reduction based on the discrete cosine transform," Translation from the artical Codifica del segnale televisivo numerica: sistemi di riduzione della ridondnza mediante l'uso della transformata coseno discrete «Elettronica e Telecomunicazioni», 1, 1989.
IC-79 Document IWP 11/7-25, Document IWP CMTT/2-103 (Italy-Spain), June 1989.
IC-80 L. Chiariglione, "Standardization of moving picture coding for interactive applications," GLOBECOM 89, IEEE Global Commun., Conf., pp. 559–563, Dallas, TX, Nov. 27–30, 1989
IC-81 J. P. Henot, "17 mbit/s algorithm for secondary distribution of 4:2:2 video signals," SPIE Visual Commun., and Image Processing, Vol. 1199, pp. 1590–1598, Philadelphia, PA, Nov. 5–10, 1989.

Image Enhancement

IE-1 W. H. Chen and S. C. Fralick, "Image enhancement using cosine transform filtering," Proc. of the Symp. on Current Math. Problems in Image Science, pp. 186–192, Monterey, CA, Nov. 1976.
IE-2 H. R. Keshavan and M. D. Srinath, "Two-dimensional interpolative models in enhancement of noisy images," National Telecommun., Conf. pp. 49:4-1 through 49:4-5, Los Angeles, CA, Dec. 1977.

Infrared Image Coding

IR-1 S. D. H. Saunders and C. J. Gillham, "Transmitting IR images at very low data rate," Proc. IEE 2nd Intl. Conf. on Image Processing and Its Applications, pp. 196–199, London, UK, June 1986.

Low-Bit Rate Coding

LBR-1 J. Anderson *et al.*, "Codec squeezes color teleconferencing through digital telephone lines," Electronics, vol. 57, pp. 113–115, Jan. 26, 1984.

LBR-2 R. Natarajan and K. R. Rao, "Design of a 64 Kbps coder for teleconferencing," SPIE's 28th Annual Intrnl. Technical Symp., San Diego, CA, vol. 504, pp. 406–413, Aug. 19–24, 1984.

LBR-3 W. H. Chen, "Scene adaptive coder," ICC '81, Intl. Conf. on Commun., Proc. pp. 22.5.1–22.5.6, Denver, CO, June 14–18, 1981.

LBR-4 L. C. Chan and P. Whiteman, "Hardware constrained hybrid coding of video imagery," IEEE Trans. Aero. Electronic Syst., vol. AES-19, pp. 71–83, Jan. 1983.

LBR-5 W. H. Chen and W. K. Pratt, "Scene adaptive coder," IEEE Trans. Commun., vol. COM-32, pp. 225–232, March 1984.

LBR-6 O. E. Bessett and W. B. Schaming, "A two dimensional discrete cosine transform video bandwidth compression system," NAECON, pp. 1146–1152, Dayton, OH, May 1980.

LBR-7 G. Eude and J. Guichard, "Hybrid transform coding for low bit rate transmission," PCS 86, Picture Coding Symp., pp. 59–60, Tokyo, Japan, April 2–4, 1986.

LBR-8 J. Guichard and G. Eude, "Intra- and interframe transform coding for moving pictures transmission," ICC 86, Intl. Conf. on Commun., pp. 381–384, Toronto, Canada, June 22–25, 1986.

LBR-9 B. Braun, "Luminance adaptive chrominance coding," ICASSP 87, Intl. Conf. Acoust., Speech, Signal Process., pp. 1075–1078, Dallas, TX, April 6–9, 1987.

LBR-10 M. Maragoudakis and J. D. Gibson, "Experiments on video teleconferencing algorithms at 56 kilobits/sec," ICASSSP 87, Intl. Conf. Acoust., Speech, and Signal Process., pp. 1071–1074, Dallas, TX, April 6–9, 1987.

LBR-11 J. Guichard and G. Eude, "Hybrid variable blocksize coding scheme based upon 3 DCTs and motion compensation techniques at 64 kbit/s," PCS 87, Picture Coding Symp., pp. 23–24, Stockholm, Sweden, June 9–11, 1987.

LBR-12 W. H. Chen and D. N. Hein, "Evaluation of various coding schemes for various size cosine transform scene adaptive coding systems," PCS 87, Picture Coding Symp., pp. 25–26, Stockholm, Sweden, June 9–11, 1987.

LBR-13 G. Kummerfeldt, F. May, and W. Wolf, "Coding television signals at 320 and 64 kbit/s," 2nd International Technical Symposium on Optical and Electro-Optical Applied Science and Engineering, SPIE, vol. 594, pp. 119–128, Cannes, France, Dec. 1985.

LBR-14 F. Sugiyama, K. Dachiku, and T. Watanabe, "64 Kbps to 1.5 Mbps color video codec with transform domain motion detection," PCS 87, Picture Coding Symp., pp. 155–156, Stockholm, Sweden, June 9–11, 1987.

LBR-15 J. Guichard, G. Eude, and J. C. Schmitt, "Hardware implementation of a 64 Kbit/s codec," PCS 87, Picture Coding Symp., pp. 181, Stockholm, Sweden, June 9–11, 1987.

LBR-16 M. W. Whybray and B. Hopper, "A compact videophone codec using two DSPs," PCS 87, Picture Coding Symp., pp. 182, Stockholm, Sweden, June 9–11, 1987.

LBR-17 A. Brandt and W. Tengler, "Hierarchical motion estimation for efficient low bit rate hybrid coding," PCS 87, Picture Coding Symp., pp. 187–188, Stockholm, Sweden, June 9–11, 1987.

LBR-18 Y. Suzuki, "A study of DSP architecture for subrate video codec," PCS 87, Picture Coding Symp., pp. 205–206, Stockholm, Sweden, June 9–11, 1987.

LBR-19 P. Gerken and H. Schiller, "A low bit-rate image sequence coder combining a progressive DPCM on interleaved rasters with a hybrid DCT technique," IEEE Journal on Selected Areas in Commun., vol. SAC-5, pp. 1079–1089, Aug. 1987.

LBR-20 S. E. Elnahas and J. G. Dunham, "Entropy coding for low bit rate visual telecommunications," IEEE Journal on Selected Areas in Commun., vol. SAC-5, pp. 1175–1183, Aug. 1987.

LBR-21 L. Chiariglione, S. Fontolan, M. Guglielmo, and F. Tomassi, "A variable resolution video codec for low bit-rate applications," IEEE Journal on Selected Areas in Commun., vol. SAC-5, pp. 1184–1189, Aug. 1987.

LBR-22 H. Gharavi, "Low bit-rate video transmission for ISDN applications," IEEE Trans. Circuits and Systems, vol. CAS-35, pp. 258–261, Feb. 1988.

LBR-23 Description of Ref. Model 5 (RM5), CCITT SGXV, Working Party XV/4, specialists group on coding for visual telephony, Document 375, Sept. 1988.

LBR-24 R. C. Nicol and N. Mukawa, "Motion video coding in CCITT SGXV—The coded picture format," GLOBECOM 88, Global Commun, Conf. pp. 992–996, Hollywood, FL, Nov.–Dec. 1988.

LBR-25 R. Plompen, Y. Hatori, W. Geuen, J. Guichard, M. Guglielmo, and H. Brusewitz, "Motion video coding in CCITT SGXV—The video source coding," GLOBECOM 88, Global Commun, Conf., pp. 997–1004, Hollywood, FL, Nov.–Dec. 1988.

LBR-26 M. Carr, J. Guichard, K. Matsuda, R. Plompen and J. Speidel, "Motion video coding in CCITT SGXV—The video multiplex and transmission coding," GLOBECOM 88, Global Commun, Conf., pp. 1005–1010, Hollywood, FL, Nov.–Dec. 1988.

LBR-27 N. Texier, J. Guichard, and G. Eude, "Hybrid transform video coding scheme for low bit rate—an improvement of the reference algorithm," PCS 88, Picture Coding Symp., pp. 8.4-1 through 8.4-2, Torino, Italy, Sept. 1988.

LBR-28 H. Gharavi, "Hybrid coding of video signals using sub-block approach," PCS 88, Picture Coding Symp., pp. 12.2-1 through 12.2-2, Torino, Italy, Sept. 1988.

LBR-29 Y. Kato, "Transformation compensation for videophone image coding," Symp., pp. 12.5-1 through 12.5-2, Torino, Italy, Sept. 1988.

LBR-30 B. Dietmar, and H. Amor, "A multi-processor concept for a narrowband picture phone," PCS 88, Picture Coding Symp., pp. 15.1-1 through 15.1-2, Torino, Italy, Sept. 1988.

LBR-31 V. Eisenhardt, G. Kummerfeldt, F. May and W. Wolf, "Architecture of a full motion 64 Kbit/s video codec," PCS 88, Picture Coding Symp., pp. 15.2-1 through 15.2-2, Torino, Italy, Sept. 1988.

LBR-32 E. Gerard, J. Guichard, and A. Plancoulaine, "A 64 Kbit/s videocodec based on transputers network," PCS 88, Picture Coding Symp., pp. 15.3-1 through 15.3-2, Torino, Italy, Sept. 1988.

LBR-33 M. Takizawa, Y. Izawa, J. Kimura and T. Fukiniki, "A consideration on hardware architecture for 64 Kb/s TV Codec," PCS 88, Picture Coding Symp., pp. 15.4-1 through 15.4-2, Torino, Italy, Sept. 1988.

LBR-34 M. Ohta, T. Omachi, and T. Koga, "Estimation for mismatch error accumulation in interframe DCT coding," PCS 88, Picture Coding Symp., pp. 5.1-1 through 5.1-2, Torino, Italy, Sept. 1988.

LBR-35 Report of the thirteenth meeting in Paris CCITT SGXV, Working Party XV/1, Specialists Group on Coding for Visual Telephony, Document no. 395R, Sept. 22, 1988.

LBR-36 G. Eude, J. Guichard, and A. Plancoulaine, "A 64 Kbit/s videocodec based on transputer network," Intl. Conf. on Commun. Systems, pp. 17.8.1 through 17.8.3, Singapore, Nov. 1988.

LBR-37 Description of Ref. Model (RM6), CCITT SGXV, Working Party XV/1, Specialists group on coding for visual telephony, Document 396, Oct. 20, 1988.

LBR-38 J. Guichard, "Motion video coding in CCITT SGXV—hardware trials," GLOBECOM 88, Global Commun., Conf., pp. 37–42, Hollywood, FL, Nov.–Dec. 1988.

LBR-39 NTT, KDD, NEC, and FUJITSU, "Accuracy of DCT calculation," Document no. 255, CCITT SGXV, Working Party XV/1, Specialists group on coding for visual telephony, Oct. 1987.

LBR-40 Report of the fifteenth meeting in Oslo CCITT SGXV, Working Party XV/1, Specialists group on coding for visual telephony, Document no. 499R, March 10, 1989.
LBR-41 S. Okubo, "Video codec standardization in CCITT study group XV," Signal Process: Image Commun., vol. 1, pp. 45–54, June 1989.
LBR-42 "Special issue on 64 Kbit/s coding of moving video," guest editor: H. G. Musmann, Signal Process: Image Commun., vol. 1, Oct. 1989.
LBR-43 2nd Intl. Workshop on 64 Kbit/s coding of moving video, Hannover, W. Germany, Sept. 4–6, 1989.
LBR-44 Report of the fifteenth meeting in Stuttgart, CCITT SGXV, Working Party XV/1, Specialists group on coding for visual telephony, Document no. 540R, June 16, 1989.
LBR-45 O. W. Kwon, J. W. Kim, J. H. Jeong, M. H. Lee, and J. S. Lee, "A software-based 64 Kbit/s video codec," ITEJ, Institution of Television Engineers Japan, pp. 469–470, Tokyo, Japan, July 1989.
LBR-46 J. S. Lee, O. W. Kwon, J. H. Kim, J. H. Jeong, and M. H. Lee, "64 Kbit/s video codec based on pipelined DSP's" PCS 90, Picture Coding Symp., pp. 000–000, March, 1990.
LBR-47 J. W. Duran and M. Kenoyer, "A PC-compatible multiprocessor workstation for video, data and voice communication," SPIE, Visual commun., and Image Processing, vol. 1199, pp. 232–236, Philadelphia, PA, Nov. 5–10, 1989.
LBR-48 J. Maeng and D. Hein, "A low-rate video coding based on DCT/VQ," SPIE, Visual Commun., and Image Processing, vol. 1199, pp. 267–273, Philadelphia, PA, Nov. 5–10, 1989.

LMS Filtering

LF-1 S. Narayan, A. M. Peterson, and M. J. Narasimha, "Transform domain LMS algorithm," IEEE Trans. Acoust., Speech, and Signal Process., vol. ASSP-31, pp. 609–615, June 1983.
LF-2 J. C. Lee and C. K. Un, "Performance of transform domain LMS adaptive digital filters," IEEE Trans. Acoust., Speech and Signal Process., vol. ASSP-34, pp. 499–510, June 1986.
LF-3 J. C. Lee and C. K. Un, "Block realization of multirate adaptive digital filters," IEEE Trans. Acoust., Speech, and Signal Process., vol. ASSP-34, pp. 105–117, Feb. 1986.

Miscellaneous

M-1 I. N. Sneddon, "Use of integral transforms," New York, NY, McGraw-Hill, 1972.
M-2 A. K. Jain, "A fast Karhunen-Loeve transform for a class of random processes," IEEE Trans. Commun. vol. COM-24, pp. 1023–1029, Sept. 1976.
M-3 K. Karhunen, "Ueber lineare methoden in der Wahrscheinlichkeitsrechnung," Ann. Acad. Sci. Fenn. Ser A.I. Math. Phys., vol. 37, 1947.
M-4 M. Loeve, "Probability theory," 2nd Ed., Princeton, NJ, Van Nostrand, pp. 478, 1960.
M-5 P. A. Devijer and J. Kittler, "Pattern recognition: A statistical approach," Englewood Cliffs, NJ, Prentice Hall, 1982.
M-6 W. D. Ray and R. M. Driver, "Further decomposition of the Karhunen-Loeve series representation of a stationary random process," IEEE Trans. Inform. Theory, vol. IT-16, pp. 663–668, Nov. 1970.
M-7 Y. Yemeni and J. Pearl, "Asymptotic properties of discrete unitary transforms," IEEE Trans. Pattern Analysis and Machine Intelligence, vol. PAMI-1, pp. 366–371, Oct. 1979.
M-8 W. B. Davenport, Jr. and W. L. Root, "An introduction to the theory of random signals and noise," New York, NY: McGraw Hill, 1958.
M-9 P. Davis, "Circulant matrices," New York, NY: Wiley, 1979.
M-10 P. Yip and B. P. Agrawal, "Theory and applications of Toeplitz matrices," CRL Report, 1979.
M-11 I. N. Sneddon, "Special functions of mathematical physics and chemistry," New York, NY: Longman Inc., 1980.

M-12 J. Pearl, "On coding and filtering stationary signals by discrete Fourier transforms," IEEE Trans. Inform. Theory, vol. IT-19, pp. 229–232, Jan. 1973.
M-13 V. I. Krylov, "Approximate calculation of integrals," New York, NY: MacMillan, 1962.
M-14 C. T. Fike, "Computer evaluation of mathematical functions," Englewood Cliffs, NJ, Prentice Hall, 1968.
M-15 G. Szego, "Orthogonal polynomials," New York, NY, AMS, 1959.
M-16 J. Pearl, "Asymptotic equivalence of spectral representations," IEEE Trans. Acoust, Speech, and Signal Process., vol. ASSP-23, pp. 547–551, Dec. 1975.
M-17 R. M. Gray, "On the asymptotic eigenvalue distribution of Toeplitz matrices," IEEE Trans. Inform. Theory, vol. IT-18, pp. 725–730, Nov. 1972.
M-18 V. Grenander and G. Szego, "Toeplitz forms and their applications," Berkeley, CA, Univ. of California Press, 1958.
M-19 H. Kitajima, "Energy packing efficiency of the Hadamard transform," IEEE Trans. Commun., vol. COM-24, pp. 1256–1258, Nov. 1976.
M-20 P. Yip and K. R. Rao, "Energy packing efficiency for the generalized discrete transforms," IEEE Trans. Commun., vol. COM-26, pp. 1257–1262, Aug. 1978.
M-21 L. D. Davisson, "Rate distortion theory and application," Proc. IEEE, vol. 60, pp. 800–808, July, 1972.
M-22 W. K. Pratt, "Generalized Wiener filtering computation techniques," IEEE Trans. Comput., vol. C-21, pp. 636–641, July 1972.
M-23 H. Reitboeck and T. P. Brodie, "A transformation with the invariance under cyclic permutation for application in pattern recognition," Information and Control, vol. 15, pp. 130–154, 1969.
M-24 H. F. Silverman, "An introduction to programming the Winograd Fourier transform algorithm (WFTA)," IEEE Trans. Acoust., Speech, and Signal Process., vol. ASSP-25, pp. 152–165, April, 1977, Corrections and an addendum, vol. ASSP-26, pp. 268, June, 1977.
M-25 B. D. Tseng and W. C. Miller, "Comments on an introduction to the Winograd Fourier transform algorithm (WFTA)," IEEE Trans. Acoust, Speech, and Signal Process., vol. ASSP-26, pp. 268–269, June 1978.
M-26 R. V. L. Hartley, "A more symmetrical Fourier analysis applied to transmission problems," Proc. IRE, vol. 30, pp. 144–150, 1942.
M-27 R. N. Bracewell, "Discrete Hartley transform," J. Optical Soc. of America, vol. 73, pp. 1832–1835, 1983.
M-28 R. N. Bracewell, "The fast Hartley transform," Proc. IEEE, vol. 72, pp. 1010–1018, Aug, 1984.
M-29 H. V. Sorenson, D. L. Jones, C. S. Burrus, and M. T. Heideman, "On computing the discrete Hartley transform," IEEE Trans. Acoust., Speech, and Signal Process., vol. ASSP-33, pp. 1231–1238, Oct. 1985.
M-30 P. Duhamel, "Implementation of 'Split-Radix' FFT algorithms for complex, real, and real-symmetric data," IEEE Trans. Acoust., Speech, and Signal Process., vol. ASSP-34, pp. 285–295, April 1986.
M-31 Document CCIR IWP 11/7 Dec. 86 (Rev. 3), Draft Amendments to Recommendation 601 (Mod I) and Reports 629-2 (Mod I) and 692 (Mod I), 1986.
M-32 R. Srinivasan and K. R. Rao, "An approximation to the discrete cosine transform for N = 16," Signal Processing, vol. 5, pp. 81–84, Jan. 1983.
M-33 L. R. Rabiner, "On the use of symmetry in FFT computation," IEEE Trans. Acoust., Speech, and Signal Process., vol. ASSP-27, pp. 233–239, June 1979.
M-34 J. W. Cooley and J. W. Tukey, "An algorithm for the machine calculation of complex Fourier series," Math of Comput., vol. 19, pp. 297–301, 1965.
M-35 J. B. Burl, "Estimating the basis functions of the Karhunen-Loeve transform," IEEE Trans. Acoust., Speech, and Signal Process., vol. ASSP-37, pp. 99–105, Jan. 1989.
M-36 D. B. Harris, J. H. McClellan, D. S. K. Chan and H. W. Schuessler, "Vector radix fast Fourier transform," Intl. Conf. on Acoust., Speech, and Signal Process., pp. 548–551, May 1977.
M-37 S. C. Pei and J. L. Wu, "Split vector radix 2-D fast Fourier transform," IEEE Trans. Circuits and Syst., vol. CAS-34, pp. 978–980, Aug. 1987.

M-38 H. R. Wu and F. J. Paoloni, "On the two-dimensional vector split-radix FFT algorithm," IEEE Trans. Acoust., Speech, and Signal Process., vol. ASSP-32, pp. 1302–1304, Aug. 1989.
M-39 H. R. Wu and F. J. Paoloni, "The structure of vector radix fast Fourier transform," IEEE Trans. Acoust., Speech, and Signal Process., vol. ASSP-37, pp. 1415–1424, Sept. 1989.

M.C. Hybrid (Transform/DPCM) Coding

MC-1 J. R. Jain and A. K. Jain, "Displacement measurement and its application in interframe image coding," IEEE Trans. Commun, vol. COM-29, pp. 1799–1808, Dec. 1981.
MC-2 W. H. Chen and D. Hein, "Motion compensated DXC system," PCS 86, Picture Coding Symp., pp. 76–77, Tokyo, Japan, April 2–4, 1986.
MC-3 M. Ohta, T. Mochizuki and T. Koga, "An adaptive hybrid coding with motion compensation and DCT," PCS 86, Picture Coding Symp., pp. 83–84, Tokyo, Japan, April 2–4, 1986.
MC-4 R. Natarajan and K. R. Rao, "Design of a 64 KBPS coder for teleconferencing," SPIE's 28th Annual Intl. Tech. Symp., vol. 575, Applications of digital image processing VII, pp. 406–413, San Diego, CA, Aug. 21–24, 1984.
MC-5 S. Ericsson, "Motion-compensated hybrid coding at 50 KB/S," ICASSP 85, Intl., Conf. on Acoust., Speech, and Signal Process., pp. 367–370, Tampa, FL, March 26–29, 1985.
MC-6 S. Kappagantula and K. R. Rao, "Motion compensated hybrid image coding," 18th Annual Asilomar Conf. on Circuits, Systems and Computers, pp. 391–395, Pacific Grove, CA, Nov. 5–7, 1984, also IEEE Trans. Commun., vol. COM-33, pp. 1011–1015, Sept. 1985.
MC-7 A. Furukawa, T. Koga, and K. Niwa, "Coding efficiency analysis for motion-compensated interframe DPCM with transform coding," GLOBECOM 85, IEEE Global Telecommun. Conf., pp. 689–693, New Orleans, LA, Dec. 2–5, 1985.
MC-8 H. Brusewitz and P. Weiss, "A video conference system at 384-Kbit/s," PCS 86, Picture Coding Symp., pp. 212, Tokyo, Japan, April 2–4, 1986.
MC-9 M. Ohta and T. Koga, "Adaptive VWL coding of transform coefficients for sub-primary rate video transmission," GLOBECOM 86, IEEE Global Telecommun. Conf., pp. 8.5.1–8.5.5., Houston, TX, Dec. 1–4, 1986.
MC-10 M. Takizawa, J. Kimura, Y. Izawa, and T. Fukiniki, "A study on motion detection method suitable for DCT," PCS 87, Picture Coding Symp., pp. 185–186, Stockholm, Sweden, June 9–11, 1987.
MC-11 T. Koga and M. Ohta, "Entropy coding for a hybrid scheme with motion compensation in sub-primary rate video transmission," IEEE Journal on Selected Areas in Commun., vol. SAC-5, pp. 1166–1174, Aug. 1987.
MC-12 Y. Kato, N. Mukawa, and S. Okubo, "A motion picture coding algorithm using adaptive DCT encoding based on coefficient power distribution classification," IEEE Journal on Selected Areas in Commun., vol. SAC-5, pp. 1090–1099, Aug. 1987.
MC-13 M. Kaneko, Y. Hatori, A. Koike, and H. Yamamoto, "Improvement of transform coding algorithm for motion compensated frame-to-frame difference signals," IEEE Global Telecommun. Conf., pp. 276–280, Houston, TX, Dec. 1–4, 1986.
MC-14 G. Kummerfeldt, F. May, and W. Wolf, "Fast algorithms of a full motion 64 Kbit/s video codec," PCS 88, Picture Coding Symp., pp. 13.8-1 through 13.8-2, Torino, Italy, Sept. 1988.
MC-15 K. Hienerwadel and J. Speidel, "Motion estimation using roughly quantized pictures," PCS 87, Picture Coding Symp., pp. 89–90, Stockholm, Sweden, June 9–11, 1987.
MC-16 T. Motizuki, S. Mizuno, and Y. Iijima, "Adaptive coefficient selection by frame difference for hybrid predictive/transform coding," PCS 87, Picture Coding Symp., pp. 157–158, Stockholm, Sweden, June 9–11, 1987.
MC-17 G. Ocylok, "A Comparison of interframe coding techniques," Intl. Conf. on Acoust., Speech, and Signal Processing, pp. 1224–1227, Boston, MA, April 14–16, 1983.
MC-18 P. Gerken and H. Schiller, "Progressive updating of unchanged picture areas in a hybrid coding system," PCS 87, Picture Coding Symp., pp. 194–195, Stockholm, Sweden, June 9–11, 1987.

MC-19 H. Brusewitz, "Filtering in the hybrid coding loop," PCS 87, Picture Coding Symp., pp. 196–197, Stockholm, Sweden, June 9–11, 1987.

MC-20 W. H. Chen and D. Hein, "Recursive temporal filtering and frame rate reduction for image coding," IEEE Journal on Selected Areas in Commun., vol. SAC-5, pp. 1155–1165, Aug. 1987.

MC-21 M. Kaneko, Y. Hatori, and A. Koike, "Improvements of transform coding algorithm for motion compensated interframe prediction errors—DCT/SQ Coding," IEEE Journal on Selected Areas in Commun., vol. SAC-5, pp. 1068–1078, Aug. 1987.

MC-22 N. Ohta, M. Nomura, and T. Fujii, "Characteristics of variable rate coding with motion compensated DCT for burst/packetized communications," paper 4-3, IEEE COMSOC International Workshop on Future Prospects of Burst/Packetized Multimedia Commun., Osaka, Japan, Nov. 22–24, 1987.

MC-23 N. Ohta, M. Nomura, and T. Fujii, "Variable rate video coding using motion compensated DCT for asynchronous transfer mode networks," Intl. Conf. on Commun., pp. 1257–1261, Philadelphia, PA, June 12–15, 1988.

MC-24 P. Strobach, D. Schutt, and W. Tengler, "Space-variant regular decomposition quadtrees in adaptive interframe coding," ICASSP 88, Intl. Conf. on Acoust., Speech, and Signal Process., pp. 1096–1099, New York, NY, April 11–14, 1988.

MC-25 G. Aartsen, R. H. J. M. Plompen, and D. E. Boekee, "Error resilience of a video codec for low bit rates," ICASSP 88, Intl. Conf. on Acoust., Speech, and Signal Process., pp. 1312–1315, New York, NY, April 11–14, 1988.

MC-26 S. Ericsson, "Fixed and adaptive predictors for hybrid predictive/transform coding," IEEE Trans. Commun., vol. COM-33, pp. 1291–1302, Dec. 1985.

MC-27 R. Srinivasan and K. R. Rao, "Predictive coding based on efficient motion estimation," ICC 88, Intl. Conf. Commun., pp. 521–526, Amsterdam, Netherlands, May 1984.

MC-28 T. Koga *et al.*, "Motion compensated interframe coding for video conferencing," NTC 81, National Telecommun. Conf., pp. G5.3.1–G5.3.5, New Orleans, LA, Nov.–Dec. 1981.

MC-29 M. Ghanbari, "Motion estimation for interframe picture coding," PCS 88, Picture Coding Symp., pp. 3.13-1 through 3.13-2, Torino, Italy, Sept. 1988.

MC-30 K. M. Yang, M. T. Sun, L. Wu, and I. Fei G. Chuang, "Very high efficiency VLSI chip-pair for full search block matching with fractional precision," Intl. Conf. Acoust., Speech, and Signal Process., pp. 2437–2440, Glasgow, Scotland, May 23–26, 1989.

MC-31 A. Artieri and F. Jutand, "A versatile and powerful chip for real time motion estimation," Intl. Conf. Acoust., Speech, and Signal Process., pp. 2453–2456, Glasgow, Scotland, May 23–26, 1989.

MC-32 T. Komarek and P. Pirsch, "VLSI architectures for block matching algorithms," Intl. Conf. Acoust., Speech, and Signal Process., pp. 2457–2460, Glasgow, Scotland, May 23–26, 1989.

MC-33 M. T. Sun and K. M. Yang, "A flexible architecture for full-search block-matching motion-vector estimation," ISCAS 89, Intl. Symp., Circuits and Systems," pp. 179–182, Portland, OR, May 9–11, 1989.

MC-34 A. Puri, H. M. Hang, and D. L. Schilling, "An efficient block matching algorithm for motion-compensated coding," Intl. Conf. Acoust., Speech, and Signal Process, pp. 1063–1066, Dallas, TX, April 1987.

MC-35 A. Puri, H. M. Hang, and D. L. Schilling, "Motion-compensated transform coding based on block motion-tracking algorithm," Intl. Conf. on Commun., pp. 136–140, Seattle, WA, June 1987.

MC-36 H. Hölzlwimmer, A. v. Brandt, and W. Tengler, "A 64 Kbit/s motion compensated transform coder using vector quantization with scene adaptive codebook," ICC 87, Intl. Conf. on Commun., pp. 151–156, Seattle, WA, June 1987.

MC-37 L. De Vos, M. Stegherr, and T. G. Noll, "VLSI architectures for the full-search blockmatching algorithm," Intl. Conf. Acoust. Speech, and Signal Process., pp. 1687–1690, Glasgow, Scotland, May 23–26, 1989.

MC-38 V. Rampa, N. Dal Degan, and A. Balboni, "VLSI implementation of a pel-by-pel motion estimator," Intl. Conf. Acoust. Speech, and Signal Process., pp. 2573–2576, Glasgow, Scotland, May 23–26, 1989.

MC-39 V. Considine and A. S. Bhandal, "Single chip motion estimator for video codec applications," IEE Third Intl. Conf. Image Process., pp. 285–289, Coventry, U.K., July 18–20, 1989.

MC-40 K. M. Yang, L. Wu, and A. Fernandez, "A VLSI architecture design for motion detection/compensation chip with full search capability," 22nd Annual Conf. on Information Sciences and Systems, pp. 695–700, Princeton, NJ, March 16, 1988.

MC-41 K. M. Yang, L. Wu, H. Chong, and M. T. Sun, "VLSI implementation of motion compensation full-search block-matching algorithm," SPIE, Visual Communications and Image Processing, vol. 1001, pp. 892–899, Cambridge, MA, Nov. 6, 1988.

MC-42 K. M. Yang, M. T. Sun, and L. Wu, "A family of VLSI designs for motion compensation block-matching algorithm," IEEE Trans. Circuits and Systems, vol. CAS-36, pp. 1317–1325, Oct. 1989.

MC-43 T. Komarek and P. Pirsch, "Array architectures for block matching algorithms," IEEE Trans. Circuits and Systems, vol. CAS-36, pp. 1301–1308, Oct. 1989.

MC-44 R. C. Kim and S. U. Lee, "A VLSI architecture for a pel recursive motion estimation algorithm," IEEE Trans. Circuits and Systems, vol. CAS-36, pp. 1291–1300, Oct. 1989.

Photovideotex

P-1 A. Leger, "Implementation of fast discrete cosine transform for full color videotex services and terminals," GLOBECOM 84, Global Telecommun. Conf., Proc. pp. 333–337, Atlanta, GA, Nov. 26–29, 1984.

P-2 K. N. Ngan and W. C. Hui, "Picture transmission for videotex," IEEE Trans. Consumer Electronics, vol. CE-31, pp. 301–310, Aug. 1985.

Performance Comparison

PC-1 M. Hamidi and J. Pearl, "Comparison of cosine and Fourier transforms of Markov-I Signals," IEEE Trans. Acoust., Speech, and Signal Process., vol. ASSP-24, pp. 428–429, Oct. 1976.

PC-2 H. Kitajima, T. Saito, and T. Kurobe, "Comparison of the discrete cosine and Fourier transforms as possible substitutes for the Karhunen-Loeve transform," Trans. of the IECE of Japan, vol. E-60, pp. 279–283, June 1977.

PC-3 H. B. Kekre and J. K. Solanki, "Comparative performance of various trigonometric unitary transforms for transform image coding," Intrnl. J. Electronics, vol. 44, pp. 305–315, 1978.

PC-4 Z. D. Wang and B. R. Hunt, "Comparative performance of two different versions of the discrete cosine transform," IEEE Trans. Acoust., Speech, and Signal Process., vol. ASSP-32, pp. 450–453, April 1984.

PC-5 T. A. Soame, "Bandwidth compression using transform techniques for image transmission systems," Marconi Review, vol. XLIII, pp. 228–240, 1980.

PC-6 A. K. Jain, "A sinusoidal family of unitary transforms," IEEE Trans. Pattern Anal., Machine Intelli., vol. PAMI-1, pp. 356–365, Oct. 1979.

PC-7 V. R. Algazi and B. J. Fino, "Performance and comparison ranking of fast unitary transforms in applications," in Proc. ICASSP, Intl. Conf. on Acoust., Speech, and Signal Process., pp. 32–35, Paris, France, 1982.

PC-8 N. C. Kim and J. K. Kim, "Behavior of generalized covariance model in picture coding," Electronics Lett., vol. 19, pp. 260–261, March 31, 1983.

PC-9 B. Mazor and W. A. Pearlman, "An optimal transform trellis code with applications to speech," IEEE Trans. Commun., vol. COM-33, pp. 1109–1116, Oct. 1985.

PC-10 R. J. Clarke, "Performance of KLT and DCT for data having widely varying values of intersample correlation coefficient," Electronics Lett., vol. 19, pp. 251–253, March, 1983.

PC-11 R. J. Clarke, "Spectral response of the discrete cosine and Walsh-Hadamard transforms," IEE Proc. Part F, Commun., Radar and Signal Process., vol. 130, pp. 309–313, June 1983.

PC-12 W. K. Cham and R. J. Clarke, "Application of the principle of dyadic symmetry to the generation of orthogonal transforms," IEE Proc. Part F, Commun., Radar, and Signal Process., vol. 133, pp. 264–270, June 1986.

PC-13 M. G. Perkins, "A comparison of the Hartley, Cas-Cas, Fourier and discrete cosine transforms for image coding," IEEE Trans. Commun., vol. COM-36, pp. 758–761, June 1988.

PC-14 R. A. Haddad and A. N. Akansu, "A new orthogonal transform for signal coding," IEEE Trans. Acoust., Speech, and Signal Process., vol. ASSP-36, pp. 1404–1411, Sept. 1988.

PC-15 P. S. Yeh, "Data compression properties of the Hartley transform," IEEE Trans. Acoust., Speech, and Signal Process., vol. ASSP-37, pp. 450–451, March 1989.

PC-16 C. Guillemot and P. Duhamel, "A new transform for image coding with reduced complexity and same performance as DCT," IEE III Intl. Conf. on Image Processing and its Applications, pp. 576–580, Univ. of Warwick, Coventry, UK, July 18–20, 1989.

Pattern Classification

PCL-1 L. T. Watson, T. J. Laffey, and R. M. Haralick, "Topographic classification of digital image intensity surfaces using generalized splines and the discrete cosine transformation," Computer Vision, Graphics, and Image Process., vol. 29, pp. 143–167, Feb. 1985.

PCL-2 Y. F. Daniyev and V. P. Ryzhov, "Signal discrimination using multi-basis representations," Telecommun., and Radio Engrg., vol. 38/39, pp. 116–117, May 1984.

PFA and DCT

PD-1 P. P. N. Yang and M. J. Narasimha, "Prime factor decomposition of the discrete cosine transform, "Intl. Conf. on Acoust. Speech, and Signal Process., pp. 772–775, Tampa, FL, March 26–29, 1985.

Printed Image Coding

PIC-1 M. Gilge, "Adaptive transform coding of four-color printed images," PCS 86, Picture Coding Symp., pp. 72–73, Tokyo, Japan, April 2–4, 1986.

PIC-2 M. Gilge, "Adaptive transform coding of four-color printed images," ICASSP 87, Intl. Conf. on Acoust., Speech, and Signal Process., pp. 1374–1377, Dallas, TX, April 6–9, 1987.

Progressive Image Transmission

PIT-1 K. N. Ngan, "Image display techniques using the cosine transform," IEEE Trans. Acoust., Speech, and Signal Process., vol. ASSP-32, pp. 173–177, Feb. 1984.

PIT-2 H. Lohscheller, "A subjectively adapted image communication system," IEEE Trans. Commun., vol. COM-32, pp. 1316–1322, Dec. 1984.

PIT-3 E. Dubois and J. L. Moncet, "Transform coding for progressive transmission of still pictures," GLOBECOM 85, IEEE Global Telecommun., Conf., pp. 348–352, New Orleans, LA, Dec. 2–5, 1985.

PIT-4 K. H. Tzou and S. E. Elnahas, "Bit-sliced progressive transmission and reconstruction of transformed images," ICASSP 86, Intl. Conf. on Acoust. Speech, and Signal Process, pp. 533–536, Tokyo, Japan, April 7–11, 1986.

PIT-5 E. Dubois and J. L. Moncet, "Encoding and Progressive transmission of still pictures in NTSC composite format using transform domain methods," IEEE Trans. Commun., vol. COM-34, pp. 310–319, March 1986.

PIT-6 K. H. Tzou and S. E. Elnahas, "An optimal progressive transmission and reconstruction scheme for transformed images," Intl. Conf. on Commun., pp. 413–418, Toronto, Canada, June 22–25, 1986.

PIT-7 L. Wang and M. Goldberg, "Progressive image transmission by multistage transform coefficient quantization," ICC 86, Intl. Conf. on Commun., pp. 419–423, Toronto, Canada, June 22–25, 1986.

PIT-8 S. E. Elnahas, R. G. Jost, J. R. Cox, and R. L. Hill, "Progressive transmission of digital diagnostic images," SPIE, Applications of digital image processing, vol. 575, pp. 48–55, San Diego, CA, Aug. 1985.

PIT-9 S. E. Elnahas, K. H. Tzou, J. R. Cox, Jr., R. L. Hill, and R. G. Jost, "Progressive coding and transmission of digital diagnostic pictures," IEEE Trans. Medical Imaging, vol. MI-5, pp. 73–83, June 1986.

PIT-10 K. H. Tzou, T. R. Hsing, and J. G. Dunham, "Applications of physiological human visual system model to image compression," Applications of Digital Image Processing VII, Proc. SPIE, vol. 504, pp. 419–424, 1984.

PIT-11 L. Wang and M. Goldberg, "Progressive image transmission by transform coefficient residual error quantization," IEEE Trans. Commun., vol. COM-36, pp. 75–87, Jan. 1988.

PIT-12 B. Chitprasert and K. R. Rao, "Human visual weighted progressive image transmission," IEEE Trans. Commun., vol. COM-38, 1990, to be published.

PIT-13 "ISO adaptive discrete cosine transform coding scheme for still image telecommunication services," ISO/TC97/SC2/WG8 N640 Rev. 1., Jan. 25, 1988.

PIT-14 J. P. Hudson, Report of the joint photographic experts group meeting," 25–27, Jan. 1988, KTAS, Copenhagen, Denmark, ISO/IEC/JTC1/SC2/WG8 N175, Rev. 2, May 1988.

PIT-15 K. Takikawa, "Fast progressive reconstruction of a transformed image," IEEE Trans. Inform. Theory, vol. IT-30, pp. 111–117, Jan. 1984.

PIT-16 B. Chitprasert and K. R. Rao, "Human visual weighted progressive image transmission," ICCS 88, Intl. Conf. on Commun., Systems, Singapore, Nov. 1988.

PIT-17 G. P. Hudson, "PICA-Photovideotex image compression algorithms, ESPRIT 86: Results and Achievements," North Holland, Elsevier Science Publishers, 1987.

PIT-18 G. P. Hudson, "Esprit 563-PICA, Photovideotex image compression algorithms-toward international standardisation," ISO/TC97/SC2/WG8 N564, Brussels, Belgium, Sept. 28–30, 1987.

PIT-19 M. Miran and K. R. Rao, "Fast progressive reconstruction of images using the DCT." (Under review).

PIT-20 T. P. Hudson and H. Yasuda, "The selection of a still picture compression technique for international standardisation," PCS 88, Picture Coding Symp., pp. 9.1-1 through 9.1-2, Torino, Italy, Sept. 1988.

PIT-21 G. Wallace, R. Vivian, and H. Poulsen, "Subjective assessment of JPEG compression techniques," PCS 88, Picture Coding Symp., pp. 9.2-1 through 9.2-2, Torino, Italy, Sept. 1988.

PIT-22 A. Leger, B. Niss, J. Vaaben, L. Chiarighione, H. Lohscheller, and S. Gicquel, "Adaptive discrete cosine transform coding scheme (ADCT) for still picture compression standard," PCS-88, Picture Coding Symp., pp. 9.3-1 through 9.3-2, Torino, Italy, Sept. 1988.

PIT-23 B. Macq and P. Delogne, "Progressive transmission of pictures by transform coding," PCS 88, Picture Coding Symp., pp. 6.4-1 through 6.4-2, Torino, Italy, Sept. 1988.

PIT-24 G. P. Hudson, H. Yasuda, and I. Sebestyen, "The international standardisation of a still picture compression technique," GLOBECOM 88, Global Commun., Conf. pp. 1016–1021, Hollywood, FL, Nov.–Dec., 1988.

PIT-25 A. Leger, J. L. Mitchell, and Y. Yamazaki, "Still picture compression algorithms evaluated for international standardisation," GLOBECOM 88, Global Commun., Conf., pp. 1028–1032, Hollywood, FL, Nov.–Dec., 1988.

PIT-26 G. K. Wallace, R. Vivian, and H. Poulsen, "Subjective testing results for still picture compression algorithms for international standardisation," GLOBECOM 88, Global Commun., Conf., pp. 1022–1027, Hollywood, FL, Nov.–Dec., 1988.

PIT-27 I. Sebestyen, C. F. Touchton, and H. Yasuda, "Application and service requirements for communication of still images," PCS 88, Picture Coding Symp., pp. 9.6-1 through 9.6-2, Torino, Italy, Sept. 1988.

PIT-28 S. E. Elnahas and E. B. Vogel, "Transform progressive coding of digitized images," Intl. Conf. on Commun. Systems, pp. 28.1.1 through 28.1.5, Singapore, Nov. 1988.

PIT-29 S. Acharya and D. R. Ahlgren, "Implementing international standards on still-frame image compression systems," SPIE, Digital image processing applications, vol. 1075, pp. 157–163, Los Angeles, CA, Jan. 1989.

PIT-30 J. L. Mitchell, W. B. Pennebaker, and C. A. Gonzales, "The standardization of color photographic image data compression," SPIE, Digital image processing applications, vol. 1075, pp. 101–106, Los Angeles, CA, Jan. 1989.

PIT-31 K. H. Tzou, "Progressive image transmission: a review and comparison," Optical Engineering, vol. 26, pp. 581–589, July 1987.

PIT-32 P. J. Burt and E. H. Adelson, "The Laplacian pyramid as a compact image code," IEEE Trans. Commun., vol. COM-31, pp. 532–540, April 1983.

PIT-33 H. Yasuda, "Standardization activities on multimedia coding in ISO," Signal Process: Image Commun., vol. 1, pp. 3–16, June 1989.

PIT-34 G. K. Wallace, "Overview of the JPEG ISO/CCITT still frame compression standard," SPIE Visual Commun., and Image Processing, Philadelphia, PA, Nov. 5–10, 1990, to be published.

PIT-35 W. B. Pennebaker and J. L. Mitchell, "Standardization of color image data compression I. Sequential coding," Electronic Imaging, 89 East, pp. 109–112, Boston, MA, Oct. 2–5, 1989.

PIT-36 J. L. Mitchell and W. B. Pennebaker, "Standardization of color image data compression II. Progressive coding," Electronic Imaging, '89 East, pp. 191–194, Boston, MA, Oct. 2–5, 1989.

PIT-37 G. K. Wallace, "Standardization of moving picture coding for interactive applications," GLOBECOM 89, IEEE Global Telecommun., Conf., pp. 564–568, Dallas, TX, Nov. 27–30, 1989.

PIT-38 R. Aravind, G. L. Cash, and J. P. Worth, "Implementing the JPEG still picture compression algorithms," SPIE Visual Commun., and Image Processing, Vol. 1199, pp. 799–808, Philadelphia, PA, Nov. 5–10, 1989.

Pattern Recognition

PR-1 Y. S. Hsu, S. Prum, J. H. Kagel, and H. C. Andrews, "Pattern recognition experiments in the Mandala/cosine domain," IEEE Trans. Pattern Anal. Machine Intell., vol. PAMI-5, pp. 512–520, Sept. 1983.

Packet Video

PV-1 N. Ohta, M. Nomura, and T. Fujii, "Characteristics of variable rate video coding with motion compensated DCT for burst/packetized communications," IEEE COMSOC Intl. Workshop on Future Prospects of Burst/Packetized Multimedia Commun., paper 4-3, Osaka, Japan, Nov. 22–24, 1987.

PV-2 M. Nomura, T. Fujii, and N. Ohta, "Layered packet-loss protection for variable rate video coding using DCT," II Intl. Workshop on Packet Video, Torino, Italy, Sept. 8–9, 1988.

PV-3 N. Ohta, M. Nomura, and T. Fujii, "Variable rate video coding using motion compensated DCT for asynchronous transfer mode networks," ICC 88, Intl. Conf. on Commun., pp. 1257–1261, Philadelphia, PA, June 1988.

PV-4 T. R. Hsing, "Video compression techniques: Networking aspects," Intl. Symp. on Circuits and Systems, pp. 223–226, Espoo, Finland, June 1988.

PV-5 L. Contin, L. Corgnier, and L. Masera, "Interframe extension of the ISO still picture coding algorithm and applications to packet network," II Intl. Workshop on Packet Video, Torino, Italy, Sept. 8–9, 1988.
PV-6 J. P. Leduc, J. P. Choffray, P. Delogne, and B. Macq, "Universal coding of transform coefficients for packet video applications," II Intl. Workshop on Packet Video, Torino, Italy, Sept. 8–9, 1988.
PV-7 Special issue on "Broadband packet communications," IEEE Journal on Selected Areas in Commun., vol. JSAC-6, Dec. 1988.
PV-8 Special issue on "Packet speech and video," IEEE Journal on Selected Areas in Commun., vol. JSAC-7, June 1989.
PV-9 Y. M. Le Pannerer, "The RACE European projects on coding," II Intl. Workshop on Packet Video, Torino, Italy, Sept. 8–9, 1988.
PV-10 R. C. Brainard and J. H. Othmer, "Television compression algorithms and transmission of packet networks," SPIE, Advances in Image Coding, vol. 1001, pp. 973–978, Cambridge, MA, Nov. 1988.
PV-11 W. Verbiest, "Video coding in an ATD environment," in New Systems and Services in Telecommun. III, G. Contraine and J. Destine (editors), pp. 249–253, North Holland, Elsevier Science Publishers, 1987.
PV-12 W. Verbiest, L. Pinnoo, and B. Voeten, "The impact of ATM concept on video coding," IEEE Journal on Selected Areas in Commun., vol. SAC-6, pp. 1623–1632, Dec. 1988.
PV-13 III Intl. Workshop on Packet Video (Visicom 90) Morristown, NJ, March 22–23, 1990.
PV-14 K. Shimamura, Y. Hayashi, and F. Kishino, "Variable-bit-rate coding capable of compensating for packet loss," SPIE, Visual Commun., and Image Proces., vol. 1001, pp. 991–998, Cambridge, MA, Nov. 1988.
PV-15 H. Holzlwimmer, "Rate control in variable transmission rate image coders," SPIE, 33rd Annual Technical Symp. on Optical and Optoelectronic Appl. Science and Engrg., vol. 1153, San Diego, CA, Aug. 1989.
PV-16 F. Kishino, K. Manabe, Y. Hayashi, and H. Yasuda, "Variable bit rate coding of video signals for ATM networks," IEEE Journal on Selected Areas in Commun., vol. JSAC-7, pp. 801–806, June 1989.

Quantization

Q-1 J. Max, "Quantizing for minimum distortion," Trans. IRE, vol. IT-6, pp. 7–12, March 1960.
Q-2 W. C. Adams and C. E. Giesler, "Quantizing characteristics for signals having Laplacian amplitude probability density function," IEEE Trans. Commun., vol. COM-26, pp. 1295–1297, Aug. 1978.
Q-3 S. P. Lloyd, "Least squares quantization in PCM," IEEE Trans. Inform. Theory, vol. IT-28, pp. 129–137, March 1982.
Q-4 K. N. Ngan, K. S. Leong, and H. Singh, "Fast convergence method for Lloyd-Max quantizer design," Electronics Letters, vol. 22, pp. 844–846, July 31, 1986.
Q-5 J. J. Y. Huang and P. M. Schultheiss, "Block quantization of correlated Gaussian random variables," IEEE Trans. Commun. Systems, vol. CS-11, pp. 289–296, Sept. 1963.
Q-6 K. H. Tzou, "A fast computational approach to the design of block quantization," IEEE Trans. Acoust., Speech, and Signal Process., vol. ASSP-35, pp. 235–237, Feb. 1987.

SCT

S-1 H. Kitajima, "A symmetric cosine transform," IEEE Trans. Comput., vol. C-29, pp. 317–323, April 1980.

Speech Coding

SC-1 R. Zelinski and P. Noll, "Adaptive transform coding of speech signal," IEEE Trans. Acoust., Speech, and Signal Process., vol. ASSP-25, pp. 229–309, Aug. 1977.

SC-2 R. Zelinski and P. Noll, "Approaches to adaptive transform speech coding at low-bit rates," IEEE Trans. Acoust., Speech, and Signal Process., vol. ASSP-27, pp. 89–95, Feb. 1979.

SC-3 R. V. Cox and R. E. Crochiere, "Real-time simulation of adaptive transform coding," IEEE Trans. Acoust., Speech, and Signal Process., vol. ASSP-29, pp. 147–154, April 1981.

SC-4 D. Malah, R. E. Crochiere, and R. V. Cox, "Performance of transform and subband coding systems combined with harmonic scaling of speech," IEEE Trans. Acoust., Speech, and Signal Process., vol. ASSP-29, pp. 273–283, April 1981.

SC-5 C. D. Heron, R. E. Crochiere, and R. V. Cox, "A 32-band subband/transform coder incorporating vector quantization for dynamic bit allocation," Intl. Conf. on Acoust., Speech, and Signal Process., pp. 1276–1279, Boston, MA., April 14–16, 1983.

SC-6 G. S. V. Rao and K. R. Rao, "Adaptive transform coding of speech signals," Intl. Conf. on Computers, Systems, and Signal Process., pp. 876–878, Bangalore, India, Dec. 10–12, 1984.

SC-7 R. E. Crochiere and J. L, Flanagan, "Current perspectives in digital speech," Intl. Conf. on Commun., Philadelphia, PA, pp. 3G.3.1–3G.3.5, June 1982.

SC-8 S., Ono and T. Araseki, "Linear transformation for low bit rate speech coding," Intl. Conf. on Acoust., Speech, and Signal Process., pp. 2391–2394, Tokyo, Japan, April 7–11, 1986.

SC-9 J. M. Tribolet and R. E. Crochiere, "Frequency domain coding of speech," IEEE Trans. Acoust., Speech, and Signal Process., vol. ASSP-27, pp. 512–530, Oct. 1979.

SC-10 T. Moriya and M. Honda, "Speech coder using phase equalization and vector quantization," Intl. Conf. on Acoust., Speech, and Signal Process., pp. 1701–1704, Tokyo, Japan, April 7–11, 1986.

SC-11 T. Moriya and M. Honda, "Transform coding of speech with weighted vector quantization," Intl. Conf. on Acoust., Speech, and Signal Process., pp. 1629–1632, Dallas, TX, April 6–9, 1987. Also IEEE Journal on Selected Areas in Commun., vol. SAC-6, pp. 425–431, Feb. 1988.

SC-12 R. Toy and W. A. Pearlman, "Backward adaptation for transform trellis coding of speech," Intl. Conf. on Acoust., Speech, and Signal Process., pp. 2209–2212, Dallas, TX, April 6–9, 1987.

SC-13 J. L. Flanagan, M. R. Schroeder, B. S. Atal, R. E. Crochiere, N. S. Jayant, and J. M. Tribolet, "Speech coding," IEEE Trans. Commun., vol. COM-27, pp. 710–737, April 1979.

SC-14 J. S. Rodrigues and L. B. Almeida, "Harmonic coding at 8 KBit/sec," Intl. Conf. on Acoust., Speech, and Signal Process., pp. 1621–1624, Dallas, TX, April 6–9, 1987.

SC-15 B. Mazor and W. A. Pearlman, "An optimal transform trellis code with applications to speech," IEEE Trans. Commun., vol. COM-33, pp. 1109–1116, Oct. 1985.

SC-16 N. Farvardin and Y. Hussain, "Adaptive block transform coding of speech based on the hidden Markov model," EUSIPCO-88, European Signal Process. Conf., Grenoble, France, Sept. 5–8, 1988.

SC-17 T. Fjallbrant and F. Mekuria, "Vector quantization of wideband transform coefficient time trajectories in speech coding," Report Li THISY-I-0917, Univ. of Linkoping, Linkoping, Sweden.

SC-18 T. Fjallbrant and F. Mekuria, "Frequency domain phase derivatives used for data reduction in an ATC system," Digital signal processing, V. Cappelini and A. G. Constantinides (eds.), pp. 289–293, Elsevier Science Publishers, North Holland, 1987.

SC-19 T. Fjallbrant, "Discrete cosine transform magnitude and phase coefficients used for vector quantization in a speech transform coder," Intl. Symp. on Circuits and Systems, pp. 1087–1089, Helsinki, Finland, June 1988.

SC-20 T. Fjallbrant, F. Mekuria, and W. Kou, "A voiced/unvoiced classified vector quantized speech transform coder implemented on a TMS 32020 signal processor," Intl. Symp. on Circuits and Systems, pp. 1333–1336, Helsinki, Finland, June 1988.

SC-21 Y. Shoham and A. Gersho, "Pitch synchronous transform coding of speech at 9.6 kb/s based on vector quantization," ICC 84, Intl. Conf. on Commun., pp. 1179–1182, Amsterdam, The Netherlands, May 14–17, 1984.

SC-22 F. S. Yeoh and C. S. Xydeas, "Split-band coding of speech signals using a transform technique," ICC 84, Intl. Conf. on Commun., pp. 1183–1187, Amsterdam, The Netherlands, May 14–17, 1984.

SC-23 T. Fjallbrant and F. Mekuria, "Vector quantization of wideband transform coefficient time trajectories in speech coding," Electronics Letters, vol. 24, pp. 773–774, June 23, 1988.

SC-24 W. Kou, T. Fjallbrant, and Y. Xiao, "Wideband transform product code vector quantization speech coder with short delays," 9th Intl. Conf. on Pattern Recognition, pp. 748–750, Rome, Italy, Nov. 1988.

SC-25 T. Fjallbrant, W. Kou, and F. Mekuria, "VQATC system for speech coding," Florence Conf. on digital signal processing, pp. 306–309, Florence, Italy, Sept. 1987.

SC-26 T. Fjallbrant, "A wide-band approach to adaptive transform coding of speech signals. A TMS 320 signal processor implementation," ISCAS 85, Intl. Symp. on Circuits and Systems, pp. 321–324, Kyoto, Japan, 1985.

SC-27 N. Farvardin and R. Laroia, "Efficient encoding of speech LSP parameters using the discrete cosine transformation," Intl. Conf. Acoust., Speech, and Signal Proces., pp. 168–171, Glasgow, Scotland, May 23–26, 1989.

SC-28 T. Fjallbrant and F. Mekuria, "A hierarchical two-level analysis structure for use in speech coding and recognition," Intl. Conf. Acoust., Speech, and Signal Proces., pp. 766–769, Glasgow, Scotland, May 23–26, 1989.

SC-29 H. S. Malvar and R. Duarte, "Transform/subband coding of speech with the lapped orthogonal transform," ISCAS 89, Intl. Symp. on Circuits and Systems, pp. 1268–1271, Portland, OR, May 1989.

SC-30 F. Mekuria, "Studies on the development and DSP implementation of vector quantization algorithms with application to speech transform coding," Linkoping studies in science and technology. Thesis No. 160, University of Linkoping, Linkoping, Sweden.

SC-31 W. Kou, J. W. Mark, T. Fjallbrant, and Y. Xiao, "Vector adaptive wideband transform product code vector quantization coding of speech signals." To be published. (Also report Li THISY, University of Linkoping, Linkoping, Sweden.)

SC-32 T. Moriya and H. Suda, "An 8 Kbit/s transform coder for noisy channels," Intl. Conf. Acoust., Speech, and Signal Process., pp. 196–199, Glasgow, Scotland, May 23–26, 1989.

SC-33 V. Cuperman, "On adaptive vector transform quantization for speech coding," IEEE Trans. Commun., vol. COM-37, pp. 261–267, March 1989.

SC-34 F. S. Yeoh and C. S. Xydeas, "A transform approach to split-band coding schemes," IEE Proc. (part F) on Commun., Radar, and Signal Process, vol. 131, pp. 56–63, Feb. 1984.

SAR Image Coding

SIC-1 T. Gioutsos and S. Werness, "Transform coding of synthetic aperture radar (SAR) images," ICASSP 87, Intl. Conf. on Acoust., Speech, and Signal Process., pp. 1370–1373, Dallas, TX, April 6–9, 1987.

Speech Recognition

SR-1 H. A. Barger and K. R. Rao, "A comparative study of phonemic recognition by discrete orthogonal transforms," Intl. Conf. on Acoust., Speech, and Signal Process., pp. 553–556, Tulsa, OK, April 10–12, 1978.

SR-2 H. A. Barger and K. R. Rao, "Evaluation of discrete transforms for use in digital speech recognition," "Computers and Elect. Engrg., vol. 6, pp. 183–197, 1979.

Surface Texture Analysis

STA-1 D. J. Mulvaney, D. E. Newland, and K. F. Gill, "A comparison of orthogonal transforms in their application to surface texture analysis," Proc. Inst. of Mechanical Engineers, vol. 200, Part C, pp. 407–414, 1986.

Transmultiplexers

T-1 M. J. Narasimha, P. P. N. Yang, B. G. Lee and M. L. Abell, "The TM 800-MI: A 60-Channel CCITT transmultiplexer," Intl. Conf. on Commun., pp. 672–674, Amsterdam, The Netherlands, May 14–17, 1984.

T-2 M. J. Narasimha, "Design of FIR filter banks for a 24-channel transmultiplexer," IEEE Trans. Commun., vol. COM-30, pp. 1506–1510, July 1982.

T-3 M. J. Narasimha and A. M. Peterson, "Design of a 24-channel transmultiplexer," IEEE Trans. Acoust., Speech, and Signal Process., vol. ASSP-27, pp. 752–762, Dec. 1979.

T-4 T. G. Marshall, Jr., "A multiple VLSI signal processor realization of a transmultiplexer," IEEE Trans. Commun., vol. COM-30, pp. 1560–1568, July 1982.

T-5 L. Gundel, "Filter bank interpretation of DFT and DCT," Eusipco-88, European Signal Process. Conf., pp. 943–946, Grenoble, France, Sept. 5–8, 1988.

T-6 H. J. Nussbaumer, "Polynomial transform implementation of digital filter banks," IEEE Trans. Acoust., Speech, and Signal Process., vol. ASSP-31, pp. 616–622, June 1983.

T-7 H. S. Malvar, "The LOT: a link between block transform coding and multirate filter banks," Intl. Symp. on Circuits and Systems, pp. 835–838, Espoo. Finland, June 7–9, 1988.

Late Additions

LA-1 K. Rose, A. Heiman, and I. Dinstein, "DCT/DST alternate transform image coding," IEEE Trans. Commun., vol. COM-38, pp. 94–101, Jan. 1990.

LA-2 M. L. Liou, "Visual telephony as an ISDN application," Commun., Magazine, vol. 28, pp. 30–38, Feb. 1990.

LA-3 M. Barbero, R. Del Pero, M. Muratori, and M. Sroppianna, "Bit-Rate Reduction Techniques Based on DCT for HDTV Transmission," ICC 90, Intl. Conference on Commun., pp. 1607–1611, Atlanta, GA, April 16–19, 1990.

LA-4 A. Fernandez, R. Ansari, D. J. LeGall, and C. T. Chen, "HDTV subband/DCT coding: analysis of system complexity," ICC 90, Intl. Conference on Commun., pp. 1602–1606, Atlanta, GA, April 16–19, 1990.

LA-5 Y. Okumura, K. Irie, and R. Kishimoto, "High quality transmission system design for HDTV signals," Intl. Conf. on Commun., pp. 1049–1053, New Orleans, LA, April 1990.

LA-6 J. Suzuki, M. Nomura, and S. Ono, "Comparative study of transform coding for super high definition images," Intl Conf. on Acoust., Speech, and Signal Process., pp. 2257–2260, Albuquerque, NM, April, 1990.

LA-7 S. S. Dixit and J. B. Nardone, "A variable bit rate layered DCT video coder for packet switched (ATM) networks," IEEE Intl. Conf. on Acoust., Speech, and Signal Process., pp. 2253–2256, Albuquerque, NM, April, 1990.

LA-8 A. Sugiyama, F. Hazu, M. Iwadare, and T. Nishitani, "Adaptive transform coding with an adaptive block size (ATC-ABS)," ICASSP 90, Int. Conf. on Acoust., Speech, and Signal Process., pp. 1093–1096, Albuquerque, NM, April, 1990.

LA-9 W. Y. Chan and A. Gersho, "High fidelity audio transform coding with vector quantization," ICASSP 90, Intl. Conf. on Acoust., Speech, and Signal Process., pp. 1109–1112, Albuquerque, NM, April, 1990.

LA-10 M. Yan, J. V. McCanny, and Y. Hu, "VLSI architectures for digital image coding," ICASSP 90, Intl. Conf. on Acoust., Speech, and Signal Process., pp. 913–916, Albuquerque, NM, April, 1990.

LA-11 P. Duhamel and C. Guillemot, "Polynomial transform computation of the 2-D DCT," ICASSP 90, Intl. Conf. on Acoust., Speech, and Signal Process., pp. 1515–1518, Albuquerque, NM, April, 1990.

LA-12 U. Totzek, F. Matthiesen, S. Wohlleben, and T. G. Noll, "CMOS VLSI implementation of the 2D-DCT with linear processor arrays," ICASSP 90, Intl. Conf. on Acoust., Speech, and Signal Process., pp. 937–940, Albuquerque, NM, April, 1990.

LA-13 M. N. Chong, J. J. Soraghan, and T. S. Durrani, "Parallel implementation and analysis of adaptive transform coding," ICASSP 90, Intl. Conf. on Acoust., Speech, and Signal Process., pp. 997–1000, Albuquerque, NM, April, 1990.

LA-14 K. Irie and R. Kishimoto, "Adaptive sub-band coder design for HDTV signal transmission," ICASSP 90, Intl. Conf. on Acoust., Speech, and Signal Process., pp. 2117–2120, Albuquerque, NM, April, 1990.

LA-15 Y. B. Yu, M. H. Chan, and A. G. Constantinides, "Low bit rate video coding using variable block size model," ICASSP 90, Intl. Conf. on Acoust., Speech, and Signal Process., pp. 2229–2231, Albuquerque, NM, April, 1990.

LA-16 F. Azadegan, "Discrete cosine transform encoding of two dimensional processes," ICASSP 90, Intl. Conf. on Acoust., Speech, and Signal Process., pp. 2237–2240, Albuquerque, NM, April, 1990.

LA-17 M. W. Marcellin, "Transform coding of images using trellis coded quantization," ICASSP 90, Intl. Conf. on Acoust., Speech, and Signal Process., pp. 2241–2244, Albuquerque, NM, April, 1990.

LA-18 D. S. Lee and K. H. Tzou, "Hierarchical DCT coding of HDTV for ATM networks," ICASSP 90, Intl. Conf. on Acoust., Speech, and Signal Process., pp. 2249–2252, Albuquerque, NM, April, 1990.

LA-19 X. Ran and N. Farvardin, "Combined VQ-DCT coding of images using interblock noiseless coding," ICASSP 90, Intl. Conf. on Acoust., Speech, and Signal Process., pp. 2281–2284, Albuquerque, NM, April, 1990.

LA-20 S. Sridharan, E. Dawson, and B. M. Goldburg, "Speech encryption using discrete orthogonal transforms," ICASSP 90, Intl. Conf. on Acoust., Speech, and Signal Process., pp. 1647–1650, Albuquerque, NM, April, 1990.

LA-21 K. M. Yang, H. Fujiwara, Y. Ishida, M. Maruyama, T. Sakaguchi, and H. Uwabu, "A flexible motion vector estimation chip for real-time video codecs," IEEE Custom Integrated Circuits Conf., pp. 17.5–17.5.4, Boston, MA, May, 1990.

LA-22 U. Sjöström, I. Defilippis, M. Ansorge, and F. Pellandini, "Discrete cosine transform chip for real-time video applications," Intl. Symp. Circuits and Systems, pp. 1620–1623, New Orleans, LA, May, 1990.

INDEX